D.G. LINDQUIST

Edited by

Richard F. Ford
Professor of Biology
San Diego State College
San Diego, California

William E. Hazen
Professor of Biology
San Diego State College
San Diego, California

Readings in Aquatic Ecology

1972 W. B. SAUNDERS COMPANY PHILADELPHIA LONDON TORONTO

W. B. Saunders Company: West Washington Square
Philadelphia, Pa. 19105

12 Dyott Street
London, WC1A 1DB

833 Oxford Street
Toronto 18, Ontario

Readings in Aquatic Ecology ISBN 0-7216-3810-4

©1972 by W. B. Saunders Company. Copyright under the International Copyright Union. All rights reserved. This book is protected by copyright. No part of it may be reproduced, stored in a retrieval system, or transmitted in any form or by any means, electronic, mechanical, photocopying, recording, or otherwise, without written permission from the publisher. Made in the United States of America. Press of W. B. Saunders Company. Library of Congress Catalog card number 77-188388.

Print No.: 9 8 7 6 5 4 3 2 1

QH
541.5
.W3
R42

INTRODUCTION

We approach the subject of aquatic ecology with the conviction that a large body of data and theories, problems and practices are common both to the ocean and to inland bodies of water. Although ecology, whether terrestrial or aquatic, comprises one subject area, practice makes bodies of water subject to an even more closely unified treatment than is given ecology as a whole.

As examples of the unity of theory in oceanography and limnology two widely quoted passages are presented here. These passages also serve to emphasize that the development of a theory of biological communities was not restricted to the biologists who worked on terrestrial plant communities. Möbius, writing of oyster beds, states (1885 translation):

> Every oyster-bed is thus, to a certain degree, a community of living beings, a collection of species and a massing of individuals, which find here everything necessary for their growth and continuance, such as suitable soil, sufficient food, the requisite percentage of salt, and a temperature favorable to their development. Each species which lives here is represented by the greatest number of individuals which can grow to maturity subject to the conditions which surround them, for among all species the number of individuals which arrive at maturity at each breeding period is much smaller than the number of germs produced at that time. The total number of individuals of all the species living together in any region is the sum of the survivors of all the germs which have been produced at all past breeding or brood periods; and this sum of matured germs represents a certain quantum of life which enters into a certain number of individuals, and which, as does all life, gains permanence by means of transmission. Science possesses, as yet, no word by which such a community of living beings may be designated; no word for a community where the sum of species and individuals, being mutually limited and selected under the average external conditions of life, have by means of transmission continued in possession of a certain definite territory. I propose the word *biocönosis* for such a community. Any change in any of the relative factors of a biocönose produces changes in other factors of the same. If, at any time, one of the external conditions of life should deviate for a long time from the ordinary mean, the entire biocönose, or community, would be transformed. It would also be transformed, if the number of individuals of a particular species increased or diminished through the instrumentality of man, or if one species entirely disappeared from, or a new species entered into, the community.

This idea was evidently known to Forbes (1887), who applied it to lakes:

> The animals of such a body of water are, as a whole, remarkably isolated—closely related among themselves in all their interests, but so far independent of the land about them that if every terrestrial animal were suddenly annihilated it would doubtless be long before the general multitude of the inhabitants of the lake would feel the effects of this event in any important way. It is an islet of older, lower life in the midst of the higher, more recent life of the surrounding region. It forms a little world within itself—a microcosm within which all the elemental forces are at work and the play of life goes on in full, but on so small a scale as to bring it easily within the mental grasp.
>
> Nowhere can one see more clearly illustrated what may be called the *sensibility* of such an organic complex, expressed by the fact that whatever affects any species belonging to it, must have its influence of some sort upon the whole assemblage. He will thus be made to see the impossibility of studying completely any form out of relation to the other forms; the necessity for taking a comprehensive survey of the whole as a condition to a satisfactory understanding of any part. If one wishes to become acquainted with the black bass, for example, he will learn but little if he limits himself to that species. He must evidently study also the species upon which it depends for its existence, and the various conditions upon which *these* depend. He must likewise study the species with which it comes in competition, and the entire system of conditions affecting their prosperity; and by the time he has studied all these sufficiently he will find that he has run through the whole complicated mechanism of the aquatic life of the locality, both animal and vegetable, of which his species forms but a single element.

These early statements, similar in tone and interest, point to the unity of the biological framework under which biological oceanography and limnology are studied, and, although the size of the ocean and its chemical composition make it complex and difficult to investigate, the problems of sampling and interpretation of data provide at least a minimum of common interests.

The individual papers were chosen for inclusion primarily on our judgment of their heuristic value. For example, more complete forms of Riley's model for phytoplankton growth exist, but the one included here is clear, relatively simple, and exemplifies the difference between an *a priori* and an *a posteriori* model. Furthermore, we have made no attempt to reproduce much of the classic literature in hydrobiology. Lindeman's seminal paper is excluded since it is by now a part of the general framework within which an ecologist works.

We are grateful to a number of persons for their suggestions and interest: to our colleagues in ecology at San Diego State College, to James T. Enright, E. W. Fager, and Michael M. Mullin of the Scripps Institution of Oceanography, and to W. T. Edmondson of the University of Washington. The faults of concept and the errors of omission and commission remain, however, entirely our own. Colleen Forbes has done most of the arduous task of preparing an index. Finally, we wish to thank the individual authors and publishers for their permission to reprint the papers included in this collection.

LITERATURE CITED

Forbes, S. A. 1887. The lake as a microcosm. Bull. Sci. Assoc. Peoria, 1887:77–87. [Reprinted with emendations in Illinois Nat. Hist. Surv. Bull., 15:537–550, 1925].

Möbius, Karl. 1877. Die Auster und die Austernwirtschaft. Berlin. [Translated 1880, The Oyster and Oyster Culture. Rept. U.S. Fish Comm., 1880:683–751.]

Contents

Part 1

PHYSIOLOGICAL AND BEHAVIORAL ECOLOGY

Unlike most scientific fields, which have relatively well-defined boundaries, ecology is an "exceptionally multidisciplinary kind of biology" (Deevey, 1964). The ecologist seeks to understand the reasons why, where, and when an organism occurs and the specific functional capabilities which allow it to remain in a state of dynamic equilibrium with its physical and biological environment. In the process he must often, of necessity, rely heavily on information and methods from allied disciplines, including particularly functional morphology, physiology, and behavior. As Bartholomew (1966) suggests, this multidisciplinary approach is both valid and necessary:

> If one attempts to attain an adequate understanding of the relation of an organism to its environment he addresses himself to a problem of such enormous complexity that he must, unfortunately, be reconciled from the outset to obtaining an incomplete answer. The task, which all scientists face, of isolation and simplification of problems is present in particularly acute form to the student of ecologically relevant physiology and behavior. He cannot reduce his problem until only a single variable remains; he cannot restrict his data to a single level of biological integration, nor, as is often the case in most other biological disciplines, even to several adjacent levels. Furthermore, he cannot limit his data gathering to techniques of any one specialty. More than most students, he must recognize that biology is a continuum. Whatever his techniques and methods, he must be a naturalist.
>
> It is a sad fact that we biologists, because of our limitations, divide ourselves into categories of specialization and then, probably in self-defense, pretend that these categories exist in the biological world. As everyone knows, organisms are functionally indivisible and cannot be split into conventional compartments of physiology and behavior, or morphology and genetics. Each of these areas of scientific specialization is but an aspect of the organism's totality, and since it is the organism with which we must deal, where can we start? Obviously, it is not just a museum specimen; and less obviously it is not an animal or plant caged in the laboratory, or observed in the field. Rather, an

organism is an interaction between a complex self-sustaining physiochemical system and the substances and conditions usually thought of as the environment. As Claude Bernard pointed out almost a century ago, organism and environment form an inseparable pair. One can be defined only in terms of the other.

The aquatic ecologist attempts to use such an integrative approach. Because of the nature of most physical systems in which his organisms occur, he is limited in his ability to make direct observations and to conduct field experiments. As a result, he has come to rely on specific laboratory studies of physiology and behavior which he attempts to design and interpret on the basis of what is observed in nature. Where possible, the more satisfactory approach of conducting comparative laboratory and field experiments is employed (see, for example, Talling in this collection).

Several of the papers included in this collection illustrate the value of this approach. Specific knowledge of the processes and requirements for photosynthesis, nutrient uptake, and growth by marine phytoplankton (Talling; Eppley and Thomas) and of grazing activity by zooplankton (Mullin) is vital in attempting to evaluate or develop predictive models for processes which occur at the population or ecosystem levels of integration (Riley; McAllister *et al.*; Dugdale).

LITERATURE CITED

Bartholomew, G. A. 1966. Interaction of physiology and behavior under natural conditions. *In* R. I. Bowman (ed.), The Galápagos: Proceedings of the Galápagos International Scientific Project of 1964. University of California Press, pp. 39–45.

Deevey, E. S., Jr. 1964. General and historical ecology. BioScience, 14:33–35.

COMPARATIVE LABORATORY AND FIELD STUDIES OF PHOTOSYNTHESIS BY A MARINE PLANKTONIC DIATOM[1]

1

J. F. Talling[2]

Division of Marine Biology, Scripps Institution of Oceanography, California

ABSTRACT

A study is described of the factors affecting measurements of phytoplankton photosynthesis by the much used field method, based upon changes of oxygen concentration in clear and darkened bottles as determined by the Winkler method. Unialgal cultures of a marine plankton diatom, *Chaetoceras affinis*, were used with short exposures of 1–3 hours in both laboratory and sea. Indications of a limitation of photosynthetic rate by inadequate CO_2 supply were only found for culture media with pH raised by prolonged photosynthesis. Otherwise photosynthetic rates were unaltered by the bicarbonate concentration over a wide range, including that of unenriched sea-water, or by cell sedimentation in unshaken cultures. A comparison of the Winkler and manometric methods of measuring oxygen production gave very similar results. High rates of photosynthesis were well maintained over the main period of each exposure, despite the development of oxygen supersaturation. Lower rates were however always found during an initial induction phase.

Photosynthetic characteristics deduced from exposures in laboratory and sea, carried out almost simultaneously and with the same algal material, were in general agreement. Data from the experiments as a whole indicated the development of unusually high photosynthetic capacities at light saturation, assessed per unit content of chlorophyll *a*. The onset of light saturation, and the development of photo-inhibition, occurred at intensities similar to those recorded for many other plankton algae. Some ecological implications of the findings are discussed.

INTRODUCTION

Knowledge of the photosynthetic activities of natural populations of planktonic algae has been obtained from experiments in both laboratory and field. Each of these approaches has limitations. The laboratory experiment may provide well-defined data on the photosynthetic characteristics of the organisms, but any deduction of the distribution of photosynthetic activity in the natural environment of sea or freshwater involves uncertain extrapolations, even when given a better knowledge of environmental variables than is usually available. The field experiment, with exposures made in the water-body itself, may appear to give more directly applicable results, and indeed has been the source of most information on the subject. Nevertheless, although the data obtained may be valid for the local conditions of the experiment, they are usually unsuitable for a quantitative analysis of the principal factors involved, and so are difficult to incorporate into a generalized body of knowledge on the photosynthetic behavior of planktonic algae. The obstacles here arise partly from the heterogeneous nature of the populations sampled, and partly from difficulties of measuring the relevant environmental factors, such as light intensity, which may fluctuate considerably during an experiment.

Comparative laboratory and field experiments, using the same well-defined algal material, provide a means of integrating the two approaches. This paper describes such studies. Photosynthesis was followed by measuring the oxygen produced by *Chaetoceras affinis* Lauder during exposures in bottles suspended at various depths in the Pacific Ocean, and was compared with results from almost simultaneous exposures under a range of light intensities in the laboratory. Particular study was made of the relation between photosynthetic rate and light intensity in these two situations. Possible limitations in the method of determining photosynthesis by exposures of algal suspensions in closed bottles were explored,

[1] Contribution from the Scripps Institution of Oceanography, New Series.

[2] Present address: Freshwater Biological Association, Ambleside, England.

Reprinted from *Limnology and Oceanography*, 5:62–77, 1960.

and several photosynthetic characteristics of the diatom used were determined. The results are relevant to similar work in which photosynthetic activity is measured from the uptake of radioactive carbon (C^{14}).

The work was carried out during the tenure of a post-doctoral Rockefeller fellowship at the Scripps Institute of Oceanography, University of California, La Jolla. Grateful acknowledgment is made of the advice and assistance received from various people, particularly Dr. F. T. Haxo, Dr. B. M. Sweeney, Dr. K. A. Clendenning, Mr. M. Neushul, Mr. E. G. Cunnison, Mrs. Anne Dodson and Mrs. Nao Belser of the Department of Marine Botany. Invaluable help in the design and construction of apparatus was given by members of the Division of Special Developments.

MATERIALS AND METHODS

Unialgal cultures of *Chaetoceras affinis* were used throughout. This species had been isolated by Dr. B. M. Sweeney from Pacific coastal water and maintained in culture at the Scripps Institution of Oceanography for several years. It was chosen because of its ease of culture and fast growth of up to at least two cell divisions per day at 15°C. The species is widely distributed in coastal and offshore waters of the Pacific and other oceans. Although pronounced morphological aberrations were not found in healthy cultures, auxospore formation occasionally developed, with corresponding changes in average cell size, during the course of these experiments. Cultures were grown, and most of the following measurements of photosynthesis made, in an enriched sea-water medium containing 75% filtered sea-water. Its composition was based upon that used for *Gonyaulax polyedra* by Sweeney and Hastings (1957), modified by the addition of 5 mg/L of silicon (added as $Na_2SiO_3 \cdot 9\ H_2O$) and of various concentrations of potassium bicarbonate as described later. The large cultures intended for measurements of photosynthesis were grown in 2.5 L flasks at 14–16°C in a constant temperature room. An illumination of approximately 5,400 lux was provided by two fluorescent lamps below the cultures. Cultures were used when they had reached densities of 19,000–46,000 cells per ml, which were still considerably lower than the maxima attainable, about 100,000 cells per ml.

Photosynthesis was followed from changes in the concentration of dissolved oxygen, determined by the Winkler method. Samples of 100 ml were titrated with approximately N/160 thiosulphate, standardized with N/10 potassium dichromate solution. Exposures were made with the algal suspension distributed in ground-glass stoppered bottles of 'Pyrex' glass, capacity about 120 ml, filled from a siphon in a darkened room. Both clear and dark bottles were used in each exposure, and the gross photosynthesis, corrected for respiration, was obtained from the difference between them. Laboratory measurements were carried out with the bottles supported horizontally in a constant temperature water-bath, and illuminated from a bank of five 'daylight' fluorescent lamps placed immediately below the glass base of the bath. Minimum spacing between fluorescent tubes in the bank and between the bank and the bottles gave maximal illumination in the bottles of 2,100–2,500 lux. Each lamp was used with two ballasts connected in parallel to increase its total output. Three lower light intensities could also be obtained in a sequence along the long axis of the water-bath, using neutral filters made from exposed photographic film mounted in glass and placed on the bottom of the bath. The neutral transmission of the filters, in the spectral region 400–700 mμ, was checked spectrophotometrically. Three bottles could be inserted above each neutral filter, in a compartment isolated optically from its neighbors by partitions of blackened plastic. Owing to the extended nature of the light source, differences of light intensity across the depth of each bottle were negligible. Intensities at the levels of the bottles, on a plane perpendicular to the direction of the light source, were measured with a selenium rectifier-cell photometer (Weston model 756) calibrated in foot-candles but here converted to meter-candles or lux. The energy content of the radiation was also measured with a thermopile, yielding an interrelation of 44 ergs/cm^2/sec per foot-candle recorded by the photo-cell or 4.1 ergs/cm^2/sec per lux.

Field exposures were made near a fixed buoy in the ocean about one km. from the shore and laboratory. Bottles were filled with algal suspension in the darkened laboratory and transported in a dark container to this station. Here they were attached by spring clips to metal rings of approximately 60 cm diameter, and the latter fastened to a rope which supported both rings and bottles horizontally at various depths in the sea. The arrangement was devised as a modification of that described by Levring (1947), and enabled up to 9 bottles to be supported horizontally at any desired depth. Dark bottles exposed at several depths yielded almost identical values for respiratory oxygen uptake, which as in dark bottles in laboratory experiments was only a small fraction (0.05–0.2) of the maximum photosynthetic rate. As the sea exposures described here were of short duration, of one to one and a half hours, and were carried out in full sunshine about midday or early afternoon, changes in the intensity of incident solar radiation were small, not exceeding ± 10%. Continuous records were taken with an Eppley thermopile ('pyrheliometer') situated on the roof of the laboratory and used to calculate the mean irradiation on a horizontal surface during each experiment. Radiant energy in the spectral region available for photosynthesis was taken as 0.46 of the total energy recorded (Talling 1957a).

Light intensities at various depths in the ocean were calculated from the surface intensity, reduced by 5% representing surface loss (estimated from the solar elevation and data in Davis 1941 and in Anderson 1952), and measurements of underwater light penetration. The latter were made with a selenium rectifier photo-cell (Weston model 856 RR) in conjunction with red, green and blue Corning glass filters. Measurements were restricted to the linear region of the current-intensity characteristic of the photo-cell, corresponding to photo-currents below 700 μa. The mean extinction midpoints of the cell-filter combinations, in their working depth ranges, were estimated as respectively 450, 525, and 625 mμ; these values were calculated from the spectral sensitivity of the cell, spectral transmission of the color filters, and color-filter action of the sea-water. Values of the percentage transmission (T) per meter were derived from measurements in each spectral region. They were compared with the families of curves obtained by Jerlov (1951, Fig. 48) to describe the variation of T with wavelength in the spectral region 400–700 mμ, from which data for the spectral regions between the experimental points were interpolated. Using the resultant curves for the spectral variation of T, and the data of Taylor and Kerr (1941, Fig. 8, curve D) for the spectral distribution of energy in the surface-incident light, the spectral distribution of energy (400–700 mμ) was calculated for the depths at which photosynthesis was measured. Planimetric integration of the areas under these spectral distribution curves (see, Sverdrup, Johnson and Fleming 1942, p. 105) enabled the total radiant energy (400–700 mμ) at each depth to be expressed as a percentage of the sub-surface intensity. Knowing the latter, the absolute irradiation at each depth could be calculated. The results were very similar to those derived by a more approximate procedure (used by Jenkin 1937 and Talling 1957a) in which the radiant energy is divided between a few spectral blocks whose contributions are summed for each depth. In these field experiments the water-column showed an optical discontinuity, with more turbid water below (Figs. 5 and 6); consequently the calculation outlined above had to be performed in two stages.

Considerable vertical stratification was also shown in the depth-distribution of temperature, which was measured during each experiment with a bathythermograph. Tests were made to determine the rate of equilibration of temperature between the contents of experimental bottles and a colder surrounding medium of flowing water. After 10 minutes exposure the remaining temperature difference was less than 1°C when the initial difference was 7°C; the latter value would not be exceeded under the conditions of the field experiments.

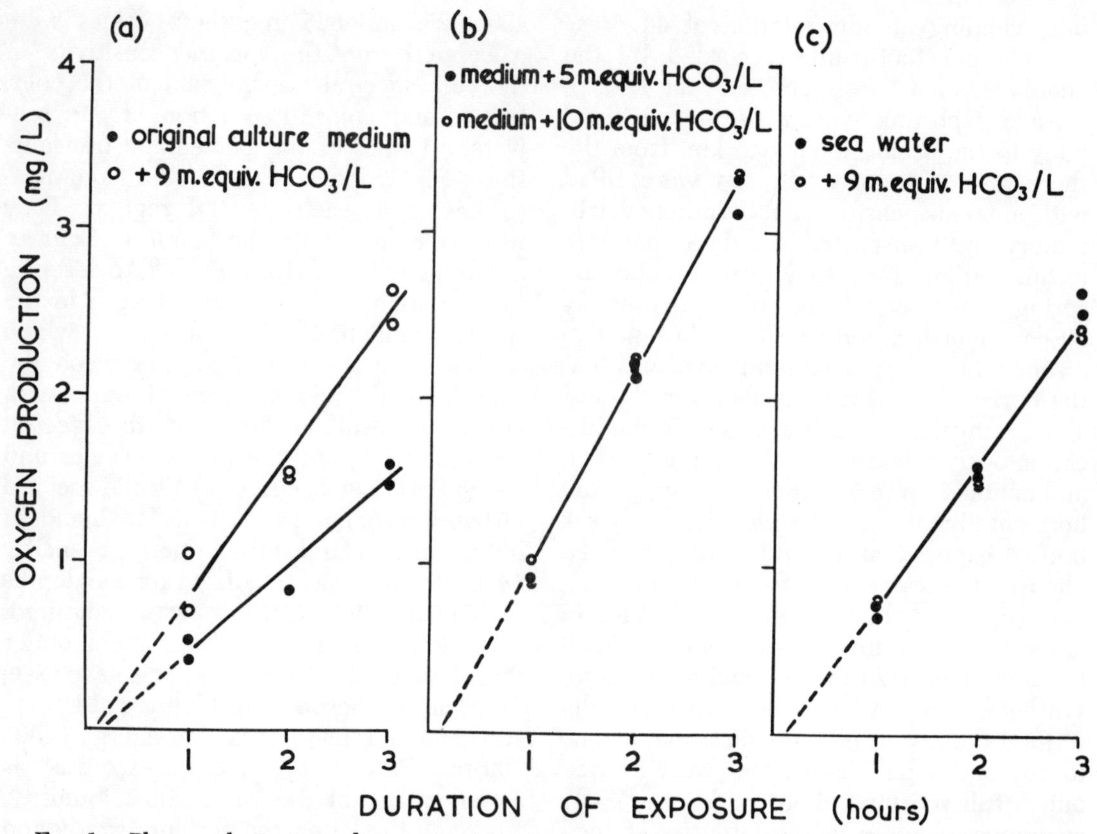

FIG. 1. Photosynthesis in media containing various supplementary additions of bicarbonate. These were made (a) to the original culture medium, (b) to the new culture medium with 5 m equiv of bicarbonate/L, (c) to cells resuspended in filtered sea water.

The densities of the algal suspensions were normally estimated from both cell counts and chlorophyll *a* determinations. Counts were made in duplicate upon volumes of 0.05 ml, counted on a normal microscope slide using the scanning procedure of Utermöhl (1936, Fig. 648). Chlorophyll *a* was estimated by the spectrophotometric method of Richards and Thompson (1952), the cells being first separated by paper filtration. Complete extraction of pigment from the cells was obtained within 3 hours.

FACTORS CONTROLLING PHOTOSYNTHESIS UNDER LABORATORY CONDITIONS

Some knowledge of the factors affecting photosynthesis under laboratory conditions was sought before attempting the analysis of photosynthetic behavior in the sea. Most measurements (in sections i–iii below) were made with high light intensities known to give complete light-saturation of photosynthesis and within a small temperature range (14–16.5°C). The variation of photosynthetic rate with temperature was not determined.

(*i*) *Composition of the medium*—It was clearly desirable to avoid determinations of photosynthesis in unfavorable media, particularly with respect to carbon dioxide and the associated carbonate-bicarbonate buffer system. Figure 1 illustrates comparative measurements of photosynthesis in a number of modifications of the culture medium previously used for the routine subculture of the diatom. The buffer capacity of the original medium was insufficient to prevent a considerable increase of pH (*e.g.* from 8.3 to 9.0) during the growth of cultures to the required density. Photosynthesis in such an alkaline medium (pH 9.0, combined CO_2 1.5–2.0 m equiv/L) is shown in Figure 1a, together with the improved rate obtained by the further addition of 9 m equiv/L of $KHCO_3$ and concomitant reduction of the initial pH to 8.5. Further experiments (Figs. 1b, 2b) showed that this response to added bicarbonate could be eliminated by the addition of 5 m equiv/L of bicarbonate after autoclaving but before inoculation of the cultures. This new culture medium with increased buffering, in which pH was maintained below 8.7, was adopted in all subsequent work. Nevertheless one experiment (Fig. 1c) showed that cells removed from an original culture by filtration, and resuspended in filtered sea-water of pH 8.3, yielded a rate of photosynthesis equal to that of similar cells resuspended in sea-water, enriched by 9 m equiv/L of bicarbonate at pH 8.1. For short exposures, therefore, in which the demands upon the carbon dioxide supply of the medium are small, sea-water without bicarbonate enrichment would appear an adequate medium. Inadequate buffering appears only after the longer exposures used during growth of the cultures to suitably high densities.

(*ii*) *Agitation of the cell suspensions*—Much criticism (*e.g.*, Emerson and Green 1934, Emerson 1935) has been levelled against measurements of photosynthesis by aquatic plants in unshaken and sealed vessels, particularly when large samples of macrophytes were used. Limitations of gaseous diffusion in unstirred media are likely to be minimal for dilute suspensions of unicellular algae, but few tests of this point are recorded in the abundant ecological literature (see, however, Talling 1957a, Doty and Oguri 1958). An experiment was therefore made to test whether rates of photosynthesis differed between unshaken suspensions in which cells were allowed to sediment and suspensions in which shaking by hand every 15 minutes maintained cells in free circulation. Continuous shaking by horizontal movements, with glass beads in the bottles, was also tested but found to be mechanically inadequate for these completely filled bottles. The experimental results (Fig. 2a) show almost identical rates of photosynthesis in shaken and unshaken bottles, despite a considerable sedimentation of cells which occurred in the latter during the three-hour exposure. Several test conditions were more severe than those likely to prevail in ecological field experiments, notably the high algal density (41.7×10^6 cells, or 0.081 mg chlorophyll *a*/L) and high photosynthetic rate (2.37 mg oxygen /L hour). The latter was the highest rate in the present experiments, being probably accelerated by the higher temperature chosen for this exposure (19.7±0.4°C, compared with 14–16.5°C in all other laboratory experiments).

Rates of photosynthesis under light saturation were also compared, using unshaken Winkler bottles and continuously shaken Warburg manometers, at 15.0–15.5°C. The manometric measurements were carried out by Dr. K. A. Clendenning as follows. All cells were derived from the same culture, of which a portion was concentrated by paper filtration to yield the higher cell densities required for the manometric measurements. The cells were centrifuged from this concentrate and resuspended in fresh growth medium with additions of bicarbonate, giving total bicarbonate enrichments of 5–25 m equiv/L at pH 8.3. The cell density so obtained was 24 times higher than that employed for the Winkler determinations. Under the conditions of the Winkler and manometric measurements, the light-saturated rate of photosynthesis was not increased by raising the bicarbonate enrichment above 5 m equiv/L. The rates of photosynthesis observed manometrically with 5 m equiv/L bicarbonate, during the first hour after induction, were within the range of the corresponding Winkler measurements shown in Figure 2b. The agreement between the two methods is noteworthy in view of the very different conditions of shaking and cell concentration which were involved. Manometric measurements of photosynthesis have been made on another species of *Chaetoceras* by Sagromsky (1943), who could not show any injurious effects of the shaking involved. A direct experimental comparison of the two methods, similar to that described above, appears to be lacking in previous work, despite the extensive literature on each method.

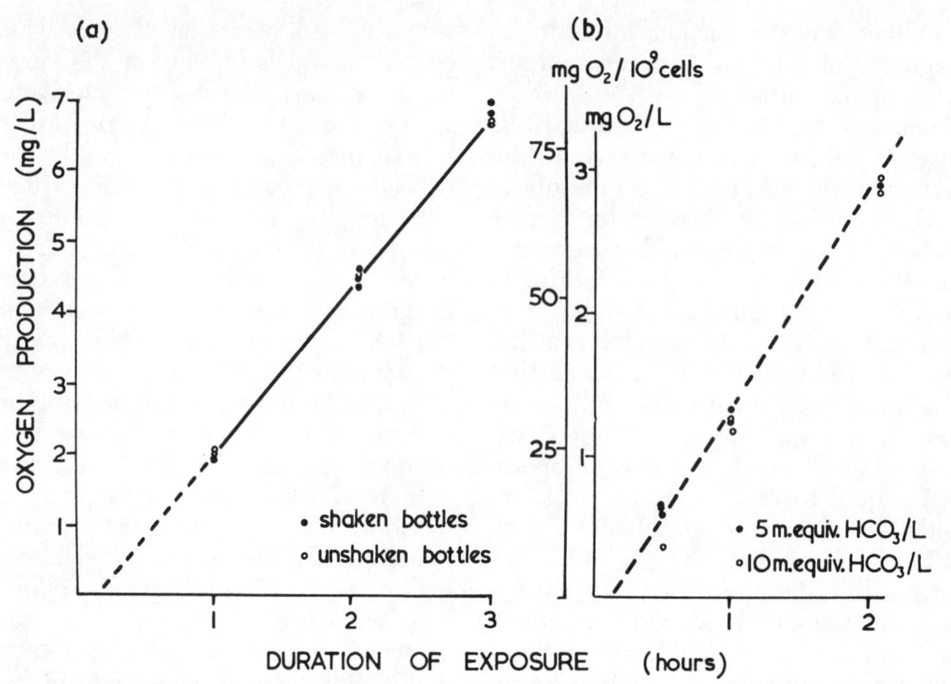

FIG. 2. Photosynthesis in shaken and unshaken cell suspensions, measured by (a) the Winkler method only (b) the Winkler and manometric methods. In (b) the mean rate obtained manometrically is expressed by the gradient of a broken line which is superimposed on Winkler determinations of oxygen production in two unshaken media.

(*iii*) *Variation of photosynthetic activity during exposures*—Measurements of photosynthesis in closed bottles are usually expressed as average rates over the periods of exposure, in which possible short-time changes of activity are not distinguished. Such changes can be followed either by repeated sub-sampling from each bottle, which is impracticable with the Winkler oxygen determination, or by use of replicate bottles which are sampled in succession. Both alternatives are inconvenient in field experiments, but successive sampling of replicate bottles has been used whenever possible in the present laboratory experiments. Results are illustrated in Figures 1, 2 and 3, from which the variation of photosynthetic rate with time can be followed within the exposure limits of 3–3¼ hours used.

In all experiments the time course of photosynthesis shows two general features. An almost constant rate is maintained during the second and third hours of each exposure, despite an increasing supersaturation with oxygen. In the first hour or half-hour the overall rate is lower, and extrapolation shows that it is *equivalent* to a lag of 6–18 minutes succeeded by the rate maintained during most of the exposure. This 'lag-equivalent' is shown by exposures in a wide range of light intensities (Fig. 3), although all these intensities were higher than the very low intensities prevailing when the bottles were being prepared. An induction phase of photosynthesis is therefore likely to occur in all exposures. The 'lag-equivalent' is also seen in exposures involving cells with different photic histories (Fig. 3).

These results clearly imply that rates measured in a single uninterrupted exposure will underestimate the rates finally realized under the same conditions. Thus in the combined laboratory and sea experiments described later with exposures of 1–1½ hours, the probable underestimation is 7–30%. This factor will diminish with increasing length of exposure; its small significance for exposures of 3 hours is illustrated in Figure 3d. Steemann Nielsen and Jensen (1957) have also described lower initial rates of photosynthesis during exposures with a natural

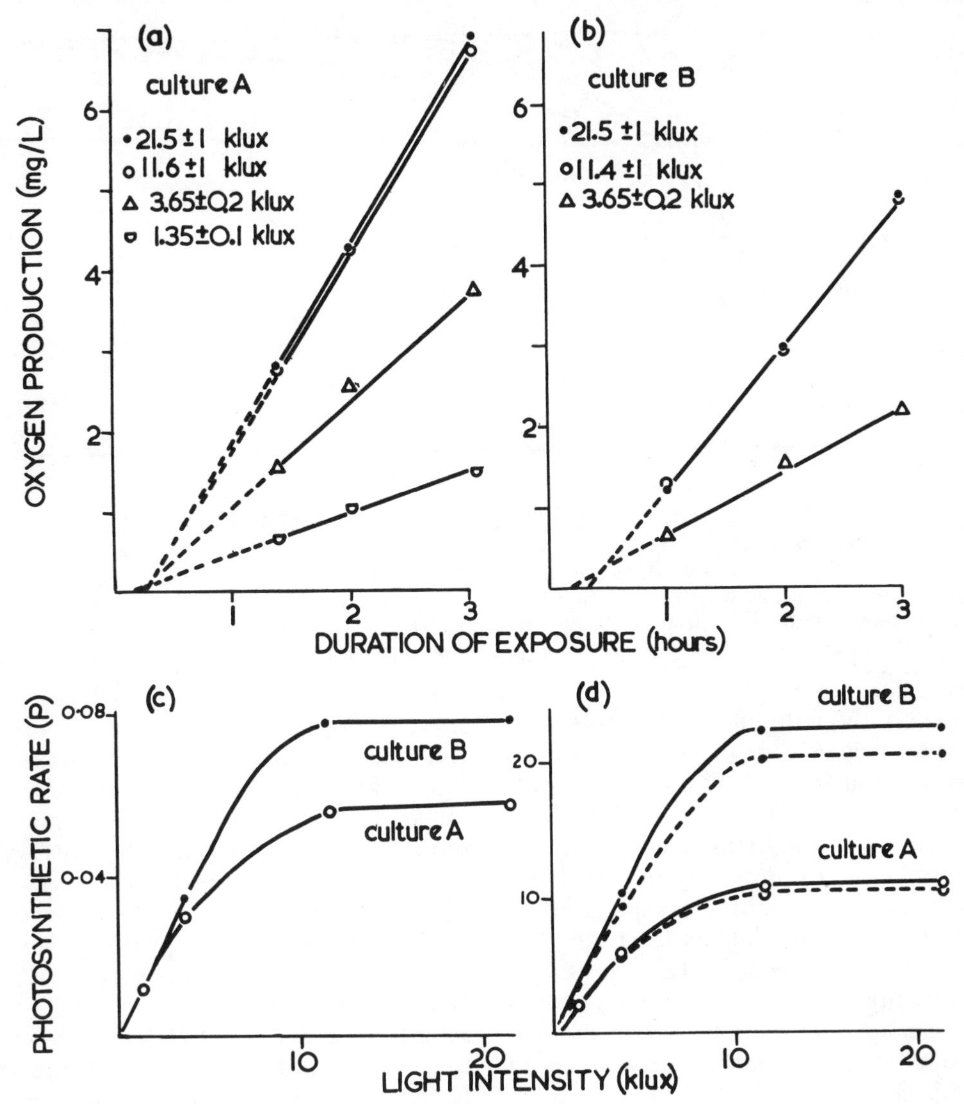

FIG. 3. Photosynthesis–time measurements for two cultures at various light intensities (above), with average gradients inserted and used to derive rate–intensity curves (below). Culture A was grown at 0.97 klux, and B at 5.4 klux, followed by darkness for 2 days. Photosynthetic rates (P) are given in mg oxygen/mm³ cell volume hour (c) and in mg oxygen/mg chlorophyll *a* hour (d). Broken lines in (d) show the overall rates of photosynthesis during the longest exposures.

population of surface plankton. They attribute this to an after-effect of light-inhibition, an explanation which is not applicable to the present results.

(*iv*) *Light intensity*—Without some knowledge of the relation between photosynthetic rate and light intensity an individual measurement of photosynthesis has little significance, as rates ranging from zero to the light-saturated capacity can be obtained at various intensities. Rates measured at light saturation (here denoted by P_{max}) allow the most direct comparison between laboratory and field experiments. The achievement of such conditions in the laboratory is described below, and the variation of rates with light intensities below saturation is discussed.

The general form of the rate-intensity relation is shown by data from laboratory experiments conducted simultaneously with exposures in the sea (Figs. 5c, 6c). Very similar curves were obtained in four other experiments, not illustrated. An upper

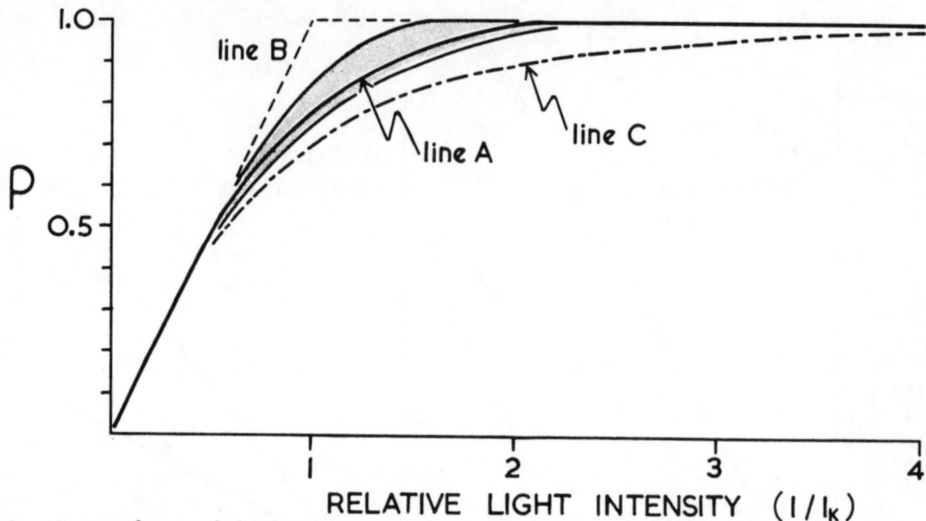

FIG. 4. Various forms of the 'intrinsic curvature' of the photosynthesis–light intensity curves below light saturation. Rates of photosynthesis are expressed as fractions (p) of the rate at light saturation, and light intensities in multiples of the I_K characteristic. The heavy line A indicates the average relation, and the stippled area the range of variation, in the present experiments. Line B shows the initial linear region continued up to the light-saturated rate: line C shows the relationship of Smith (1936, equation 1) here expressed as $p\ (1-p)^{-1/2} = I/I_K$.

'plateau' on the rate-intensity curves provides good evidence that light-saturation is fully realized at the highest intensities used, 24,000–27,000 lux. The onset of saturation in relation to the light intensity scale can be indicated by the intensity at which extrapolations of the initial linear gradient and final saturation 'plateau' intersect (see Fig. 5c; also Kok 1956a, b, Talling 1957a). This intensity, here denoted by I_K, varied little (between 4,600–6,800 lux) in five experiments, all at temperatures between 14 and 16.5°C.

Considerable interest attaches to the form of the 'intrinsic curvature'—the curvature property independent of the absolute scales used—shown by the rate-intensity curves at intermediate light intensities. With P_{max} and I_K variables, it can be used to summarize much of the rate-intensity characteristic. It has received little attention in ecological work, often being completely neglected (*e.g.* by Ryther 1956). Modifications associated with optically dense suspensions (Rabinowitch 1951, p. 1008) are unlikely to be appreciable with the relatively thin suspensions used in this and most other ecological work. The description of such intrinsic curvature is possible either by the application of a mathematical relation (*e.g.* Kok 1956b, Talling 1957a), or by plotting curves with standard scales for the light-saturated rate (P_{max}) and the initial linear gradient. The latter approach is the more flexible and is illustrated in Figure 4. The same figure (line C) shows the form of curvature implied by the relation proposed by Smith (1936, equation 1) and used in several previous studies (*e.g.* Winokur 1948, Talling 1957a). The approach to light-saturation so derived is clearly more gradual than the experimental data indicate. Similar divergence is shown in some, but not all, of the data of Talling (1957a) obtained from studies of a freshwater plankton diatom. Steemann Nielsen and Jensen (1957) have stated that the equation was inapplicable to their rate-intensity curves for marine phytoplankton, but the type of curvature appears variable and the attainment of complete light-saturation is sometimes not established. Other equations, such as that of the rectangular hyperbola, have been tested here with even less success. Consequently the curvature relation is probably best expressed in a standardized graphical form, as in Figure 4, although errors can easily arise in estimating the initial gradient of the curves.

A further experiment was designed to ob-

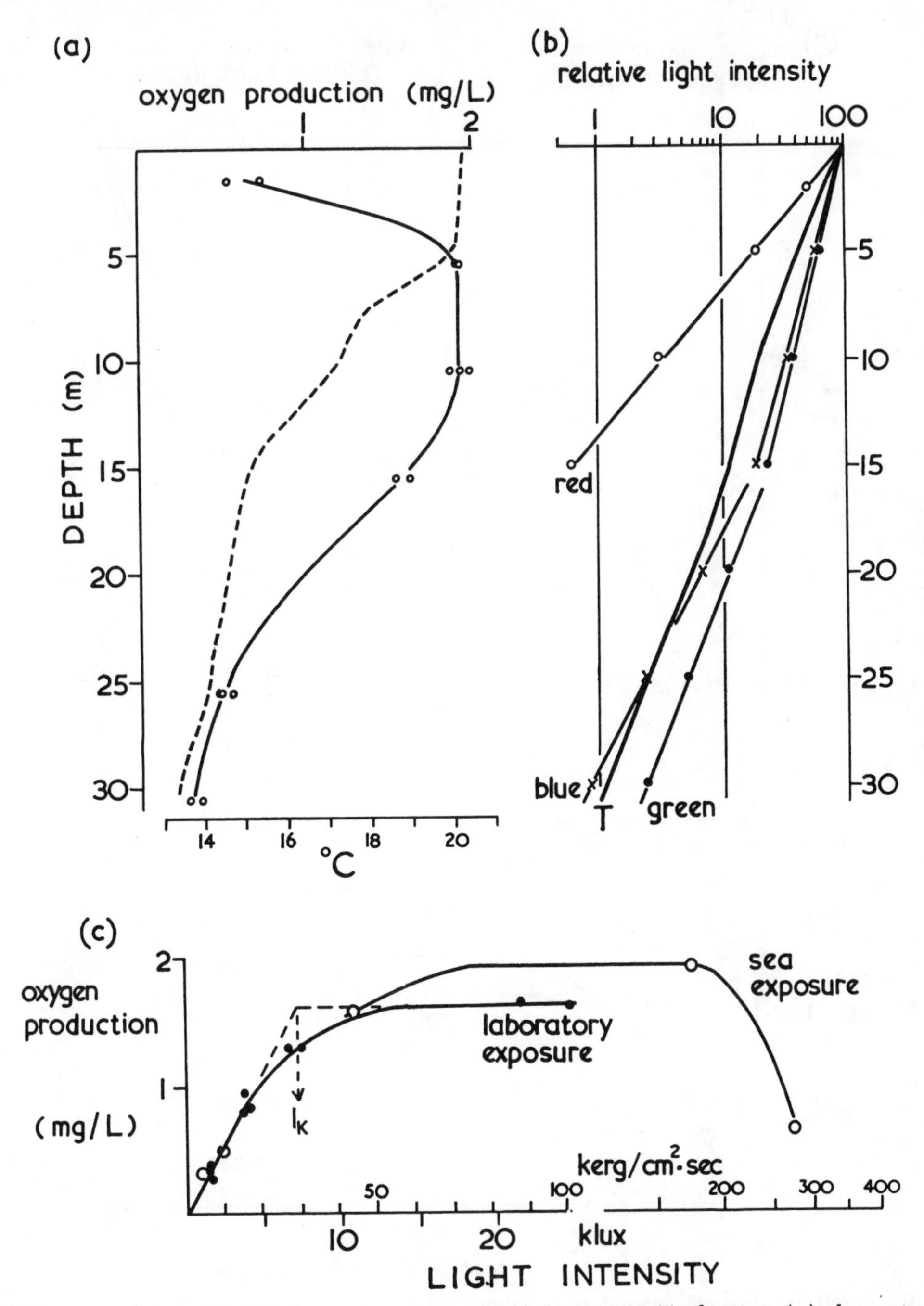

FIG. 5. Combined field and laboratory experiment of 5 Aug., 1957, showing (a) the variation of photosynthetic rate (solid line) and temperature (broken line) with depth, (b) light penetration measured for three spectral regions and of total energy (T) in the region 400–700 mμ, (c) the variation of photosynthetic rate with light intensity (higher intensities shown on a logarithmic scale) obtained from both laboratory and sea exposures.

tain data on the rate-intensity characteristic from exposures in which a time-sequence of determinations enabled the effects of the initially lower rates to be eliminated. The same experiment included a comparison of two sets of cells given different pre-treatments, one being grown at a low intensity of 970 lux, and the other at the normal cul-

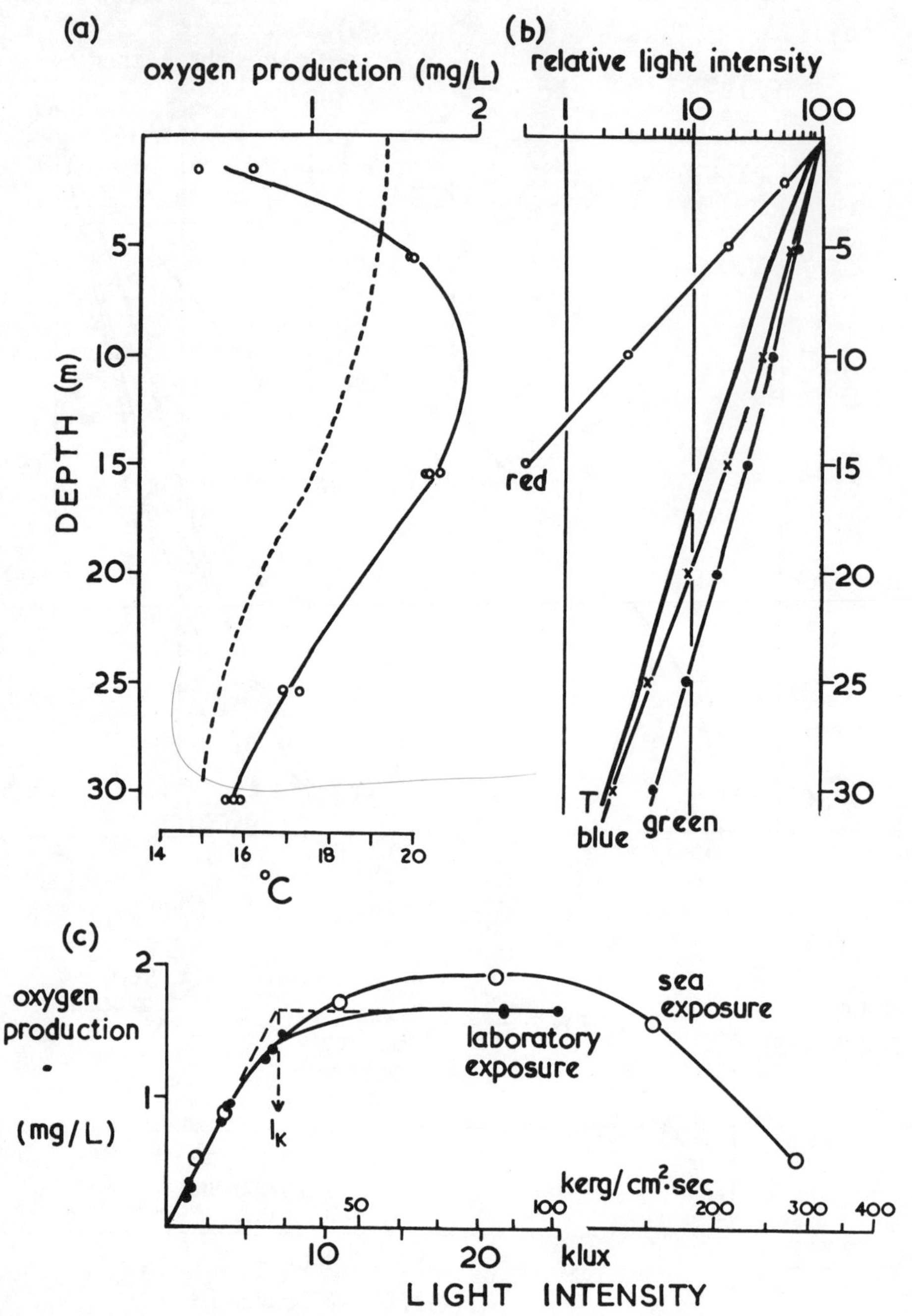

FIG. 6. Combined field and laboratory experiment of 7 Oct., 1957. Representation as in Figure 5.

ture intensity of 5,400 lux but then kept in darkness for two days. Figure 3 illustrates both the original oxygen determinations and the derived rate-intensity characteristics. The latter do not show any appreciable modification in the intensities at which light saturation developed. Similar behavior by cultured populations of several planktonic algae, grown under different types of illumination, is described by Ryther (1956). However, marked modifications have been found in several natural populations, when the characteristics of algae obtained from various depths have been compared (Steemann Nielsen and Hansen 1959, and unpublished experiments of the author).

PHOTOSYNTHESIS IN THE SEA AND LABORATORY COMPARED

Rate-intensity characteristics have been calculated for two experiments, in which simultaneous field and laboratory exposures with the same algal material were made. Figures 5 and 6 illustrate the variation of photosynthetic rate with depth, with the vertical gradient of light intensity expressed for three spectral regions and for the entire visible spectrum (400–700 mμ). From these data the corresponding rate-intensity curves are derived, each depth being replaced by the equivalent light intensity expressed as an energy flux for the 400–700 mμ spectral region. For each experiment direct comparison can be made with the rate-intensity curves obtained in the parallel laboratory exposure with fluorescent illumination. Intensities of the latter are also expressed in terms of energy flux, using the factor noted above; the limits of the spectral region involved (illustrated by Withrow and Withrow 1956, Fig. 3–15) were close to the limits of 400 and 700 mμ adopted in calculating light intensities in the sea. The rate-intensity curves obtained from the sea exposures suffer from various complicating influences connected with the vector distribution and spectral modification of underwater light (see below), so that their correlation with laboratory data is, *a priori*, uncertain. An important difficulty was the presence of vertical gradients of temperature in the sea (see Figs. 5 and 6). The laboratory temperatures, 14.9–16.4°C, were chosen to lie near the centers of the ranges occurring in the photic zone of the sea. Experiments in which the sea surface temperature exceeded 20.5°C are not illustrated.

The variation of photosynthetic rate with depth is of a form familiar from many other studies, both in the sea and freshwater. Much of it can be deduced from the rate-intensity characteristic measured in the laboratory, considered in relation to the almost exponential decrease of photosynthetically active radiation with depth in the sea. The initial linear region of the rate-intensity characteristic corresponds to the lower 'hollow' part of the rate-depth curve; the succeeding 'plateau' region at light-saturation is also reflected, in an abbreviated form, in the depth curve. A partial inhibition of photosynthesis is present at the higher light intensities near the sea surface, which were not attainable with the fluorescent lighting used in the laboratory exposures. The general form of the inhibition, and the intensities under which it developed, resemble those recorded in several other field studies with planktonic diatoms (Jenkin 1937, Talling 1957a). Nevertheless the extent of the inhibition may well depend on the spectral composition of the underwater radiation, particularly in the blue (Kok 1957a) and ultra-violet regions. Consequently the general validity of a calculation of the depth-distribution of photo-inhibition in nature, based on measurements in sunlight reduced by 'neutral' filters (Ryther 1956, Steemann Nielsen 1958, p. 45), is not established. Even direct field measurements of the inhibition may be influenced by the absorption of ultra-violet radiation in the glass walls of the experimental vessels (Gessner and Diehl 1951, Gessner 1955, p. 415). This factor requires further investigation.

Quantitative comparison of the photosynthetic behavior in laboratory and sea exposures can be based upon (i) the rates obtained at light saturation (P_{max}) or (ii) the initial gradients of the rate-intensity curves ($= P_{max}/I_K$). These features have contrasted responses to two factors (Rabinowitch 1951, 1956), temperature and spectral composition of the illumination, which differ between exposures in laboratory and sea. Temperature differences are likely to alter (i) but not (ii), but for spectral differences this situation is reversed. Both characteristics are displayed in Figures 5 and 6, which show considerable agreement between photosynthetic behavior in laboratory and sea. Rates at light-saturation are slightly higher in the sea exposures, but the differences probably result from the higher temperatures present in the upper layers of the sea. As such rates measured in the laboratory have been shown to be uninfluenced by cell sedimentation in unshaken bottles, the limitation of rates measured in field experiments by this factor is very unlikely. At lower light intensities the gradients of the rate-intensity relationship agree between both types of exposure. This similarity, with the small differences between the light-saturated rates, determines the correspondence between I_K values measured in laboratory and sea.

SOME PHOTOSYNTHETIC CHARACTERISTICS OF *CHAETOCERAS AFFINIS*

Although not the main object of these experiments, some descriptive information was accumulated on the photosynthetic characteristics of the diatom used. It is discussed below in relation to the few other comparable data which are available for the ecologically important algae of the marine and freshwater plankton.

The onset of light saturation of photosynthesis in relation to light intensity is an important determinant of the depth-distribution of photosynthetic rates; it can be characterized by a single intensity the I_K value. Its variation in natural populations of phytoplankton is poorly understood, but includes the effects of a pronounced, though often neglected temperature dependence. The present laboratory experiments yield I_K values between 4,600 and 7,700 lux, or an estimated equivalent of 19–32 kerg/cm^2 sec, at 14–16·5°C. The results of the sea experiments are similar but involve a greater temperature range. Wider comparison is often difficult, as other work with planktonic algae has frequently involved light fields with different spectral qualities and the use of different instruments or means of calculation for estimating light intensities. There is a general similarity with the values indicated for planktonic algae, particularly diatoms, at comparable temperatures (*e.g.* Jenkin 1937, Sagromsky 1943, Ryther 1956, Talling 1957a). The most conspicuous divergencies are shown by the high saturation intensities obtained for the surface phytoplankton of tropical oceans by Steemann Nielsen (1952) and Steemann Nielsen and Jensen (1957), using fluorescent illumination and an indirect means of assessing the effective light intensities.

The maximum photosynthetic capacity of a unit quantity of algae, measured under light-saturation, is another characteristic useful for analytical and comparative purposes. Three indices of algal quantity were used here: (i) cell numbers, (ii) cell volume, calculated from cell numbers and cell dimensions, and (iii) chlorophyll *a* content. The first is unsuitable as a general measure owing to the variation of cell size, which also contributed to changes of chlorophyll *a* content per cell. Contents measured for 11 healthy cultures grown at 5,400 lux varied from 1.0–3.6 mg chlorophyll *a*/10_9 cells; a single culture grown at 970 lux gave the highest value of 4.6 mg/10^9 cells. Rates based upon cell volume and particularly chlorophyll *a* content have been much used in other studies of phytoplankton photosynthesis and productivity, although the experimental conditions have often been ill-defined especially as regards media, temperature, and light-saturation. The present laboratory experiments, carried out at 14–16°C using adequately buffered media and cells grown at 5,400 lux, yield rates of 29–32 mg 0_2mg chlorophyll *a*/hour for the four longer experiments in which the initially lower rates can be eliminated (Figs. 1, 2 and 3). These values are much greater than those commonly reported for either cultured or natural populations of algae. Ryther and Yentsch (1957) have based a method of calculating photosynthetic production in the sea upon an average value of 3.7 g C/g chlorophyll *a*/hour, which is equivalent, using their value of 1.25 for the photosynthetic quotient, to 12.3 mg O_2/mg chlorophyll *a*/hour. The present experiments provide fewer data for rates based upon cell volume, owing to uncertainties in estimating volume for geometrically complex cells. The values obtained from two laboratory experiments, under the conditions listed above, are 0.048 and 0.073 mg 0_2/mm^3/hour; in the first figure the effect of the initial 'lag' is eliminated. These rates are also higher than most values previously published for natural populations of phytoplankton (Verduin 1956), although higher values are not unknown (Talling 1957c).

Little is known concerning the variation of photosynthetic behavior between cells with different histories, which can be important in other algae (*e.g.* Sorokin 1958). An experiment already described (Fig. 3) compared the behavior of cells grown at a low light intensity of 970 lux with that of cells grown at a higher intensity of 5,400 lux and then kept in darkness for two days. These differing pre-treatments had little effect on the I_K characteristic, but light-saturated rates of photosynthesis were lower in cells cultured at the lower intensity whether expressed in terms of cell numbers, cell volume, or chlorophyll *a* content. The relatively vigorous activity of cells after two days in darkness is ecologically significant, as such dark conditions must often befall cells circulating in nature. A similar response has been obtained with a freshwater plankton diatom (Talling 1957a).

DISCUSSION

Most of the numerous recent measurements of photosynthesis of phytoplankton populations have involved attempts to duplicate, in field experiments, conditions prevailing in a natural water-body during a certain time. As even the field experiments described here involve simplified conditions, such as uniform algal material and short exposures, many factors affecting the populations in nature are not represented. Examples include the possible differentiation of photosynthetic capacities between cells circulating at various depths or photosynthesising at different times of the day. Nevertheless such simplification allows a better resolution of many important factors and permits direct tests of the application of laboratory data to field conditions.

Within these limits, the results generally support the value of measurements based on exposures in closed bottles even when, as in field experiments, effective shaking of the cell suspensions is not possible. The conditions controlling the appearance of secondary effects during long exposures (*cf.* also Ryther and Vaccaro 1954, Vollenweider 1956, Bauer 1957, Rodhe 1958) are not explored, but the photosynthetic rates were shown to be well maintained in exposures of up to three hours. They appeared only slightly modified by lower initial rates, in which induction phenomena were presumably involved. The effectiveness of the method of following photosynthesis by the Winkler oxygen estimation is particularly noteworthy in view of the relatively high algal densities and photosynthetic rates involved. It has been suggested (Steemann Nielsen and Jensen 1957) that uptake of iodine by unsaturated algal constituents may render the Winkler method unusable with dense algal populations. This consequence is not shown in the present experiments or many others described in the literature. The measurements here also show no indications of ill-effects due to the supersaturation with oxygen which resulted from high photosynthetic rates. Similar results were obtained by Vinberg (1934), using the Winkler method with freshwater phytoplankton. In general a marked supersaturation is undesirable, due to its possible effects on bubble-formation and rates of respiration (Gessner and Pannier 1958a,b). Recent unpublished work by the author demonstrates that the Winkler oxygen determination can also be used to detect photosynthetic changes as low as 0.02 mg/L, about two orders of magnitude lower than the changes involved in the present experiments.

Experiments which combine both laboratory and field exposures have relevance for the various recent attempts to use laboratory data for estimating photosynthetic productivity in nature (Ryther 1956, Ryther and Yentsch 1957, Rodhe, Vollenweider and Nauwerck 1958, Cushing 1958). A principal difficulty here is the comparison of light intensities measured in the two situations, particularly in view of the progressive spectral modification of underwater light with depth, which cannot be reproduced by a single color filter as used by Cushing (1957, 1958). A full comparison would require evaluation of the effects of differing qualities of vector and spectral distribution of radiation in laboratory and sea, the latter considered in relation to the action-spectrum of photosynthesis. The present assessment of effective light intensities, by the total energy flux of photosynthetically active radiation, is only an approximation but has value for interpreting field experiments under different conditions. Here it has given

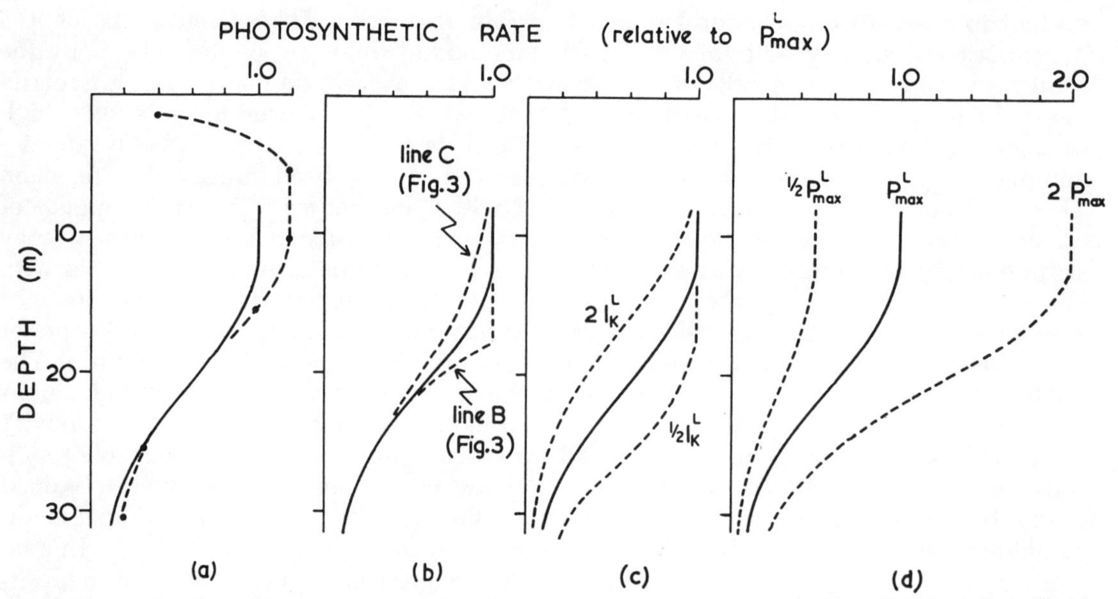

FIG. 7. The depth distribution of photosynthetic rate (full lines) calculated from the laboratory rate–light intensity curve, and the intensity–depth curve, for 5 Aug., 1957. With this depth distribution are compared (a) the photosynthetic rates measured *in situ*, (b) the depth distribution of rates calculated from two variants of the rate–intensity curve (see Fig. 4), (c) the depth distributions calculated for given changes in the I_K characteristic, and (d) for the same changes in the P_{max} characteristic. All photosynthetic rates are expressed relative to the laboratory value at light saturation ($P_{max}{}^L$).

general agreement between rate-intensity characteristics of photosynthesis measured in the laboratory and at sea, so providing some support for the use of such laboratory data in calculating the vertical distribution of photosynthetic rates in nature. Such a calculation is illustrated in Figure 7. Because of the approximations involved, the agreement should be regarded as a working correlation rather than the outcome of a physically and physiologically adequate comparison. However, the type of calculation appears more complete than that involved in any previous comparison of laboratory and field experiments with planktonic algae. Thus the correlative relationship established by Steemann Nielsen (1952) does not include a direct comparison of specific light intensities or their effects. Ryther (1956) has described and compared rate-intensity characteristics of photosynthesis measured in the sea and in sunlight reduced by graded neutral filters. His estimates of underwater light intensities in foot-candles are based upon a single vertical extinction coefficient expressing the diminution in the response of a photo-cell with depth, and suffer from limitations discussed by Lund and Talling (1957) and Strickland (1958).

For many ecological purposes, estimates are needed of the total photosynthetic production below unit area. This can be expressed by the area enclosed by a photosynthesis-depth profile, and can be related to the individual variables controlling the distribution of photosynthetic rate with depth (Talling 1957b). When this depth distribution includes a well-developed region with light-saturation, as illustrated in Figures 5 and 6, it can easily be shown (Fig. 7) that the area mentioned is much more sensitive to variation in the maximum rate (P_{max}) than to similar variation in the onset of light-saturation (expressed by I_K). This feature is expressed in the equations derived by Talling (1957b) to relate the areal production to the P_{max} and I_K variables. With these conditions the experimental errors in determining I_K are less significant than errors in P_{max}, for which measurement of light intensity is not required. The obvious uncertainties in the measurement of I_K, therefore, are unlikely to affect strongly the final calculation of areal production.

Errors or uncertainties in the measurement of I_K, or of the initial gradient of the rate-intensity relation, will tend to become more serious in this connection when the surface intensity (I_0) is low and approaches I_K, and photosynthesis decreases rapidly with depth from the water surface. Another factor affecting the area under the photosynthesis-depth profile, and hence the areal production, is the intrinsic curvature of the rate-intensity curve. However when the effects of the most extreme curvature types of Figure 4 were tested, using a graphical construction of photosynthesis-depth profiles, the changes induced were small (Fig. 7b).

On present knowledge, field experiments involving exposures *in situ* still provide the most direct estimation of photosynthetic productivity by natural populations of phytoplankton. With suitable design, they can also allow the observed behavior to be interpreted in terms of a limited number of variables, from which photosynthetic behavior in other field situations can be estimated (Manning and Juday 1941, Edmondson 1956, Talling 1957a, b). The application of laboratory measurements involves still wider extrapolation, but has been used in recent work with radio-carbon in view of the saving of time and greater ease of measurement. Nevertheless the significance of the measurements is often uncertain, partly on purely physiological grounds such as the unknown effects of temperature on the P_{max} and I_K variables and the incomplete attainment of light-saturation, and partly from uncertainties from the widespread application of formulae (*e.g.* those of Steemann Nielsen 1952, p. 132; Ryther 1956, Ryther and Yentsch 1957) without sufficient testing of the validity of the basic assumptions for local conditions. Other attempts to simplify field conditions of exposure, using photosynthetic rates measured only near the water surface, are still more unsatisfactory in view of the unknown significance of photo-inhibition in such exposures. A need exists for the greater development of analytical studies of photosynthetic productivity by phytoplankton, broadly based upon both laboratory and field experiments, and involving short exposures under conditions better defined than in most previous field experiments.

REFERENCES

ANDERSON, E. R. 1952. Energy-budget studies. U.S. Geol. Surv., Circ. **229:** 71–119.

BAUER, G. 1957. Die Ausnützung der Sonnenergie durch das Phytoplankton eines eutrophen Sees. Inaugural-Dissertation, Universität München, 77 pp.

CUSHING, D. H. 1957. Production of carbon in the sea. Nature, Lond., **179:** 876.

———. 1958. Some experiments using the ^{14}C technique. Rapp. Cons. Explor. Mer., **144:** 73–75.

DAVIS, F. J. 1941. Surface loss of solar and sky radiation by inland lakes. Trans. Wis. Acad. Sci., Arts & Lett., **33:** 83–93.

EDMONDSON, W. T. 1956. The relation of photosynthesis by phytoplankton to light in lakes. Ecology, **37:** 161–174.

EMERSON, R. 1935. The effect of intense light upon the assimilatory mechanism of green plants, and its bearing on the carbon dioxide factor. Cold Spring Harbor Symp. Quant. Biol., **3:** 128–137.

——— AND L. GREEN. 1934. Manometric measurements of photosynthesis in the marine alga *Gigartina*. J. Gen. Physiol., **17:** 817–843.

GESSNER, F. 1955. Hydrobotanik. I. Energiehaushalt. Deutscher Verlag der Wissenschaften, Berlin, 517 pp.

——— AND A. DIEHL. 1951. Die Wirkung natürlicher Ultraviolettstrahlung auf die Chlorophyllzerstörung von Planktonalgen. Arch. Mikrobiol., **15:** 439–453.

——— AND F. PANNIER. 1958a. Der Sauerstoffverbrauch der Wasserpflanzen bei verschiedenen Sauerstoffspannungen. Hydrobiologia, **10:** 323–351.

———. 1958b. Influence of oxygen tension on respiration of phytoplankton. Limnol. Oceanogr., **3:** 478–480.

JENKIN, P. M. 1937. Oxygen production by the diatom *Coscinodiscus excentricus* Ehr. in relation of submarine illumination in the English Channel. J. Mar. Biol. Assoc., U.K., **22:** 301-343.

JERLOV, N. G. 1951. Optical studies of ocean waters. Rep. Swedish Deep-Sea Expd., **3:** 1–59.

KOK, B. 1956a. On the inhibition of photosynthesis by intense light. Biochim. Biophys. Acta, **21:** 234–244.

———. 1956b. Photosynthesis in flashing light. Biochim. Biophys. Acta, **21:** 245–248.

———. 1957. Report on some recent results at Wageningen. In: Gaffron, H. (ed.), Research in Photosynthesis, pp. 353–365. Interscience Publishers, New York. xiv + 524 pp.

LEVRING, T. 1947. Submarine daylight and the photosynthesis of marine algae. Goteborgs

VetenskSamh. Handl., Sjatte Folj. B, **5**(6): 1–90.

LUND, J. W. G., AND J. F. TALLING. 1957. Botanical limnological methods with special reference to the algae. Bot. Rev., **23**: 489–583.

MANNING, W. H., AND R. E. JUDAY. 1941. The chlorophyll content and productivity of some lakes in north-eastern Wisconsin. Trans. Wis. Acad. Sci., Arts & Lett., **31**: 377–410.

RABINOWITCH, E. I. 1951. Spectroscopy and fluorescence of photosynthetic pigments; Kinetics of photosynthesis. Photosynthesis and related processes, II, Part 1. Interscience Publ., New York pp. xi + 603–1208.

———. 1956. Kinetics of photosynthesis (cont'd). Addenda to I and II, Part 1. Photosynthesis and related processes, II, Part 2. Interscience Publ., New York. pp. xvi + 1209–2088.

RICHARDS, F. A., AND T. G. THOMPSON. 1952. The estimation and characterisation of plankton populations by pigment analyses. II. A spectrophotometric method for the estimation of plankton pigments. J. Mar. Res., **11**: 156–172.

RODHE, W. 1958. The primary production in lakes: some results and restrictions of the ^{14}C method. Rapp. Cons. Explor. Mer, **144**: 122–128.

———, R. A. VOLLENWEIDER, AND A. NAUWERCK. 1958. The primary production and standing crop of phytoplankton. pp. 299–322. *In* 'Perspectives in marine biology,' ed. A. A. Buzzati-Traverso. Univ. of California Press, Berkeley. xvi + 621 pp.

RYTHER, J. 1956. Photosynthesis in the sea as a function of light intensity. Limnol. Oceanogr., **1**: 61–70.

——— AND R. F. VACCARO. 1954. A comparison of the oxygen and ^{14}C methods of measuring marine photosynthesis. J. Cons. Int. Explor. Mer, **20**: 25–34.

——— AND C. S. YENTSCH. 1957. The estimation of phytoplankton production in the ocean from chlorophyll and light data. Limnol. Oceanogr., **2**: 281–286.

SAGROMSKY, H. 1943. Die Bedeutung des Lichtfaktors für den Gaswechsel planktontischer Diatomeen und Chlorophyceen. Planta, **33**: 299–339.

SMITH, E. L. 1936. Photosynthesis in relation to light and carbon dioxide. Proc. Nat. Acad. Sci., Wash., **22**: 504–511.

SOROKIN, C. 1958. The effect of the past history of cells of *Chlorella* on their photosynthetic capacity. Physiol. Plant., **11**: 275–283.

STEEMANN NIELSEN, E. 1952. The use of radioactive carbon (C^{14}) for measuring organic production in the sea. J. Cons. Int. Explor. Mer, **18**: 117–140.

———. 1958. Experimental methods for measuring organic production in the sea. Rapp. Cons. Explor. Mer, **144**: 38–46.

——— AND E. A. JENSEN. 1957. Primary oceanic production. The autotrophic production of organic matter in the oceans. Scientific Results of the Danish Deep-Sea Expedition round the World 1950–52, Galathea Report I: 49–136. Copenhagen.

——— AND V. HANSEN. 1959. Light adaptation in marine phytoplankton populations and its interrelation with temperature. Physiol. Plant., **12**: 353–370.

STRICKLAND, J. D. H. 1958. Solar radiation penetrating the ocean. A review of requirements, data and methods of measurement, with particular reference to photosynthetic productivity. J. Fish. Res. Bd. Canada, **15**: 453–493.

SVERDRUP, H. U., M. W. JOHNSON AND R. H. FLEMING. 1942. The oceans, their physics, chemistry and general biology. Prentice-Hall, New York. x +1060 pp.

SWEENEY, B. M., AND HASTINGS, J. W. 1957. Characteristics of the diurnal rhythm of luminescence in *Gonyaulax polyedra*. J. Comp. Cell. Comp. Physiol., **49**: 115–128.

TALLING, J. F. 1957a. Photosynthetic characteristics of some freshwater plankton diatoms in relation to underwater radiation. New Phytol., **56**: 29–50.

———. 1957b. The phytoplankton population as a compound photosynthetic system. New Phytol., **56**: 133–149.

———. 1957c. Diurnal changes of stratification and photosynthesis in some tropical African waters. Proc. R. Soc. B, **147**: 57–83.

TAYLOR, A. H., AND G. P. KERR. 1941. The distribution of energy in the visible spectrum of daylight. J. Opt. Soc. Amer., **31**: 3–8.

UTERMÖHL, H. 1936. Quantitative Methoden zur Untersuchung des Nannoplanktons. *In*: Aberhalden Handb. Biol. ArbMeth., **9**(2): 1879–1937.

VERDUIN, J. 1956. Energy fixation and utilization by natural communities in western Lake Erie. Ecology, **37**: 40–50.

VINBERG, G. G. 1934. Versuch zum Studium der Photosynthese und der Atmung des Seewassers. Zur Frage ueber Bilanz des organischen Stoffes. (Mitteilung I). (In Russian: German summary.) Arb. Limnol. Sta. Kossino, **18**: 5–24.

VOLLENWEIDER, R. A. 1956. Das Strahlungsklima des Lago Maggiore und seine Bedeutung für die Photosynthese des Phytoplanktons. Mem. Ist. Ital. Idrobiol. Marchi, **9**: 293–362.

WINOKUR, M. 1948. Photosynthesis relationships of *Chlorella* species. Amer. J. Bot., **35**: 207–214.

WITHROW, R. B., AND A. P. WITHROW. 1956. Generation, control, and measurement of visible and near-visible radiant energy. *In* Hollaender, A. (ed.), Radiation biology, vol. **3**. Visible and near-visible light, pp. 125–258. McGraw-Hill Co., New York, Toronto, London. viii + 765 pp.

COMPARISON OF HALF-SATURATION CONSTANTS FOR GROWTH AND NITRATE UPTAKE OF MARINE PHYTOPLANKTON[1,2]

2

Richard W. Eppley and William H. Thomas

Institute of Marine Resources, University of California, La Jolla, California 92037

SUMMARY

Growth rates and rates of nitrate uptake by N-depleted cells were measured for an oceanic diatom, Chaetoceros gracilis, *and a neritic diatom,* Asterionella japonica, *as functions of nitrate concentration of the medium. Both growth and N-uptake rates appeared to be hyperbolic with nitrate concentration and could be fit to an equation of Michaelis-Menten form:*

$$v = \frac{V_m S}{K_s + S}$$

where v *is rate,* V_m *is the maximum rate,* S *is nitrate concentration, and* K_s *is the half-saturation constant.* K_s *values for uptake and growth were similar if not identical for each species. Uptake experiments can provide a presumptive measure of* K_s *for growth, thought to be an ecologically significant characteristic of a species.*

INTRODUCTION

Variation in the rate of uptake of nitrate by marine phytoplankton with nitrate concentration follows a hyperbola that can be fitted by equations such as the Langmuir adsorption isotherm or the Michaelis-Menten equation (*5,8,14*). Under certain conditions phytoplankton specific growth rate is also hyperbolic with NO_3^- concentration (*23,24*) but not always (*3*). The relative ability of an organism to take up low levels of NO_3^- can be evaluated from the hyperbolae by calculating the maximum uptake velocity (V_m) or maximum specific growth rate (μ_m) and the half-saturation constant K_s (analogous to the Michaelis constant for an enzyme reaction).

The variation in specific growth rate with nutrient concentration is probably an important characteristic of species living in low nutrient environments, such as phytoplankton in the sea, and may determine, in part, the geographical and seasonal distribution of species. Coastal phytoplankton communities show K_s values > 1 μM for nitrate uptake while oceanic communities have a lower K_s, ~ 0.2 μM (*14*). Much of the emphasis in this work has been on NO_3^- and NH_4^+ uptake since nitrogen is thought to be the nutrient most often limiting the size of phytoplankton crop and its specific growth rate (*23–25*, and references cited therein). Silicate, phosphate, and certain vitamins and trace metals may be growth-rate limiting in fresh water and possibly in the sea, although the latter is not well substantiated. Dugdale (*5*) has given hypothetical examples of interspecies competition based upon assumed values of μ_m and K_s, and Eppley *et al.* (*9*) give experimental data for 4 species at 2 levels of light intensity (irradiance). The distribution of *Skeletonema costatum* and *Coccolithus huxleyi*, in 1967, in coastal waters off La Jolla could be accounted for largely by their differences in K_s for NO_3^- and NH_4^+ and their growth responses to irradiance (*6*).

The recent work, cited above, has been sufficiently favorable to the hypothesis that K_s for NO_3^- and NH_4^+ uptake and μ_m for N-limited growth influence species distribution as to stimulate more detailed analysis. The present point of interest concerns the comparison of different kinds of measurements of the kinetic parameters. For example, the technique of MacIsaac & Dugdale (*14*) employed 24-hr incubations of samples of natural communities with ^{15}N-labeled NO_3^- and NH_4^+. Caperon (*1–3*) studied NO_3^--limited chemostat cultures of *Isochrysis galbana.* Thomas measured the specific growth rate of cultures (*23*) and natural communities (*24*) and the other measured short-term uptake of NO_3^- and NH_4^+ with N-depleted culture (*9*).

Several lines of reasoning suggest that half-saturation constants for uptake should exceed those for growth. And it has long been known that uptake can be separated in time or uncoupled from cell division, as shown by very high initial rates of NH_4^+ uptake by N-deficient cells (e.g., *10,20*) compared to the growth rate. Thus uptake velocity will vary with the nitrogen nutrition of the cells. Maximum velocity estimates are best taken from specific growth

[1] *Received April 14, 1969; revised July 23, 1969.*

[2] Work supported by Grant GB-5541 from the National Science Foundation and by Contract AT 11-1 (GEN 10) P.A. 20 with the U.S. Atomic Energy Commission.

Reprinted from *Journal of Phycology*, 5:375–379, 1969.

rate measurements of important species *in situ* or under conditions which closely match those prevalent in the sea.

A question important for plankton ecology is which method, if any, gives half-saturation constants pertinent to phytoplankton growth in the sea. In this paper we report work with 2 species of marine phytoplankton which suggest similar half-saturation constants for growth and for NO_3^- uptake as functions of NO_3^- concentrations. Reasons are discussed for expecting different estimates from the 2 methods and a possible explanation is offered for the similar values observed. We thank Mrs. Anne N. Crayne, Mrs. Jane N. Rogers, and Mr. Edward Renger for expert technical assistance and Miss Elizabeth Fuglister for computer analysis of the data.

MATERIAL AND METHODS

Chaetoceros gracilis was obtained in May 1958 from a tropical oceanic upwelling area (*22*). *Asterionella japonica* was obtained in June 1964 from a coastal upwelling area west of southern Baja California (24°16.5′N latitude, 111°32.5′W longitude). Both cultures were routinely maintained at both room temperature in a north window and at 21 C and a constant illuminance of 500 ft-c. The medium used for maintenance was enriched 75% seawater (*19*). Both diatoms are in unialgal culture and *C. gracilis* is axenic. Attempts to obtain bacteria-free cultures of *Asterionella* were unsuccessful; the algae could be freed of bacteria by washing and antibiotic treatment, but subsequent transfers produced misshapen, slowly growing cells. Other workers have been unable to obtain bacteria-free cultures of this alga (*13*; E. J. F. Wood, personal communication).

For experiments on the effects of NO_3^- on the growth rate of *C. gracilis*, the inoculum was depleted of nitrogen by growing it for 4 days in a basal enriched seawater medium containing no added nitrogen. This medium was prepared from nutrient-poor, tropical, surface seawater collected during a March 1967 EASTROPAC cruise; the medium contained as enrichments to 75% seawater: 10 μM K_2HPO_4, 100 μM Na_2SiO_3, trace metals in the concentrations recommended by Taylor (*21*) for his medium CF-1, and 0.2 mg/liter iron, added as ferric citrate (*16*). One-liter flasks containing 750 ml of medium to which NO_3^- was added at varying concentrations up to 10 μM were inoculated with depleted cells. The cultures were then incubated at 26 C and an illuminance of 800–1100 ft-c (0.04–0.065 ly/min) in a glass-bottomed water bath. This temperature and light intensities were those previously found optimal for cultures of *C. gracilis* containing an excess of nutrients (*22*). Light was supplied continuously by cool-white fluorescent lamps. Cells were counted each day with a Model B Coulter Counter.

Depleted cells were not used as an inoculum in experiments with *A. japonica*. Instead, cells in the exponential phase of growth were washed aseptically with basal medium using the filtration apparatus of Soli (*17*). The washed cells were then resuspended in basal medium and inoculated into 750-ml cultures containing varying amounts of NO_3^- (to 25 μM). The basal medium for *Asterionella* consisted of 75% nutrient-poor seawater enriched with 200 μM K_2HPO_4, 1000 μM Na_2SiO_3, vitamins and trace metals in the concentrations recommended by Guillard & Ryther (*11*) for their medium "f," and 0.5 mg/liter iron added as the citrate. Nitrate-containing cultures of *Asterionella* were incubated at the same temperature and light intensity as *C. gracilis*; such conditions were optimal for *Asterionella* (*13*). Daily measurements were made of cell volume per volume of culture using a Model J plotter connected to the Coulter Counter (*15*) since *Asterionella* is a chain-forming diatom. Preliminary experiments showed that growth rates calculated from cell volumes were the same as those calculated from microscopic cell counts.

Cell numbers or volumes were plotted against time on semi-logarithmic paper so that exponential growth rates could be calculated. Rates are expressed as doublings per day.

In experiments with both *Chaetoceros* and *Asterionella*, nutrients other than NO_3^- were present in excess and both rates of growth and yield of cells were functions of the initial NO_3^- in the medium. Nitrate in the media was calculated from the NO_3^- present in basal seawater media [as determined by chemical analysis (*18*)], plus the NO_3^- added.

Uptake measurements. Cells were inoculated into 6 liters of enriched seawater containing 25 μM nitrite as the N source and the cultures were incubated at 18 C with fluorescent lighting (~ 700 ft-c with a 12/12 hr light-dark cycle) until all the added NO_2^- was depleted from the medium. The resulting crop of cells contained ~ 50 μg chlorophyll *a*/liter as determined by chlorophyll fluorescence in 90% acetone extracts (*27*). At 9 AM on the day of an experiment, 1.5 liters of N-depleted culture was supplemented with KNO_3 to a concentration of 1 μM. By 1 PM this NO_3^- was consumed. (Without this preincubation NO_3^- uptake was not linear with time over the 30–90 min period of measurement.) Aliquots of 200 ml each were then supplemented with KNO_3 at 6 different concentrations in the range 1 to 20 μM and the subcultures were incubated a further 15–90 min. The times were chosen to allow measurable nitrate uptake but less than 50% nitrate depletion. The cells were then removed by filtration (Whatman GF/C glass fiber filters) and nitrate was determined on duplicate 100 ml samples of filtrate (*18*). NO_3^- uptake was calculated as the difference between initial and final values. The analytical method for NO_3^- involves conversion of NO_3^- to nitrite by passing samples through a column of copper-coated cadmium metal. We noted that the degree of reduction in our columns varied slightly with NO_3^- concentration and our estimates of NO_3^- uptake include this correction.

RESULTS

Growth experiments. Figure 1 shows a typical hyperbolic plot of growth rate and initial NO_3^- concentration (S) for an *Asterionella* experiment and a transformation of the data (S/μ vs. S) so that the maximum growth rate μ_m and the half-saturation constant K_s could be estimated.

These constants and their 95% confidence limits for growth experiments with organisms are summarized in Table 1.

For *C. gracilis*, K_s agrees with a value that might be calculated by extrapolation of μ at high S values from earlier experiments with this organism (*23*). The new data are more meaningful, however, because by using NO_3^--poor tropical seawater, we were able to measure rates at much lower NO_3^- concentrations than those achieved previously.

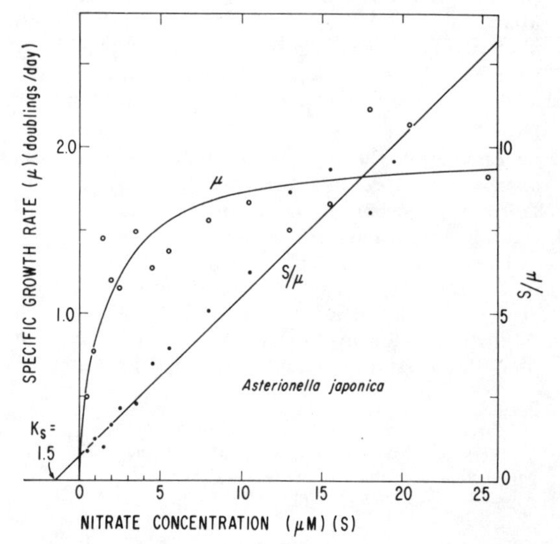

FIG. 1. Specific growth rate of *Asterionella japonica* as a function of nitrate concentration. The S/μ vs. S regression line was used to calculate K_s and μ_m.

TABLE 1. *Half-saturation constants* (K_s) *(in μmoles/liter), maximum velocity of growth* (μ_m), *and nitrate uptake* (V_m) *with 95% confidence limits of the estimates.*

		K_s		Doublings/day	
Organism	Exp.	Growth	Uptake	μ_m	V_m
Asterionella	1	1.5 ± 1.4	1.3 ± 0.5	1.97 ± 0.24	7.25 ± 0.32
japonica	2	1.2 ± 1.2	0.7 ± 0.3	1.90 ± 0.18	6.62 ± 0.37
Chaetoceros	1	0.2 ± 0.2	0.3 ± 0.5	3.22 ± 0.18	11.8 ± 1.4
gracilis	2		0.1 ± 0.2		12.0 ± 0.3

Uptake experiments. Rates of NO_3^- uptake were calculated in units equivalent to those in the growth experiments as doublings of cell N per day as:

$$v = \left[\frac{\log_2 10}{\text{uptake period (days)}}\right] \times$$

$$\log_{10}\left[\frac{\mu\text{moles cell-N} + \mu\text{moles } NO_3^- \text{ uptake}}{\mu\text{moles cell-N}}\right].$$

Initial cell N was taken as 25 μmoles/liter, the concentration of NO_2 added. An example is shown in Fig. 2 along with a linear plot, s/v vs. S, where v is the uptake rate and S the NO_3^- concentration initially present in the medium. Estimates of maximum velocity and half-saturation constants and their 95% confidence limits are given in Table 1.

The uptake velocities of the N-depleted cells exceed those required for growth as expected from earlier observations on NH_4^+ uptake rate (*10,20*). However, the half-saturation constants are, if only because of the large confidence limits, indistinguishable from those for growth.

It is interesting to compare the half-saturation constant for NO_3^- uptake with the results of a NO_3^--limited continuous culture of *Isochrysis galbana* (*1–3*). NO_3^- concentration in the chemostat vessel at steady state was not very dependent upon dilution rate (growth rate) and averaged 0.32 μM (*1*). This value compares favorably with our results with *Isochrysis*, where K_s values (for uptake) and their 95% confidence limits were 0.09 ± 0.23 and 0.13 ± 0.24, in 2 experiments (*9*). However, a graph of N per cell vs. μ was a hyperbola in the chemostat experiments (*3*) and the half-saturation constant calculated with the assumption of a constant N per cell (the value noted in batch cultures at the stationary phase) exceeded 2 μM (*1*).

DISCUSSION

Droop (*4*), Caperon (*3*), and earlier evidence (see ref. *7*) suggest that growth rate may not always vary with external nutrient concentration, but may depend upon the nutrient content of the cell. Thus graphs of growth rate vs. cell content of nutrient often appear to be hyperbolas. Droop (*4*) gives a useful equation relating these variables in the steady state:

$$\text{nutrient content of cell} = \frac{\text{uptake rate}}{\text{cell division rate}} \quad (1)$$

It is easily shown from Eq. (1), with hypothetical values for growth and uptake taken from hyperbolic rates vs. concentration curves, that nutrient/cell will be constant during the exponential growth, regardless of what may happen to external nutrient concentrations, if the half-saturation constants are equal for growth and uptake. However, if the latter constant for uptake exceeds that for growth, then nu-

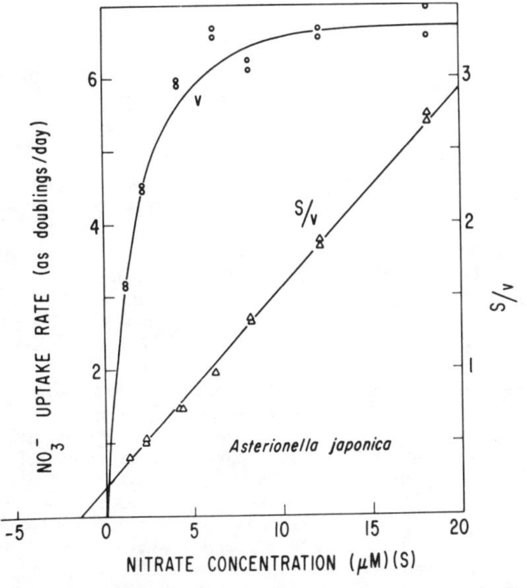

FIG. 2. Nitrate uptake rate of *Asterionella japonica* vs. nitrate concentration. Uptake rate is calculated as the doubling rate of cell nitrogen. The S/v vs. S regression line was used to calculate K_s and V_m.

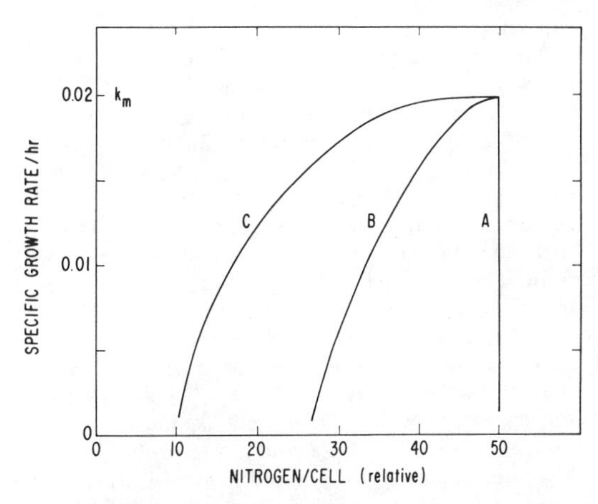

FIG. 3. Results of different K_s for growth and uptake of nitrate. Variation in specific growth rate with cell nitrogen content calculated from Eq. (1) with the following assumptions. In all cases specific growth rate varied with nitrate concentration (S) with $\mu_m = 0.02$ hr^{-1} and $K_s = 1.0$ μM and $\mu = \frac{\mu_m S}{K_s + S}$. In line A uptake rate is assumed to follow the equation $v = \frac{V_m S}{K_s + S}$ with $V_m = 1$ and $K_s = 1$ μM; in line B, $K_s = 2$ μM, and in line C, $K_s = 5$ μM.

trient/cell will be greater with high external nutrient concentration than with low. Growth rate plotted against nutrient/cell will then be roughly hyperbolic with a finite intercept on the abscissa, as was observed by Droop & Caperon. A hypothetical example is shown in Fig. 3. Note on this graph that the greater the difference between half-saturation constants for uptake and growth, the smaller is the nutrient/cell intercept when growth rate is extrapolated to zero. (The nutrient/cell intercept may provide a measure of difference between uptake and growth half-saturation constants.) Apparently the difference between the constants for uptake and growth need not be large (a factor of 2 or less) for growth rate vs. nutrient/cell plots to be of the observed form, although Droop (*4*) calculated a 20-fold difference in his study of vitamin B_{12} with *Monochrysis lutheri.*

A second model for growth and uptake leads also to a larger half-saturation constant for uptake than for growth, given a higher maximum velocity for uptake than growth. If uptake rate is assumed to be hyperbolic with nutrient concentration and is growth rate-limiting at low nutrient concentration but ceases to be limiting at high nutrient levels, then μ will increase hyperbolically until the maximum value is reached. If the maximum value is sufficiently small, growth rate will no longer appear to be hyperbolic with NO_3^- concentration. However, the difference between V_m for uptake and μ_m for growth would determine the magnitude of difference between the half-saturation constant for uptake and the nutrient concentration allowing a half-maximal growth rate. Growth rate variation with concentration of phosphate fit such a model with *Chaetoceros gracilis* (*26*). If this model is generally valid then growth rate must be limited by NO_3^- concentration over a broad range (in our experiments) in order, first, to give an apparent hyperbola with NO_3^- concentration and second to give a similar estimate of half-saturation constant to that of the uptake experiments.

The agreement between estimates of half-saturation constant must also be regarded as surprising because of the large differences in V_m and μ_m. The high V_m for NO_3^- uptake may be due to an increase in the concentration of NO_3^- absorption sites in the N-depleted cells rather than the rate constant for the reaction relieving the site (see ref. *2*). Alternatively, there may be temporary imbalance between rates of cellular influx and efflux of NO_3^- as was argued in a study of potassium uptake by barley plants (*12*).

The K_m for the enzyme NO_3^- reductase in the diatom *Ditylum brightwellii* is about 100-fold greater than the K_s for NO_3^- uptake (*8*). This observation suggests that NO_3^- uptake rate at low NO_3^- concentration may be important in controlling cellular NO_3^- levels which presumably in turn influence the rate of N assimilation and growth. Evidently, close coupling of uptake rate and growth rate is lacking as in N-deficient cells where uptake rate may exceed growth rate 3–4-fold (Table 1) and where the rate of NO_3^- uptake could quickly elevate cellular NO_3^- concentrations sufficient to saturate NO_3^- reductase. NADH–NO_3^- reductase is at very low levels in *D. brightwellii* grown on nitrite and in N-deficient cells but is present in cells growing with NO_3^- as the N source (*8*).

Asterionella japonica is a coastal diatom whereas this isolate of *Chaetoceros gracilis* is from the open sea. It is interesting that K_s values (either uptake or growth) for *Asterionella* are higher than those for *Chaetoceros.* This observation adds further weight to the hypothesis that species distribution may be controlled by differences in K_s between coastal and offshore species and by differences in nutrient concentration between coastal and offshore environments.

CONCLUSIONS

There are several possible explanations for the agreement between our estimates of half-saturation constant for NO_3^- uptake and growth: (1) differences are hidden in the statistical uncertainty of the measurements; (2) N/cell may be relatively constant at the *observed* growth rates; (3) differences in maximum velocity may be unimportant if the high V_m for uptake represents a temporary imbalance between NO_3^- influx and efflux or if it is due to a temporary increase in the concentration of cellular nitrate absorption sites in the N-depleted cells; (4) growth rate may be limited by NO_3^- concentration over a fairly wide range of NO_3^- concentration because of large differences in the half-saturation constant for uptake and intracellular NO_3^- reduction. It is an understatement to comment that our data do not allow a choice among these alternatives. However, NO_3^- uptake measurements can provide a presumptive measure of the half-saturation constant for growth. This constitutes a significant advantage, since such measurements of uptake rate are much more easily carried out than growth rate determinations. Our conclusion implies that the half-saturation constants determined in NO_3^- uptake experiments with natural communities of phytoplankton (*14*) are valid.

REFERENCES

1. CAPERON, J. 1965. The dynamics of nitrate limited growth of *Isochrysis galbana* populations. Ph.D. dissertation. Univ. Calif., San Diego. 71 pp.
2. ——— 1967. Population growth in micro-organisms limited by food supply. *Ecology* **48**:713–22.
3. ——— 1968. Population growth response of *Isochrysis galbana* to nitrate variation at limiting concentrations. *Ecology* **49**:866–72.
4. DROOP, M. R. 1968. Vitamin B_{12} and marine ecology. IV. The kinetics of uptake, growth and inhibition in *Monochrysis lutheri. J. Mar. Biol. Assoc. U.K.* **48**:689–733.
5. DUGDALE, R. C. 1967. Nutrient limitation in the sea: dynamics, identification, and significance. *Limnol. Oceanogr.* **12**:685–95.
6. EPPLEY, R. W. 1969. The ecology of plankton off La Jolla, California in the period April through September, 1967. IV. Relationship of phytoplankton species distribution to the depth distribution of nitrate. Submitted to *Bull. Scripps Inst. Oceanogr.*
7. ——— & STRICKLAND, J. D. H. 1968. Kinetics of marine phytoplankton growth. *In* Droop, M. R. & Ferguson Wood, E. J. [eds.], *Advances in Microbiology of the Sea,* Academic Press, London, 23–62.
8. EPPLEY, R. W., COATSWORTH, J. L., & SOLORZANO, LUCIA. 1969. Studies of nitrate reductase in marine phytoplankton. *Limnol. Oceanogr.* **14**:194–205.
9. EPPLEY, R. W., ROGERS, JANE N., & MCCARTHY, JAMES J.

1969. Half-saturation constants for uptake of nitrate and ammonium by marine phytoplankton. *Limnol. Oceanogr.* (in press).

10. FITZGERALD, G. P. 1968. Detection of limiting or surplus nitrogen in algae and aquatic weeds. *J. Phycol.* **4**:121–6.
11. GUILLARD, R. R. L. & RYTHER, J. H. 1962. Studies of marine planktonic diatoms. I. *Cyclotella nana* Hustedt and *Detonula confervacea* (Cleve). Gran. *Can. J. Microbiol.* **8**:229–39.
12. JOHANSEN, C., EDWARDS, D. G., & LONERAGAN. J. F. 1968. Interactions between potassium and calcium in the absorption by intact barley plants. II. Effects of calcium and potassium concentration on potassium absorption. *Plant Physiol.* **43**:1722–6.
13. KAIN, J. M. & FOGG, G. E. 1958. Studies on the growth of marine phytoplankton. I. *Asterionella japonica* Gran. *J. Mar. Biol. Assoc. U.K.* **37**:397–413.
14. MACISAAC, J. J. & DUGDALE, R. C. 1969. The kinetics of nitrate and ammonium uptake by natural populations of marine phytoplankton. *Deep-Sea Res.* **16**:415–22.
15. PARSONS, T. R. 1965. An automated technique for determining the growth rate of chain-forming phytoplankton. *Limnol. Oceanogr.* **10**:598–602.
16. RODHE, W. 1948. Environmental requirements of freshwater plankton algae. *Symb. Bot. Upsalienses* **10**:1–149.
17. SOLI, G. 1964. A system for isolating phytoplankton organisms in unialgal and bacteria-free culture. *Limnol. Oceanogr.* **9**:265–8.
18. STRICKLAND, J. D. H. & PARSONS, T. R. 1968. A practical handbook of seawater analysis. *Fish. Res. Bd. Canada Bull. No. 167*, 311 pp.
19. SWEENEY, B. M. & HASTINGS, J. W. 1957. Characterization of the diurnal rhythm of luminescence in *Gonyaulax polyedra*. *J. Cell Comp. Physiol.* **49**:115–28.
20. SYRETT, P. J. 1962. Nitrogen assimilation. *In* Lewin, R. A. [ed.], *Physiology and Biochemistry of Algae*, Academic Press, N.Y., 171–88.
21. TAYLOR, W. R. 1964. Inorganic nutrient requirements for marine phytoplankton organisms. *Symp. Exptl. Mar. Ecology Grad. School Oceanogr. Univ. Rhode Island Occ. Publ. No. 2*, 17–24.
22. THOMAS, W. H. 1966. Effects of temperature and illuminance on cell division rates of three species of tropical oceanic phytoplankton. *J. Phycol.* **2**:17–22.
23. ——— 1967. The nitrogen nutrition of phytoplankton in the northeastern tropical Pacific Ocean. *Proc. Intern. Conf. Tropical Oceanogr. Stud. Tropical Oceanogr. Miami* **5**:280–9.
24. ——— 1969. Effect of nitrate and ammonium concentration on chlorophyll increases in natural tropical Pacific phytoplankton populations. Submitted to *Limnol. Oceanogr.*
25. ——— 1969. On nitrogen deficiency in tropical oceanic phytoplankton: Photosynthetic parameters in rich and poor water. Submitted to *Limnol. Oceanogr.*
26. ——— & DODSON, ANNE N. 1968. Effects of phosphate concentration on cell division rates and yield of a tropical oceanic diatom. *Biol. Bull.* **134**:199–208.
27. YENTSCH, C. S. & MENZEL, D. W. 1963. A method for the determination of phytoplankton chlorophyll and phaeophytin by fluorescence. *Deep-Sea Res.* **10**:221–31.

3 SOME FACTORS AFFECTING THE FEEDING OF MARINE COPEPODS OF THE GENUS *CALANUS*[1]

Michael M. Mullin
Biological Laboratories, Harvard University

ABSTRACT

The grazing rates of 4 species of *Calanus* were measured on several species of marine phytoplankton, both singly and in mixtures. Grazing rates varied inversely with phytoplankton concentration, the duration of the experiment, and the age of the phytoplankton cultures used.

Calanus spp. generally removed large cells at higher rates than small when feeding on a mixture of two species of phytoplankton. Because of this selectivity and the difference in volume between large and small cells, the large cells contributed by far the greater fraction of the total volume of food ingested by the copepods, even when considerably less abundant than the smaller cells. Comparison of the grazing rates of the copepodite stage V and adult female animals of *C. finmarchicus*, *C. glacialis*, and *C. hyperboreus* on a mixture of 7 species of diatoms demonstrated further the importance of large cells. Marked quantitative differences in rate of food intake were found between the different developmental stages of each species and between the different species of copepods, and some qualitative differences in selective feeding were also indicated.

INTRODUCTION

The food relationships of copepods of the genus *Calanus* have been investigated by several workers. Early studies were made of the gut contents (Dakin 1908; Lebour 1922) and feeding mechanism (Esterly 1916; Cannon 1928; Lowndes 1935) of *Calanus*, and some measurements of rates of feeding were made as early as 1887 (Hensen, cited in Dakin 1908).

Organic matter is present in sea water as dissolved molecules, colloidal material, particulate detritus, bacteria, nanoplankton, "net" phytoplankton, and animals. Marshall and Orr (1955a) have concluded that *Calanus* is best considered as a filter-feeding herbivore, deriving most of its nutrition from phytoplankton (including nanoplankton). Even so, a great variety of forms of organic matter is potentially available as food for *Calanus*, because of the heterogeneity of phytoplankton communities. It is therefore necessary to determine whether *Calanus* acts as an indiscriminate filterer, removing all types of particles with equal efficiency, or whether feeding is selective, removing some food organisms preferentially to others. Various workers have noted the ability of *Calanus* to reject unwanted particles which had been filtered out of suspension (Esterly 1916; Cannon 1928) and to remove organic matter preferentially out of a natural suspension containing much inorganic particulate material (Corner 1961). Harvey (1937), using diatoms of three sizes, concluded that the larger diatoms were preferentially removed from a suspension of two species by *Calanus* during feeding. He found that factors other than cell size also influenced selective feeding, as the diatom *Chaetoceros* was not eaten, apparently because of its spinous processes. On the other hand, Marshall and Orr (1955b) found no major differences in feeding on those species of phytoplankton tested that were large enough to be effectively filtered at all.

In addition to possible variability in feeding due to the composition of the phytoplankton community, variability may arise with each species of *Calanus* if the developmental stages feed at different rates or show differences in selectivity. Quantitative differences in feeding of the developmental stages of *Calanus* have been demonstrated by Marshall and Orr (1956) and Gauld (1951). Similarly, the various species of the genus may show quantitative or

[1] Contribution No. 1299 from the Woods Hole Oceanographic Institution.

Reprinted from *Limnology and Oceanography*, 8:239–250, 1963.

qualitative differences in feeding behavior. Such differences may be of evolutionary as well as ecological significance, especially where species of the same genus occur sympatrically.

In view of these potential sources of variability, it was the purpose of the present investigation to study the feeding of 4 species of *Calanus*, to determine the degree of applicability of a single value for the feeding rate, and to clarify the interactions between various phytoplankton and zooplankton populations in the natural environment.

In considering feeding activity, two factors are involved, namely, the grazing rate and the rate of food intake. The *grazing rate*, also called the filtering rate, is defined as the rate at which water is "swept clear" of food organisms by the copepods. It is measured as the volume of water containing the number of cells removed per unit time per copepod, assuming that all the cells are removed from water passing through the filtering mechanism. If filtration is not 100% efficient, the actual rate at which water is filtered must be greater than the experimentally computed grazing rate. The *rate of food intake* is measured as the number, or biomass, of food organisms ingested per unit time, and depends both on the grazing rate and on the concentration and size of the food organisms. This concept is equivalent to the "feeding rate" as defined by Rigler (1961), but the term "rate of food intake" is less easily confused with the term "filtering rate."

The experiments reported here were carried out at Scripps Institution of Oceanography with the advice of Dr E. W. Fager, and at the Woods Hole Oceanographic Institution and the Biological Laboratories, Harvard University, with the advice and assistance of Drs G. L. Clarke and R. J. Conover. The contributions of these and other workers, especially Drs B. M. Sweeney and R. R. L. Guillard, are gratefully acknowledged. During the course of the work, the investigator was supported by a fellowship from the National Science Foundation. N. S. F. Grant 8339 provided for working time aboard ships from Woods Hole.

METHODS

The copepods studied were *Calanus helgolandicus*, collected off the California coast from Scripps Institution, and *C. finmarchicus*, *C. glacialis*, and *C. hyperboreus*, collected in Cape Cod Bay and the Gulf of Maine and studied at Woods Hole. Some *C. hyperboreus* collected from deep water ($>$ 1,000 m) south of Cape Cod (39°36′ N, 71°06′ W) were also used. Animals were captured with plankton nets, sorted to species and developmental stage under a dissecting microscope, and maintained in filtered sea water at constant temperature in the manner described by Conover (1960), except that no antibiotics were used. Various species of marine phytoplankton, grown in unialgal cultures, were added as food. With these methods, copepods could be readily maintained for experimental use in the laboratory for over 4 months.

The *Calanus* from the California coast is here referred to the species *helgolandicus* on morphological grounds alone, as the true status is unestablished. The three species from the Gulf of Maine were distinguished on the basis of total body length. The boreal species *C. finmarchicus* (females 2.76–4.30 mm in length) is frequently the dominant copepod in the offshore waters of the Gulf of Maine. The somewhat larger *C. glacialis* (females 4.42–5.24 mm) and the still larger *C. hyperboreus* (females 6.28–8.49 mm) are primarily arctic species, and are much less abundant during most seasons. *C. glacialis* has been distinguished from *C. finmarchicus* only recently (Jaschnov 1955, cited in Grainger 1961). All these populations may be augmented by influx of animals from more northern waters carried into the Gulf by currents (Bigelow 1924; Grainger 1961; Jaschnov 1961).

Grazing rates were measured with the species of phytoplankton shown in Table 1, both singly and in mixtures. All of these grew as single cells in culture. In addition, one experiment was run with the chain-

TABLE 1. *Phytoplankton cells, volumes in μ^3, computed from linear dimensions*

1. *Ditylum brightwellii*	1.2×10^5
2. *Rhizosolenia setigera*	7.6×10^4
3. *Gymnodinium nelsoni*	7.5×10^4
4. *Striatella unipunctata*	6.9×10^4
5. *Gonyaulax polyedra*	3.0×10^4
6. *Thalassiosira fluviatilis*	2.3×10^3
7. unknown cryptomonad	3.6×10^2
8. *Cyclotella nana*	2.1×10^2

forming diatom *Asterionella japonica.* Suspensions of the phytoplankton to be tested were made up by addition of known amounts of cultures to filtered, pre-cooled sea water, and divided between 12 screwcap glass jars of 950-ml capacity, leaving a 50-ml air space in each. Variance tests on counts of cells showed that this method did not introduce any bias between the jars. The cultures used were generally less than 10 days old. Copepods which had been kept in sea water without food for at least 24 hr prior to the experiment were added to some of the jars (3 to 18 copepods per jar). At least 3 jars in each experiment contained no animals and served as controls. A minimum of 24 animals was used in each experiment, although usually the number was greater than this.

In order to keep the phytoplankton cells in suspension during an experiment, the jars were strapped to a wheel rotating at 0.2 to 0.5 rev/min. As rotation alone was insufficient to maintain a uniform suspension of the larger cells, each jar was equipped with a small magnetic stirring bar which was activated for a few seconds during each revolution of the wheel. No difference in grazing rates was found between animals in stirred and unstirred jars, when small cells which did not settle out were used. Experiments were run in the dark, and at the temperature at which the animals had been maintained in the laboratory (12°C at Scripps, 5–6°C at Woods Hole). These temperatures were close to those of the water masses from which the animals were collected.

The concentrations of the cells in control and experimental jars after an experiment were determined either with an electronic Coulter Counter Model A (Hastings, Sweeney, and Mullin 1962), or by microscopic examination of samples concentrated by settling, depending on the experiment. Direct comparison between Coulter Counter counts and visual counts of the same samples showed good agreement. Microscopic examination showed that the frustules of even the largest diatoms were either ingested or thoroughly crushed during feeding, so that since fecal pellets were excluded, suspensions of cells which had been grazed by the animals could be satisfactorily counted electronically. In one experiment performed with the help of Dr R. Lasker, chains of *Asterionella japonica* tagged with C^{14} were used, and grazing was measured as removal of activity from the suspension, after exclusion of fecal pellets. Grazing rates were computed as the volume of water "swept clear" of cells per day per copepod (Gauld 1951). The concentrations of cells in the control jars at the end of the experiment were taken as the initial concentrations for the experiment in computing the grazing rates. This procedure corrected for possible changes in concentrations due to factors other than grazing. Such changes were small under the conditions of these experiments. Rates of food intake were computed from the grazing rates and the initial concentration or biomass of the food organisms (Cushing 1958).

RESULTS

Grazing on single species of phytoplankton

Because of the wide range of values reported for the grazing of *Calanus* in the laboratory (reviewed by Jørgensen 1955), it was deemed advisable to estimate the effects of the experimental conditions on grazing. As many investigators have recognized, such tests are necessary to extrapolate from laboratory measurements to grazing under field conditions, although such extrapolation may still be open to question.

Experiments with *C. helgolandicus* demonstrated that up to 18 animals could be

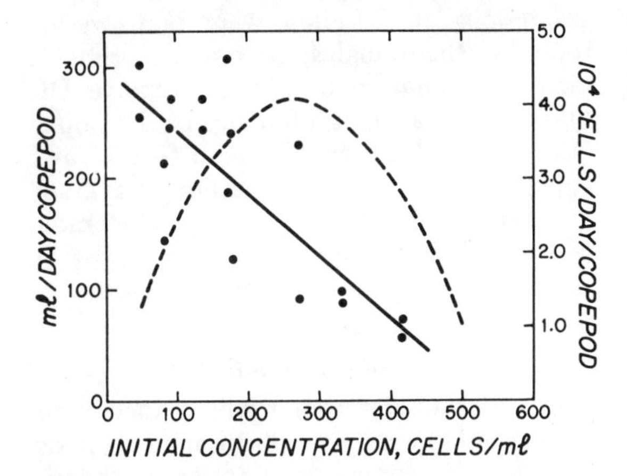

FIG. 1. Grazing by female *Calanus hyperboreus* on various concentrations of *Ditylum brightwellii* (three experiments).
Left ordinate and solid line—grazing rate.
regression equation: $\bar{Y}_X = -0.56X + 300$
correlation coefficient: $r = -0.791$
Right ordinate and broken line—rate of intake of cells.

used per 950-ml jar without decreasing the grazing rate per animal. No significant difference was found between the 950-ml jars and 475-ml jars with half the number of animals per jar (Wilcoxon signed rank test). Thus, dependence of grazing rate on container size, found by Cushing (1959) for containers up to 500 ml, was not demonstrated for these larger containers.

A series of 24-hr experiments tested the effects of the initial concentration of food on grazing. The grazing rate of female *C. helgolandicus* varied inversely with the initial concentration of *Ditylum* over the range of concentrations tested (50–500 cells/ml). The linear regression of grazing rate $\bar{Y}_X$ on initial cell concentration X was given by $\bar{Y}_X = -0.192X + 129.88$. The regression for similar experiments using *Gonyaulax* was non-linear (linearity test from Dixon and Massey 1957, p 197), but showed an inverse relationship of the same sort between 250 and 800 cells/ml. Figures 1 and 2 show the results of similar experiments in which female *C. hyperboreus* fed on *Ditylum* or *Thalassiosira*. In each case, the regression of grazing on initial concentration was linear, and from the correlation coefficients, it is highly improbable that grazing and phytoplankton concentration are independent ($p < 0.0005$). Changes in the concentration of the large *Ditylum* had a much greater effect on grazing than changes in the small *Thalassiosira*.

Rigler (1961) concluded that above a certain critical concentration of cells at which the grazing rate of *Daphnia* began to be limited by the ability to ingest or digest and excrete food, the grazing rate decreased but the rate of food intake remained constant. At concentrations less than this critical level, the grazing rate was constant. This has not been found for *Calanus* in the present experiments. With increasing concentrations of cells, the intake of food by *Calanus* increased to a maximum value and then decreased again (broken lines in Figs. 1 and 2), while the grazing rate decreased steadily. In computing the points for these curves, values for the grazing rate at each concentration were taken from the regression lines. Translating cell numbers into cell volumes, the concentration of food at which the greatest rate of ingestion occurred was 3.3×10^7 μ^3/ml for *Ditylum* and 9.2×10^6 μ^3/ml for *Thalassiosira*.

The grazing rate also decreased with time, up to 24 hr. Figure 3 shows the grazing rates which were maintained by female *C. helgolandicus* during various intervals of three experiments, graphed as percent-

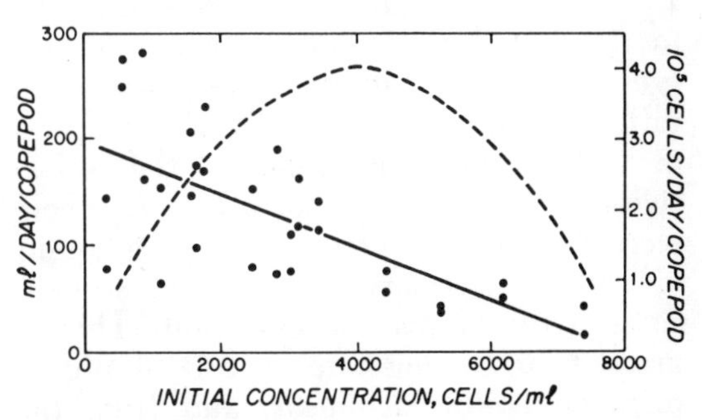

FIG. 2. Grazing by female *Calanus hyperboreus* on various concentrations of *Thalassiosira fluviatilis* (four experiments).
Left ordinate and solid line—grazing rate.
regression equation: $\bar{Y}_X = -0.0246X + 197$
correlation coefficient: $r = -0.696$
Right ordinate and broken line—rate of intake of cells.

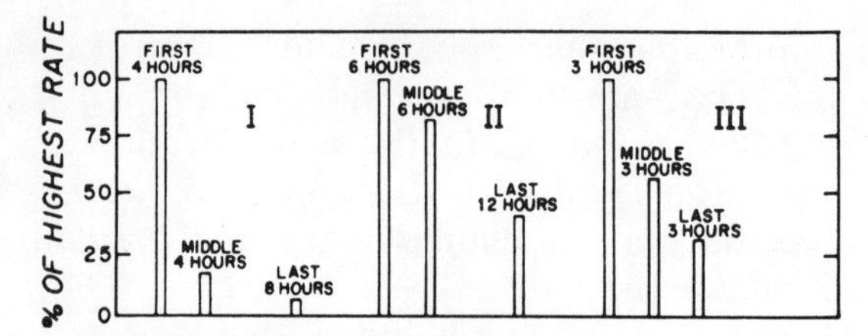

FIG. 3. Relative grazing rates of female *Calanus helgolandicus* at various intervals during three experiments.

I. 16-hr experiment; feeding on *Asterionella;* 100% = 255.6 ml/day/copepod.
II. 24-hr experiment; feeding on *Ditylum;* 100% = 363.9 ml/day/copepod.
III. 9-hr experiment; feeding on *Ditylum;* 100% = 246.9 ml/day/copepod.

ages of the highest rate found during each experiment. It is apparent that the copepods fed at a high rate when initially placed in the phytoplankton suspension, and then gradually reduced their feeding. The reduction of grazing rate cannot be due to changes in the concentrations of cells present, since grazing rate would be expected to increase with time as the cell concentrations are reduced, rather than decrease.

The age of the phytoplankton culture used in making up the suspensions of cells may also affect the grazing rate, as shown in Figure 4. This graph combines the results of two experiments, each run for 18 hr. All the suspensions were initially close to 2,330 cells/ml (maximum deviation 13%), and all grazing rates were corrected to this initial concentration, using the regression line in Figure 2. No significant difference in size of the cells from cultures of various ages was found, so that selectivity for cell size was not a factor. There were no great differences in the amounts of culture necessary to make up suspensions of the proper concentration. The regression of grazing rate on age of phytoplankton culture is linear, and from the correlation coefficient, it is highly improbable that the two are independent ($p < 0.0005$). Conover (1956) found that senescent cultures of the diatom *Skeletonema* depressed the filtering rates of *Acartia*.

Finally, the grazing rates of male *C. helgolandicus* on *Ditylum* were ⅛ to ⅒ of those of the females; in one experiment, male *C. finmarchicus* did not remove *Ditylum* at all, although females of the population showed a grazing rate of 80 ml/day/copepod on the same suspension. Raymont and Gross (1941) found the grazing rates of male *C. finmarchicus* to be 1/15 to 1/40 of those of the females of that population.

Selective grazing on mixtures of phytoplankton

In order to determine which species of phytoplankton may be most important as food for the copepods, a series of experiments was run to measure feeding on mixtures of two species, and the rate of grazing on each species was determined separately (Table 2). Each pair of values represents a separate experiment. Because the copepods fed on a mixture, any difference in the grazing rates could be attributed to a selective effect of feeding, rather than to some other factor. In experiments with *C. helgolandicus*, *C. finmarchicus*, and *C. hyperboreus*, the larger species of phytoplankton was removed more efficiently than the smaller from the mixture, except for two cases involving the dinoflagellate *Gymnodinium* (Table 2, column II). The preferential removal of large cells

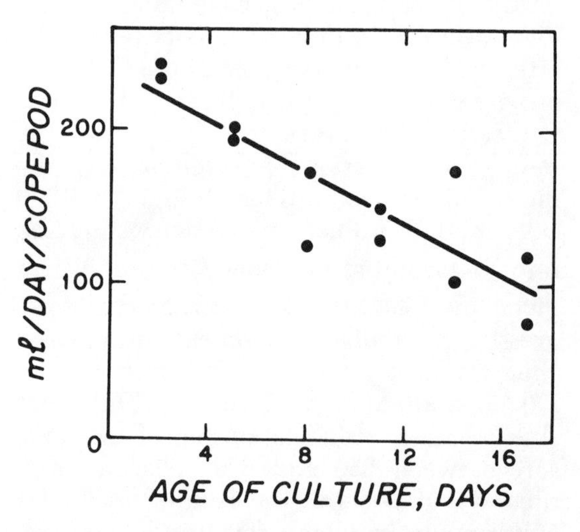

FIG. 4. Grazing rates of female *Calanus hyperboreus* on suspensions of *Thalassiosira* made up with cultures of various ages (two experiments).
regression equation: $\bar{Y}_X = -8.29X + 240.2$
correlation coefficient: $r = -0.857$

TABLE 2. I. *Phytoplankton, designated by numbers from Table 1*
II. *Grazing rates, computed for each species in the mixture*
III. *Percentage of the volume of available food represented by each species which would be ingested per day per copepod at the observed grazing rates*
IV. *Percentage contributed by each species to the total volume of food ingested during each experiment, calculated from the observed grazing rates and the following cell concentrations: For species #1–#5, 50 cells/ml; for species #6, 1,000 cells/ml; for species #7–#8, 5,000 cells/ml*

I	II	III	IV
	C. finmarchicus		
#1	118.2 ml/day/copepod	13.1%	74%
#2	64.1 "	7.1%	26%
#1	38.9 "	4.4%	75%
#3	21.0 "	2.3%	25%
#1	59.4 "	6.6%	80%
#4	26.4 "	2.9%	20%
#1	25.2 "	2.8%	79%
#6	18.1 "	2.0%	21%
#1	82.9 "	9.4%	90%
#8	54.2 "	5.6%	10%
#3	5.1 "	0.6%	12%
#4	40.0 "	4.4%	88%
#3	20.3 "	2.2%	50%
#5	50.8 "	5.6%	50%
#6	164.4 "	18.7%	85%
#7	36.4 "	3.7%	15%

I	II	III	IV
	C. helgolandicus		
#1	123.0 ml/day/copepod	13.6%	88%
#5	67.8 "	7.5%	12%
	C. hyperboreus		
#1	197.8 ml/day/copepod	22.1%	65%
#2	167.6 "	18.6%	35%
#1	61.4 "	6.7%	92%
#6	25.6 "	2.9%	8%
#2	329.0 "	36.5%	72%
#6	211.0 "	23.4%	28%
#6	158.0 "	17.5%	100%
#8	0.0 "	0.0%	0%

occurred even when the large cells were from older cultures than the small cells. Higher grazing rates on large cells than on small were usually found also in simultaneous comparisons of grazing on different species of phytoplankton in unialgal suspensions. However, the differences were less striking than those reported by Harvey (1937).

The population of the larger species generally suffered a relatively greater loss in biomass due to this selective grazing, as shown by calculation of the percentage of the available volume of food represented by each species in the mixture which would be ingested per day per copepod when grazing at the observed rates (Table 2, column III).

Computation of the fraction contributed by each species in the mixture to the total volume of food ingested, corrected to standard cell concentrations to make the different experiments comparable, demonstrates the importance of the larger cells in the nutrition of the copepods, even when the large cells are numerically much less abundant than smaller cells (Table 2, column IV). The cell concentrations actually used were generally close to the standard values chosen for computing the total ingested volume of food, and were picked to give a rough approximation of the relative abundances of natural populations. If these approximations are in error, it is probably in underestimating the relative abundance of the nanoplankton (*Thalassiosira*, *Cyclotella*, and the cryptomonad), and therefore underestimating somewhat the fraction they contribute to the total food ingested.

The importance of the size of the food organism, independent of its species, was demonstrated by the fact that *C. helgolandicus* preferentially removed the longer chains of cells in a suspension of *Asterionella* (significance by a rank sum test of Ansari and Bradley 1960). Further, *C. hyperboreus* showed strongly preferential removal of the largest cells in a suspension of *Rhizosolenia setigera*, in which about 10% of the cells were the broad type (diameter about 4 times that of the other cells) arising from auxospores.

Preferential removal of large particles

can also extend to animal food. In two experiments, female *C. finmarchicus*, *C. glacialis*, and *C. hyperboreus* all removed *Artemia* nauplii (volume approximately 10^7 μ^3) at significantly higher rates than *Ditylum* from a mixture. In one experiment, female *C. hyperboreus* had slightly higher grazing rates on eggs of the sea urchin *Strongylocentrotus* (volume approximately 10^6 μ^3/egg) than on *Ditylum*, although this difference was not significant.

"Conditioning" C. *helgolandicus* to *Ditylum* or *Gonyaulax* by feeding only one or the other for two weeks had no significant effect on the rate at which either was grazed (rank sum test). Marshall and Orr (1955b) reached similar conclusions.

In order to test the effect of grazing on a complex mixture of phytoplankton, a suspension of 5 species of diatoms was made up in which the small species far outnumbered the large, but the biomass of the large species exceeded that of the small, as may be the case under oceanic conditions. Grazing rates of *C. finmarchicus*, *C. glacialis*, and *C. hyperboreus*, copepodite stage Vs and females, were determined for each species of diatom after 18 hr. The animals were all taken in the same deep net tows in the Gulf of Maine (43°22′ N, 67°41′ W) in late July 1961. These observations were later extended to include two larger diatoms, *Coscinodiscus concinnus* (volume 6.9 × 10^6 μ^3) and *Coscinodiscus* sp. 2 (1.15 × 10^6 μ^3). Animals were collected at 41°39′ N, 65°29′ W in late March 1962, and grazing rates on a mixture of the two *Coscinodiscus* spp. plus *Ditylum* were determined after 18 hr. In order to combine these results with those of the previous experiment, each of the grazing rates of each copepod was multiplied by the factor necessary to make the grazing rates of that copepod on *Ditylum* the same in the two experiments. The total available biomass of food was approximately the same in the two sets of experiments.

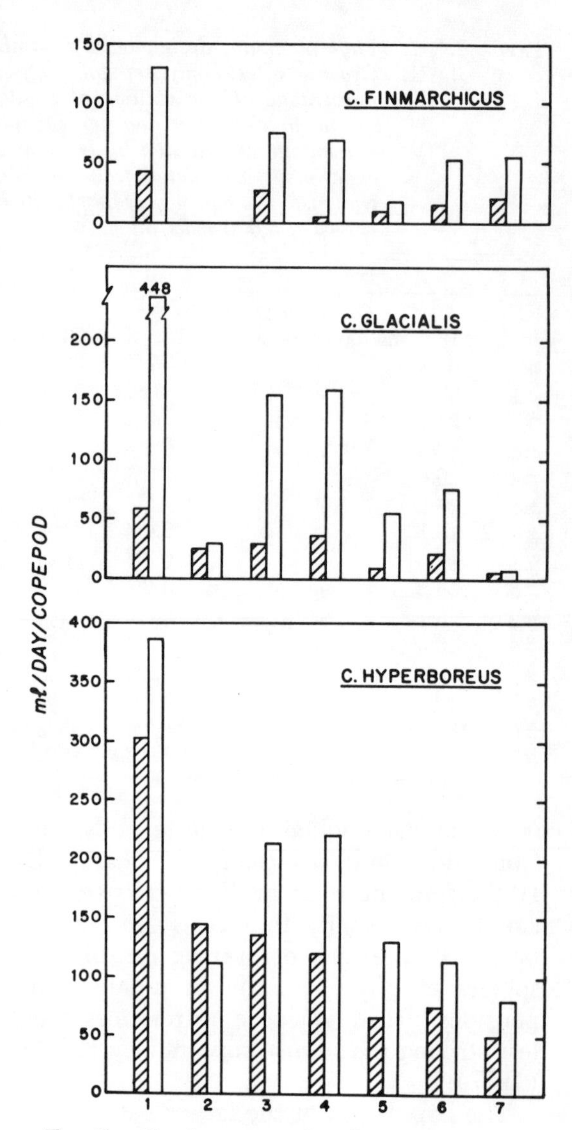

FIG. 5. Grazing rates of *Calanus* spp. on various diatoms in a mixture.
shaded bars—copepodite stage Vs
open bars—adult females

Column	Diatom sp.	Initial concentration (cells/ml)	Initial biomass (cell volume) (× 10^6 μ^3/ml)
1	*C. concinnus*	0.6	4.1
2	*C.* sp. 2	0.9	1.04
3	*Ditylum*	47.4	5.6
4	*Rhizosolenia*	53.8	4.1
5	*Striatella*	51.0	3.5
6	*Thalassiosira*	460.0	1.05
7	*Cyclotella*	1,467.0	0.3

In the combined experiments, the net phytoplankton (all except *Thalassiosira* and *Cyclotella*) constituted 7.5% of the total number of cells, less than the value of 9% found for Vineyard Sound by Yentsch and Ryther (1959). The actual concentrations of cells were considerably higher than those generally reported for the off-

TABLE 3. I. *Grazing rates computed from the total number of cells of all kinds removed*
II. *Rates of intake of food as total volume ingested*
III. *Grazing rates computed from the total volume of cells of all kinds removed*

	I	II	III
C. finmarchicus stage V	19.9 ml/day/copepod	4.06 × 10^8 μ^3/day/copepod	20.6 ml/day/copepod
" " female	54.9 "	13.8 "	70.1 "
C. glacialis stage V	15.1 "	6.72 "	34.1 "
" " female	34.0 "	37.1 "	188.2 "
C. hyperboreus stage V	56.0 "	29.8 "	151.2 "
" " female	91.4 "	44.3 "	224.5 "

shore waters of the Gulf of Maine, but all except *Cyclotella* were within the ranges reported for the maximum numbers of diatoms during spring blooms in these waters (Bigelow, Lillick, and Sears 1940; Gran and Braarud 1935).

The results of these experiments are summarized in Figure 5. The relative rates at which the different copepods grazed the various species of phytoplankton in the mixture are consistent with a series of preliminary experiments comparing the feeding of the three species on each of the diatoms separately (the two species of *Coscinodiscus* were not tested). The particular species of phytoplankton present apparently have some effect, since both *Coscinodiscus* sp. 2 and *Striatella* were grazed at unexpectedly low rates. Otherwise, all the copepods had higher mean grazing rates on the net phytoplankton than on the nanoplankton. The tendency for the larger diatoms to be removed preferentially to the smaller was particularly marked in the female *C. glacialis*.

The importance of the larger diatoms, in spite of their numerical scarcity, is further emphasized when the total volume of food ingested per day is computed (Table 3, column II), and the fraction of this contributed by each species of diatom is determined. For all the copepods, the nanoplanktonic *Thalassiosira* and *Cyclotella* together constituted less than 6% of the total volume of food ingested from this particular mixture of cells. Because of their low efficiency in filtering *Cyclotella*, the two stages of *C. glacialis* swept clear a smaller volume of water, in terms of removal of cells of all kinds, than the comparable stages of *C. finmarchicus*, but ingested a larger volume ot food because of higher efficiency in removing the large cells (Table 3, columns I and II). This demonstrates that grazing rates computed simply from the number of cells removed from a heterogeneous mixture do not give the most useful information. A more realistic calculation of the grazing rate is based on the removal of biomass of food. Rates of this sort, computed from the biomass represented by all the cells initially present and the rate of intake of food, are shown in Table 3, column III. Such grazing rates indicate more accurately the relative efficiencies of the different copepods in feeding on the whole mixture of phytoplankton.

Column III of Table 3 also illustrates the influence on grazing of the developmental stage of the copepod, as well as its absolute body size. The females of each species had higher grazing rates than the stage V copepodites of the next larger species, but are smaller in body size. This suggests that when food is plentiful, females show greater feeding relative to their size than do the stage V animals. As has already been pointed out, male *Calanus* show little or no feeding.

DISCUSSION

Grazing rates of 150 to 350 ml/day/copepod were measured for female *C. helgolandicus* feeding on dilute suspensions of *Ditylum* or *Asterionella* for 4 to 8 hr. Harvey's (1937) values for grazing by *Calanus* on *Ditylum* and some measurements by Cushing (1959) and Conover,

Marshall, and Orr (1959) fall within this range, but most previous workers have recorded maximum values much lower than this. The cell concentrations and short time periods used in the present experiments may represent a closer approximation to the natural situation than some of these earlier measurements, especially if the copepod populations undergo a diurnal vertical migration into and out of a phytoplankton-rich zone (Gauld 1953). The somewhat lower grazing rates of *C. finmarchicus*, which is similar in size to *C. helgolandicus*, may be due in part to the lower temperatures at which experiments with the former were run.

Many previous measurements of the grazing and respiratory rates of *Calanus* have led to the conclusion that the copepods cannot meet their metabolic requirements by feeding on natural concentrations of phytoplankton at the low grazing rates measured under laboratory conditions (reviews by Clarke 1939; Marshall and Orr 1955a). It has often been stated that μ-flagellates, which are seldom sampled quantitatively and leave no recognizable residue in the fecal pellets, must make up the difference. However, the results of the present investigation show that unless these tiny cells are extremely abundant, they can contribute little to the total nutrition of *Calanus*. Corner (1961) has found that female *C. helgolandicus* can readily meet their metabolic requirements at 10°C by grazing at relatively low rates (10–36 ml/day/copepod, computed from removal of weight of organic matter) on natural suspensions of phytoplankton for long periods of time. The grazing rate of female *C. finmarchicus* at 5°C on the 7 species of diatoms reported above was considerably higher than this, when calculated either in terms of cells or of biomass removed from suspension (Table 3). A possible explanation is that the biomass of food available per unit volume of water in Corner's experiments was roughly twice that in the total mixture of cells used in the present experiments.

Gauld (1951) found that the filtering rate of *Calanus* on *Chlamydomonas* was independent of the concentration of cells, and this has frequently been assumed in constructing models of interactions between phytoplankton and zooplankton populations (Fleming 1939; Harvey, *et al.* 1935; Riley, *et al.* 1949; Steele 1956). The dependence of grazing rate on concentration of cells, demonstrated in the present experiments, does not necessarily introduce an important source of error in such models unless feeding rates on phytoplankton populations which differ greatly in density are to be compared. A depression of grazing rate by very high concentrations of cells was found by Marshall and Orr (1955a) for *Calanus*, and by Conover (1956) for *Acartia*.

From experiments demonstrating a depression of the grazing rate of *Daphnia* at high concentrations of cells, and from other evidence, Ryther (1954) suggested the presence of a toxic or inhibitory substance in *Chlorella* cells which depressed feeding when released from ingested cells in the animal's gut. The dependence of grazing rate on phytoplankton concentration and on the duration of the experiment, demonstrated here for *Calanus*, could be explained as decreased feeding activity as the animal becomes sated, or as reduced efficiency in filtering or ingesting cells in dense suspensions or when the gut is full. Rigler (1961) has put forward this hypothesis of "overloading" as a result of his experiments on the feeding of *Daphnia* on yeast, and from a reevaluation of Ryther's data. However, the data presented here for *Calanus* do not entirely agree with Rigler's analysis for *Daphnia*, particularly since the rate of food intake by *Calanus* decreases above a certain concentration of cells. If overloading merely reduces the efficiency of filtration or ingestion, one would expect the rate of food intake to become constant at high concentrations, rather than decreasing. All 4 species of phytoplankton tested—*Ditylum*, *Gonyaulax*, *Asterionella*, and *Thalassiosira*—have shown effects on the copepods which could be attributed to the presence of inhibitory substances, so that Ryther's critical experiment comparing the grazing rates of *Daphnia* fed on *Chlorella* (toxic) with those fed on bacteria (non-toxic) could not be repeated. Ryther concluded that the inhibitory property was most pronounced in senescent *Chlorella* cells, and the inverse relationship between grazing rate and the age of the phytoplankton culture would seem to confirm this finding for *Thalassiosira*. Size of the senescent cells or amount of culture medium used in making up the suspensions cannot explain this observation, although changes could have occurred in the culture medium as a result of the growth of the cells, rather than within the cells themselves. Until further studies of senescent cells and their culture medium are made, the presence of an inhibitory substance produced within phytoplankton cells is at best hypothetical.

The demonstration of selective grazing by the various species of *Calanus* has important implications for the specific zooplankton and phytoplankton populations in a given area. In the relatively unstructured environment occupied by planktonic communities, the possibilities for separation of ecological niches may be rather limited (Hutchinson 1951, 1961; Pennak 1957). One might suppose, on the basis of the so-called competitive exclusion principle (Hardin 1960), that closely related, sympatric species of filter-feeding zooplankton would tend to feed most efficiently on different sizes of food organisms, and thus fit into slightly different niches (Slobodkin 1960, p 150). Hutchinson (1951) has suggested that sympatric species of congeneric copepods are often separated by size differences, probably related to differences in selectivity of food, and therefore reducing interspecific competition. This hypothesis has been supported by work of Comita and Comita (1957) and Cole (1961) on fresh-water diaptomid populations. The present study of *Calanus* indicates a greater relative reliance by larger species or the larger developmental stages of each species on large cells for food. The extent to which this tendency can, in itself, constitute an effective separation of niches is open to question, however. The wide range of sizes of cells on which all the copepods can feed is perhaps more striking than the differences in selectivity between the species. Fryer (1957) found that although closely related carnivorous fresh-water cyclopoids showed some divergence in food preference, this tendency was not found among the herbivorous species, so that several related herbivores could fill the same feeding niche.

It may also be noted that Grainger's (1961) measurements of *C. finmarchicus* and *C. glacialis* do not show evidence for character displacement (Brown and Wilson 1956) of body size in regions where the ranges of these two species overlap. Although several environmental factors influence body size, character displacement in sympatric populations might be expected if the relationship between body size and food preference is valid, if the competitive exclusion principle is applicable to feeding in this case, and if gene flow between local populations is not so rapid as to mask the phenomenon on the local level.

In regard to the effects of grazing on the qualitative composition of the phytoplankton community, Riley, *et al.* (1949) and Margalef (1958) have pointed out that even a slight preferential removal of certain species could have considerable effect if maintained for a number of weeks. From the present experiments, it appears that populations of large-celled species will usually experience a greater loss of biomass to grazing by *Calanus* than small-celled populations, even if the populations are initially equal in total biomass, because of the preferential removal of the larger cells. Similarly, a chain-forming species might experience heavier predation than a single-celled species of similar cell size. Evidence demonstrating a succession of dominant diatoms from large to small species in the waters of Georges Bank as the spring bloom is dissipated is given by Sears (1941). Such a succession would be expected if larger cells are removed preferentially by *Calanus finmarchicus*, the dominant copepod, although other causes are of course possible. Factors other than cell volume, such as cell shape, the presence of spines or a thick frustule, or some chemical factor making the cell "distasteful," might also affect selective grazing on the natural phytoplankton community. Since all three species of *Calanus* can feed on organisms as large as *Artemia* nauplii, the possible importance of microzooplankton in the diet of the older stages must also be considered.

Possible sources of variability in feeding behavior, such as the concentration of the phytoplankton, the length of time the animals feed, the species composition of the phytoplankton community, and the species and stage of the dominant copepods, illustrate some of the difficulties in assigning a single grazing rate for mixed zooplankton on a complex phytoplankton community. Approximations of this sort are frequently necessary in the construction of manageable models for oceanic productivity, but considerable information concerning the interactions of the species present may be lost in the process.

REFERENCES

Ansari, A. R., and R. A. Bradley. 1960. Rank-sum tests for dispersions. Ann. Math. Statistics, **31**: 1174–1189.

Bigelow, H. B. 1924. Plankton of the offshore waters of the Gulf of Maine. Bull. U. S. Bur. Fish., **40**, document no. 968, 509 pp.

———, L. C. Lillick, and M. Sears. 1940. Phytoplankton and planktonic protozoa of the offshore waters of the Gulf of Maine. I. Numerical distribution. Trans. Amer. Phil. Soc., **31**: 149–191.

Brown, W. J., Jr., and E. O. Wilson. 1956. Character displacement. Systematic Zool., **5**: 49–64.

Cannon, H. G. 1928. On the feeding mechanism of the copepods *Calanus finmarchicus* and *Diaptomus gracilis*. Brit. J. Exp. Biol., **14**: 131–144.

Clarke, G. L. 1939. The relation between diatoms and copepods as a factor in the productivity of the sea. Quart. Rev. Biol., **14**: 60–64.

Cole, G. A. 1961. Some calanoid copepods from Arizona with notes on congeneric occurrences of *Diaptomus* species. Limnol. Oceanogr., **6**: 432–442.

Comita, G. W., and J. J. Comita. 1957. The internal distribution patterns of a calanoid copepod population, and a description of a modified Clarke-Bumpus plankton sampler. Limnol. Oceanogr., **2**: 321–336.

Conover, R. J. 1956. Biology of *Acartia clausi* and *A. tonsa*. Bull. Bingham Oceanogr. Coll., **15**: 156–233.

———. 1960. The feeding behavior and respiration of some marine planktonic Crustacea. Biol. Bull., **119**: 399–415.

———, S. M. Marshall, and A. P. Orr. 1959. Feeding and excretion of *Calanus finmarchicus* with reference to the possible role of zooplankton in the mineralization of organic matter. Woods Hole Oceanogr. Inst. Ref. 59-32 (*unpublished manuscript*).

Corner, E. D. S. 1961. On the nutrition and metabolism of zooplankton. I. Preliminary observations of the feeding of the marine copepod, *Calanus helgolandicus* (Claus). J. Mar. Biol. Ass. U. K., **41**: 5–16.

Cushing, D. H. 1958. The effect of grazing in reducing the primary production: A review. Rapp. Cons. Intern. Expl. Mer, **144**: 149–154.

———. 1959. On the nature of production in the sea. Fish. Invest. London, Ser. II, **22**: 1–40.

Dakin, W. J. 1908. Notes on the alimentary canal and food of the Copepoda. Intern. Rev. Hydrobiol. Hydrogr., **1**: 772–782.

Dixon, W. J., and F. J. Massey, Jr. 1957. Introduction to statistical analysis. New York. McGraw-Hill. 488 pp.

Esterly, C. O. 1916. The feeding habits and food of pelagic copepods and the question of nutrition by organic substances in solution in the water. Univ. California Publ. Zool., **16**: 171–184.

Fleming, R. H. 1939. The control of diatom populations by grazing. J. Cons. Intern. Expl. Mer, **14**: 1–20.

Fryer, G. 1957. The food of some freshwater cyclopoid copepods and its ecological significance. J. Animal Ecol., **26**: 261–286.

Gauld, D. T. 1951. The grazing rate of planktonic copepods. J. Mar. Biol. Ass. U.K., **29**: 695–706.

———. 1953. Diurnal variations in the grazing of planktonic copepods. J. Mar. Biol. Ass. U.K., **31**: 461–474.

Grainger, E. H. 1961. The copepods *Calanus glacialis* Jaschnov and *Calanus finmarchicus* (Gunnerus) in Canadian arctic-subarctic waters. J. Fish. Res. Bd. Canada, **18**: 663–678.

Gran, H. H., and T. Braarud. 1935. A quantitative study of the phytoplankton in the Bay of Fundy and the Gulf of Maine (including observations on hydrography, chemistry, and turbidity). J. Biol. Bd. Canada, **1**: 279–476.

Hardin, G. 1960. The competitive exclusion principle. Science, **131**: 1292–1298.

HARVEY, H. W. 1937. Note on selective feeding by *Calanus*. J. Mar. Biol. Ass. U.K., **22**: 97–100.

———, L. H. N. COOPER, M. V. LEBOUR, AND F. S. RUSSELL. 1935. Plankton production and its control. J. Mar. Biol. Ass. U.K., **20**: 407–441.

HASTINGS, J. W., B. M. SWEENEY, AND M. M. MULLIN. 1962. Counting and sizing of unicellular marine organisms. Ann. New York Acad. Sci., **99**: 280–289.

HUTCHINSON, G. E. 1951. Copepodology for the ornithologist. Ecology, **32**: 571–577.

———. 1961. The paradox of plankton. Amer. Nat., **95**: 137–147.

JASCHNOV, V. A. 1961. Water masses and plankton. I. Species of *Calanus finmarchicus s.l.* as indicators of definite water masses. Zool. Zhurn., **40**: 1314–1334.

JØRGENSEN, C. B. 1955. Quantitative aspects of filter feeding in invertebrates. Biol. Rev. Cambridge Phil. Soc., **6**: 412–442.

LEBOUR, M. V. 1922. The food of plankton organisms. J. Mar. Biol. Ass. U.K., **12**: 644–677.

LOWNDES, A. G. 1935. The swimming and feeding of certain calanoid copepods. Proc. Zool. Soc. London, **1935**: 687–715.

MARGALEF, R. 1958. Temporal succession and spatial heterogeneity in phytoplankton. *In* Perspectives in marine biology, Buzzati-Traverso [Ed.], Univ. California Press, pp. 323–349.

MARSHALL, S. M., AND A. P. ORR. 1955a. The biology of a marine copepod. Edinburgh. Oliver & Boyd. 188 pp.

———, AND ———. 1955b. On the biology of *Calanus finmarchicus*. VIII. Food uptake, assimilation, and excretion in adult and stage V *Calanus*. J. Mar. Biol. Ass. U.K., **34**: 495–529.

———, AND ———. 1956. On the biology of *Calanus finmarchicus*. IX. Feeding and digestion in the young stages. J. Mar. Biol. Ass. U. K., **35**: 587–603.

PENNAK, R. W. 1957. Species composition of limnetic zooplankton communities. Limnol. Oceanogr., **2**: 222–232.

RAYMONT, J. E. G., AND F. GROSS. 1941. On the feeding and breeding of *Calanus finmarchicus* under laboratory conditions. Proc. Roy. Soc. Edinburgh, **61**: 267–287.

RIGLER, F. H. 1961. The relation between concentration of food and feeding rate of *Daphnia magna* Straus. Canadian J. Zool., **39**: 857–868.

RILEY, G. A., H. STOMMEL, AND D. F. BUMPUS. 1949. Quantitative ecology of the plankton of the western North Atlantic. Bull. Bingham Oceanogr. Coll., **12**: 1–169.

RYTHER, J. H. 1954. Inhibitory effects of phytoplankton upon the feeding of *Daphnia magna* with reference to growth, reproduction, and survival. Ecology, **35**: 522–533.

SEARS, M. 1941. Notes on the phytoplankton on Georges Bank in 1940. J. Mar. Res., **4**: 247–257.

SLOBODKIN, L. B. 1961. Preliminary ideas for a predictive theory of ecology. Amer. Nat., **95**: 147–153.

STEELE, J. H. 1956. Plant production on the Fladen Ground. J. Mar. Biol. Ass. U.K., **35**: 1–33.

YENTSCH, C. S., AND J. H. RYTHER. 1959. Relative significance of the net phytoplankton and nanoplankton in the waters of Vineyard Sound. J. Cons. Intern. Expl. Mer, **24**: 231–238.

4

FEEDING, ASSIMILATION AND RESPIRATION RATES OF *DAPHNIA MAGNA* UNDER VARIOUS ENVIRONMENTAL CONDITIONS AND THEIR RELATION TO PRODUCTION ESTIMATES

By D. W. SCHINDLER*

Department of Zoology, Oxford University

INTRODUCTION

Although a number of studies, both terrestrial and aquatic, have provided quantitative estimates of production, few workers have identified or analysed the factors which affect the efficiency of energy transfer from one trophic level to the next or the partitioning of available energy among different species at the same trophic level.

This paper describes a laboratory study designed primarily to single out environmental factors which have a possible bearing on production for further study under field conditions. Field production estimates will be dealt with in later publications.

It is well known that the standing crop (S) at any time t_2 is an exponential function of the standing crop at time t_1

$$S_{t_2} = S_{t_1} \, e^{(A-R)(t_2-t_1)} \tag{1}$$

where A and R are instantaneous assimilation and respiration rates per unit standing crop (for a discussion see Clarke, Edmondson & Ricker 1946). Since assimilation rates have never been determined for any species under a wide variety of environmental conditions, equation (1) has never been used to calculate production. Assimilation and feeding rates for filter-feeding zooplankton can, however, be estimated easily by feeding them photosynthetically tagged radioactive algae. This method has been widely used for some years, but most workers have concentrated their efforts upon investigating the relationship between food concentration (usually expressed in units of cells per unit volume with no allowance made for cell size or shape) and zooplankton feeding rates (F) (usually expressed as millilitres of water filtered per animal per unit time; an artificial unit of doubtful value). The relationship between A and F will be discussed later in this paper, in light of the results of the series of experiments described here.

In this paper the effects of the following environmental factors upon assimilation, feeding and respiration of *Daphnia magna* Straus were tested: light, motion, crowding, animal weight and reproductive state. The effects of food concentration, food energy content and different species of algae on feeding and assimilation rates were also examined.

MATERIALS AND METHODS

A. *Assimilation*

Carbon-14 was used for tagging all radioactive diets, as first done by Rodina & Troshin (1954) and Marshall & Orr (1955) for determining zooplankton feeding rates. The technique used was similar to that of Bourne (1959). Animals were allowed to graze for

* Present address: Freshwater Institute, Fisheries Research Board of Canada, 501 University Crescent, Winnipeg 19, Manitoba, Canada.

Reprinted from *Journal of Animal Ecology*, 37:369–385, 1968.

a period in a radioactive food suspension. They were then transferred to a like concentration of non-radioactive food to clear their guts of undigested radioactive algae. Calories of food assimilated (C_a) were assumed to equal

$$C_a = (C_e)(\gamma_a) \qquad (2)$$

where C_e is the calories per unit radioactivity of food, and γ_a is the radioactivity per animal after the determination, corrected for background, counter efficiency and self-absorption of β emissions due to sample thickness. The length of an assimilation experiment is not critical, except that respiration of assimilated ^{14}C will begin after 16–24 h, after which the method no longer measures true assimilation (see Fig. 1). Malovitskaya & Sorokin (1961b) give further information on respiration of assimilated ^{14}C.

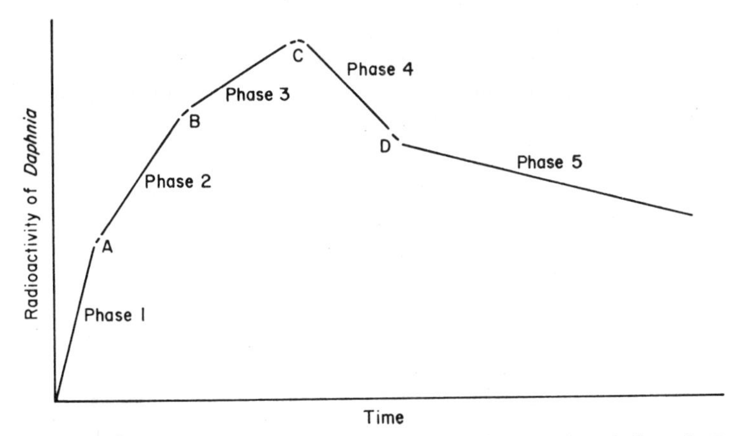

FIG. 1. A schematic diagram of ^{14}C uptake by *Daphnia* fed on labelled diets. A, Egestion of labelled diet begins; B, respiration of $^{14}CO_2$ begins; C, animals removed from radioactive diet; D, animal clears its gut of radioactive diet. In this study, point A occurred 30–60 min after introduction of the radioactive diet, depending on food concentration. Point B occurred at least 16 h after the experiment was begun. Point D followed point C by 30–60 min. Phase 1, ^{14}C ingested; Phase 2, ^{14}C ingested and egested; Phase 3, ^{14}C ingested, egested and respired; Phase 4, ^{14}C egested and respired; Phase 5, ^{14}C respired.

Steps in testing the effect of varying environmental conditions were as follows.

1. An initial set of assimilation experiments was run at 20° C using 5·0 mg of *Chlorella*/l. The *Chlorella* used was from vigorously growing cultures having a calorific value of 5·272 cal/mg. *Daphnia* of 2·0 mm in length (0·063 mg) were used in these determinations. Experiments were run in the dark, and bottles were kept motionless.

2. Assimilation was determined at several temperatures, food concentrations, food calorific values, and for several types of diet and animal sizes. Results of these experiments were tested against the initial set by analysis of variance. The effects of light and motion were also examined. If a change of any factor caused a change in assimilation rate, the relationship was investigated in more detail and included in a multiple regression analysis with other causal factors.

A detailed description of the conditions tested is included in the Results section.

Cultures of *Chlorella vulgaris* Beyerinck were grown at several combinations of light intensity and nitrogen level, as outlined in Table 1. All cultures were kept at 25° C in a plexiglass cupboard, with illumination provided by a fluorescent light. Light intensities were determined with a Weston Master IV light meter. Five hundred and 1000 ml conical

Table 1. *Conditions for* Chlorella *cultures*

Basic medium

KCl	0·1 g/l
$MgSO_4$	0·1 g/l
KH_2PO_4	0·1 g/l

Nitrate as in column (2) below.

(1) Culture	(2) $Ca(NO_3)_2$ (g/l)	(3) Illumination (ft-candles)	(4) ^{14}C (μc/l)	(5) Culture age when used (days)
A*	0·1	150	5	7
B	0·01	250	5	7
C	0·001	1000	15	14
D	0·0001	2500	20	21

* Also used for *Chlamydomonas* and *Anabaena* cultures.

Culture	μg/cell $\times 10^{-3}$	cal/mg dry weight
A	0·126	5·204–5·272
B	0·118	6·006
C	0·109	7·154
D	–	Unsuccessful

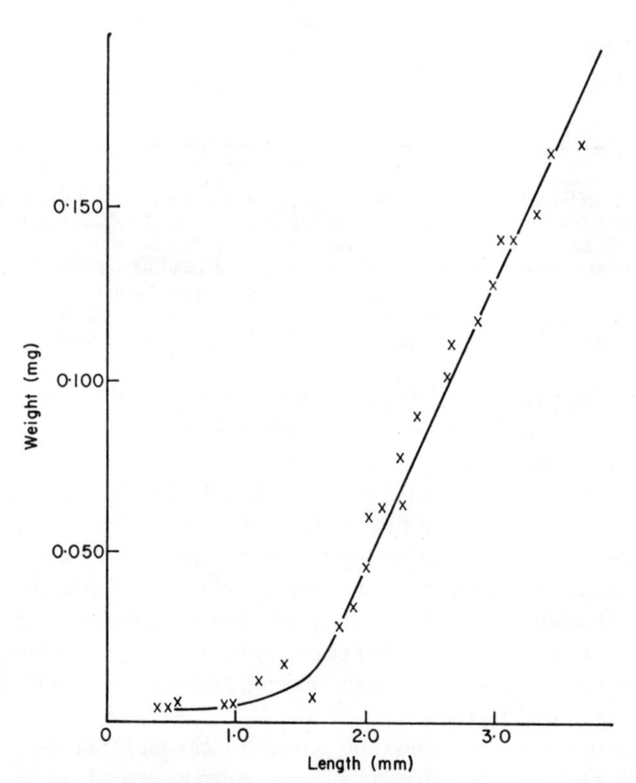

FIG. 2. Length–dry weight relationships for *Daphnia magna*. The curve shown has been drawn by eye. A log–log curve fits the data with $r^2 = 0·96$, but results for small animals (under 1 mm in length) did not fit the curve. Richman (1958) obtained similar results.

flasks were used as culture vessels, with the desired amount of $Na_2{}^{14}CO_3$ injected when the cultures were inoculated. More radioactive carbonate and longer growing periods were needed in cultures with nitrogen deficiencies as shown in Table 1. (See also Spoehr & Milner 1949.) All algae used for this study were obtained from the National Type Collection of Algae, the Botany School, Cambridge. Low-calorie radioactive diets were prepared by feeding large aquaria of *Daphnia* on photosynthetically tagged mixed algae, then collecting and screening detritus from the bottom of the tanks for feeding and assimilation experiments. These will be referred to as 'detritus' diets in the ensuing discussion. When tanks were left for 3 weeks after the last feeding and *Daphnia* populations were large, detritus diets as low as 2·0 cal/mg were obtained.

Daphnia were first sorted into size categories by screening them through nylon mesh. They were then re-sorted several times with an eyedropper. Sub-samples were occasionally checked for size using a stereoscopic microscope fitted with an ocular micrometer. As a result, the average length of animals used for any set of assimilation determinations had a coefficient of variation of less than 5%. Previous authors have used groups of animals varying by nearly 50% in length, which would mean a two- to three-fold difference in weight. It is therefore not surprising that there is little agreement among the results of different investigators.

Twenty-five to 125 animals were used for each assimilation determination. These were placed in a 250 ml reagent bottle containing a non-radioactive diet similar to that to be tested and allowed 3–5 h to adjust to the conditions to be tested. McMahon (1962) and McMahon & Rigler (1963) found that starved *Daphnia* fed at a higher rate, but that adjustment to new conditions took place within minutes; therefore the acclimatization period in this study was probably unnecessarily long.

After the acclimatization period, the non-radioactive diet suspension was siphoned off, and the bottles were refilled with a radioactive diet suspension of the same concentration, energy content and temperature. Bottles were tightly capped to avoid air bubbles; otherwise some *Daphnia* became trapped at the air–water interface and were unable to swim properly. The end of the siphon was covered with a tiny sack of fine-meshed nylon. Some radioactive suspension was allowed to proceed up the siphon, and any *Daphnia* stuck to the screen were dislodged by back-flushing. Very few *Daphnia* (roughly one in 500) were injured in this way.

After 4–6 h, the radioactive suspension was siphoned off and replaced once again with a non-radioactive suspension. A period of 1½ h was allowed for the *Daphnia* to clear their guts of radioactive algae, after which the animals were screened from the culture by pouring it through a piece of nylon mesh, and transferred to a bowl of millipore-filtered culture medium where any injured animals were quickly removed. Healthy animals were then narcotized with ether, measured with a microscope and ocular micrometer, and counted into a small glass homogenizer. They were then ground in a small quantity of distilled water and transferred quantitatively to planchets, using tiny pipettes for the transfer. The homogenizer was rinsed several times with distilled water, and rinses were also added to the planchets. A drop of Teepol, a commercial detergent, was added to each planchet to facilitate even spreading and to eliminate 'creep' on drying. Planchets were dried at 60° C, and their activities were determined with a Mullard end-window geiger tube and an I.D.L. (Isotope Developments, Ltd) scaler. All samples were timed for 4000 counts when possible. Counts were corrected for background and counter efficiency. Self-absorption was assumed to be zero, since the layer on each planchet was so thin as to be invisible to the naked eye. Monakov & Sorokin (1960,

1961) and Malovitskaya & Sorokin (1961a, b) did not crush their animals before counting, but applied standard self-absorption corrections. This technique was tried early in the present study, but counting variances were high and erratic. Homogenized samples had both higher activities and smaller counting variances.

Food concentrations in mg/l dry weight were determined by first filtering several aliquots of diet suspension on to tared millipore HA (0·45 μ) membrane, drying these at 60° C and weighing them, then diluting the algal suspension with membrane-filtered culture water to obtain the desired concentration. Food radioactivity was determined by counting several 2 ml aliquots filtered on to millipore membranes, correcting these for background, counting efficiency, and self-absorption. Samples of an identical non-radioactive diet were concentrated by centrifuging and dried at 60° C for calorific analysis.

B. *Feeding*

Feeding determinations were started in the same manner as assimilation experiments, but were terminated 20–30 min after introduction of the radioactive diet. None of the ingested radioactive algae had been egested at that time. The rate of increase in radioactivity of several groups of animals fed under identical conditions was tested over a period of time, and the first inflection point in the curve was taken to be the time when egestion began. (See the schematic Fig. 1.) Passage times agreed well with those found by Bourne (1959) and McMahon (1962), ranging from 30 to 60 min.

After removal from the radioactive suspension, animals were killed immediately (while their guts still contained food), homogenized and counted as described for the assimilation determinations.

C. *Respiration*

Oxygen consumption was measured in 250 ml glass-stoppered reagent bottles, each containing twenty-five to 125 *Daphnia*, depending on animal size and the prescribed experimental conditions. Bottles were filled with millipore-filtered culture water (containing no algae). Several control bottles, containing no *Daphnia*, were also prepared. After 6–12 h, oxygen was fixed by the Winkler method (Welch 1948) and a 200 ml aliquot was titrated with 0·005 N sodium thiosulphate and a burette graduated in 0·02 ml. A series of experiments of varying length showed that respiration rates of *Daphnia* were relatively constant for up to 24 h; therefore it was assumed that O_2 depletion and CO_2 build-up in the bottles did not affect respiration.

D. *Calorific analysis*

Calorific values of *D. magna* and algae were determined in a Parr model 1411 oxygen bomb calorimeter, which gives good results with samples of from 30 to 100 mg. Because of the small size of algal and some *Daphnia* samples, the burning of filter material was necessary (see Comita & Schindler 1963).

E. *Other corrections*

The possibility that animals were respiring assimilated ^{14}C was checked as follows: three sets of assimilation determinations were begun as usual at 20° C, with food of 5 mg/l and 5·272 cal/mg, using animals 2 mm in length. After 6 h, the *Daphnia* in these experiments were transferred to bottles of filtered culture water and allowed to swim for 16–24 h, after which all $^{14}CO_2$ was precipitated as $BaCO_3$ by adding 5% $BaCl_2$.

Planchets prepared from $BaCO_3$ precipitate showed no activity; therefore no correction for respiratory loss was necessary.

Determinations were also made using culture water from which all particles had been removed by millipore filtering, to test whether *Daphnia* could assimilate ^{14}C directly from the medium. The results of these tests were also negative. Nauwerck (1959) reports similar findings for *Diaptomus*.

Assimilation by *Daphnia magna* may therefore be calculated directly from equation (2).

D. magna weights were taken from a length–weight table constructed by sorting non-ovigerous animals into length categories, in lots of thirty to 500, drying them at 60° C, and weighing. Lengths were taken as the distance from the anterior-most tip of the head to the point of inflection at the base of the shell spine. Measurements were made with a microscope and ocular micrometer (see Fig. 2).

RESULTS

A. *Assimilation, feeding and respiration rates*

The effect of environmental factors on assimilation, feeding and respiration rates is given in Table 2. Factors exerting a significant effect were tested in more detail and

Table 2. *The effect of environmental factors on assimilation, feeding and respiration rates of* Daphnia magna

	Effect on:		
Factor	Assimilation	Feeding	Respiration
Light intensity (1000 ft-candles *v.* dark)	NS	NS	NS
Motion* (rotated at 1 rpm *v.* motionless)	NS	NS	NS
Crowding (one *Daphnia*/2 ml *v.* one *Daphnia*/10 ml)	NS	NS	NS
Reproductive state (ovigerous *v.* non-ovigerous)	S†	NS	NS
Diet species (*Chlorella, Chlamydomonas, Anabaena*)	S‡	NS	Not tested
Food concentration (1–10 mg/l)	S	S	Not tested
Food energy content (2–5 cal/mg)	S	S	Not tested
Animal weight (of animals 1–4 mm in length)	S	NS	S
Water temperature (10° *v.* 20° C)	S	NS	S

S = Significant; NS = not significant.

* The variance of the rotated set was significantly higher but there was no significant difference between means. Rotation of samples has been extremely popular among previous investigators of feeding or assimilation as a means of keeping food suspended, even though no previous author appears to have tested its possible effect. It was therefore examined here.

† Animals carrying six to eight eggs or embryos each assimilated at a significantly higher rate than the non-ovigerous control group. Animals carrying one to three eggs or embryos each assimilated at a higher rate than controls, but not significantly so. This factor was not included in the multiple regression analyses.

‡ Assimilation rates with *Chlamydomonas* and *Chlorella* were not significantly different, but *Anabaena* was assimilated at a much lower rate. The poor assimilability of Myxophyceae is well documented. See Edmondson (1957) for a review, also Malovitskaya & Sorokin (1961b).

subjected to multiple regression analysis. The results of linear multiple regression analyses are set out in Tables 3–5. The linearity of the regression analyses was tested graphically by the method of Ezekial & Fox (1959), by plotting the residuals of each partial regression equation (the difference between measured values and those predicted by the partial regression line). These results are given for assimilation in Fig. 3 (a–c), for respiration in Fig. 4, and for feeding in Fig. 5 (a and b). The residuals for food concentration and food energy content showed some departure from linearity with

Table 3. *Results of the multiple linear regression analysis relating* Daphnia magna *assimilation (A in cal/mg of animal/h) to environmental factors*

$A =$	$0{\cdot}0286E$	$+\ 0{\cdot}0038T$	$+\ 0{\cdot}00310C$	$-\ 0{\cdot}1444W$	$-\ 0{\cdot}1405$
S.E. of partial regression coefficient	0·00156	0·00047	0·00051	0·0254	
t value of partial regression coefficient	18·23 HS	8·12 HS	6·10 HS	5·68 HS	
Standard partial regression coefficient	0·98	0·44	0·28	0·26	
Units	cal/mg	°C	mg/l	mg dry weight	

Coefficient of multiple determination, $R^2 = 0{\cdot}84$; $F = 32{\cdot}3$ HS; d.f. = 4 and 76.

E, food energy content; *T*, water temperature; *C*, food concentration; *W*, animal weight; S.E. = standard error; H.S. = highly significant.

Table 4. *Results of the multiple linear regression analysis relating* Daphnia magna *respiration (R in μl/mg of animal/h) to environmental factors*

$R =$	$0{\cdot}293T$	$-\ 4{\cdot}275W$	$+\ 0{\cdot}882$
S.E. of partial regression coefficient	0·008	0·97	
t value of partial regression coefficient	33·26 HS	4·39 HS	
Standard partial regression coefficient	0·97	0·13	
Units	°C	mg dry weight	

Coefficient of multiple determination, $R^2 = 0{\cdot}95$; $F = 6{\cdot}9$ HS; d.f. = 2 and 57.

S.E. = Standard error; H.S. = highly significant.

Table 5. *Results of the multiple linear regression analysis relating* Daphnia magna *feeding (F in cal/mg of animal/h) to environmental factors*

Initial form:

$F =$	$0{\cdot}0019C$	$+\ 0{\cdot}0352E$	$-\ 0{\cdot}0561W$	$-\ 0{\cdot}0079T$	$-\ 0{\cdot}0719$
S.E. of partial regression coefficient	0·0011	0·0036	0·0577	0·0106	
t value of partial regression coefficient	16·21 HS	9·51 HS	0·97 NS	0·74 NS	
Standard partial regression coefficient	0·74	0·51	0·04	0·04	
Units	mg/l	cal/mg	mg dry weight	°C	

Coefficient of multiple determination, $R^2 = 0{\cdot}85$; $F = 0{\cdot}55$ NS; d.f. = 4 and 76.

Final form:

$F =$	$0{\cdot}0019C$	$+\ 0{\cdot}0352E$	$-\ 0{\cdot}0958$
S.E. of partial regression coefficient	0·0011	0·0030	
t value of partial regression coefficient	16·26 HS	11·67 HS	
Standard partial regression coefficient	0·74	0·53	
Units	mg/l	cal/gm	

Coefficient of multiple determination, $R^2 = 0{\cdot}92$; $F = 136{\cdot}1$ HS; d.f. = 2 and 78.

S.E. = Standard error; H.S. = highly significant; N.S. = not significant.

respect to both assimilation and feeding, but more data would be needed to determine the exact nature of the relationship, as results are quite variable. Temperature could not be included in the test for linearity since only two experimental temperatures were used,

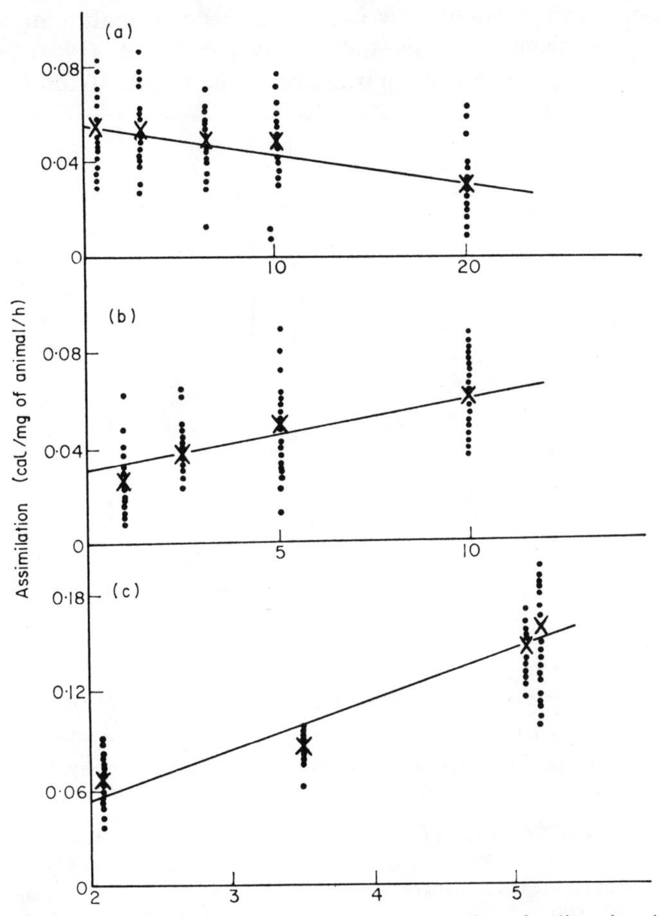

FIG. 3. A test of the assimilation partial regression equations for linearity. (a) Animal weight (W) (mg); (b) food concentration (C) (mg/l); (c) food energy content (E) (cal/mg). Small marks denote residuals, plotted from the partial regression line; large $\times$ denotes the average of a group of residuals.

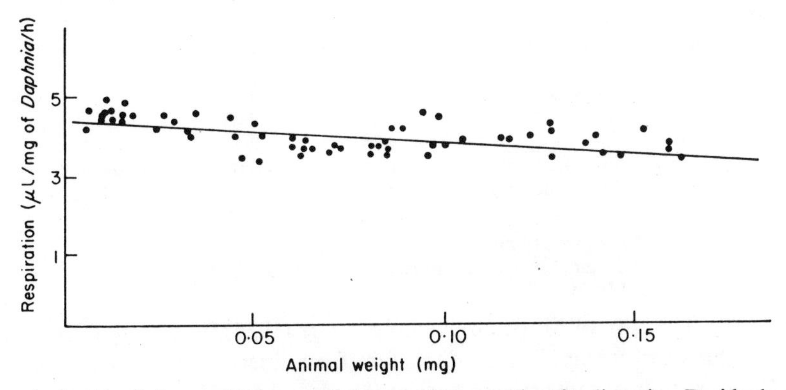

FIG. 4. A test of the respiration partial regression equation for linearity. Residuals are denoted as in Fig. 3.

allowing only linear computations. The lack of further temperature data is probably the greatest single shortcoming of this study. Comita & Comita (1964) found a linear relationship between respiration and temperature in the region 10–20° C for *Mixodia-*

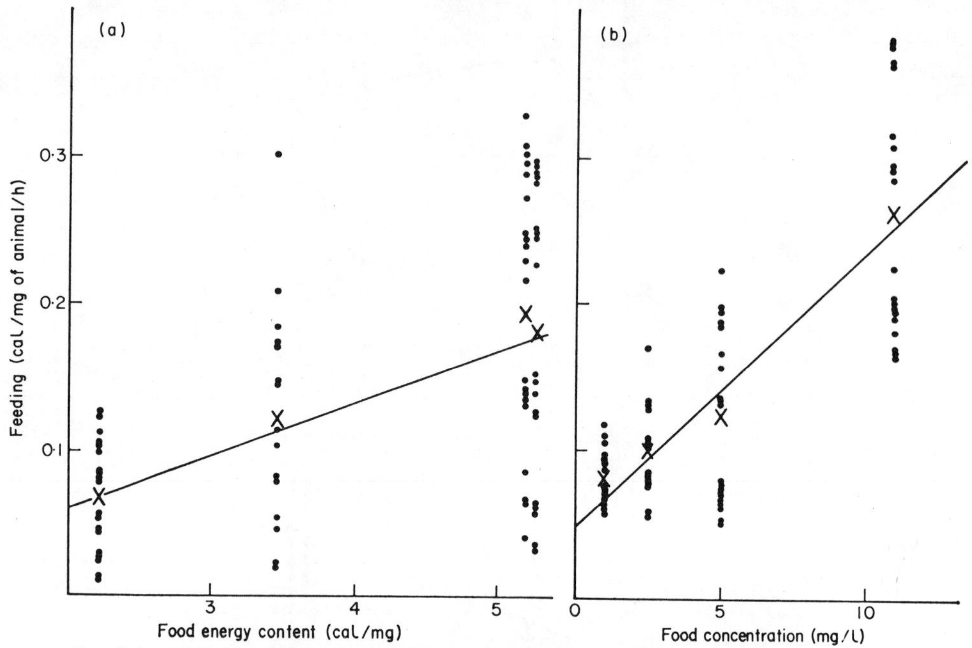

FIG. 5 (a and b). A test of feeding partial regression equations for linearity. Residuals are denoted as in Fig. 3.

Table 6. *Calorific values of* Daphnia magna *and experimental diets*

No. of det.	Animals/det.	*Daphnia* length (mm)	mg/det.	cal/mg	Filler used (mg)
1	2000	1·0	12·52	4·582	30·52‡
2	500	2·0	32·24	4·901	–
2	400	2·5	37·65	4·890	–
2	200	2·5*	29·45	4·925	–
1	Unknown	Eggs and embryos	12·02	4·909	36·65‡
1	Unknown	Carapaces	44·22†	4·855	–

No. of det.	Type of diet	mg/det.	cal/mg	Filler used (mg)
1	*Chlorella*	25·14	5·204	–
1	*Chlorella*	29·36	5·272	–
1	*Chlamydomonas*	21·16	5·105	–
1	Detritus I	37·24	3·455	–
1	Detritus II	29·97	2·155	–
1	*Chlorella* (N-deficient)	14·15	6·006	24·15‡
1	*Chlorella* (N-deficient)	6·14	7·154	31·52‡

* Ovigerous animals.
† Original weight of entire animals = 631·71 mg.
‡ Millipore HA membrane.

ptomus laciniatus. The data of Marshall, Nichols & Orr (1935) for *Calanus* show the more typical curvilinear relationship, but the departure from linearity is slight at temperatures under 20° C. No author has investigated assimilation and feeding over a range of several temperatures.

B. *Calorific values*

Juvenile and adult *Daphnia magna* had very similar calorific values, contrary to the findings of Richman (1958) for *Daphnia pulex*. All calorific determinations for *D. magna* were in the range 4·852–4·925 cal/mg as compared with a range of 4·059–5·075 cal/mg for Richman's *D. pulex*. Results of calorific determinations on *D. magna* and experimental diets are summarized in Table 6. As a result of these determinations, a calorific value of 4·9 cal/mg will be assumed for *D. magna* throughout this paper.

C. *Production by* Daphnia magna

Final assimilation and respiration equations are set out below.

$$A = 0{\cdot}0286E + 0{\cdot}0038T + 0{\cdot}0031C - 0{\cdot}1444W - 0{\cdot}1405 \quad (3)$$

$$R = 0{\cdot}293T - 4{\cdot}27W + 0{\cdot}882 \quad (4)$$

where A is expressed in cal/mg of animal/h, and R is expressed in μl/mg of animal/h.

Both equations may be transformed into energy units by dividing by 4·9 cal, the calorific content of 1 mg of *D. magna*. Multiplying both equations by 24/4·9, and multiplying the respiration equation by 0·005, the number of calories respired per ml of oxygen consumed, gives both assimilation and respiration in units of cal/cal of animal/day. These expressions are set out below.

$$A = 0{\cdot}1401E + 0{\cdot}0186T + 0{\cdot}0152C - 0{\cdot}7076W - 0{\cdot}6884 \quad (5)$$

$$R = 0{\cdot}0072T - 0{\cdot}1046W + 0{\cdot}0216 \quad (6)$$

Subtracting equation (6) from equation (5) gives a single expression for production by *D. magna*

$$(A - R) = 0{\cdot}1401E + 0{\cdot}0114T + 0{\cdot}0152C - 0{\cdot}6030W - 0{\cdot}7100 \quad (7)$$

It is probable that food energy content is only a crude measure of the nutritive properties of a diet. This is suggested by the results of determinations using *Chlorella* of 6·006 and 7·154 cal/mg, which gave lower assimilation rates in cal/mg of animal/unit time than *Chlorella* of lower calorific value. These results were not included in the regression expression, since this inclusion would have introduced a strong degree of non-linearity which could not be approximated by any simple mathematical expression. Data are shown in Fig. 6.

Calorific values as high as 6 cal/mg have never been recorded for primary producers in nature, and of all species tested, only *Chlorella* was capable of reaching this level in the laboratory.

Fig. 7 compares the results of assimilation determinations with those of Malovitskaya & Sorokin, and Monakov & Sorokin. Since these authors gave no data on food energy content and animal size, it was assumed that the calorific value of their diets was 5·0 cal/gm, since healthy algal cultures were used; and that animals were 2 mm in length. The solid line in Fig. 7 represents a solution to the partial regression equation found for food concentration and assimilation in this study, with $E = 5{\cdot}0$, $W = 0{\cdot}063$, and $T = 15°$ C. Results obtained in this study were considerably higher than those of the above-mentioned authors. The difference may be due to self-absorption in the preceding studies, since the above authors determined the radioactivity of whole animals, with a standard correction for self-absorption of β emissions. This procedure was tried in the early stages of this study, but radioactivities were very low and the self-absorption correction varied even for animals of the same size. The reduced assimilation found

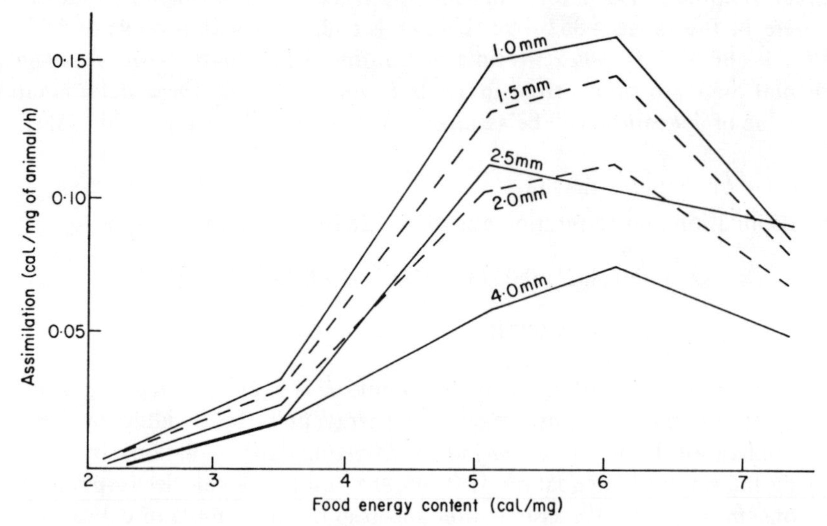

FIG. 6. Variation in the assimilation of *Daphnia magna* as related to animal length and food energy content. Diets of 2–4 cal/mg were of detritus collected from *Daphnia* tanks. Diets over 6 cal/mg were of *Chlorella* grown at high light intensity with nitrogen deficiency.

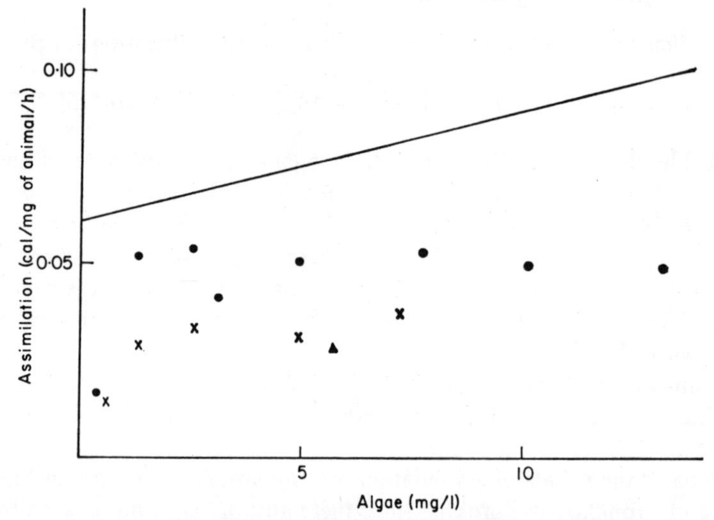

FIG. 7. A comparison of *Daphnia magna* assimilation rates with values found by previous authors. ×, *Daphnia pulex* fed upon *Chlorococcus* (Monakov & Sorokin 1961); ●, *Daphnia longispina* fed upon *Chlorococcus* (Monakov & Sorokin 1961); ▲, *Daphnia longispina* fed upon *Scenedesmus* (Malovitskaya & Sorokin 1960). The solid line represents a solution to equation (3) with $T = 15°$ C, $E = 5{\cdot}0$ cal/mg and $W = 0{\cdot}030$ mg/animal. The experiments of other authors were all run at 15° C.

by other authors at food concentrations under 1 mg/l is also of interest. No determinations using less than 1 mg/l were done in this study, since the seston content of natural ponds in which *Daphnia magna* is found seldom drops to such low levels.

D. *Testing production under controlled conditions*

Glass jars of approximately 3 litre capacity were filled with *Chlamydomonas* obtained from tanks in the Zoology Department courtyard at Oxford University. The calorific value of these algae was 5·050 cal/mg. They were concentrated by centrifuging or diluted with filtered culture water as required to obtain a food density of 5 mg/l. Twenty-five *Daphnia* 1·5 mm in length were put into each jar. All were kept at 20±2° C. Every 24 h,

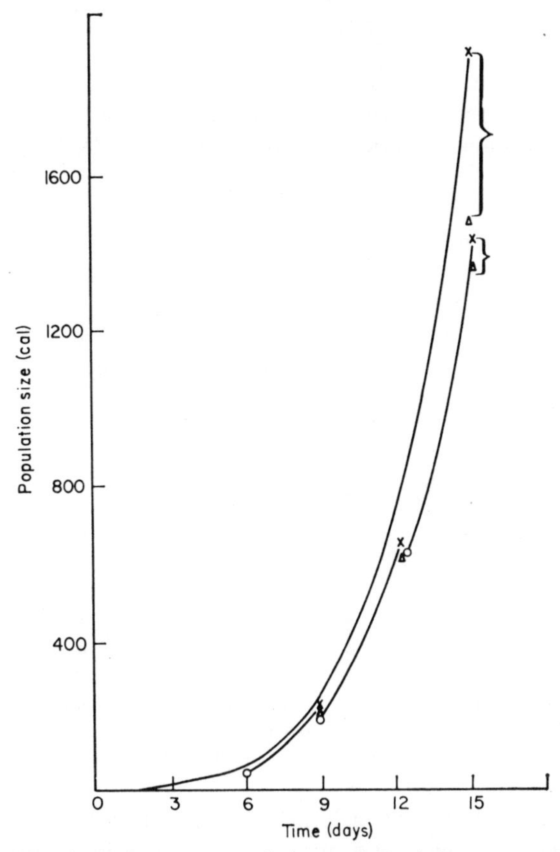

FIG. 8. Production by laboratory populations of *Daphnia magna* under controlled conditions. The distance between × and △ represents the difference between measured production and that predicted from equation (7). Upper curve = $Se^{(A-R)t}$ with t = 15 days; lower curve = $Se^{(A-R)t}$ with t = 3 days. ○, Initial standing crop of live animals; ×, initial standing crop plus production during the interval (t) as predicted from equation (7); △, production actually measured (= growth and numerical increase in live animals, plus dead animals and cast carapaces). S = Standing crop; t = interval between samplings.

cultures were poured through a No. 20 net, cast carapaces and dead animals were removed and frozen, and live animals were put back into new medium. On every third day, dead animals and cast carapaces were sorted into depression slides, measured and counted. The weight of animals was determined from Fig. 2. Fig. 8 gives the results of these experiments. The agreement between results as predicted from equation (1) and those actually measured is quite good when calculations are corrected for population size at 3-day intervals, but the equation tends to overestimate production when corrections are less frequent. This is largely due to the fact that some of the population is lost through death and cast carapaces, and is not capable of further production. This error tends to build up exponentially and may be quite large if the interval between samplings is large. The upper, smooth curve in the figure represents the equation:

$$(A-R) = 0{\cdot}1401E + 0{\cdot}0115T + 0{\cdot}0152C - 0{\cdot}7303 \qquad (8)$$

which does not include the variable for weight. Results will vary only a few per cent as a result of this omission.

DISCUSSION

Assimilation rates were sensitive to a number of environmental factors which did not affect feeding rates, or affected them to a lesser degree. These were the following.

1. *Temperature.* Assimilation rates were markedly affected, while feeding rates were

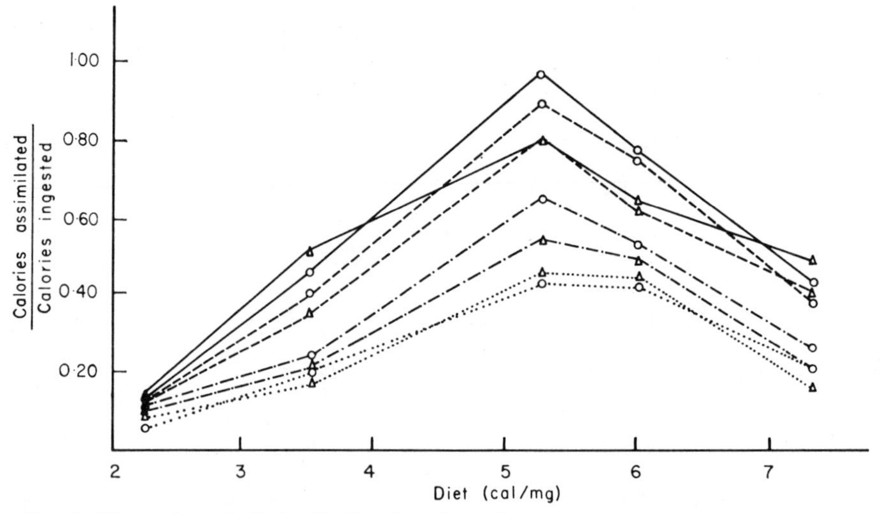

FIG. 9. The ratio assimilation/feeding for diets of several calorific values. △, 2·5 mm animals; ○, 1·0 mm animals; ———, 1 mg of diet/l; – – – –, 2·5 mg of diet/l; — · —, 5·0 mg of diet/l; · · · · · ·, 10·0 mg of diet/l.

unaffected. Assimilation rates, being largely dependent upon chemical digestive processes, would be expected to increase as the temperature is raised. Feeding in zooplankton is highly connected to normal swimming movements, and as a result may be more strongly influenced by density, viscosity and turbulence of the medium, by wave length and direction of light, and by food distribution rather than being dependent directly upon temperature. Further investigation is needed.

2. *Food energy content and concentration.* Assimilation of detritus and nutrient-starved algae by *D. magna* was poorer than the assimilation of algae grown in normal culture. Feeding rates were not so strongly affected, although the effect was a significant one. The variation in the ratio assimilation/feeding with food energy content is given in Fig. 9. It appears from these results that *Daphnia* may defaecate and reingest food several times, probably with a somewhat lower assimilation rate at each reingestion. Recently Johannes & Satomi (1966) have shown that marine zooplankton may reingest their own faeces, assimilating these with high efficiencies in some cases. Assimilation efficiency decreases with increased food concentration if all other factors are equal (see Fig. 9), resulting in the defaecation of faecal pellets with high nutritional values

when food densities are high. Analyses of the seston of a small, turbid Minnesota lake made by the author revealed that the energy content of suspended particles between 0·45 μ and 0·07 mm in diameter never exceeded 4·0 cal/mg dry weight, falling to values as low as 0·8 cal/mg in late winter. Healthy algal cultures are usually 4·5–5·3 cal/mg. Healthy algae seldom made up more than a few per cent of the natural seston, and during 18 months of observation, algae only exceeded 50% of the total seston on one occasion, during a spring diatom bloom. Silt, scraps of algal cells, pieces of zooplankton carapace and unidentifiable organic particles made up the remainder of the seston. Laboratory observations revealed that planktonic Copepoda and Cladocera in the lake consumed particles of different origin non-selectively, although there was some selection for size and shape of particles. The classic idea of defaecated material being passed to the next trophic level or to a separate decomposer food chain does not appear to be valid in turbid freshwater lakes or in the sea.

3. *Animal weight.* Assimilation rates (cal/mg of animal/h) decreased slightly as body size increased. Feeding rates were unaffected. Brooks & Dodson (1965) present evidence that filtration rate is roughly proportional to the square of body length for a number of zooplankton species. Since weight is more related to the cube of body length, the rate

Table 7. *Efficiency of growth and reproduction by* Daphnia pulex *recalculated from tables* 11 *and* 12 *of Richman* (1958)

Food concentration* (mg/l)	Growth/Assimilation (young)	Growth and reproduction / Assimilation (adult)
0·0062	0·55	0·56
0·0124	0·58	0·68
0·0186	0·57	0·70
0·0248	0·59	0·73

* Calculated using Richman's figure of $0{\cdot}248 \times 10^{-6}$ mg/*Chlamydomonas* cell and cell concentrations of 25 000, 50 000, 75 000 and 100 000 cells/ml.

of filtration per unit should decrease with increased body size. Increased body size also enables animals to ingest larger food particles, however (see the discussion by Brooks & Dodson 1965), so that the actual rate of ingestion should be a function of both body size and food particle size. It is therefore doubtful whether generalized conclusions may be drawn from the feeding rate experiments outlined here.

4. *Reproductive state.* The higher assimilation rates of animals carrying large broods are inexplicable, and evidence in this paper is largely circumstantial. The data of Richman (1958) however, demonstrate a higher production efficiency for *D. pulex* populations at high food concentrations, when most of the energy used for production is channelled into reproduction. Table 7 gives a summary of (growth + reproduction)/assimilation derived from Richman's tables 11 and 12. This ratio will be called the production efficiency throughout the remainder of this paper. The efficiency of adult animals rises with increased food concentration and the resulting increase in the channelling of energy into reproduction. The production efficiency of young animals does not rise with increased food concentration, and is similar to the production efficiency of adults at low food concentration, where only a small fraction of the energy assimilated is used for reproduction. A higher metabolic efficiency appears to affect both growth and reproduction in populations where reproduction is the chief method of production, as Richman's respiration data reveal no significant difference between the maintenance

requirements per unit weight of ovigerous and non-ovigerous adults. A study similar to the removal experiments of Slobodkin (1959) would provide an invaluable check on the efficiency of ovigerous *v.* non-ovigerous animals, if the following two removal categories were used: (a) juveniles plus non-ovigerous adults, and (b) ovigerous adults.

The dependence of feeding upon food energy content is inexplicable. It is probable, however, that the size of food particles is much less homogeneous at lower food energy contents where fully-formed algae form a small proportion of the food supply and that large scraps of diffuse organic matter, zooplankton carapace and allochthonous material may hamper feeding in such situations. The relationship between feeding rate and food concentration has been tested by a large number of previous authors, but results are difficult to compare, since little attention has been given to the size of food particles or to food concentration by weight.

The direct measurement of assimilation and respiration may be done quite simply under field conditions, using a procedure similar to the suspended bottle technique used by Richman (1966) for measuring feeding rates of zooplankton. Such experiments would allow the direct evaluation of the effect of environmental conditions upon zooplankton production. More traditional methods for measuring the effect of environment upon animal populations have involved monitoring birth rates, death rates and individual growth rates. Such procedures have not been used for zooplankton production work to date, though recent work has shown that zooplankton hatching rates are generally dependent only upon temperature (Elster 1954; Eichhorn 1957; Edmondson 1965; Schindler 1966) and that death rates may be calculated from temperature-dependent birth rates and accurate population data (Hall 1964). No author has yet included measurements of individual growth in such a study, and correlations of birth or death rates with environmental factors have been generally low (see Edmonson 1965; Edmondson, Comita & Anderson 1962). The direct measurement of assimilation and respiration under field conditions should provide useful information on such problems where individual growth, hatching and death rates are impossible to measure.

ACKNOWLEDGMENTS

I am grateful to the following persons for assistance provided: to Dr Peter Brunet for advice on the technique used in this study, and for criticizing several drafts of the manuscript; to Mr Charles Elton, F.R.S., for several helpful criticisms on the work as part of a D.Phil. thesis, and for furnishing writing space at the Bureau of Animal Population; to Dr K. H. Mann, for several valuable discussions of the method; to Professor A. J. Cain and Professor F. C. Evans, for many stimulating discussions while the work was in its early stages; to Professor G. W. Comita, who kindly allowed me to do the bomb calorimetry at North Dakota State University and arranged financial support for that portion of the study, as well as for many discussions of zooplankton energy relations; to Dr D. F. Mayers and Dr D. Peterson for assistance in programming the data for the English Electric KDF9 and the IBM 1620 computers at Oxford and at North Dakota State University; to Professor R. L. Edwards, for several helpful comments on the manuscript; to Mr E. D. Le Cren, for an exhaustive review of the paper and many helpful criticisms. This study is part of a thesis presented for the Doctor of Philosophy degree at Oxford University. The calorimetry at North Dakota State University was financed by N.S.F. Grant No. 16415 to Professor Comita, and the North Dakota Institute for Regional Studies.

Work at Oxford was done whilst on a Rhodes Scholarship.

SUMMARY

1. Assimilation and feeding rates of *Daphnia magna* were determined using ^{14}C-tagged diets. The effects of temperature, animal size, food energy content and concentration, light, motion, crowding and reproductive state were examined.

Light, motion and crowding did not affect either assimilation or feeding in the ranges tested. Assimilation rates increased with higher temperature and decreased as animal size increased. Neither temperature nor animal size affected feeding rates significantly.

Food concentration and energy content affected both assimilation and feeding rates significantly.

2. Respiration of *D. magna* was measured using the Winkler method. Respiration rates increased as temperature increased and decreased as animal size increased.

3. Equations for assimilation, feeding and respiration were determined by multiple regression analysis. The linearity of these relationships was tested. Equations were tested by comparing production as predicted from regression analysis with actual production in laboratory populations.

REFERENCES

Bourne, N. F. (1959). *The determination of carbon transfer from* Chlorella vulgaris *Beyerinck to* Daphnia magna *Straus using radioactive carbon* (C^{14}) *as a tracer.* Ph.D. thesis, University of Toronto, Ontario.

Brooks, J. L. & Dodson, S. I. (1965). Predation, body size, and composition of plankton. *Science, N.Y.* **150,** 28–35.

Clarke, G. L., Edmondson, W. T. & Ricker, W. E. (1946). Mathematical formulation of biological productivity. *Ecol. Monogr.* **16,** 336–7.

Comita, G. W. & Comita, J. J. (1964). Oxygen uptake in *Mixodiaptomus laciniatus* Lill. *Memorie Ist Ital. Idrobiol.* **17,** 151–66.

Comita, G. W. & Schindler, D. W. (1963). Calorific values of microcrustacea. *Science, N.Y.* **140,** 1394–6.

Edmondson, W. T. (1957). Trophic relations of the zooplankton. *Trans. Am. microsc. Soc.* **76,** 225–45.

Edmondson, W. T. (1965). Reproductive rates of planktonic rotifers as related to food and temperature in nature. *Ecol. Monogr.* **35,** 61–111.

Edmondson, W. T., Comita, G. W. & Anderson, G. C. (1962). Reproductive rate of copepods in nature and its relation to phytoplankton population. *Ecology,* **43,** 625–34.

Eichhorn, R. (1957). Zur Populationsdynamik der calanoiden Copepoden im Titisee und Feldsee. *Arch. Hydrobiol.* **24,** Suppl., 186–246.

Elster, H. J. (1954). Uber die Populationsdynamik von *Eudiaptomus gracilis* Sars und *Heterocope borealis* Fischer im Bodensee-Obersee. *Arch. Hydrobiol.* **20,** Suppl., 546–614.

Ezekial, M. & Fox, K. A. (1959). *Methods of Correlation and Regression Analysis.* New York.

Hall, D. J. (1964). An experimental approach to the dynamics of a natural population of *Daphnia galeata mendotae.* *Ecology,* **45,** 94–112.

Johannes, R. E. & Satomi, M. (1966). Composition and nutritional value of fecal pellets of a marine crustacean. *Limnol. Oceanogr.* **11,** 191–7.

Malovitskaya, L. M. & Sorokin, Yu. I. (1961a). On the feeding of some species of diaptomids (Copepoda, Calanoida) on bacteria. *Dokl. Akad. Nauk SSSR,* **136,** 948–50. A.I.B.S. translation.

Malovitskaya, L. M. & Sorokin, Yu. I. (1961b). An experimental study of the feeding of *Diaptomus* (Crustacea, Copepoda) using C^{14}. *Trudÿ Inst. Biol. Vodokhran.* **4,** 262–73 (in Russian).

Marshall, S. M., Nicholls, A. G. & Orr, A. P. (1935). On the biology of *Calanus finmarchicus.* Part 6. Oxygen consumption in relation to environmental conditions. *J. mar. biol. Ass. U.K.* **20,** 1–27.

Marshall, S. M. & Orr, A. P. (1955). Experimental feeding of the copepod *Calanus finmarchicus* (Gunner) on phytoplankton cultured with radioactive carbon (14C). *Deep Sea Res.* **3,** Suppl., 110–14.

McMahon, J. W. (1962). The feeding behaviour and feeding rate of *Daphnia magna* Straus. *Can. J. Zool.* **43,** 603–11.

McMahon, J. W. & Rigler, F. H. (1963). Mechanisms regulating the feeding rate of *Daphnia magna* Straus. *Can. J. Zool.* **41,** 321–32.

Monakov, A. V. & Sorokin, Yu. I. (1960). An experimental investigation of *Daphnia* nutrition using C^{14}. *Dokl. Akad. Nauk SSSR,* **135,** 1516–18. A.I.B.S. translation.

Monakov, A. V. & Sorokin, Yu. I. (1961). Quantitative data on the feeding of *Daphnia*. *Trudӯ Inst. Biol. Vodokhran.* **4,** 251–61 (in Russian).

Nauwerck, A. (1959). Zur Bestimmung der Filtrierrate limnischer Planktontiere. *Arch. Hydrobiol.* **25,** Suppl., 83–101.

Richman, S. (1958). The transformation of energy by *Daphnia pulex*. *Ecol. Monogr.* **28,** 273–91.

Richman, S. (1966). The effect of phytoplankton concentration on the feeding rate of *Diaptomus oregonensis*. *Verh. int. Verein. theor. angew. Limnol.* **16,** 392–8.

Rodina, A. G. & Troshin, A. S. (1954). The use of labelled atoms in the study of feeding in aquatic animals. *Dokl. Akad. Nauk SSSR,* **98** (2), 297–300.

Ryther, J. H. (1954). Inhibitory effects of phytoplankton upon the feeding of *Daphnia magna*, with reference to growth, reproduction, and survival. *Ecology,* **35,** 522–33.

Schindler, D. W. (1966). *Energy relations at three trophic levels in an aquatic food chain.* D.Phil. thesis, Oxford University.

Slobodkin, L. B. (1959). Energetics in *Daphnia pulex* populations. *Ecology,* **40,** 232–43.

Spoehr, H. A. & Milner, A. W. (1949). The chemical composition of *Chlorella*: effect of environmental conditions. *Pl. Physiol., Lancaster,* **24,** 120–49.

Welch, P. S. (1948). *Limnological Methods.* Philadelphia.

THE SWIMMING RHYTHM OF THE SAND BEACH ISOPOD *EURYDICE PULCHRA*

5

D. A. Jones [1] and E. Naylor

Department of Zoology, University College of Swansea, Swansea, Wales

Abstract: *Eurydice pulchra* is shown to have an endogenously controlled swimming rhythm of tidal frequency; in the laboratory this is expressed by animals kept in sea water alone but not by animals given sand, from which they rarely emerge. On the beach the animals evidently rely on being washed from the sand by the rising tide, the endogenous component of the rhythm ensuring that they swim for up to 5–6 h before reburrowing in the sand in a restricted zone between M.T.L. and H.W.N. The rhythm is synchronized by wave action and reinforced by variations in hydrostatic pressure. The number swimming is affected by light, fewer animals swimming by day since well fed animals are photonegative and reburrow immediately they emerge into the surf.

Introduction

Eurydice pulchra Leach (Crustacea: Isopoda) is found intertidally in clean sand beaches most abundantly between M.T.L. and H.W.N. (Jones & Naylor, 1967). It is a predaceous isopod (Jones, 1968) which leaves the sand with the incoming tide, feeds actively during high tide, and returns to the sand on the ebb (Elmhirst, 1932; Watkin, 1942; Salvat, 1966). The present work was initiated to investigate how the zonation is maintained in relation to the tidal rhythm of swimming, to assess the role of external and internal factors in the control of the swimming rhythm, and to determine any seasonal changes in this behaviour.

Material and Methods

To determine the pattern of tidal excursions into the surf by *E. pulchra*, samples were taken from the water's edge at hourly intervals on both flood and ebb tides using a surf plankton net of the kind described by Colman & Segrove (1955) but fitted in this instance with a single net. In laboratory experiments spontaneous swimming activity was initially recorded by counting at hourly intervals throughout the day and night the number of animals swimming above mid-depth in a sea water aquarium. In other experiments, activity was recorded automatically using a water-filled Perspex tube of 1 cm diameter which was formed into a closed circle about 20 cm across and in which the animals could swim continuously and unimpeded in one direction. Sea water circulated slowly through the tube and swimming activity was monitored when the animals swam through a dim beam of light focussed onto a photo-electric cell (Williams & Naylor, 1967). All experiments, unless otherwise indicated, were conducted at a fairly constant temperature in a dark room, lit by dim red light. Temper-

[1] Present address: Marine Science Laboratories, Menai Bridge, Anglesey.

Reprinted from *Experimental Marine Biology and Ecology*, 4:188–199, 1970.

atures ranged from 18–21 °C throughout the course of all the experiments and varied by no more than 1 °C during the course of any one experiment. Most experiments ran for several days and results have been analyzed as form estimates for selected whole-hour frequencies.

The Rhythm in the Field

Fig. 1 shows the results of monthly surf samples from which it can be seen that most *Eurydice* were taken at the water's edge during two periods of 2–3 h, one just before and one just after the time of high tide. This pattern was repeated in most

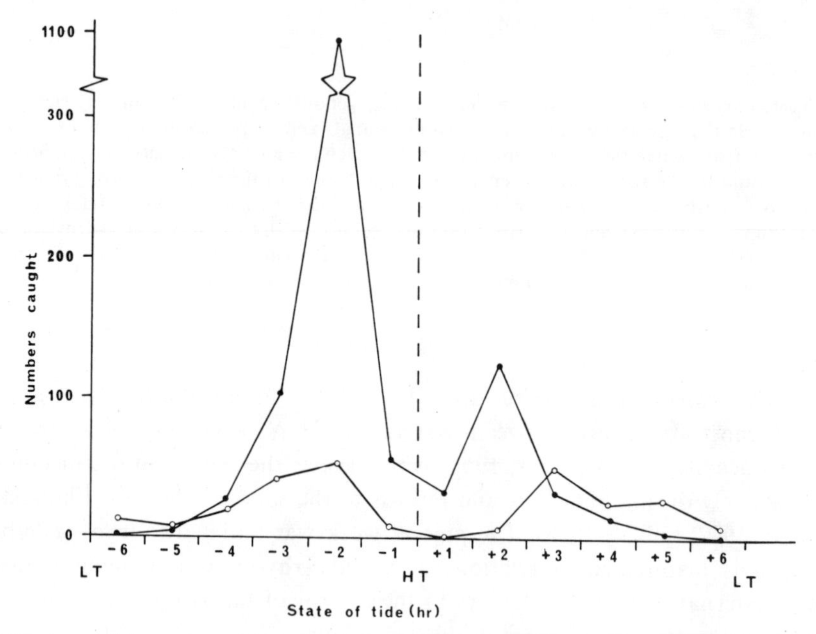

Fig. 1. Mean monthly catches of *Eurydice pulchra* in standard 25 m surf plankton hauls at Rotherslade, S. Wales; samples were taken hourly at the water's edge as the tide rose and fell throughout the day and were carried out once each month during 1966; ● March to November hauls, ○ January, February and December hauls.

months of the year, the peaks being least distinct in winter when most of the population migrated below tidemarks. The results of a more detailed sampling over a 24 h period in summer (Fig. 2) show that the numbers caught during night-time high tides were greater than during the day. All these results agree with evidence from sand samples (Jones, unpublished observations) that *Eurydice pulchra* occupies a distinct zone between M.T.L. and H.W.N., peak numbers of animals being taken as the tide flows and ebbs through the zone of their greatest abundance.

Reference to Figs 1 and 2 shows that *Eurydice* was sparse in surf plankton samples obtained at the highest levels of the beach, *i.e.* higher up the shore than the zone where they are abundant in the sand. Investigations were carried out to determine whether this was due to animals reburrowing at the time of high tide and then swimming again on the ebb, as has been reported for *Synchelidium* (Enright, 1963). Accordingly, just before high tide a number of cast iron plates (20 × 20 cm) were laid behind the rising tide with a view to preventing the re-emergence of any isopods which might have temporarily burrowed; no animals were trapped beneath such plates, so evidently some other explanation must be sought to account for the fall in number of *Eurydice* at the water's edge at high tide. One possibility is that isopods 'maintain station' immediately above the sand from which they are washed out, but this cannot always be so because it was sometimes possible to obtain large numbers

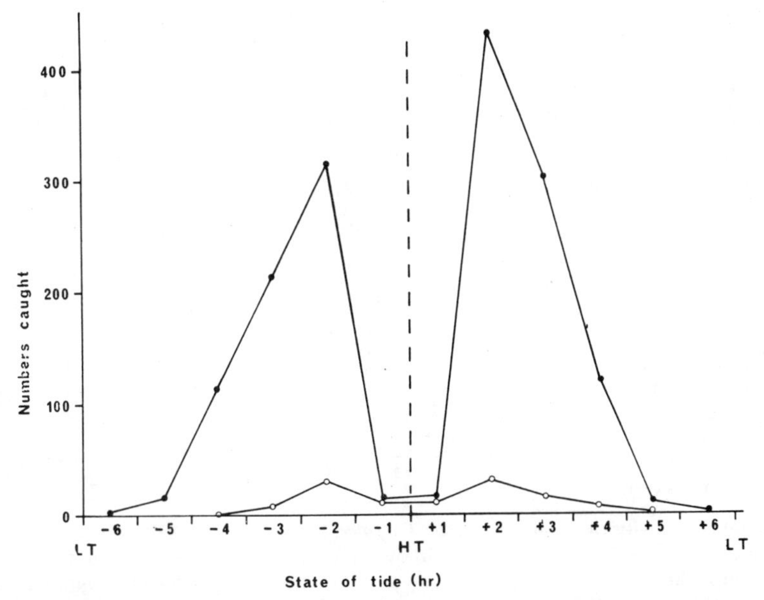

Fig. 2. Total *Eurydice pulchra* in standard 25 m surf plankton hauls taken hourly throughout two consecutive tidal cycles in summer; ● night hauls, o day hauls.

of animals in surf even at high spring tide. In any event, from these observations and the laboratory experiments below, it seems clear that *Eurydice* probably swims throughout the whole high tide period. Perhaps the most likely explanation of the low catches at high tide is that at high water mark, where the slope of the beach increased, boulders were more abundant and surf sampling was much less efficient than lower down the beach.

The Rhythm in the Laboratory

A particularly striking feature of the laboratory experiments was that whenever sand was provided in the aquarium the isopods promptly burrowed and never reappeared. This experiment was repeated many times using fresh sand from the beach containing undisturbed animals and adding sea water at the time of high tide in the field; at no time under these conditions was *E. pulchra* observed to leave the sand. Only mechanical disturbance of the sand by shaking or stirring caused the animals to swim and it seems likely that in the field wave action must wash them out.

Despite this important exogenous feature of the rhythm there is also a pronounced endogenous component, since animals kept in an aquarium tank without sand exhibited a tidal rhythm of swimming. This is well illustrated in Fig. 3a which gives the numbers swimming in such a tank during a period of 30 sec every hour over a period of four days. Peaks of swimming activity in these freshly collected animals clearly coincided with the times of high tide which they would have experienced on the shore. Automatic recording of activity confirmed the existence of a tidal rhythm,

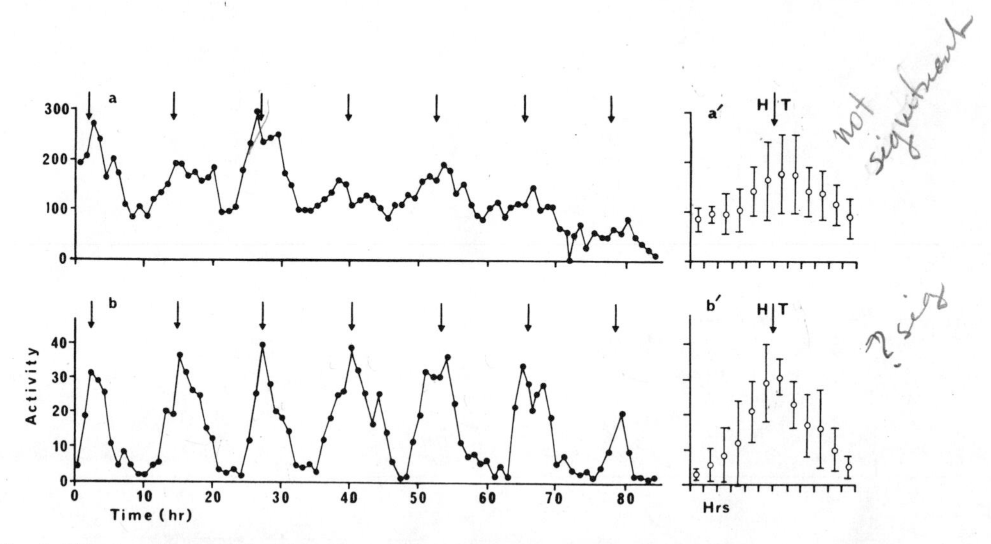

Fig. 3. Hourly activity of freshly collected *Eurydice pulchra* during 4 days in constant laboratory conditions: a, 50 isopods in an aquarium without sand, showing numbers swimming above mid-tank depth during a 30 sec observation period every h; b, 5 isopods in a photocell actograph, plotting total numbers swimming past light source per h; arrows indicate 'expected' times of high tide; a′, b′ (note expanded time-scale) are form estimates plotting mean and standard deviations of hourly activity values for the 6 h periods before and after 'expected' high tide times.

and showed that the swimming period continued for about 5 h centred around the time of high tide (Fig. 3b). These results also show that in the laboratory *Eurydice* was more or less equally active at the times of day and night high tides, whereas on the shore they swim much less by day than by night (see Fig. 2). The rhythm more or less disappeared after about 10 days under constant conditions.

Effects of External Factors

The major problems which arise from these observations are, 1) to determine the mode of synchronization of the rhythm, 2) to explain the difference between the laboratory rhythm, in which successive activity peaks may be of equal height (Fig. 3b), while in the field rhythmic activity was greater at night than during the day

(Fig. 2). Under constant conditions the *Carcinus* walking rhythm (Naylor, 1958) shows tidal peaks of activity which are alternately large and small, suggesting an endogenous twice-tidal or circadian component in addition to a tidal component, but in *Eurydice* it seems most likely that increased nocturnal activity (or decreased daytime activity) in the field is related to some external factor, the most obvious being light.

Light

In studying this factor attempts were made to entrain arhythmic *Eurydice* in laboratory aquaria with normal day–night cycles; however, after 7 days under such conditions animals transferred to actographs under constant conditions showed no evidence of a persistent circadian rhythm (Fig. 4). Such rhythmicity as is evident in Fig. 4 could be a residual tidal periodicity reflected by the two small peaks about 12 h apart in the 24 h form estimate. There appears, therefore, to be no endogenous circadian component in the rhythm of *E. pulchra* and there is no evidence that light is an important synchronizer of the rhythm.

On the other hand, *E. pulchra* does show important behavioural responses to light. Fraenkel (1961) showed that photopositive individuals attracted to a lamp at night, later became photonegative in the laboratory. Present observations confirm this, and also show that in any population, whether from the sand or surf plankton, both photonegative and photopositive animals are present at the same time. Moreover, experiments indicate that these phototaxic responses appear to be related to the nutritional state of the animal (Table I), animals with a full gut being photonegative and starved

TABLE I

Light responses of 42 starved and 44 recently fed *E. pulchra*. (χ^2 tests comparing observed results with hypothetical random occurrence of negatively phototaxic and positively phototaxic animals, give $P < 0.001$).

Nutritional condition	Animals used: Positively phototaxic	Animals used: Negatively phototaxic
Starved 14 days	42	0
Fed 1 h	8	36

TABLE II

Light responses of *E. pulchra* taken in surf and sand samples during daytime. (χ^2 test comparing observed surf hauls with expected hauls based upon ratio of negatively phototaxic and positively phototaxic animals in sand gives $P < 0.001$).

Day samples	Animals caught: Positively phototaxic	Animals caught: Negatively phototaxic	Ratio
Surf	544	61	8.9 : 1
Sand	168	88	1.9 : 1

animals photopositive. Of particular interest is that whereas nighttime surf samples contained both fully fed and starved individuals, daytime hauls contained very few fully fed photonegative animals (Table II).

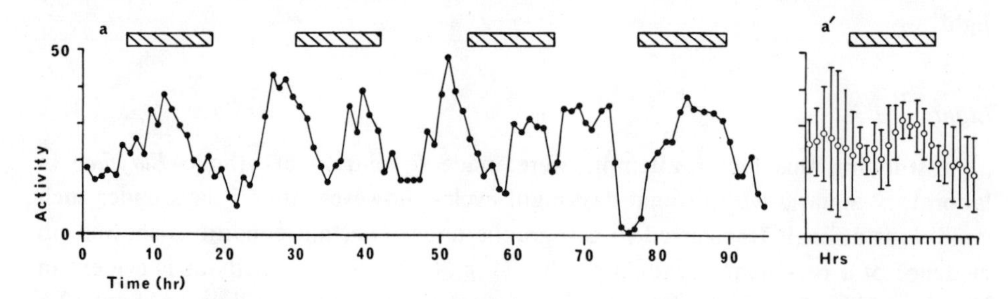

Fig. 4. Total hourly activity of 5 *Eurydice pulchra* after 7 days in a 12 h light : 12 h dark regime and then placed in a photocell actograph in constant dim light: shaded areas are 'expected' dark periods; a′ is a form estimate giving mean and standard deviations of hourly activity values through a 24 h period centred on each 'expected' dark period.

Wetting

Immersion on the rising tide probably does not act to synchronize the rhythm, particularly since it has been shown that immersion in the laboratory fails to bring the animals out of the sand.

Wave action

As has been indicated already, wave action and the associated mechanical stimuli would seem to be important in bringing about the active phase of these isopods both in the field and laboratory, and this factor has already been cited as the main synchronizer for another sand-beach isopod, *Exocirolana chiltoni* (Enright, 1965) and possibly also for the amphipod, *Synchelidium* (Enright, 1963). In the present work, wave action was simulated by stirring the animals among sand and Fig. 5a shows that by subjecting arhythmic animals to stirring for 30 min at 12 h intervals for 4 days, persistent rhythmicity of approximately tidal periodicity could be induced with large peaks of activity generally occurring around the 'expected' time of stirring. The small rise in activity at the time of otherwise low activity in Fig. 5a′ appears to be due to a few sporadic outbursts of activity. These are seen in Fig. 5a and are reflected in the particularly high standard deviations of values contributing to the means at −6 and +6 h in Fig. 5a′. An arhythmic group which was unstirred, but otherwise handled in the same manner as experimental animals, was included as a control (Fig. 5c). Mean hourly activity values in the controls (Fig. 5c′) showed little divergence from the overall mean, indicating lack of tidal rhythmicity in the data.

In other experiments it was found (Fig. 5b, b′) that a single period of stirring for as short a time as 1 h in the absence of sand partially re-established a rhythm in arhythmic *Eurydice pulchra*. Here a weak rhythm appeared with peaks occurring after the 'expected' times of stirring and persisted for a few cycles. The phasing of activity was slightly different depending upon whether the animals were stirred once for 1 h or repeatedly at tidal intervals, but in both types of experiment there are indications of a rhythm of tidal frequency in the results. Evidently swirling water can re-establish a tidal rhythm in arhythmic *Eurydice*, which suggests that wave action is an important synchronizer in the field.

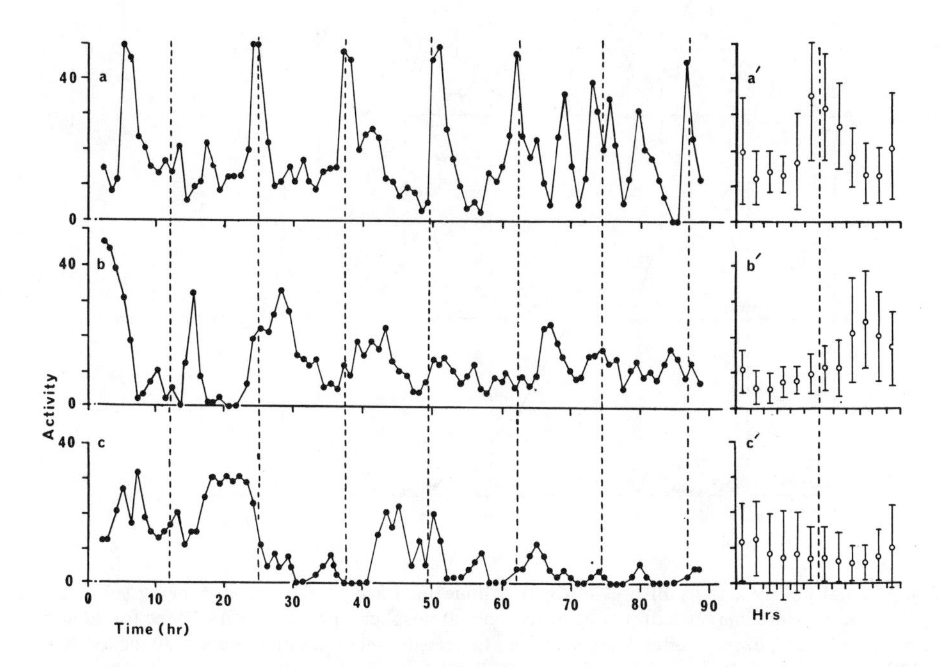

Fig. 5. Total hourly activity of 5 previously arhythmic *Eurydice pulchra*: a, after being stirred for 30 min every 12 h for 4 days; b, after stirring for 1 h only; activity of 5 control animals shown in c; vertical dotted lines at 12.4 h (tidal) intervals from time when experimental animals were last stirred; a′–c′ are form estimates as in Fig. 3.

Pressure

Although *E. pulchra* is pressure sensitive when swimming (Knight-Jones & Morgan, 1966), it has been shown that increases in pressure equivalent to wave heights of up to 1 m failed to evoke any swimming reaction from animals buried in the sand (Singarajah, 1966). In the present work, when free-swimming arhythmic *Eurydice* without sand were subjected to cyclic pressure fluctuations equivalent to 5 m of water every 30 sec for 30 min at 12 h intervals for 5 days, rhythmic activity of approximately tidal periodicity was induced with peak activity occurring before the 'expected' time of increased pressure (Fig. 6a). Similar experiments using pressure fluctuations of the order of 1 m of water resulted in sporadic peaks in the subsequent actograph record but there is little evidence of a rhythm of tidal frequency resulting from this treatment (Fig. 6b). It seems likely then that pressure cycles, while not the main synchronizer, have some synchronizing effect on animals, perhaps by reinforcing the effects of wave action.

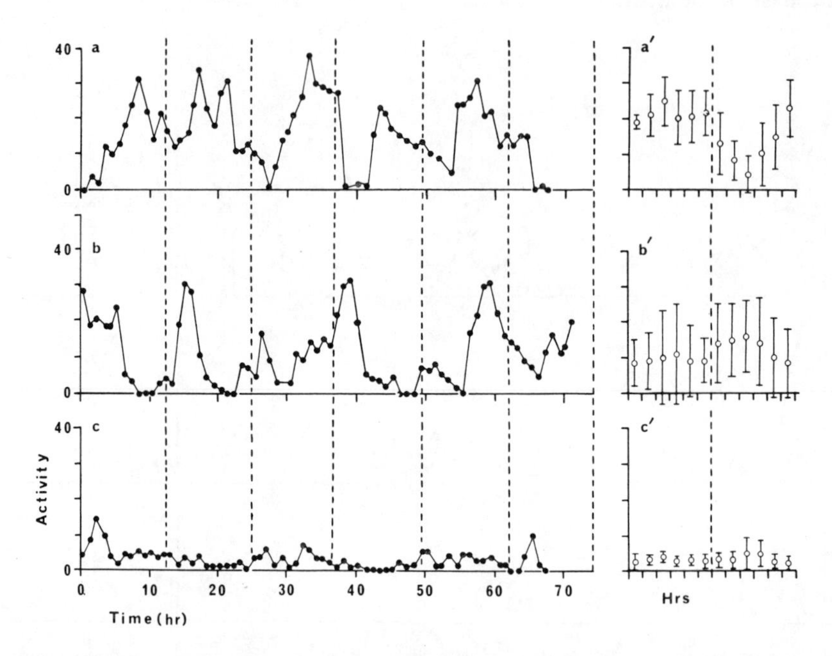

Fig. 6. Total hourly activity of previously arhythmic *Eurydice pulchra*: a, after being subjected to hydrostatic pressure increases of 5 m of water from atmospheric pressure every 30 sec for 30 min at 12 h intervals for 5 days; b, after being subjected to pressure increases of 1 m every 30 sec for 30 min at 12 h intervals for 5 days; c, activity of untreated controls; vertical dotted lines and a′–c′ as in Fig. 3.

Temperature

The effects of cyclically varying temperatures were not investigated, but in view of observations that chilling induces a rhythm of tidal frequency in shore crabs and fish which have lost their rhythmicity in the laboratory (Naylor, 1963; Gibson, 1967) *Eurydice* was subjected to low temperatures. Chilling rhythmic *Eurydice* to 5 °C for periods as short as 3 h caused a complete rephasing of the rhythm, activity occurring shortly after the animals were returned to normal temperatures and recurring at 12.4 h intervals thereafter (Fig. 7a). A rhythm of approximately tidal periodicity was also induced by subjecting arhythmic *Eurydice* to periods of chilling at 5 °C (Fig. 7b) activity again occurring after the 'expected time of chilling'. These results conform closely to those obtained with the shore crab, *Carcinus* (Naylor, 1963) and seem to confirm the presence of an endogenous tidal component in the activity rhythm of the isopod, as in the crab.

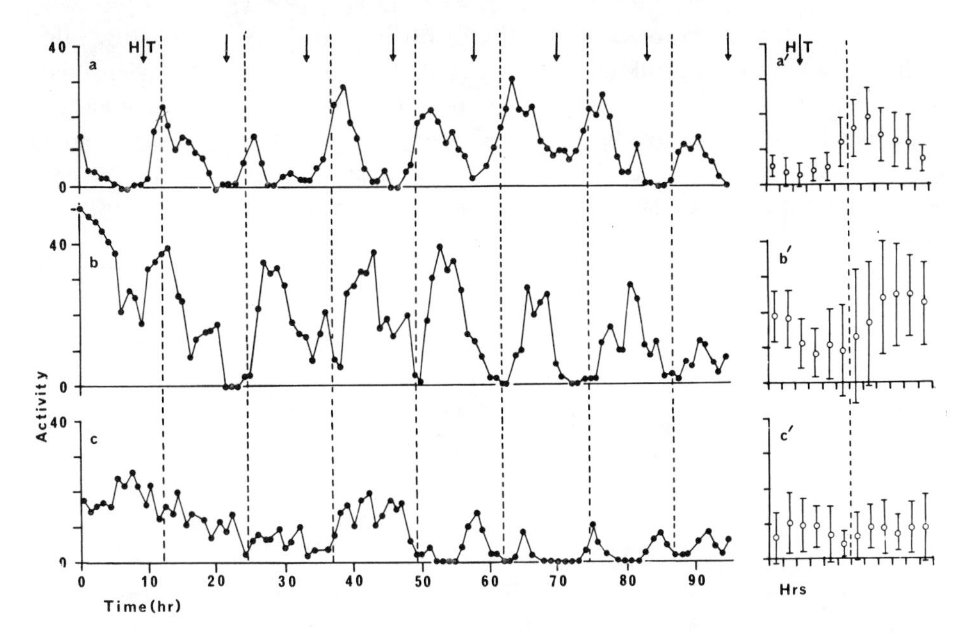

Fig. 7. Total hourly activity of 5 *Eurydice pulchra* after being chilled to 5 °C for 3 h: a, rhythmic animals fresh from the shore at 15 °C; b, previously arhythmic animals; c, arhythmic unchilled controls; arrows in 'a' indicate 'expected' times of high tide; dotted lines at 12.4 h intervals from end of chilling of experimental animals; a′–c′ as in Fig. 3.

Discussion

The possible relationships between external physical factors and the endogenous rhythm in *E. pulchra* are summarized in Fig. 8. Wave action, the only factor apparently capable of initiating the active swimming phase on the beach, both washes the animals out of the sand and acts as the main synchronizer. Once out of the sand variations in hydrostatic pressure then act as a reinforcement, indicating as in *Carcinus* (Williams & Naylor, in press) that more than one environmental variable is involved in synchronization. Additional evidence emphasizing the dependence of *Eurydice* upon wave action comes from field observations upon depth of burrowing (Jones, unpublished observations); these show that few *Eurydice* burrow deeper than the average depth disturbed by wave action and that the numbers of animals caught in surf is linearly related to the wave height recorded at the time of sampling. Fig. 8 also suggests how the light responses and hunger state combine to modulate the endogenous swimming rhythm, so explaining the differences in numbers caught in the surf by day and by night. The smaller numbers taken by day would be explained since, even if most animals were washed out from the sand by the rising tide, those which were fully fed and therefore photonegative (see p. 193) would promptly reburrow. At night the total population whether starved or fed would remain swimming, since the absence of light would not select against photopositive or photonegative individuals.

Similar phototaxic responses in relation to feeding have been observed by other authors in a number of planktonic crustacea (Clarke, 1932; Lucas, 1936; Singarajah, Moyse & Knight-Jones, 1967) but this appears to be the first observed instance where, under natural conditions, the light response operates directly to suppress the amount of activity dependent upon an endogenous component of a tidal rhythm. This may have important ecological significance since predation by intertidal fish will probably

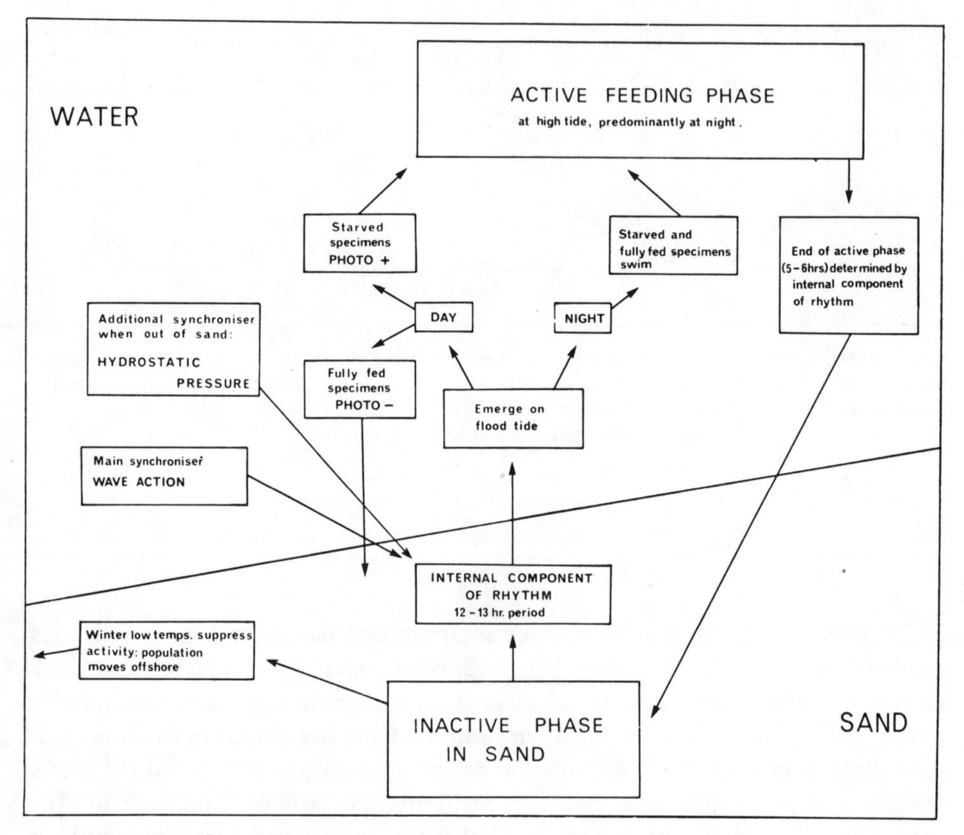

Fig. 8. Summary of interplay between environmental variables and the endogenous component of tidal periodicity in *Eurydice pulchra*.

be greater during the hours of daylight, while intertidal plankton, which forms the main diet of *Eurydice* (Jones, 1968), is more abundant at night; indeed there is evidence to suggest that in the clearer Atlantic waters of southwest France, intertidal *Eurydice* show a nocturnal rhythm only, activity on daytime high tides being completely suppressed (Salvat, 1966).

Varying light responses in relation to satiation and starvation are also of interest in relation to observations on the behaviour of ovigerous *Eurydice* which are often reported to remain buried in the sand throughout incubation without feeding. They have seldom been caught in hauls taken during daytime (Bacesco, 1940; Soika, 1955;

Salvat, 1966) or taken with lights at night (Fage, 1933); this was found not to be entirely true in the present study for, although no ovigerous females were taken in day plankton, they were found in surf plankton at night. Moreover, it is already known (Jones, 1968) that these ovigerous females are capable of feeding and the present investigations show that ovigerous females of *E. pulchra* retain an activity rhythm throughout incubation. The absence from tidal waters during the day is probably related to their incubatory condition, the developing embryos leaving little room for expansion of the parent gut during feeding (see, Jones 1968) so that they become quickly satiated and hence are photonegative for much of the incubatory period.

In view of the observation that *Eurydice* will not emerge from the sand unless stirred by mechanical action it remains to discuss the significance of the endogenous rhythm. One important function of the rhythm appears to be in phasing the end of the swimming period with the falling tide and so helping to maintain the zonation of *Eurydice*. The species is active for about 5–6 h during high tide and reburrows at approximately the level from which it emerged (M.T.L.–H.W.N.), thus avoiding being carried out to sea on the ebb tide (Fig. 8). Some confirmation of this is seen in their behaviour during spring tides when they are washed out from the sand relatively earlier and correspondingly re-enter the sand at a higher level than at neaps (Jones, unpublished observations). Thus, the tidal rhythm appears to have a role in controlling the zonation of *E. pulchra* on the beach.

Whilst small variations in temperatures do not appear to influence the rhythm, it has been shown that temperatures of 5 °C inhibit activity and a subsequent return to room temperature rephases the tidal rhythm. Since *Eurydice* collected in winter are arhythmic, low temperatures at that time probably disrupt the normal tide-synchronized rhythm which maintains zonation. This disruption, together with the general lethargy shown by the isopods at low temperatures, is no doubt correlated with the concentration of the population offshore in winter, where there are less extensive temperature variations. In spring, the return to higher temperatures, together with the depth regulating response to increased pressure (Knight-Jones & Morgan, 1966) could lead to a population movement onshore where substratum choice (Jones, unpublished observations) would ensure that these isopods return to their appropriate zone on the beach, and wave action would then result in eventual resynchronization of the rhythm.

Acknowledgements

We are grateful to Professor E. W. Knight-Jones for the provision of laboratory facilities, to Mr. C. Cole who constructed the actographs, and to N.E.R.C. for financial support.

References

Bacesco, M., 1940. Les Mysidacés des eaux roumaines (étude taxonomique, morphologique, biogéographique et biologique). *Annls scient. Univ. Jassy*, T. 26, pp. 453–803.

Clarke, G. L., 1932. Quantitative aspects of the phototropic signs in *Daphnia*. *J. exp. Biol.*, Vol. 9, pp. 180–211.

COLMAN, J. S. & F. SEGROVE, 1955. The tidal plankton over Stoupe Beck Sands, Robin Hood's Bay (Yorkshire, North Riding). *J. Anim. Ecol.*, Vol. 22, pp. 445–462.

ELMHIRST, R., 1932. Quantitative studies between the tide-marks. *Glasg. Nat.*, Vol. 10, pp. 52–62.

ENRIGHT, J. T., 1963. The tidal rhythm of activity of a sand beach amphipod. *Z. vergl. Physiol.*, Bd 46, S. 276–313.

ENRIGHT, J. T., 1965. Entrainment of a tidal rhythm. *Science, N.Y.*, Vol. 147, pp. 864–867.

FAGE, L., 1933. Pêches planctoniques à la lumière effectuées à Banyuls-sur-Mer et à Concarneau. 3, Crustacea. *Archs Zool. exp. gén.*, T. 76, pp. 228–232.

FRAENKEL, G., 1961. A new type of negative phototaxis observed in a marine isopod *Eurydice. Physiol. Zoöl.*, Vol. 34, pp. 228–232.

GIBSON, R. N., 1967. Experiments on the tidal rhythm of *Blennius pholis*. *J. mar. biol. Ass. U.K.*, Vol. 47, pp. 97–116.

JONES, D. A. & E. NAYLOR, 1967. The distribution of *Eurydice* (Crustacea: Isopoda) in British waters, including *E. affinis* new to Britain. *J. mar. biol. Ass. U.K.*, Vol. 47, pp. 373–382.

JONES, D. A., 1968. The functional morphology of the digestive system in the carnivorous intertidal isopod *Eurydice*. *J. Zool.*, Vol. 156, pp. 363–376.

KNIGHT-JONES, E. W. & E. MORGAN, 1966. Responses of marine animals to changes in hydrostatic pressure. *Oceanogr. Mar. Biol. Ann. Rev.*, Vol. 4, pp. 267–299.

LUCAS, E. E., 1936. On certain inter-relations between phytoplankton and zooplankton under experimental conditions. *J. Cons. perm. int. Explor. Mer*, Vol. 11, pp. 343–362.

NAYLOR, E., 1958. Tidal and diurnal rhythms of locomotory activity in *Carcinus meanus* (L). *J. exp. Biol.*, Vol. 35, pp. 602–610.

NAYLOR, E., 1963. Temperature relationships of the locomotor rhythm of *Carcinus*. *J. exp. Biol.*, Vol. 40, pp. 669–679.

SALVAT, B., 1966. *Eurydice pulchra* Leach (1815) *Eurydice affinis* (Hansen 1905) Isopodes Cirolanidae, taxonomie, éthologie, écologie, répartition verticale, et cycle reproducteur. *Act. Soc. linn. Bordeaux*, T. 103, Sér. A, pp. 1–77.

SINGARAJAH, K. V., 1966. Some aspects of the behaviour of plankton animals. Ph.D. Thesis, University of Wales, 91 pp.

SINGARAJAH, K. V., J. MOYSE & E. W. KNIGHT-JONES, 1967. The effect of feeding upon the phototactic behaviour of cirripede nauplii. *J. exp. Biol. Ecol.*, Vol. 1, pp. 144–153.

SOIKA, G. A., 1955. Ethologie, écologie, systématique et biogéographie des *Eurydice* s. str. *Vie Milieu*, T. 6, pp. 38–52.

WATKIN, E. E., 1942. The macrofauna of the intertidal sand of Kames Bay, Millport, Buteshire. *Trans. Roy. Soc. Edinb.*, Vol. 60, pp. 543–561.

WILLIAMS, B. G. & E. NAYLOR, 1967. Spontaneously induced rhythm of tidal periodicity in laboratory-reared *Carcinus*. *J. exp. Biol.*, Vol. 47, pp. 229–234.

WILLIAMS, B. G. & E. NAYLOR, in press. Synchronization of the locomotor tidal rhythm of *Carcinus*. *J. exp. Biol.*

Part 2

SMALL-SCALE DISTRIBUTION AND SAMPLING PROBLEMS

Terrestrial and aquatic ecologists are concerned both with the structure and dynamics of a population and with its distribution in space. Studies of distribution for a given species may span the broad range from consideration of the dispersion pattern of individuals within the habitat to ecological aspects of its geographical distribution.

Small-scale distribution, or pattern of dispersion, which is a measure of the spatial relationships of individuals in a given area, is a subject of particular interest to those concerned with specific effects of the physical and biological environment on individuals and populations and with problems of sampling error involved in censusing populations (see, for example, Cassie, 1959; Grieg-Smith, 1964; Pielou, 1969). Plants and animals exhibit three basic types of small-scale distribution, the relationships of which can be represented as a continuous spectrum:

Uniform ———— Random ————Aggregated

In practice, statistical methods are used to determine the position of an observed small-scale distribution pattern on this continuum. Evidence from a wide variety of terrestrial and aquatic habitats has shown that both random and uniform (even) spacing of individuals are quite uncommon patterns compared to aggregation. An aggregated distribution pattern is characterized by densities of individuals which are markedly higher than the overall mean density in some parts of the area sampled and markedly lower in others. This effect is the result of the tendency for individuals to occur in groups which are often called aggregations, clumps, or patches. Thus, the terms clumped, patchy, and contagious also are used in referring to an aggregated small-scale distribution pattern.

Dispersion patterns of organisms may be considered quantitatively by using a device for sampling areas or volumes to obtain replicate density estimates, or by using nearest neighbor distance measurements obtained by a plotless sampling technique. The latter method is of limited value in most aquatic situations where conditions generally dictate the use of volumetric samplers. The observed data are compared statistically with those expected on the basis of a random dispersion pattern. These expected values can be determined from the terms of the Poisson distribution, which has the form

$$p_n = \frac{\lambda^n e^{-\lambda}}{n!}$$

where n is the number of individuals per sampling unit, λ is the mean number of individuals per sampling unit (mean density), e is the base of the natural logarithms, and p_n is the probability or expected frequency of occurrence for n individuals per sampling unit. The expected numbers of sampling units containing 0, 1, 2, 3, . . ., n individuals are each obtained by multiplying the expected relative frequency (p_n) for each n value by the total number of sampling units; they may then be compared by a chi-square test to determine if the observed pattern, as reflected by a frequency distribution of density data, conforms to a random pattern or departs from it in the direction of uniformity or aggregation.

Another useful property of the Poisson distribution, which allows an alternate method of analysis, is that the variance equals the mean. Thus, using a statistical significance test, one can compare an observed variance-to-mean ratio based on replicate density estimates with the expected Poisson ratio of 1.0 to determine whether the observed data fit or depart from a random dispersion pattern.

While it is useful within the context considered previously, the Poisson distribution provides an inadequate representation of the aggregated dispersion patterns exhibited by most terrestrial and aquatic forms. Other distributions, particularly the negative binomial, provide more meaningful descriptions of the small-scale distribution continuum. In the negative binomial, the probability or expected relative frequency of there being n individuals in a sampling unit is expressed as

$$p_n = \frac{k(k+1)\ldots(k+n-1)(k\lambda)^n}{n!\,Q^{k+n}}$$

where k is a fitted constant which reflects degree of aggregation, n is the number of individuals per sampling unit, λ is the mean number of individuals per sampling unit, and $Q = 1 + \lambda/k$. Several different models based on the negative binomial have been developed to take into consideration special biological conditions related to dispersion pattern.

Hutchinson (1953) suggests that there are five major categories of factors which influence the type of pattern individuals of a species exhibit in a given area. Vectorial patterns are those induced by light, currents, bottom sediment conditions, and other external environmental factors. Reproductive patterns result from such processes as parental care of the young and asexual reproduction by budding, with retention of these new individuals nearby, as in sea anemones and many other forms. Social patterns are produced by such effects as territorial behavior, which tends to cause uniform spacing, and by schooling or swarming to reduce predation, causing an aggregated pattern. Competition and, presumably, other interactions between species may induce coactive patterns. In the absence of any specific physical or biological effects, random processes alone may produce a stochastic or chance pattern. While this classification is a useful means of considering the causes and advantages of a particular dispersion pattern, it does not take into account the fact that several different factors, acting separately or in combination, may produce the pattern observed.

Specific information about small-scale distribution can be useful in studies concerned with the relationships between density and factors, such as predation, which affect mortality and other population characteristics. For example, if predation mortality rates for two crustacean populations of equal size are related to density, one cannot necessarily assume that the effects of predator activity will be the same in the two populations if the individuals of one are randomly or uniformly distributed while those of the other occur in widely spaced aggregations.

With the exception of intertidal areas and small bodies of fresh water, most aquatic habitats present the ecologist with unique and difficult sampling problems (Holme, 1965; UNESCO, 1968). In most instances, he does not have direct access to the habitat and organisms he wishes to study and must rely on the inefficient process of manipulating his sampling devices from the surface in a dynamic, three-dimensional system. Under these conditions, maintaining a balance between adequacy and cost of sampling is a much more pressing dilemma than in terrestrial work. An even more serious problem confronting the aquatic ecologist is that usually in nature he cannot make detailed observations of the organisms he is studying, their reactions to the sampling gear, or how the gear itself performs. As a result, such observations must be made instead under laboratory conditions in which an attempt is made to duplicate critical features of the natural environment. Papers by Fleminger and Clutter, Mullin, and Schindler in this collection provide good examples of both the shortcomings and the successful application of this approach. Recently, the use of self-contained diving equipment has enabled ecologists to make visual observations and to apply slightly modified terrestrial sampling methods in detailed studies on the shallow continental shelf, which formerly was accessible only by remote sampling techniques (see, for example, Fager in this collection).

The four papers included in this part reflect the fact that in aquatic, as in terrestrial, ecology a wide variety of factors must be taken into account in the design and successful application of both sampling equipment and sampling programs. Specific knowledge of the interacting effects of abundance, dispersion pattern, and behavior is critical in choosing the appropriate sampler type, size, and operating method, and in determining the number and placement of these sampling units. All of these factors, in turn, are critical to evaluations of sampling error.

LITERATURE CITED

Cassie, R. M. 1959. An experimental study of factors inducing aggregation in marine plankton. New Zealand J. Sci., 2:339–365.

Grieg-Smith, P. 1964. Quantitative Plant Ecology (2nd ed.). Butterworths, London, 256 pp.

Holme, N. A. 1965. Methods of sampling the benthos. Adv. Mar. Biol., 2:171–260.

Hutchinson, G. E. 1953. The concept of pattern in ecology. Proc. Acad. Nat. Sci. (Phila.), 105:1–12.

Pielou, E. C. 1969. An Introduction to Mathematical Ecology. Wiley-Interscience, New York. 286 pp.

UNESCO. 1968. Zooplankton Sampling. Monographs in Oceanographic Methodology. 2. Imprimeries Populaires, Geneva. 174 pp.

6 AVOIDANCE OF TOWED NETS BY ZOOPLANKTON[1]

Abraham Fleminger and Robert I. Clutter

Scripps Institution of Oceanography, La Jolla, University of California, San Diego, and U.S. Bureau of Commercial Fisheries, Biological Laboratory, La Jolla, California

ABSTRACT

Experiments were conducted to determine the effect of behavior on the accuracy of sampling populations composed of several species of marine copepods and mysids held in a large, enclosed seawater pool. Plankton nets having mouth areas of 1,600, 800, and 400 cm^2 were towed in replicate sets during midday and midnight under three light conditions in plankton populations of different densities. Statistical treatment of the results showed that smaller nets were more effectively avoided and that the degree of avoidance varied among species. Estimates of the peripheral escape zone were obtained. More mysids, but not more copepods, were caught in darkness than in light, and avoidance tended to be less in denser populations. Time of day had no effect on avoidance behavior. Escape mechanisms and the possible effect of population density are discussed.

INTRODUCTION

Evaluation of plankton populations in the sea requires sampling and thus involves uncertainty that may affect both precision and accuracy. Contagious distribution of organisms decreases precision, while sampling gear performance and the reactions of the organisms affect the accuracy of estimates of population densities.

The proposition that zooplankton avoid nets is not new; for example, Franz (1910) suggested that plankton capable of seeing an oncoming net in daylight would avoid capture, Mackintosh (1934) observed euphausiids avoiding a net near the surface, Bary et al. (1958) suggested that plankters may avoid nets if there is a zone of pressure preceding the net, and Hansen and Andersen (1962) contended that nets are grossly inefficient for catching copepods. By contrast, Winsor and Clarke (1940) implied that avoidance was not a problem in that they concluded that a 12.7-cm-diameter net was as reliable as a 75-cm net for sampling the copepod species that they considered.

The work reported here, part of a continuing study of the causes of variability in plankton sampling, deals only with the accuracy of net sampling for zooplankton as affected by the behavior of the animals themselves. These experiments were conducted on captive populations of zooplankton to eliminate several problems adversely affecting field collections: 1) the effects of vertical migration, 2) large-scale patchiness in the distribution of organisms, and 3) variation in vessel speed and drift. Moreover, it allowed for factorial experimental design wherein light conditions, time of sampling, and course of sampling could be precisely repeated.

We conclude that zooplankton can avoid capture to such an extent that accuracy is affected and that both visual and mechanical disturbances can be involved.

This research was partially supported by funds from National Science Foundation Grants G-19417 and G-7141, and by Scripps Tuna Oceanography Research (U.S. Bureau of Commercial Fisheries Contract 14-17-0007-28). The manuscript was read and criticized by E. W. Fager, M. B. Shaefer, J. D. Isaacs, E. H. Ahlstrom, and D. G. Chapman. W. D. Clarke aided in formulating the study and installing the lighting system. R. Schwartzlose furnished logistic aid. We thank them all for their valued help.

METHODS

The experiments were conducted in a large rectangular concrete pool (Fig. 1) that had been exposed to seawater for

[1] A contribution of Scripps Institution of Oceanography, conducted under the Scripps Institution of Oceanography Marine Life Research Program, a component of the California Cooperative Oceanic Fisheries Investigations sponsored by the Marine Research Committee of the State of California.

Reprinted from *Limnology and Oceanography*, 10:96–104, 1965.

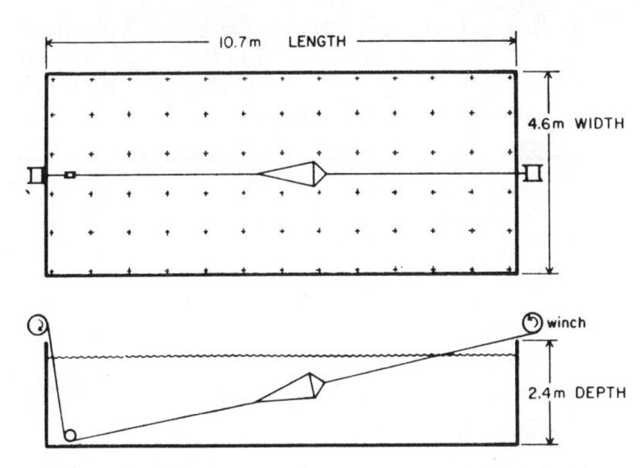

FIG. 1. Diagram of experimental pool illustrating (above) arrangement of overhead lights (+) and (below) method of towing nets.

several years. It was enclosed by an essentially lighttight superstructure; interior illumination was provided by 78 100-w incandescent lamps placed 3.04 m (10 ft) above the water surface at 0.91-m (3-ft) intervals and controlled by rheostats.

The pool was filled with unfiltered seawater pumped through an opaque plastic pipe from an intake 300 m offshore, about 3 m below the sea surface at mean tide level and 3 m above the sea floor. Thus, it was possible to introduce experimental zooplankton populations without directly handling the animals. Populations of different densities were obtained by allowing the water to overflow at the surface under full illumination for different time periods. Tests showed that significantly more animals entered the pool than were lost in the outflow under these conditions. The plankton was allowed to acclimate for one week preceding the start of experimentation. The animals remained vigorous; delicate fish larvae and chaetognaths survived, and copepods reproduced.

Three conical plankton nets (Fig. 2) having mouth areas of 1,600, 800, and 400 cm^2, and proportional in all other dimensions, were used in the study. The towing bridles were constructed of three strands of 1.7-mm-diameter plastic-coated steel cable; forward extension of each bridle was equal to the mouth diameter of the net to which it was attached. To overcome the effect of any variation in vertical distribution of the animals, net tows were made obliquely from the bottom of the pool to the surface of the water (Fig. 1). A constant towing rate of 30 cm (1 ft)/sec was used. The line of traverse was kept straight by the use of a restraining line. The hauling line was made of 2.4-mm (3/32-inch) hydrographic wire. Since the tows were short and the nets were washed thoroughly as well as being towed backward to the starting position after each haul, clogging was not a

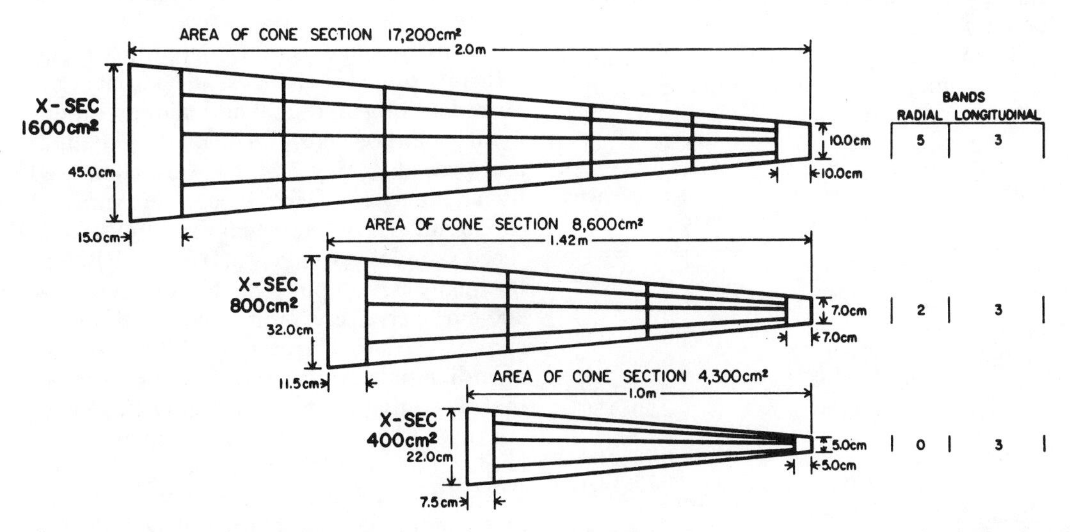

FIG. 2. Diagram of experimental plankton nets used in study, showing construction specifications. Filtering material of 333-μ aperture monofilament nylon, collars and bands of dacron.

TABLE 1. *Basic design for each of the six sampling experiments*

Net	Full light	Reduced light	Dark
1,600 cm²	2 trials	2 trials	2 trials
800 cm²	2 trials	2 trials	2 trials
400 cm²	2 trials	2 trials	2 trials

problem. The nets were left in the starting position on the bottom for 10 min before each tow, and 10 additional minutes were allowed after changes in light intensity. Actual filtration efficiency was not determined. The three nets strained theoretical maxima of 0.8, 0.4, and 0.2 m^3 of water per haul.

Six experiments were carried out according to the balanced design shown in Table 1, full light intensity was about 550 lux at the water surface, and reduced light intensity about 22 lux. Three experiments were carried out on 5–6 July 1961 and three on a different population on 13–14 July 1961. The 13–14 July population was made larger to test the possible effect of population density on avoidance behavior. Two midday (1200–1600 hours) and one midnight (2200–0200 hours) experiments were carried out on each population. All three light conditions were imposed in each experiment. The night series were included to test for possible endogenous diurnal changes in activity of the animals.

The animals in the total of 108 samples were examined and counted under a stereomicroscope at 16× magnification.

TABLE 2. *Species and size ranges of zooplankton included in the analyses*

Species	Body length (mm)
Copepoda	
Acartia clausi	0.80–1.10
Acartia tonsa	0.90–1.20
Calanus helgolandicus	2.50–3.50
Corycaeus anglicus	0.85–1.10
Oithona spp.	0.70–0.90
Paracalanus parvus	0.75–1.00
Mysidacea	
Metamysidopsis elongata	1.50–5.50

TABLE 3. *Sums of actual and adjusted* numbers of copepods and mysids caught in four experiments, including two midday and two midnight experiments*

Group	Actual catches		Adjusted catches	
	Day	Night	Day	Night
Copepoda				
5–6 July	2,580	2,545	3,565	3,187
13–14 July	16,233	15,599	25,315	22,664
Total	18,813	18,144	28,880	25,851
Mysidacea				
5–6 July	74	57	121	76
13–14 July	560	608	856	975
Total	634	665	977	1,051

* Sums computed after data for 800- and 400-cm² nets were adjusted to number of individuals per 0.8 m³ for each sample.

RESULTS

The animals included in this analysis and their size ranges are listed in Table 2. Two other, less abundant, copepods were taken, along with a chaetognath and fish larvae, but their occurrence was too sporadic to include them in the analysis. Two species of mysids occurred in the samples; the second species, *Mysidopsis californica* (juveniles), was not distinguished from *Metamysidopsis elongata* in the sample counts, but it never composed more than about 5% of the mysid population.

It was first necessary to determine if significant diurnal changes in activity occurred or if the midnight and midday experiments could be considered equal. The night and preceding day samples were compared by an analysis of variance. The sums of the numbers of copepods and mysids in these samples are given in Table 3. Before the analysis, sample data for the 800- and 400-cm² nets were multiplied by 2 and 4, respectively, to correspond to the number of individuals per 0.8 m^3 of water strained, that is, equivalent to the catch of the 1,600-cm² net, and all values were normalized by being transformed to logarithms (Barnes 1952). Complete partition of the variance was not made, but neither copepods nor mysids showed any evidence of diurnal

TABLE 4. *Mean adjusted catch per haul of six replicate hauls for each net and light condition, listed by species for the two populations*

Population	Net	Light condition	*Acartia clausi*	*Acartia tonsa*	*Calanus helgolandicus*	*Corycaeus anglicus*	*Oithona* spp.	*Paracalanus parvus*	Mysidacea
5–6 July	1,600 cm²	Full	229.3	88.7	22.8	3.7	22.7	32.7	12.8
		Reduced	198.2	77.3	22.3	2.3	13.0	24.7	7.8
		Dark	151.0	62.3	20.5	3.2	2.7	22.7	13.3
	800 cm²	Full	70.7	40.0	12.3	1.7	16.7	8.0	6.3
		Reduced	95.3	49.0	13.7	1.7	6.7	7.3	6.7
		Dark	78.3	46.0	20.3	0.7	2.0	10.7	10.0
	400 cm²	Full	34.0	22.7	6.0	0.7	3.3	2.0	2.0
		Reduced	51.3	20.0	8.7	0.0	5.3	1.3	2.0
		Dark	29.3	10.0	8.7	0.0	0.0	2.7	4.7
13–14 July	1,600 cm²	Full	281.5	1,001.5	7.2	123.7	156.2	228.5	17.5
		Reduced	276.7	933.3	2.5	88.3	123.3	217.5	30.3
		Dark	286.7	967.5	3.5	87.5	175.8	156.7	154.8
	800 cm²	Full	126.7	605.7	5.7	34.3	72.0	45.7	15.7
		Reduced	124.0	496.3	3.7	31.0	100.7	38.7	19.3
		Dark	156.7	685.7	4.3	39.7	79.3	49.3	123.7
	400 cm²	Full	70.7	673.3	2.0	14.7	118.7	37.3	16.0
		Reduced	105.3	828.0	4.7	18.7	118.7	37.3	28.0
		Dark	80.7	495.3	7.3	29.3	107.3	36.0	71.3

changes in activity. Therefore, the two mid-day experiments and the one midnight experiment could be treated as replicates.

Table 4 shows, by species, the adjusted mean number of individuals per haul for each replicate set of samples.

The mean numbers of mysids per haul, partitioned in various ways, for all six experiments are given in Table 5. The results of analysis of variance of the data are given in Table 6; the unpartitioned residual variance was used for the error term in calculating variance ratios. As expected, the population sampled on 13–14 July yielded a significantly larger mean catch than the population sampled on 5–6 July. The replicates were not significantly different, indicating no inconsistencies in sampling technique. Changes in light intensity had a significant effect on sampling for mysids; many more were caught in darkness. The two smaller nets took significantly fewer animals per theoretical unit volume sampled than did the largest net. Interaction between population size and net size was significant, suggesting that avoidance tended to be less in the denser population. The significant interaction between population size and light condition was due to a higher ratio of catches in the dark to catches in the light in the larger population.

Similar data for copepods are given in Tables 7 and 8. The differences between means for species were expected because of the known differences in abundance of

TABLE 5. *Mean values of actual number of mysids caught in 108 hauls*

	Mean number per haul
Populations	
5–6 July	5.3
13–14 July	34.5
Nets	
1,600 cm²	39.4
800 cm²	15.1
400 cm²	5.2
Light conditions	
Full light	7.6
Reduced light	9.8
Darkness	42.3
Replicates	
First trial	18.2
Second trial	21.6
Third trial	17.8
Fourth trial	20.1
Fifth trial	23.3
Sixth trial	18.6

TABLE 6. *Analysis of variance for mysids caught in 108 hauls**

Source of variation	Degrees of freedom	Variance ratio†	Probability higher value by chance
Populations	1	148.2	<0.0005
Nets	2	11.8	<0.0005
Light conditions	2	30.8	<0.0005
Replicates	5	1.3	0.25–0.50
Population–net interaction	2	6.7	<0.005
Population–light interaction	2	11.5	<0.0005
Net–light interaction	4	0.4	0.75–0.90

* Data for 800- and 400-cm² nets adjusted to number of individuals per 0.8 m³ for each sample and all data transformed to logarithms before analysis.

† Mean square for stated source divided by residual mean square of 0.513 (89 degrees of freedom).

TABLE 7. *Mean values of actual number of copepods caught in 108 hauls*

	Mean number per haul
Species	
Acartia clausi	102.3
Acartia tonsa	255.8
Calanus helgolandicus	6.6
Corycaeus anglicus	21.0
Oithona spp.	40.0
Paracalanus parvus	44.0
Populations	
5–6 July	143.5
13–14 July	796.0
Nets	
1,600 cm²	1,019.6
800 cm²	265.0
400 cm²	124.6
Light conditions	
Full light	494.0
Reduced light	460.5
Darkness	454.7
Replicates	
First trial	538.4
Second trial	512.1
Third trial	424.9
Fourth trial	410.3
Fifth trial	474.3
Sixth trial	461.4

these species in the populations. Based on catches, the population sampled on 13–14 July was significantly larger than the 5–6 July population, as intended. The six means for replicates were not significantly different from one another. Unlike the case of the mysids, changes in light intensity had no significant effect on sampling copepods. Like that of the mysids, the two smaller nets took significantly fewer animals per unit volume of water strained than the largest net. This apparent avoidance differed among species, as indicated by the significant variance ratio for interaction between species and net size. A separate analysis has shown that even the two species of *Acartia* reacted differently from each other. The significant interaction between population size and net size suggests that avoidance tended to be less in the denser populations.

Since the analysis for copepods is a simultaneous treatment of six species, it does not show the contribution of individual species to the variations among net catches. The apparent reaction of each species was tested separately by analysis of variance of the adjusted catches (number per 0.8 m³ of water sampled) of the three nets (Table 9). In 25 of 36 comparisons, the smaller nets took significantly fewer animals than expected, the most notable exceptions being *Calanus helgolandicus* (13–14 July) and *Oithona* spp. (5–6 July). In 5 of 12 comparisons, the catch per unit volume of the smallest net was significantly less than that of the intermediate net. In no case did the catch of a smaller net significantly exceed that of a larger one.

We cannot precisely evaluate the distances moved by the test animals in avoiding capture without detailed knowledge of their escape behavior. Nevertheless, the apparent minimum peripheral escape zone can be calculated by comparing the catches of the smaller nets with those of the larger net. The width of this peripheral escape zone, here called Δr, is determined by assuming that the "effective" sampling areas

TABLE 8. *Analysis of variance for copepods caught in 108 hauls**

Source of variation	Degrees of freedom	Variance ratio†	Probability higher value by chance
Species	5	140.7	<0.0005
Populations	1	449.9	<0.0005
Nets	2	239.8	<0.0005
Light conditions	2	0.7	0.50–0.75
Replicates	5	0.6	0.50–0.75
Species–net interaction	10	15.7	<0.0005
Species–light interaction	10	0.8	0.50–0.75
Population–net interaction	2	3.7	0.025
Net–light interaction	4	1.3	0.25–0.50

* Data for 800- and 400-cm² nets adjusted to number of individuals per 0.8 m³ for each sample and all data transformed to logarithms before analysis.

† Mean square for stated source divided by residual mean square of 1.06 (606 degrees of freedom).

TABLE 9. *Results of analyses of variance for differences within species of copepods of numbers caught per 0.8 m³ with nets of three sizes**

Species	Population	Nets (cm²) compared		
		1,600–800	1,600–400	800–400
Acartia clausi	5–6 July	+	+	+
	13–14 July	+	+	+
Acartia tonsa	5–6 July	+	+	+
	13–14 July	+	+	0
Calanus helgolandicus	5–6 July	+	+	+
	13–14 July	0	0	0
Corycaeus anglicus	5–6 July	+	+	0
	13–14 July	+	+	0
Oithona spp.	5–6 July	0	0	0
	13–14 July	+	+	0
Paracalanus parvus	5–6 July	+	+	+
	13–14 July	+	+	0

* Employing the method of Tukey (1953). The symbol + indicates that the catch of the larger net was significantly greater, and the symbol 0 indicates no significant difference at the 5% confidence level.

of the smaller nets are proportional to their catches, that is,

$$\pi\hat{r}^2 = \pi r_L^2 \frac{C_s}{C_L},$$

where

$\hat{r}$ = effective radius of the smaller (800- or 400-cm²) net,
r_L = radius of 1,600-cm² net,
C_s = catch of smaller (800- or 400-cm²) net,
C_L = corresponding catch of 1,600-cm² net.

By definition,

$$\Delta r = r_s - \hat{r},$$

where

Δr = minimum width of peripheral escape zone,
r_s = actual radius of the smaller (800- or 400-cm²) net.

Therefore,

$$\Delta r = r_s - r_L(C_s/C_L)^{1/2}.$$

This calculation was made for the two smaller nets for each species population independently, giving the results summarized in Table 10. Combining nets and populations, the mean values for Δr (cm) for each species are: *Acartia clausi*, 5.4; *Acartia tonsa*, 3.8; *Calanus helgolandicus*, 1.6; *Corycaeus anglicus*, 6.8; *Oithona* spp., 3.6; and *Paracalanus parvus*, 7.2. The calculated minimum width of the peripheral escape zone is not significantly correlated with body length (*see* Table 2). The rank correlation statistic (Tate and Clelland 1957) is $\tau = 0.13$; the probability that a higher absolute value may occur by chance is greater than 20%.

DISCUSSION

In these experiments, the larger nets caught more copepods and mysids per theoretical unit volume of water filtered than the smaller nets. We have no reason to believe that filtration efficiency differed among the nets under the conditions of the experiment. Since the largest net used presumably did not sample with 100% efficiency, the interpretation of significant differences in catches is probably conservative.

Vulnerability to capture under full and reduced light conditions differed significantly between mysids and copepods. For mysids, but not for copepods, apparent avoidance was greater in light than in darkness. This could be attributed to dissimilar visual acuity. The mysids have well-developed compound eyes, whereas the eyes of the copepods are capable of light perception but probably not of image formation. Smaller catches of mysids in light could have resulted from better perception of the moving net, towline, and bridle. Alternatively, they could have resulted from changes in the distribution of the animals preceding the actual tow. In a lighted medium, the mysids habitually remain within 5–20 cm of the bottom and also shun foreign objects. If the area around the net was evaded by mysids that otherwise crowded close to the bottom of the pool, an oblique tow would fail to sample representatively the densest stratum of the population. This may not have occurred in darkness, either because the immobile net was not avoided or because of the tendency for the animals to become more dispersed vertically in darkness.

Although the effect of bioluminescence was not determined, the moving nets were visible as glowing cones during all of the tows made in darkness. This may have affected dark sampling for both copepods and mysids.

Since all samples were taken with nets preceded by bridles and a long towing line, it is possible that the animals merely reacted to tactile or vibrational stimuli, or, alternatively, to pressure waves formed ahead of the moving nets. According to Brook and Woodward (1956), the reaction of copepods to rapidly accelerating currents is a function of hydrostatic pressure change. Both mysids and copepods are responsive to changes in hydrostatic pressure (Knight-Jones and Quasim 1955; Rice 1961, 1962), but the response times may not be short enough to allow the animals to use this cue in avoiding sampling devices. Enright (1962) found that single square pressure waves must include at least a full second of increased pressure to produce appreciable response in an amphipod.

Zooplankton may be oriented initially in any direction relative to the trajectory of a sampling device and still disperse away from the path of the device so long as a gradient in intensity of the cue, or of the reaction to it, occurs in the vicinity of the trajectory. However, avoidance will be increased if the animals are capable of oriented movement.

Storch (1929) and Lowndes (1935) describe escape swimming in calanoids as a darting movement produced by repeated backward flexure of the biramous thoracic legs. Lowe (1935) has presented morphological evidence of a system of giant nerve fibers in *Calanus* that supply the muscles used in rapid leaping, escape movements. She argues convincingly that a stimulus applied differentially to the paired sensory receptors (thought to be the first antennae) will cause this darting movement to be directed away from the source of stimulation. We have observed a variety of calanoid species exhibiting darting movements away from the mouth of a pipette being used to capture them. We have also observed that the mysids used in this study are capable of well-oriented escape movements.

The lack of significant correlation between copepod body length and the estimated width of the peripheral escape zone does not support the report by Lukjanova (1940) that when different species are compared, the velocities of movement of marine planktonic copepods are proportional to the cube roots of their lengths. *Calanus helgolandicus* was the largest copepod sampled in our study, but it gave the least evidence of avoidance of all six copepod species. However, the degree of avoidance may not be directly related to swimming capability but to a general reaction pattern characteristic of the species. We have found that *Calanus finmarchicus* may show little response to displacement by a current that is simultaneously strongly opposed by the copepod *Temora longicornis.* Knight-Jones and Quasim (1955) failed to obtain clear pressure responses in *Calanus* while at the same time obtaining them from another copepod, *Caligus rapax*, and Rice (1962) found the responses of *Calanus helgolandicus* to pressure to be unpredictable.

The swimming velocities of marine copepods have been estimated to be between 0.7 cm/sec and 12 cm/sec by Welsh (1933), Hardy and Bainbridge (1954), and Nishigawa, Fukuda, and Inoue (1954). The highest of the estimated velocities are probably sufficient to allow significant avoidance of nets among disturbed animals, but the observations included cruising velocities and therefore are not applicable directly to escape behavior. Lukjanova (1940) reported more directly applicable maximum velocities of up to 15 cm/sec for *Anomalocera*, and Lowndes (1933) observed a rate of 20 cm/sec for *Diaptomus gracilis*. By applying the velocity data of Storch (1929), Lochhead (1961) calculated that speeds for each jump in excess of the 20 cm/sec reported for *Diaptomus* should certainly be possible. Recent analysis of slow-motion films (P. E. Smith, Bureau of Commercial Fisheries Biological Laboratory, La Jolla, personal communication) has shown that, when being attacked by fish, the copepod *Labidocera acutifrons* is capable of directed movement at a rate of 80 cm/sec (230 body lengths/sec) over distances of at least 15 cm. At this rate, a copepod would take only 0.1 sec to pass through the broadest peripheral escape zone estimated in this study (*see* Table 10).

The sampling results show that relatively more animals were taken in the smaller nets than in the larger net, and in the denser population than in the less dense population. This suggests that interindividual interference may have restricted their movements. If we assume that 1) the animals were randomly distributed in the pool and 2) physical contact was necessary for interference to occur, we find that the mean free path of the animals is on the order of 25 m or more, even in the denser population. The probability of contact in a path 16 cm long (the radius of the intermediate size net) would thus be < 0.007 in the denser population. Thus, the animals either experienced effective contact by other than tactile means, or their aggregation increased the probability of contact.

It is concluded that accuracy in sampling can be affected significantly by the behavior of the animals themselves. To collect representative cross sections of the zooplankton community, or to make accurate estimates of the population structure of any given species, it will be necessary to use sampling devices with minimal escape cues, and with mouth areas or towing velocities, or both, that are adequate to overcome the escape capabilities of the animals sought. This will require further background studies on the general behavior and locomotor capabilities of representative species of the plankton community.

TABLE 10. *Estimates of width of peripheral escape zone (Δr) for copepods*

Species	Population	Net (cm^2)	Δr (cm)
Acartia clausi	5–6 July	800	5.6
		400	6.0
	13–14 July	800	5.0
		400	4.9
Acartia tonsa	5–6 July	800	3.7
		400	5.7
	13–14 July	800	3.4
		400	1.9
Calanus helgolandicus	5–6 July	800	2.6
		400	4.4
	13–14 July	800	0.3
		400	0.5
Corycaeus anglicus	5–6 July	800	6.4
		400	8.1
	13–14 July	800	6.6
		400	5.9
Oithona spp.	5–6 July	800	3.0
		400	5.7
	13–14 July	800	4.2
		400	1.4
Paracalanus parvus	5–6 July	800	5.9
		400	7.9
	13–14 July	800	8.5
		400	6.4

REFERENCES

BARNES, H. 1952. The use of transformations in marine biological statistics. J. Conseil, Conseil Perm. Intern. Exploration Mer, **18**: 61–71.

BARY, B. M., J. G. DE STEFANO, M. FORSYTH, AND J. VAN DEN KERKHOF. 1958. A closing, high speed plankton catcher for use in vertical and horizontal towing. Pacific Sci., **12**: 46–59.

BROOK, A. J., AND W. B. WOODWARD. 1956. Some observations on the effects of water inflow on the plankton of small lakes. J. Animal Ecol., **25**: 22–35.

ENRIGHT, J. T. 1962. Responses of an amphipod to pressure changes. Comp. Biochem. Physiol., **7**: 131–145.

Franz, V. 1910. Phototaxis und Wanderung. Nach Versuchen auf Jungfischen und Fischlarven. Intern. Rev. ges. Hydrobiol., **3**: 306–334.

Hansen, V. K., and K. P. Andersen. 1962. Sampling the smaller zooplankton. Rappt. Proces-Verbaux Reunions, Conseil Perm. Intern. Exploration Mer, **153**: 39–47.

Hardy, A. C., and R. Bainbridge. 1954. Experimental observations on the vertical migration of plankton animals. J. Marine Biol. Assoc. U.K., **33**: 409–448.

Knight-Jones, E. W., and S. Z. Quasim. 1955. Responses of some marine plankton animals to changes in hydrostatic pressure. Nature, **175**: 941–942.

Lochhead, J. H. 1961. Locomotion, p. 313–364, v. 2. *In* T. H. Waterman [ed.], The physiology of Crustacea. Academic, New York and London.

Lowe, E. 1935. On the anatomy of a marine copepod, *Calanus finmarchicus* (Gunnerus). Trans. Roy. Soc. Edinburgh, **58**: 561–603.

Lowndes, A. G. 1933. The feeding mechanism of *Chirocephalus diaphanus* Prévost, the fairy shrimp. Proc. Zool. Soc. London, **1933**: 1093–1118.

———. 1935. The swimming and feeding of certain calanoid copepods. Proc. Zool. Soc. London, 1935: 687–715.

Lukjanova, W. S. 1940. On the maximum velocities of marine plankters. Compt. Rend. (Doklady) Acad. Sci. URSS N.S., **28**: 641–644.

Mackintosh, N. A. 1934. Distribution of the macroplankton in the Atlantic sector of the Antarctic. Discovery Rept., **9**: 65–160.

Nishigawa, S., M. Fukuda, and N. Inoue. 1954. Photographic study of suspended matter and plankton in the sea. Bull. Fac. Fisheries, Hokkaido Univ., **5**: 36–40.

Rice, A. L. 1961. The responses of certain mysids to changes in hydrostatic pressure. J. Exp. Biol., **38**: 391–401.

———. 1962. Responses of *Calanus finmarchicus* (Gunn.) to changes of hydrostatic pressure. Nature, **194**: 1189–1190.

Storch, O. 1929. Die Schwimmbewegung der Copepoden, auf Grund von Mikro-Zeitlupenaufnahmen analysiert. Verhandl. Deut. Zool. Ges., **33**: 118–129.

Tate, M. W., and R. C. Clelland. 1957. Nonparametric and shortcut statistics. Interstate Printers and Publishers, Danville, Illinois.

Tukey, J. W. 1953. Some selected quick and easy methods of statistical analysis. Trans. N.Y. Acad. Sci. Ser. 2, **16**: 88–97.

Welsh, J. H. 1933. Light intensity and the extent of activity of locomotor muscles as opposed to cilia. Biol. Bull., **65**: 168–174.

Winsor, C. P., and G. L. Clarke. 1940. A statistical study of variation in the catch of plankton nets. J. Marine Res., **3**: 1–34.

SMALL-SCALE SPATIAL DISTRIBUTION IN OCEANIC ZOOPLANKTON[1]

7

Peter Howard Wiebe[2]

Scripps Institution of Oceanography, University of California, San Diego, La Jolla 92037

ABSTRACT

A Longhurst-Hardy plankton recorder was used near Guadalupe Island (Baja California) to obtain information on the small-scale horizontal patchiness of a number of zooplankton species. On one occasion, a daytime series of 11 horizontal recorder tows was taken forming a grid over an area 500 m square at 90-m depth (total number of samples, 459). On another, a single long nighttime horizontal tow was taken at 20 m (82 samples).

A method has been developed to measure spatial structure in terms of patch size and shape, number and distribution of patches in the grid area, concentration of individuals in each patch, and densities of background individuals. The patches were approximately circular with a median patch radius ranging from 13.6 to 15.6 m for the species on the grid and 38.4 to 73.1 m for the night tow. Estimates of the numbers of patches in the grid area ranged from 183 to 222. They were distributed randomly. Average densities of individuals in the patches over the grid varied from 2.6 to 5.1 times the background densities. Although the abundances of different species showed a significant degree of concordance, the relationship between the same species pair on different tows and between different species pairs was quite variable.

The degree of aggregation of the less abundant species was associated with their density; the lower the density the less frequently significant departures from a random expectation were observed.

INTRODUCTION

Investigators studying the oceanic distributions of zooplankton must obtain information from samples taken at stations usually separated by huge distances (tens to hundreds of kilometers), although their interest may be focused on populations and communities on a range of scales from less than a meter to hundreds of meters (Cassie 1963). Implicit in these studies are the assumptions that the samples represent, quantitatively, population and community parameters in the parcel of water sampled and that the parcel sampled contains numbers and kinds of organisms representative of the area the station is to represent (Taft 1960). The extent to which these assumptions can be accepted depends on the various errors associated with the sampling method, and a knowledge of its accuracy and precision is required. Such knowledge is also necessary for testing and using theoretical ecological models such as those discussed by Riley (1963).

In most studies of field sampling error, the emphasis has been on the statistical confidence placed on the numerical data from a single net tow. However, with existing sampling techniques it is impossible to observe significant population growth or decline of an abundant species unless the numbers of individuals change by at least a factor of 2, and sometimes the factor must be 5 or 10 or larger.

The role of small-scale patchiness as a component of sampling error has been emphasized (Hardy 1936, 1955; Wiebe and Holland 1968). However, there is little information about the actual spatial structure of planktonic populations. Plankton pump studies (Barnes 1949; Barnes and Marshall 1951; Anraku 1956; Cassie 1959*a*, *b*, 1960) have shown organisms to be

---

[1] Based in part on a dissertation submitted in partial fulfillment of the requirements for the Ph.D. degree at the University of California, San Diego (Scripps Institution of Oceanography). This work was supported by National Science Foundation Grant GB-2861 and the Marine Life Research Program (Scripps Institution of Oceanography's part of the California Cooperative Oceanic Fisheries Investigations, sponsored by the Marine Research Committee of the State of California).

[2] Present address: Woods Hole Oceanographic Institution, Woods Hole, Massachusetts 02543.

Reprinted from *Limnology and Oceanography*, 15:205–217, 1970.

clumped or aggregated more frequently than randomly or evenly distributed. Other studies (Tonolli 1949; Tonolli and Tonolli 1960) have furnished additional evidence that aggregations occur on a wide variety of scales, but only two (Cushing 1953; Cushing and Tungate 1963) have provided any specific information about the dimensions of single or multispecies aggregations and the changes in their spatial structure over time and space.

This research was designed to obtain information about the size, shape, distribution, and density of patches of zooplankton species in the open ocean on smaller scales (1–10 m up to hundreds of meters) than previously obtained with nets.

I gratefully acknowledge the assistance of Drs. J. McGowan and E. W. Fager in the planning and execution of this research. Drs. C. B. Miller and B. W. Frost also gave considerable help.

METHODS AND MATERIAL

The lee of Guadalupe Island (28° 54′ N lat, 118° 16′ W long) was selected as the study area because the identity and gross patterns of distribution and abundance of many of the species were known. It can be an area of relatively slow currents and little turbulence—conditions necessary to conduct the grid study described below. Most of the fieldwork was done during the day on 16 August 1966. Supplementary data were gathered on the night of 2 March 1967.

Grid study—16 August 1966

To study the structure of patchiness on scales as small as 5–10 m, ideally, three-dimensional information is needed. However, available sampling devices permit information to be gathered, at best, in only two, so that variations in the third dimension must be held constant. Since many species of zooplankton migrate vertically, sampling was limited to the period of day when this was minimal. A grid of contiguous samples taken on a horizontal plane was used to produce an approximately synoptic two-dimensional view of the dis-

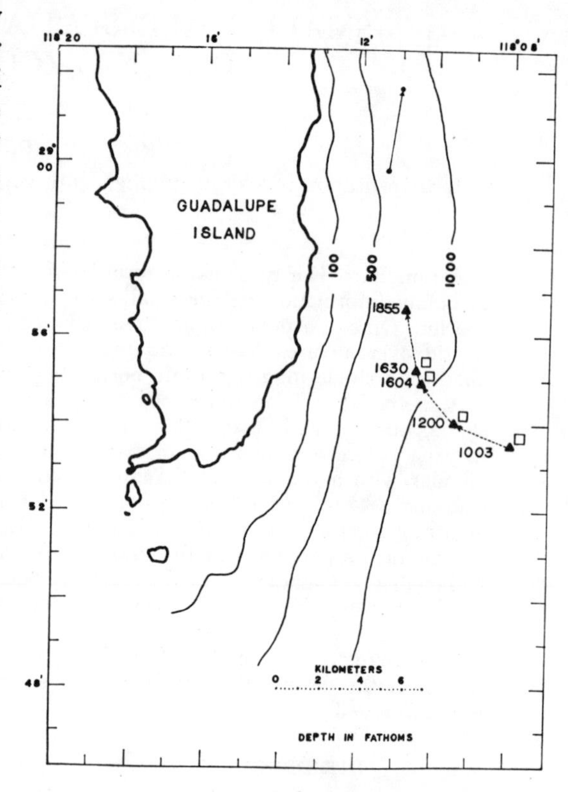

FIG. 1. Positions of the grid of horizontal net tows (open squares) and the parachute drogue (closed triangles). The grid study began at 1003 hours on 16 August 1966 and was completed at 1630. The closed circles (upper right) mark the beginning and end locations of the night horizontal tow which began at 2208 hours and was completed at 2255.

tribution and abundance of the species present at sampling depth.

A modified version of the discrete plankton recorder system described by Longhurst et al. (1966) was used. The modifications include the use of a 12-v power supply that is hooked in series to run a Rustrak chart recorder and the plankton recorder motor, Bongo nets (71-cm diam, 5.04-m overall length) with a conical filtering surface of 0.505-mm nylon gauze (McGowan and Brown 1966), and a modified Bongo net frame to form the nucleus from which the supporting frame was constructed. The hoops to which the nets are attached were extended anteriorly 43.2 cm to reduce the disturbances created by the supporting frame members and the tow

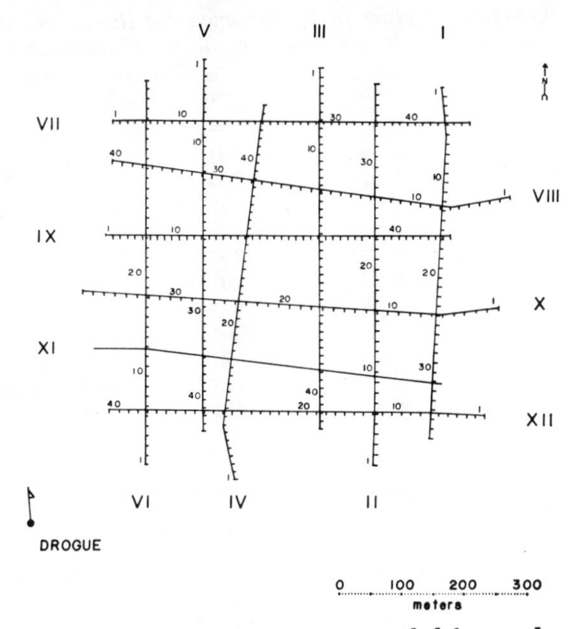

FIG. 2. Grid configuration compiled from radar ranges and bearings using the drogue as the reference point. Roman numerals indicate the order in which the tows were taken. They were started at the end of the line closest to the numeral. Only the horizontal segment of each tow is illustrated. Discrete samples are indicated by short lines perpendicular to the tow paths. The length of line between adjacent marks represents the scaled distance traveled by the net for that sample. The first and every tenth sample on a tow are numbered. There was a failure in the plankton recorder system on tow XI.

TABLE 1. *Temperature data (°C) from bathythermograph recordings taken during the grid study*

Depth (m)	Time (hours)						
	1240	1400	1435	1500	1525	1610	1640
0	20.7	21.0	20.5	21.2	21.6	21.3	20.9
5	20.5	20.6	20.4	20.7	20.5	20.7	20.5
10	20.3	20.3	20.1	20.4	20.1	20.4	20.2
15	20.0	20.1	20.0	20.2	20.1	20.2	20.0
20	19.8	20.0	19.8	20.0	19.9	20.0	19.8
25	19.3	19.8	19.4	19.6	19.7	19.8	19.5
30	18.8	19.1	18.9	18.9	19.2	19.0	19.1
35	18.6	18.7	18.6	18.7	18.7	18.8	18.8
40	18.5	18.6	18.5	18.6	18.5	18.6	18.6
45	18.4	18.5	18.4	18.5	18.5	18.5	18.4
50	18.4	18.3	18.2	18.4	18.5	18.4	18.3
55	18.2	18.1	17.8	18.2	18.3	18.2	18.2
60	17.8	17.7	17.2	17.8	18.1	17.8	17.7
65	17.3	17.2	16.8	17.2	17.4	17.3	17.1
70	16.8	16.6	16.4	16.6	16.8	16.8	16.7
75	16.5	16.2	16.3	16.2	16.5	16.4	16.3
80	16.2	15.8	16.1	15.9	16.2	16.2	16.1
85	15.9	15.5	15.9	15.7	16.0	16.0	15.8
90	15.6	15.2	15.6	15.5	15.6	15.6	15.5
95	15.3	15.1	15.2	15.1	15.2	15.2	15.2
100	15.0	15.0	14.8	15.1	14.8	15.1	14.9
105	14.9	14.7	14.4	15.0	14.3	15.0	14.5
110	14.3	13.8	14.1	14.4	13.9	14.9	14.2
115	13.9	13.5	13.6	13.7	13.6	14.3	13.9
120	13.1	13.0	13.3	13.2	13.4	13.5	13.6

wire. The tow bar and bridle are mounted on top of the frame, posterior to the hoops, and the bridle is attached to the end of the tow wire. Thus, the net openings are clear of any towing device.

Before the main series of horizontal tows, an oblique recorder tow to 250 m was taken to find a relatively dense and discrete layer of animals. A depth of 90 m was selected because a dense layer of *Limacina inflata* was located between 75 and 110 m.

A parachute drogue (Volkman, Knauss, and Vine 1956) was placed at 90 m to identify the parcel of water and to serve as the reference point for navigation. During the study, the drogue moved in a northwesterly direction and on completion of the grid it was about 5.5 km from the starting point (Fig. 1).

The square grid (approx 500 m on a side) was formed by 12 net tows, each composed of a long horizontal segment at sampling depth bracketed by two short oblique segments. Only samples from the horizontal segments were used in the analysis (Fig. 2). Each of these took an average of 10.5 min to complete (range 9.3–12.0 min). Ship speed was about 2.8 km/hr. The gauze in the plankton recorder was stepped through at 15-sec intervals giving a distance traveled by the net for individual samples of 11.7 to 15.9 m. The grid started at 1000 hours and finished at 1633 (PDT). A failure in the plankton recorder system on tow XI caused the loss of most of the samples, and data from this tow have been omitted. Bathythermograph recordings were taken at varying points along the grid (Table 1); only minor changes in temperature structure took place. Larger scale changes possible over the distance the drogue moved were not observed, and this suggests that the same body of water was sampled by the grid tows.

Night horizontal tow—2 March 1967

This consisted of a single horizontal plankton recorder tow (N.H.T.) taken to provide information about night spatial structure. The sampling device was towed at about 20-m depth; the gauze tape was stepped through the recorder at 30-sec intervals, giving a length for individual samples of 38.6 m. Ship speed during the tow was about 4.6 km/hr.

Laboratory analysis and sources of error in the counts

The zooplankton species used in the analysis were selected on the basis of their frequency in the conventional cod-end sample, their morphology as it related to presumed visual acuity and mobility, feeding type, absolute size, and, in some cases, sex.

All individuals of the identified species were counted, and the daytime grid and N.H.T. counts were standardized to numbers per 5 and 8.5 m^3 respectively. Use of the entire sample minimized but did not eliminate laboratory counting error; individuals were occasionally skipped or counted more than once. The average error associated with a single count was estimated at ±2.3 (95% confidence) for counts greater than 4 (Wiebe 1968).

A more serious error might occur if the plankton in passing through the net and recorder were alternately stalled and moved (a result of variation of flow through the net or of the animals themselves hanging up in the mesh temporarily), causing artificially high and low concentrations to occur along the gauze tape. If stalling mechanisms were important, concentrations of all species and subgroups of species should show a high degree of concordance. In addition, peak values (a sample is considered a peak when it is larger than the values on either side) should show significant agreement between species in both number and placement on any grid tow line. The presence of significant values for either would imply that mechanical biasing has taken place or that the animals were abundant or rare in the same parcels of water at the time of capture.

TABLE 2. *Values of concordance coefficient* W. *The coefficients are based on 37 to 47 samples*

Tow No.	All species	Subgroup I*	Subgroup II†
I	0.43‡	0.49‡	0.57‡
II	0.57‡	0.72‡	0.57‡
III	0.21	0.55‡	0.29
IV	0.40‡	0.47‡	0.56‡
V	0.33‡	0.70‡	0.31
VI	0.32‡	0.46‡	0.41
VII	0.38‡	0.54‡	0.44
VIII	0.46‡	0.60‡	0.51‡
IX	0.46‡	0.52‡	0.58‡
X	0.59‡	0.71‡	0.62‡
XXII	0.37‡	0.68‡	0.37
Mean	0.41	0.59	0.48

* *Limacina inflata, Sagitta pseudoserratodentata, Corycaeus flaccus* female.

† *Euphausia* furcilia, *Euphausia* calyptopis, *Stylocheiron* furcilia.

‡ Values significantly greater than zero ($p < 0.05$).

To test the degree of agreement among the concentrations of individuals of the six most frequent species along each tow line, a ranking procedure and concordance test of Kendall (1955) were used. The abundance values for each of k species in n samples on each grid tow line were ranked from smallest to largest with tied values scored by the midrank method. The ranks for each of the n samples were summed and a concordance coefficient, W, based on the sum of the squares of the deviations from the mean sum was calculated. Concordance coefficients were also calculated for two subgroups of three species each (Table 2). The species in subgroup I are quite varied morphologically and might be expected to show somewhat less concordance than those in subgroup II, which are structurally similar. Significance of the concordance coefficients was tested by chi-square. In most cases, the W values were significantly greater than zero; high and low concentrations of all species tended to co-occur more often than expected by chance. The matrix of W values (Table 2) was used in a two-way analysis of variance (Tukey 1953) to reduce the effect of variability between tows on group means. The mean concordance values for the species in subgroup I was significantly higher

($p < 0.05$) than that of either all the species or of subgroup II. The other two means were not different ($p > 0.05$). There should have been no difference in means if the stalling mechanism had acted independently on the species. The fact that the species of subgroup I, which differed greatly in body structure, had a mean value significantly greater than both other mean values suggests that the co-occurrence of high and low numbers of individuals of different species does represent their distribution in the water.

Evaluation of the peak position on the grid lines also supported this interpretation. On each tow line, the abundance values of each of the six species in each of the n samples (except the end values) were examined to see if they were peaks. The number of peaks per sample (0–6) was calculated and compared with that expected if the alignment of peaks occurred randomly. In only 2 cases out of the 11 did the observed values deviate significantly ($p < 0.01$) from the expected. Abundance peaks of different species are apparently distributed independently, although the species tend to be positively correlated in terms of samples in which they have abundance above their median. In addition, counts of *L. inflata* taken with the same apparatus (Miller 1969) indicated that the vertical structure on the down and up portions of the 4 oblique hauls was essentially the same. Had stalling been an important factor, the vertical structure for a species would have been displaced downward on the down portion of the hauls compared with the up portion. There was no evidence of this.

TABLE 3. *List of zooplankton forms counted in the grid tows*

Copepoda
- *Candacia bipinnata*
- *Candacia ethiopica*
- *Corycaeus flaccus* female*
- *Corycaeus flaccus* male
- *Sapphirina intestinalis* female
- *Sapphirina intestinalis* male
- *Sapphirina metallina*

Euphausiacia
- *Euphausia calyptopis* larvae*
- *Euphausia* furcilia larvae*
- *Sytlocheiron* furcilia larvae*

Chaetognatha
- *Pterosagitta draco*
- *Sagitta enflata*
- *Sagitta minima*
- *Sagitta pseudoserratodentata**

Thecosomata
- *Cavolinia inflexa*
- *Clio pyramidata*
- *Desmopterus pacificus*
- *Limacina inflata**

* Indicates forms occurring in 50% or more of the samples on a tow line and therefore considered abundant. The others are considered common.

PATCHINESS STRUCTURE GRID—ANALYSIS AND RESULTS

A single horizontal plane was not sampled (the horizontal segments ranged in avg depth from 78 to 102 m). Therefore, the grid design permits only a composite picture of the two-dimensional patch structure to be developed from the separate analysis of each tow. The analysis has been divided into two sections. The first deals with species usually present in 50% or more of the samples on a tow line, and the second considers the less common forms (Table 3). The grid data have been given (Wiebe 1968).

Abundant species

As a first step, Fisher's index of dispersion (Fisher 1958) was calculated for the six numerically dominant forms on each tow line. The values showed significant departure ($p < 0.05$) from randomness in the direction of overdispersion in 64 of 66 cases. The general implication is that the individuals of each species are distributed in aggregations or patches (Greig-Smith 1964).

Various mathematical models could be fitted to the data, however, even in those models that use parameters supposedly related directly to aspects of the patch structure, it is by no means certain that the parameters are adequate measures of the real structure. Therefore, another approach was used.

It is assumed that the clumps can be discriminated and will be adequately described if measures of the following parameters can be obtained for each species: 1) patch size and shape; 2) number and distribution of patches in the area (bounded by the outer grid tow lines); and 3) concentration of individuals in each patch and densities of background individuals (i.e., individuals outside the patches). Essential to this evaluation is a decision as to what constitutes a patch—subjective, since no clear natural demarcation line has been observed in these data. A patch was therefore defined as a concentration of individuals exceeding the central value in a data set. The data are not normally distributed, so the median was used as the indicator of central tendency rather than the arithmetic mean. A sample will thus be considered as having been taken from a patch if it exceeds the median value on its tow line.

If many of the aggregations are on a scale equal to or smaller than the sampling interval (11.7 to 15.9 m), some of the structural information will be biased because the patch size will be overestimated and concomitantly the densities of individuals in a patch (on a per unit volume basis) underestimated. However, because the plankton recorder takes an integrative sample over the sampling interval, the sequence of patches along a tow should remain unaltered. Tests have suggested (Wiebe 1968) that a small amount of bias is present and that the estimates of patch size and densities in a patch are probably maxima and minima respectively.

Using this criterion for identifying a patch, we can estimate the number of patches a tow line crosses by counting the number of single values or sets of values above the median with adjacent values below the median. From plots of abundance versus distance (e.g., Fig. 3), the length of the tow line across each patch can be measured. The spatial structure across the grid in terms of numbers and size of concentrations of individuals above

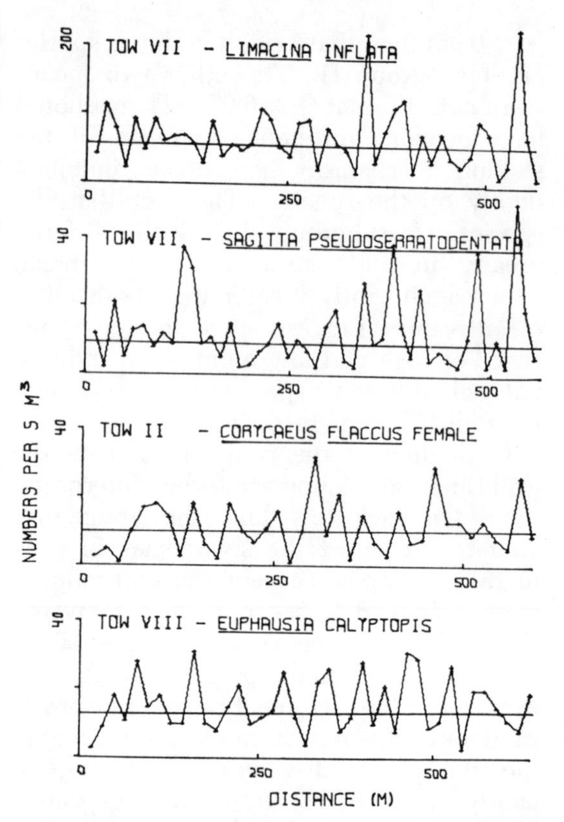

FIG. 3. Examples of the plots of abundance versus distance used to estimate patch sizes. These show some of the most extreme changes in abundance observed over short distances on the grid. The median value is also plotted.

the median value is remarkably similar for all species (Table 4). With the possible exception of *L. inflata* on tow XII, no large-scale changes in abundance occurred. If it is assumed that the patches are circular, the average size and size range can be easily calculated from the measured lengths of the tow line across patches. The median of these lengths is an estimate of the median chord length (cl) of a line dropped at random on a circle and related to radius (r) by:

$$cl = 2(r^2 - r^2/4)^{\frac{1}{2}} = r\sqrt{3}.$$

Median r is obtained by dividing cl by $3^{\frac{1}{2}}$ and the average area of a patch is πr^2.

Number of patches encountered is related to number of patches as follows. An estimate of the per cent of an area covered by an entity on a line transect is the per cent of transect line covered by the entity

TABLE 4. *Results of the statistical analysis of patch structure*

	Median No. patches/ tow line	95%*	Median length tow line across patch (m)	95%*	Median patch radius	95%*	Median patch area (m^2)	95%*	Area covered (%)	Median No. patches in grid area†	95%*
		11		28		16.2		821			341
Limacina inflata	9		25.0		14.4		655		43.2	218	
		6		20		11.5		419			174
		11		33		19.1		1,244			301
Corycaeus flaccus female	8		26.5		15.3		735		46.2	207	
		6		22		12.7		507			133
		11		30		17.3		943			306
Sagitta pseudoser-ratodentata	9		27.0		15.6		763		46.9	203	
		7		22		12.7		507			164
		10		30		17.3		943			255
Euphausia calyptopis	7		26.0		15.0		708		42.8	200	
		5		23		13.3		554			150
		11		29		16.7		880			253
Euphausia furcilia	9		23.5		13.6		579		38.8	222	
		7		22		12.7		507			146
		10		30		17.3		942			255
Stylocheiron furcilia	9		26.0		15.0		708		39.1	183	
		5		22		12.7		507			137

* 95% limits (Tate and Clelland 1957, Table D).
† Area bounded by the outer grid tow lines was 330,500 m^2.

(Brown 1954). Once the area covered by an entity (such as a plankton patch) is known, the density can be estimated by dividing the area by the average entity size. Here, dividing area by median patch area gives the estimated number of patches in the grid area (Table 4). Distribution of the patches of these six species showed no significant departures from randomness.

As a last step in analyzing the spatial structure, the densities of individuals inside and outside the patches were determined (the background density was the average number of individuals not in the patches). To compare the results between tow lines and between species, the densities were expressed as a ratio, patch density : background density (Table 5). When the background density was zero, no value was calculated. A two-way analysis of variance indicated that the differences between tow lines were nonsignificant (= 0.50) and that

TABLE 5. *Ratios of patch density per 5 m^3 to background density per 5 m^3*

	Grid tow line											Row mean*
	I	II	III	IV	V	VI	VII	VIII	IX	X	XII	
Limacina inflata	2.796	3.287	8.909	3.046	3.258	1.950	2.723	2.410	1.837	2.525	1.808	3.332
Sagitta pseudoserratodentata	2.749	3.398	2.789	2.974	2.935	2.929	3.794	3.301	4.882	3.576	3.033	3.076
Corycaeus flaccus female	2.310	3.164	2.464	2.479	3.344	1.862	4.332	2.253	2.589	2.617	2.674	2.574
Euphausia furcilia	6.397	4.820	3.576	2.885	6.233	3.622	3.820	3.975	†	4.436	3.375	4.349
Euphausia calyptopis	3.019	3.759	2.588	2.915	4.347	4.500	†	2.592	5.186	2.743	2.207	3.186
Stylocheiron furcilia	6.121	4.702	6.655	3.745	3.547	6.111	4.845	4.272	2.662	3.999	7.029	5.131
Column mean	3.899	3.855	4.547	3.007	3.944	3.496		3.104		3.316	3.354	

* Values from tows VII and IX were omitted from calculation of row mean values.
† On this tow line, the background density was 0./5 m^3.

TABLE 6. *Concordance coefficients* W *calculated for all pairs of abundant species**

	I	II	III	IV	V	VI	VII	VIII	IX	X	XII
					Limacina inflata						
Sagitta pseudoserratodentata	**+0.824**	**+0.805**	+0.608	**+0.682**	+0.657	−0.478	**+0.746**	+0.517	**+0.675**	**+0.738**	+0.604
Corycaeus flaccus female	**+0.744**	**+0.763**	+0.535	**+0.710**	+0.665	**+0.682**	**+0.783**	+0.544	+0.673	**+0.798**	+0.520
Euphausia furcilia	+0.663	**+0.728**	+0.522	+0.526	−0.452	+0.612	**+0.750**	+0.684	−0.437	**+0.790**	**+0.747**
Euphausia calyptopis	+0.536	+0.728	+0.556	**+0.718**	+0.604	+0.586	−0.457	+0.679	+0.528	+0.676	**+0.728**
Stylocheiron furcilia	+0.573	**+0.794**	+0.559	+0.573	**+0.827**	+0.568	+0.560	+0.650	−0.452	**+0.732**	+0.545
					Sagitta pseudoserratodentata						
Corycaeus flaccus female	**+0.704**	**+0.780**	**+0.780**	**+0.697**	+0.650	+0.635	**+0.797**	+0.748	+0.631	**+0.836**	**+0.722**
Euphausia furcilia	+0.540	**+0.748**	**+0.756**	+0.673	+0.568	+0.645	+0.607	+0.594	−0.399	**+0.752**	+0.667
Euphausia calyptopis	+0.562	**+0.709**	**+0.733**	**+0.776**	+0.639	+0.591	−0.358	**+0.781**	+0.643	**+0.758**	**+0.746**
Stylocheiron furcilia	+0.615	**+0.710**	+0.684	**+0.687**	+0.644	+0.575	−0.417	+0.558	−0.412	**+0.810**	+0.661
					Corycaeus flaccus female						
Euphausia furcilia	+0.611	**+0.797**	**+0.752**	+0.614	+0.500	+0.644	+0.656	+0.596	−0.398	**+0.739**	+0.557
Euphausia calyptopis	+0.683	**+0.850**	**+0.830**	**+0.789**	**+0.713**	+0.571	−0.406	**+0.708**	−0.493	**+0.751**	+0.686
Stylocheiron furcilia	+0.623	**+0.717**	**+0.709**	**+0.800**	+0.646	+0.588	+0.574	+0.516	+0.530	**+0.735**	+0.599
					Euphausia furcilia						
Euphausia calyptopis	+0.621	**+0.765**	**+0.730**	+0.653	+0.562	+0.562	−0.391	**+0.716**	**−0.313**	+0.662	**+0.735**
Stylocheiron furcilia	−0.467	**+0.705**	+0.679	+0.567	+0.555	+0.602	−0.464	+0.658	+0.501	**+0.724**	+0.618
					Euphausia calyptopis						
Stylocheiron furcilia	−0.481	+0.681	+0.649	**+0.684**	+0.594	+0.510	−0.449	+0.634	+0.498	+0.647	+0.680

* Positive association ($p < 0.50$) is indicated by a plus sign; negative association ($p > 0.50$) by a minus sign. Numbers in boldface indicate significant positive ($p < 0.05$) or negative ($p > 0.95$) association depending on the sign.

the mean ratios for the species differed significantly ($p < 0.005$). This appears to be related to the overall density of the species: As density increases, the mean ratio tends to decrease.

Since most of the parameters of the spatial structure did not differ significantly between species, it is of interest to know whether the patches are single or multispecies structures. Use of Kendall's concordance coefficient, W, showed that the three species, *L. inflata*, *Sagitta pseudoserratodentata*, and *Corycaeus flaccus* female, agreed in their relative abundances to a significantly greater degree than did the three types of euphausiid larvae or all six species together. However, all but 1 of the 11 W values calculated for the six species combined were significant. To look more closely at the relations between pairs of species, the calculation of concordance coefficients was extended to all possible pairs of the six species on each tow (Table 6). W can also be used to indicate negative associations when it is calculated for species pairs. Significant positive or negative association is assumed when the χ^2 value is less than the $p = 0.05$ or greater than the $p = 0.95$ levels. Of the 165 coefficients, only 18 were negative, a result not expected by chance alone ($p < 0.0005$), indicating that the high and low concentrations of individuals of all species tend to co-occur more often than expected by chance and that the patches are multispecies structures. However, the values in Table 6 suggest that the degree or intensity of the association between species pairs is not constant over the grid nor is it the same for different pairs of species.

Common species

The 11 species considered common were listed in Table 3. Only *Sapphirina intestinalis* male, *Desmopterus pacificus*, and *Candacia ethiopica* did not occur on every tow line. The index of dispersion indicated marked differences between species as to the number of lines on which they showed significant aggregation. For example, *C. flaccus* male was significantly aggregated on 10 of the 11 tow lines, *Candacia bipinnata* was aggregated on 5, and *Sapphirina metallina* on only 1. The differences apparently are associated with differences in the densities of the species (Fig. 4). As the

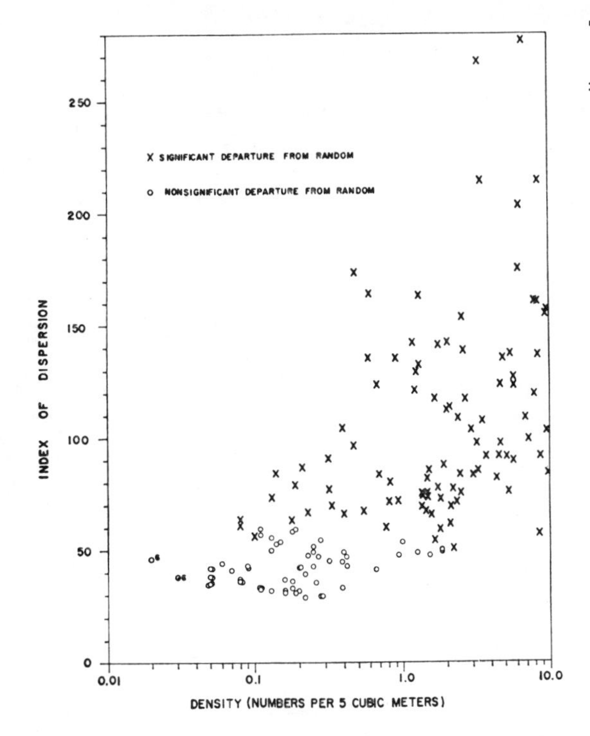

FIG. 4. Plot of density of the common species versus the index of dispersion to illustrate the apparent association between departure from randomness of distribution and density.

density of individuals decreased the distributions appeared to approach randomness. Cassie (1963) has pointed out that apparent tendencies toward random distributions with a decrease in density can mean that the statistical techniques are not sensitive enough at low densities to detect any nonrandomness present.

The index of dispersion does not take into account the sequence in which the values were drawn. A test which does consider the sequence of the observations, and to a lesser extent their value, was used on those data sets that had more than one positive value and in which the index of dispersion did not indicate significant aggregation. This is a generalized runs test (Wallis and Roberts 1956). Out of 46 cases in which it was applied, in only one was the sequence of values significantly ($p < 0.05$) nonrandom. This could have been a chance occurrence (expected by chance, 2.3). The distributions of the individuals of species in the densities below

TABLE 7. *List of zooplankton species counted in the night horizontal tow*

Copepoda
Candacia bipinnata
Euchaeta media
*Heterorhabdus papilliger**
*Pleuromamma abdominalis**
*Pleuromamma gracilis**
Undeuchaeta intermedia
Euphausiacia
Euphausia eximia
*Euphausia recurva**
Nematobrachion flexipies
Chaetognatha
*Sagitta bierii**
*Sagitta enflata**
Sagitta hexaptera
Sagitta minima
*Sagitta pseudoserratodentata**
Thecosomata
Limacina inflata

* Indicates the species occurring in 50% or more of the samples and therefore considered abundant. The others are considered common.

0.5 individual per 5 m^3 do not deviate significantly from random.

NIGHT HORIZONTAL TOW—ANALYSIS AND RESULTS

The species counted in the N.H.T. samples are listed in Table 7. Although this series of samples was taken in the same area as the grid tows, it was 6.5 months later and the species composition had changed. For example, *L. inflata* was the numerically dominant form on the grid, but it was relatively unimportant in the night tow; *Pleuromamma gracilis*, which was not observed in the day samples, was the most abundant species in the night samples. As a result of the change in plankton composition between the two sampling periods, an exact comparison of patch structure between night and day at the species level is not possible. However, general similarities and dissimilarities can be noted. Analyses are presented in Wiebe (1968).

Abundant species

The index of dispersion values for the seven frequent species indicate that six were significantly aggregated. The excep-

TABLE 8. *Results of the statistical analysis of patch structure*

	No. of patches	Median length tow line across patch (m)	95%	Median patch radius	95%	Median patch area (m^2)	Area covered (%)	Patch density : background density
Pleuromamma gracilis	9	110.4	278 58	67.7	160 33	14,499	49.0	3.11
Pleuromamma abdominalis	19	95.5	110 53	55.2	64 31	9,572	56.4	4.03
Sagitta pseudoserratodentata	14	79.4	166 39	45.8	96 22	6,590	48.9	2.13
Sagitta enflata	13	66.7	133 25	38.4	77 15	4,633	53.1	3.04
Sagitta bierii	13	126.5	150 55	73.1	86 32	16,787	50.9	2.95
Euphausia recurva	13	105.8	159 51	61.2	92 29	11,767	49.3	4.05

tion, *Heterorhabdus papilliger*, although occurring in over 50% of the samples, had a low mean density. The generalized runs test also indicated a nonsignificant ($p = 0.37$) departure from random for this species.

For the six species showing significant patchiness, estimates of median patch radius were made as before. The results (Table 8) suggest the following. First, as with the grids, the estimates of median patch radius for the species do not differ significantly (i.e., the median radius for each species is within the 95% limits of all other species). Second, the night estimates are all larger than the day values. One factor responsible for this increase is the larger sample size; patch sizes smaller than the tow length cannot be discriminated. Another factor that might lead to higher patch size estimates from the N.H.T. is the presence of larger scale trends in abundance. This is most clearly seen in the plot of abundance versus distance (Fig. 5) for *P. gracilis*, which exhibits an increase in abundance midway along the tow line.

TABLE 9. *Concordance coefficients* W *calculated for all pairs of abundant species**

	Sagitta pseudoserratodentata	*Euphausia recurva*	*Heterorhabdus papilliger*	*Pleuromamma abdominalis*	*Sagitta bierii*	*Pleuromamma gracilis*
Sagitta enflata	**+0.7563**	+0.6020	+0.5484	+0.5356	+0.5888	**−0.3634**
Sagitta pseudoserratodentata		**+0.6413**	+0.5701	+0.5561	**+0.7374**	−0.3906
Euphausia recurva			+0.5477	+0.5044	+0.5818	−0.4117
Heterorhabdus papilliger				+0.5917	+0.5645	+0.5348
Pleuromamma abdominalis					+0.6260	**+0.6646**
Sagitta bierii						+0.5336

* Positive association ($p < 0.50$) is indicated by a plus sign; negative association ($p > 0.50$) by a minus sign. Numbers in boldface indicate significant positive ($p < 0.05$) or negative ($p > 0.95$) association depending on the sign.

There are less distinct, but noticeable, large-scale changes in the abundances of *S. pseudoserratodentata*, *Sagitta enflata*, and *Euphausia recurva*. The trend for these three species is inverse to that for *P. gracilis* and represents changes in abundance that are sharp enough to alter the patch size estimates appreciably, indicating that a larger scale of patchiness is present in the area for these species.

Concordance coefficients were calculated for all species pairs (Table 9). As expected, *P. gracilis* was negatively associated with *S. enflata*, *S. pseudoserratodentata*, and *E. recurva*. All other combinations of species pairs (18) showed positive associations. The probability of 3 negative values out of 21 is $p < 0.005$.

Common species

The index of dispersion was calculated for the eight common species. Half show significant ($p < 0.05$) departure from randomness. Those indicating randomness were further treated with the generalized runs test. The results of this test also indicated nonsignificant ($p > 0.05$) departures from random. As with the grid, low density of individuals was associated with apparent randomness of distribution of individuals.

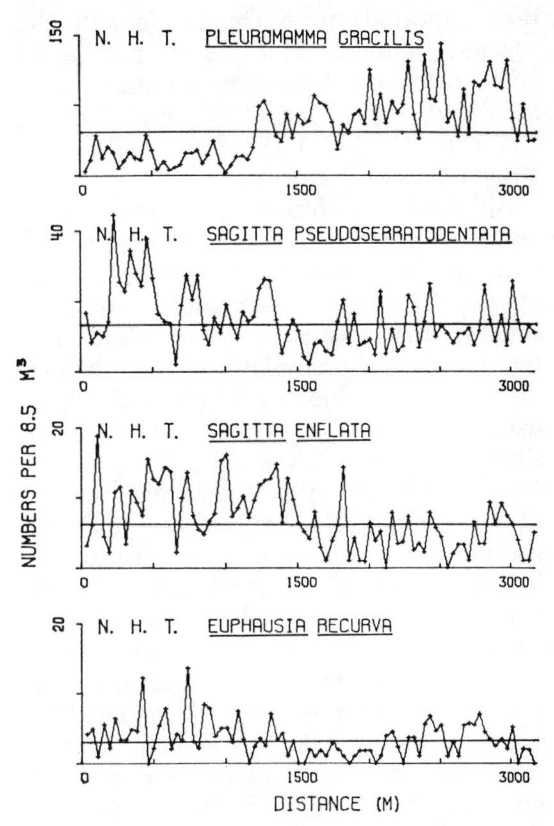

FIG. 5. Plots of abundance versus distance illustrating the presence of a larger scale of patchiness on which is superimposed the smaller scale structure observed on the night horizontal tow. The median value is also plotted.

DISCUSSION

All of the abundant and many of the common species in both the day and night samples exhibited significant, nonrandom changes in abundance over short horizontal distances (the minimum distance detectable was about 12 m). The numerical values determined for the patch parameters of the separate species show a remarkable degree of similarity, although the species are morphologically and ecologically quite diverse. This result was unexpected and much effort has been devoted to reducing the possibility of an artifact due to the mechanics of the sampling gear or to the interaction of the spatial structure and sampling scale. The second alternative, that the structure is really on a smaller scale than the sample length and the resultant measured structure is a function of the latter, has not been discredited beyond doubt. But even if most of the patches were smaller than the estimates indicate, the patch sizes for different species would still be within the range of 1 to 15 m. Winsor and Clarke (1940) reached a similar conclusion, although they made no effort to determine the spatial structure of their plankton. Winsor and Walford (1936) had noted earlier that the variability of abundances of different types of organisms taken in paired net tow samples did not differ appreciably, but they argued that different organisms would be expected to exhibit disparate amounts of variability, so the organisms were most likely randomly distributed. Bernhard and Rampi (1965) working in the Mediterranean Sea found intensive horizontal patchiness on about the same scale as was found here. The structure appeared quite similar for all the species examined. Much larger patches of zooplankton (30 to 50 km in radius) were identified by Cushing (1953) and Cushing and Tungate (1963) although smaller scale structures may have existed.

Although no physical parameters were monitored simultaneously with the collection of the samples, the fact that the abundant organisms have similar structure suggests that the causal mechanisms may be fundamentally physical. That is, the animals may be subjected to physical aggregating mechanisms similar to those in surface waters (Langmuir 1938; Owen 1966), although such subsurface mechanisms have not yet been postulated. Or the species may be reacting to small-scale inhomogeneities of physical factors (e.g., temperature, salinity, light, etc.), avoiding parcels of water with unfavorable properties and collecting in more favorable parcels. It is, however, difficult to visualize such parcels as being permanent or constant. In addition, evidence for such inhomogeneities (LaFond 1963) indicates they are of small magnitude compared with the range of values the animals are exposed to in performing vertical migrations. The bathythermograph data from the grid show that a migrating organism would have been subjected to a temperature change of at least 5C in swimming from 90 m to the surface. The temperature change in moving horizontally across the grid would have been at the most 0.4C. Hardy and Bainbridge (1951) and Knight-Jones and Morgan (1966) have shown that zooplankton in the laboratory are sensitive to very small changes in pressure even though the changes they experience in the ocean can be orders of magnitude larger. A similar sensitivity to other factors would provide the possibility of reacting to small changes in the physical environment. Information about the relationship between environmental parameters and subsurface horizontal distributions of individuals is meager. Cassie (1959*a*, *b*, 1960) found that changes in the abundance of numerically dominant zoo- and phytoplankton species 5 to 10 m below the surface in the Wellington Harbor area were correlated with changes in temperature and salinity. He concluded that physical factors can make as substantial a contribution as biological interactions to the distribution pattern. Barnes and Marshall (1951) found that variations in salinity with depth (down to 10 m) and with time were similar to variations in the abundance of four different species of copepod nauplii and suggested that population changes were associated with different water masses that maintained identity while the populations developed and that the association of a number of species was maintained. But Bernhard and Rampi (1965) concluded that biological interactions were more important in determining the spatial structures in their species. In both the daytime and the night tow samples, different coefficients of association were observed between different pairs of species. In all cases on the grid, the associations between the same species pair changed in magnitude from tow to tow. Apparently, although the species may be reacting to, or regulated by, a common set of factors both biological and physical in origin, the manner of accomplishing this may differ.

The comparison made here indicates an increase in patch size at night, but the abundant species in the night tow were still strongly aggregated horizontally. In addition, the sample size used could not discriminate smaller scale structures. These data cannot therefore be interpreted as indicating a marked diurnal change in horizontal spatial structure.

The biological significance of this spatial structure is not readily apparent and probably will not become so until the patch-forming mechanisms are better known. On this scale (10–15 m), the data suggest that important interactions between species abundances and the biological-physical environment are taking place.

REFERENCES

ANRAKU, M. 1956. Some experiments on the variability of horizontal plankton hauls and on the horizontal distribution of plankton in a limited area. Bull. Fac. Fisheries, Hokkaido Univ. **7**: 1–16.

BARNES, H. 1949. On the volume measurement of water filtered by a plankton pump, with some observations on the distribution of planktonic animals. J. Marine Biol. Assoc. U.K. **28**: 651–661.

———, AND S. M. MARSHALL. 1951. On the variability of replicate plankton samples and some application of "contagious" series to the statistical distribution of catches over restricted periods. J. Marine Biol. Assoc. U.K. **30**: 233–263.

BERNHARD, M., AND L. RAMPI. 1965. Horizontal microdistribution of marine phytoplankton in the Liqurian Sea. Botan. Gothoburgensia **3**: 13–24.

BROWN, D. 1954. Methods of surveying and measuring vegetation. Commonw. Bur. Pastures Field Crops Bull. 42. Hurley, Berks.

CASSIE, R. M. 1959*a*. An experimental study of factors inducing aggregation in marine plankton. New Zealand J. Sci. **2**: 339–365.

———. 1959*b*. Micro-distribution of plankton. New Zealand J. Sci. **2**: 398–409.

———. 1960. Factors influencing the distribution of plankton in the mixing zone between oceanic and harbour waters. New Zealand J. Sci. **3**: 26–50.

———. 1963. Microdistribution of plankton. Oceanog. Marine Biol. (H. Barnes [ed.]) **1**: 223–252.

CUSHING, D. H. 1953. Studies on plankton populations. J. Conseil, Conseil Perm. Intern. Exploration Mer **19**: 3–22.

———, AND D. S. TUNGATE. 1963. Studies on a calanus patch. I. The identification of a calanus patch. J. Marine Biol. Assoc. U.K. **43**: 327–337.

FISHER, R. A. 1958. Statistical methods for research workers, 13th ed. Hafner Publ. Co., New York. 356 p.

GREIG-SMITH, P. 1964. Quantitative plant ecology. Butterworths. 256 p.

HARDY, A. C. 1936. Observations on the uneven distribution of oceanic plankton. Discovery Rept. **11**: 511–538.

———. 1955. A further example of the patchiness of plankton distribution. Papers Marine Biol. Oceanog. Deep-Sea Res. **3**(Suppl.): 7–11.

———, AND R. BAINBRIDGE. 1951. Effect of pressure on the behavior of decapod larvae (Crustacea). Nature **167**: 354–355.

KENDALL, M. G. 1955. Rank correlation methods, 2nd ed. Griffen, London. 196 p.

KNIGHT-JONES, E. W., AND E. MORGAN. 1966. Responses of marine animals to changes in hydrostatic pressure. Oceanog. Marine Biol. (H. Barnes [ed.]) **4**: 267–300.

LAFOND, E. C. 1963. Detailed temperature structure of the sea off Baja California. Limnol. Oceanog. **8**: 417–425.

LANGMUIR, I. 1938. Surface motion of water induced by wind. Science **87**: 119–123.

LONGHURST, A. R., A. D. REITH, R. E. BOWER, AND D. L. R. SEIBERT. 1966. A new system for the collection of multiple serial plankton samples. Deep-Sea Res. **13**: 213–222.

MCGOWAN, J. A., AND D. M. BROWN. 1966. A new opening-closing paired zooplankton net. Scripps Inst. Oceanogr. Ref. 66-23. 56 p.

MILLER, C. B. 1969. Some environmental consequences of vertical migration. Ph.D. thesis, Univ. Calif., San Diego. 230 p.

OWEN, R. W., JR. 1966. Small-scale, horizontal vortices in the surface layer of the sea. J. Marine Res. **24**: 56–66.

RILEY, G. A. 1963. Theory of food chain relations in the ocean, p. 438–463. *In* M. N. Hill [ed.], The sea, v. 2. Interscience.

TAFT, B. A. 1960. A statistical study of the estimation of abundance of sardine (*Sardinops caerulea*) eggs. Limnol. Oceanog. **5**: 245–264.

TATE, M. W., AND R. C. CLELLAND. 1957. Nonparametric and shortcut statistics. Interstate Printers, Danville, Ill. 171 p.

TONOLLI, V. 1949. Struttura spaciale del popolamento meso planctico. Eterogeneita della densita dei popolamenti orrizontali e sua variazione in funzione della quota. Mem. Ist. Ital. Idrobiol. **5**: 191–289.

———, AND L. TONOLLI. 1960. Irregularities of distribution of plankton communities, considerations and methods, p. 137–143. *In* A. A. Buzzati-Traverso [ed.], Perspectives in marine biology. Univ. Calif. Press, Berkeley.

TUKEY, J. W. 1953. Some selected quick and easy methods of statistical analysis. Trans. N.Y. Acad. Sci. **16**: 88–97.

VOLKMAN, G., J. KNAUSS, AND A. VINE. 1956. The use of parachute drogues in the measurement of subsurface ocean currents. Trans. Am. Geophys. Union **37**: 573–577.

WALLIS, W. A., AND H. V. ROBERTS. 1956. Statistics: A new approach. Free Press, Glencoe, Ill. 646 p.

WIEBE, P. H. 1968. Zooplankton patchiness and its effects on sampling error: a field and computer model study. Ph.D. thesis, Univ. Calif., San Diego. 184 p.

———, AND W. R. HOLLAND. 1968. Plankton patchiness: effects on repeated net tows. Limnol. Oceanog. **13**: 315–321.

WINSOR, C. P., AND G. L. CLARKE. 1940. A statistical study of variation in the catch of plankton nets. J. Marine Res. **3**: 1–34.

———, AND L. A. WALFORD. 1936. Sampling variations in the use of plankton nets. J Conseil, Conseil Perm. Intern. Exploration Mer **11**: 190–204.

8 A COMPUTER MODEL STUDY OF ZOOPLANKTON PATCHINESS AND ITS EFFECTS ON SAMPLING ERROR[1]

Peter H. Wiebe

Woods Hole Oceanographic Institution, Woods Hole, Massachusetts 02543

ABSTRACT

A computer model using estimates of patch structure parameters determined from an ocean study was developed to study the effects of spatial structure on the error of replicate net tows. The precision of nine replicate tows increases with increasing patch size, regardless of the position of the patch centers, and the largest net provides the most precise estimates. Lengthening the tow also results in significant increases in precision (up to a factor of 2), not related directly to increases in the volume of water filtered, but apparently to the increase in the probability of sampling the "right" number of patches. When spatial structure is homogeneous, increasing tow length results in a larger reduction of sampling error than does a corresponding increase in net diameter. In general, better estimates of species proportions are obtained when the numbers of individuals per species are inequitably distributed and the degree of patch overlap is the greatest.

INTRODUCTION

In this paper I shall examine the interactions of the patch structure variables, size, distribution, and density (i.e., numbers of individuals per volume) with net size and tow length, and their relation to the precision of net tows. The results of these experiments can be used to qualify only one of the assumptions basic to survey samples (Taft 1960; Wiebe 1970); that is, that plankton samples taken with nets represent quantitatively population and community parameters in the parcel of water sampled. The term *parcel* refers to the water body that would be sampled if a series of replicate tows were taken at a geographic location. None of the work reported here relates to the other basic assumption: that the parcel itself contains numbers and kinds of organisms representative of the general area of the station.

My approach to the study of sampling error associated with the patchiness of zooplankton has been to perform experiments using a synthetic system or model in which only one error-producing component was present. I used the model approach because it did not appear possible to develop a field sampling design that would allow the effects associated with other error components to be reduced, eliminated, or controlled in such a way that they could with certainty be extracted from the experimental results.

Preliminary work (Wiebe and Holland 1968) has shown that the size and distribution of patches significantly affects both the accuracy and precision of estimates of single species abundance. A comparison of tows of the same length taken with nets of 25-, 100-, and 200-cm diam showed that, in general, estimates from the largest net were the most accurate and precise. In this study, two experiments were conducted with a modified version of the model. The first is an extension of the preliminary work, the results of which apply to single species populations. The second examines the aspect of zooplankton sampling dealing with the estimation of relative abundances of species. In both cases, I am concerned primarily with two questions: How does the spatial structure affect the precision of net tow estimates? How much error should be expected from this error component?

[1] Contribution No. 2490 of the Woods Hole Oceanographic Institution. This work was supported by Atomic Energy Commission Contract AT(30-1)-3862 (Ref. No. NYO-3862-32); National Science Foundation Grant GB 2861; and the Marine Life Research Program, the Scripps Institution of Oceanography's part of the California Cooperative Oceanic Fisheries Investigations, which are sponsored by the Marine Research Committee of the State of California.

Reprinted from *Limnology and Oceanography*, 16:29–38, 1971.

TABLE 1. *Values of variables in model experiment 1*

Variable	Fixed value
Box length (x)	5,000 m
Box width (y)	50 m
Box height (z)	50 m
No. of patches	250
Position of patch centers	Given by 3 different sets of 750 random numbers* each
Patch sizes	7.5, 15.0, 25.0, 35.0, 42.5 m/side
Individuals per patch	10,000
Net sizes	25-, 100-, 200-, 400-cm diam
Length of tow (x)	500, 2,000, 5,000 m; midpoint of all tows at midpoint of box length, i.e., 2,500 m
Starting positions of the 9 replicate tows (y, z)	12.5, 12.5; 25.0, 12.5; 37.5, 12.5; 12.5, 25.0; 25.0, 25.0; 37.5, 25.0; 12.5, 37.5; 25.0, 37.5; 37.5, 37.5

* The random numbers were obtained from a random number generating function programmed into the CDC 3600 computer located at the University of California, San Diego.

Answers to these questions are necessary if we are to undertake ecological studies requiring quantitative estimates of the number of species or subspecific forms, the sizes of their populations, and changes in these.

Although the model is based on information about the structure of plankton patchiness obtained from the field (Wiebe 1970), the necessary assumptions and simplifications made mean that the conclusions drawn from the experiments presented below may not apply to nature.

I acknowledge and appreciate the liberal assistance of Drs. J. A. McGowan and E. W. Fager in the planning and execution of this research. Drs. C. B. Miller and J. T. Enright also gave considerable help through discussions and constructive criticism.

MODEL DESCRIPTION AND EXPERIMENTAL DESIGN

The model has been previously described (Wiebe and Holland 1968) and, except for one substantial modification in the method of capturing individuals, was unchanged for these experiments.

The simulated water body is rectangular with variable dimensions. Inside are placed cubic patches of variable size, position, number, and density. Nets are passed through the volume and individuals are captured from the patches in the following way: When a patch lies in the path of a net tow, the volume of water in the patch sampled by the net is calculated. The average density in a patch times the volume sampled from the patch equals the average number (m) of individuals expected to be caught in a large number of trials. Since the individuals are distributed randomly in the patches, a certain amount of variation about this average is expected. To introduce this source of variation into the model, the individuals are assumed to be distributed as a Poisson series with a mean, m. The frequency distribution of this series is given by

$$p(x) = \frac{e^{-m} m^x}{x!},$$

where x is the number of individuals in a sample. Starting with $x = 0$, the $p(x)$ terms are cumulated until the sum is greater than or equal to a previously generated random number between 0.0 and 1.0. The value of x at this point is used as the number of individuals captured by the net in that particular patch. For m greater than 30, a normal approximation to the Poisson is used to select the number of individuals captured. The method consists of selecting a random normal number, r_n, from a population with a mean equal to 0 and a variance equal to 1. Then the number captured is set equal to $m + r_n m$. The total catch for a net tow is the sum of the individuals caught in the separate patches along the net tow path.

Experimental design 1

Table 1 presents the values given the variables used in experiment 1. The values associated with the parameters of patch structure were derived from the field study, and a series of nine "replicate" tows was run for each possible combination of five patch sizes, three random placements of their centers, four net sizes, and three tow lengths (Wiebe 1970). For any series of replicate tows all the patches in the model volume were the same size. From the net tows, a total of 1,620 ($9 \times 5 \times 3 \times 4 \times 3$) estimates of abundance were obtained. The estimated total number of individuals in the water body was made from the number of individuals caught on each net tow. Coefficients of variation ($s/\bar{x}$) were used as a measure of the precision (the deviation of a set of values from their mean value) of each set of replicate tows. To assess the effect of the spatial structure on net tow estimates, an analysis of variance was done using the precision measurements as data. In addition, an overall coefficient of variation was calculated from the counts of all tows taken for each of the 12 combinations of net size and tow length. These were computed to obtain estimates of the amount of error arising from the interaction between the distribution of the organisms and different sized nets and tow lengths.

Experimental design 2

In experiment 2, a community of 10 species was placed in the model with the individuals of each species again placed in patches. The values given the model variables for this experiment are presented in Table 2. Two possible extreme distributions of single species patches exist. The patches of different species may overlap 100% (the patches are perfect multispecies structures), or they may be independently and randomly distributed with respect to each other. Both cases were treated in the experiment. When the patches of all species overlapped 100%, the patches of the different species were positioned by a single set of random numbers. When the patches were independent, their positions were determined by different sets of random numbers, one for each species. The distribution of individuals among the species was based on two extreme situations found in actual samples from the California Current, where a comparatively complete census of zooplankton species has been made. The relative proportions of individuals of the 10 most abundant species at the stations at which individuals were most and least equitably distributed among species (California Cooperative Oceanic Fisheries Investigations Cruise No. 5804, stations 100.40 and 80.60 respectively) were used to determine the numbers of individuals of each species in the model (Table 2). The total number of individuals in the model volume was based on the sum of the average numbers of individuals of the 10 most abundant species caught in a series

TABLE 2. *Values of variables in model experiment 2 that were changed from experiment 1*

Variable	Fixed value
Patches concordant:	
No. of patches	250
Position of patch centers	Given by 1 set of 750 random numbers
Patches discordant:	
No. of patches	250
Position of patches	Given by 10 different sets of 750 random numbers; one set for each species
Patch sizes	7.5, 25.0, 42.5 m/side
Proportions of individuals in species 1, 2, 3, 4, 5, 6, 7, 8, 9, 10:	
Equitable case	0.201, 0.162, 0.138, 0.121, 0.079, 0.069, 0.066, 0.066, 0.055, 0.043
Nonequitable case	0.774, 0.047, 0.044, 0.038, 0.027, 0.024, 0.018, 0.011, 0.009, 0.008
Net sizes	25-, 100-, 400-cm diam
Lengths of tow	500, 2,000 m
Starting position of the 6 replicate tows	12.5, 12.5; 25.0, 12.5; 37.5, 12.5; 12.5, 37.5; 25.0, 37.5; 37.5, 37.5

TABLE 3. *Experiment 1, analysis of variance results; data were coefficients of variation*

Source of variation	SS	df	MS	Significance level
Nets	0.588	3	0.1959	$p < 0.0005$
Tow length	4.239	2	2.1193	$p < 0.0005$
Random distribution	0.329	2	0.1644	$p < 0.0005$
Patch size	5.250	4	1.3139	$p < 0.0005$
Residual	2.952	168	0.0176	

of net tows taken in the Guadalupe Island area on 2 March 1967. At that time there were about 50 individuals of these species per cubic meter. This value times the volume of water in the model (12.5×10^6 m^3) equals 6.25×10^8, which was used as the total number of individuals in the model.

A series of six replicate tows was run for each combination of three net sizes, two net tow lengths, three patch sizes, two sets of proportions of the 10 species, and two arrangements of relative patch position. As in experiment 1, the total number of individuals of a species in the water body was estimated from the number of individuals of that species caught on each net tow. For each replicate tow, these values were summed and used to calculate estimates of the proportion of the total individuals represented by each species. To measure the accuracy of the estimates (i.e., the deviation of the observed value from the known), the "percentage similarity" index, S, used by Whittaker and Fairbanks (1958) was used:

$$S = 100 \times (1.0 - 0.5 \times \sum_{i=1}^{n} |P_i - OP_i|),$$

where P_i and OP_i are the known and observed proportions of the ith species, and n is the total number of species. S equals 100 when the estimated proportions are exactly the same as those in the model and approaches 0 as the similarity between the estimated and actual model proportions diminishes. S is assigned a zero when no patches are sampled on a tow. An analysis of variance was done using the percent similarities as data.

TABLE 4. *Overall mean coefficient of variation for each variable category in experiment 1*

Net size (cm diam)	
25	0.458
100	0.340
200	0.331
400	0.312
Tow length (m)	
500	0.568
2,000	0.310
5,000	0.203
Patch size (m/side)	
7.5	0.682
15.0	0.391
25.0	0.258
35.0	0.255
42.5	0.246
Random distribution	
21	0.322
29	0.420
30	0.339

RESULTS

Experiment 1

The results of the analysis of variance are summarized in Table 3. The mean estimates of the total number of individuals in the model obtained by the four sizes of nets differed significantly in precision ($p < 0.0005$) with the largest net giving the best estimates. However, it is clear from Table 4 that the improvement in precision was not linear with increasing net size. The greatest improvement came when the net diameter was increased from 25 to 100 cm (26% in precision). A proportional increase in net diameter from 100 to 400 cm only improved the precision 8.2%.

The estimates of total numbers of individuals in the model changed more dramatically with increasing tow length. The mean estimates based on all sizes of nets (Table 3) pooled according to tow length increased significantly ($p < 0.0005$) in precision with increased tow length. The

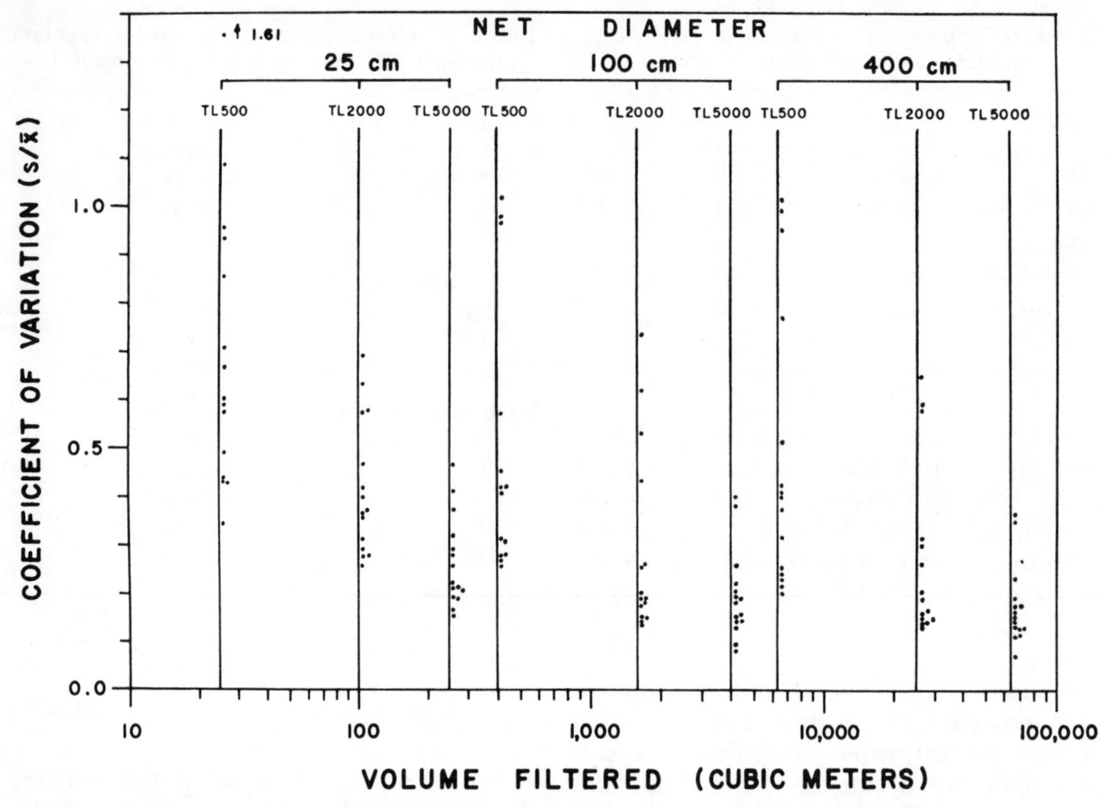

FIG. 1. Plot of volume of water filtered vs. the coefficient of variation to illustrate that, on a unit volume increase basis, greater improvement in the precision occurs by increasing tow length than by increasing net size.

greatest improvement took place with the first increase in tow length. With an increase from 500 to 2,000 m, the precision was almost doubled; a further increase to 5,000 m had relatively less effect.

Increasing either the size of net or length of tow results in an increase in the volume of water sampled. However, the plot of precision of the sample estimates versus volume filtered (Fig. 1) clearly shows that the increases in precision are not simply due to the increase in sample volume. For example, the 25-cm-diam net towed 5,000 m provided better estimates than did the 400-cm-diam net towed 500 m, although the former net filtered only 248 m³ and the latter 6,280 m³.

It is also apparent from Table 3 that the size of the patches and their position plays an important role in precision of the estimates. In general, the precision increased significantly with increasing patch size (Table 4). However, as was the case in the preliminary study, with one of the random distributions (21) significant overestimates of the known model population were often made when the patches were greater than 15 m/side. The magnitude of the overestimates decreased with increasing length of tow and net size.

The average coefficients of variation for all net size–tow length combinations are presented in Table 5. The 95% confidence limits associated with a single observation, x_i, can be calculated as:

$$x_i \pm x_i \times 1.96 \times s/\bar{x}.$$

For example, if 1,000 individuals were caught by the 400-cm-diam net towed 2,000 m, the 95% limits would be 1,773 to 277, while the limits for the same net towed 500 m would be 2,397 to 0.

TABLE 5. *Mean coefficients of variation for the combinations of net size and tow length used in sampling the model populations. Each coefficient based on 135 tows*

Tow length (m)	Net size (cm diam)			
	25	100	200	400
500	0.8387	0.7878	0.7788	0.7130
2,000	0.4789	0.4120	0.4135	0.3943
5,000	0.3190	0.2729	0.2671	0.2469

Experiment 2

The results of the analysis of variance are summarized in Table 6. The overall mean values for each variable category are given in Table 7. There is a trend of increasing mean percent similarity, S, with increasing net size, significant at the $p = 0.10$ level. Increasing tow length resulted in a much more significant ($p < 0.0005$) improvement of the net tow estimates of S. This again suggests that it is not the increase in the volume of water filtered that is responsible for the improved estimates.

The size of the patches, the degree of overlap of patches of different species, and the distribution of individuals per species significantly ($p < 0.0005$) influenced the net tow estimates. As patch size decreased, the values of S became progressively smaller. When the single species patches were independently positioned and 7.5 m/side, values as low as 36% were observed when the individuals were equitably distributed among the species and 15% when they were inequitably distributed. When the patches of different species were superpositioned, zero values of S were observed

TABLE 6. *Experiment 2, analysis of variance results**

Source of variation	SS	df	MS	Significance level
Concordance	1.241	1	1.2409	$p < 0.0005$
Equitability	0.176	1	0.1757	$p = 0.0005$
Nets	0.076	2	0.0381	$p = 0.10$
Tow length	0.664	1	0.6643	$p < 0.0005$
Patch size	2.466	2	1.2328	$p < 0.0005$
Residual	5.887	424	0.01388	

* Data were indexes of percentage similarity of observed and known relative proportions of species.

TABLE 7. *Overall mean percent similarity value for each variable category in experiment 2*

Overlap of patches	
Concordant	0.959
Discordant	0.852
Equitability	
Equitable	0.885
Inequitable	0.925
Net size (cm diam)	
25	0.892
100	0.901
400	0.923
Tow length (m)	
500	0.866
2,000	0.944
Patch size (m/side)	
7.5	0.799
25.0	0.950
42.5	0.967

with the smallest patches, because it was possible to tow the net in particular model regions and not sample a patch. However, in all cases where patches were sampled, better estimates of the species' relative proportions resulted with the multispecies patches. In general, I got higher values of S when the distribution of individuals was inequitable.

DISCUSSION

The increases in precision with increasing net size and tow length are not entirely explained by increases in the volume filtered, so other possible explanations must be considered. These appear to involve two factors related to patch size; namely, the probability of sampling the "right" number of patches and the probability of capturing the correct number of individuals once the net is inside a patch. When the patches are large ($\geqslant 25.0$ m/side) and nets of different size are towed along the same path, they all usually pass through the same number of patches (i.e., the probability of hitting a patch is independent of net size). The larger nets provide more precise estimates than the smaller ones because the variability of the estimates is a function of the numbers caught. The individuals in a patch (in the model) are randomly distributed (Poisson) and, therefore, the variance equals the mean. The coefficient of variation is then $\sqrt{\bar{x}}/\bar{x}$ and this decreases nonlinearly as $\bar{x}$ increases. This may explain why there is a much greater improvement in precision in going from a 25-cm-diam net to one of 100 cm as compared to the change from 100 to 400. With a constant number of individuals per patch and the consequent reduction in density within patches as they were made larger, it was even possible for the smallest net to pass through some patches ($\geq$25 cm/side) and not capture a single individual. When patches were smallest, the density of individuals, and therefore the value of $\bar{x}$, was increased to a point where estimates of concentrations within patches were more similar in precision for the different net sizes. However, the probability of meeting a patch increases with net size and when the patches are small, it is more likely that the right number of them would be sampled by a larger net than by a smaller one. A graphical illustration of this is presented in Fig. 2. The volume of water effectively sampled by the net is here defined as the volume surrounding the net tow path in which a patch center must be located to be sampled by the net. The formula used to calculate the volume effectively sampled is:

$$V = \pi \times (r_n + r_p)^2 \times L + \tfrac{4}{3}\pi \times r_p^3,$$

where r_n equals the net radius, r_p equals the mean patch radius, and L equals the tow length. I have assumed that the patches are spheres rather than cubes to simplify the calculations. For a given tow length, the larger net effectively samples a larger volume of water and would be expected to provide better estimates, since an increase in volume increases the chances of sampling the right number of patches to provide a correct estimate of the model population. It is also clear that if $r_p \gg r_n$, as was the case in the model, extending the tow length has more effect in increasing the volume effectively sampled than

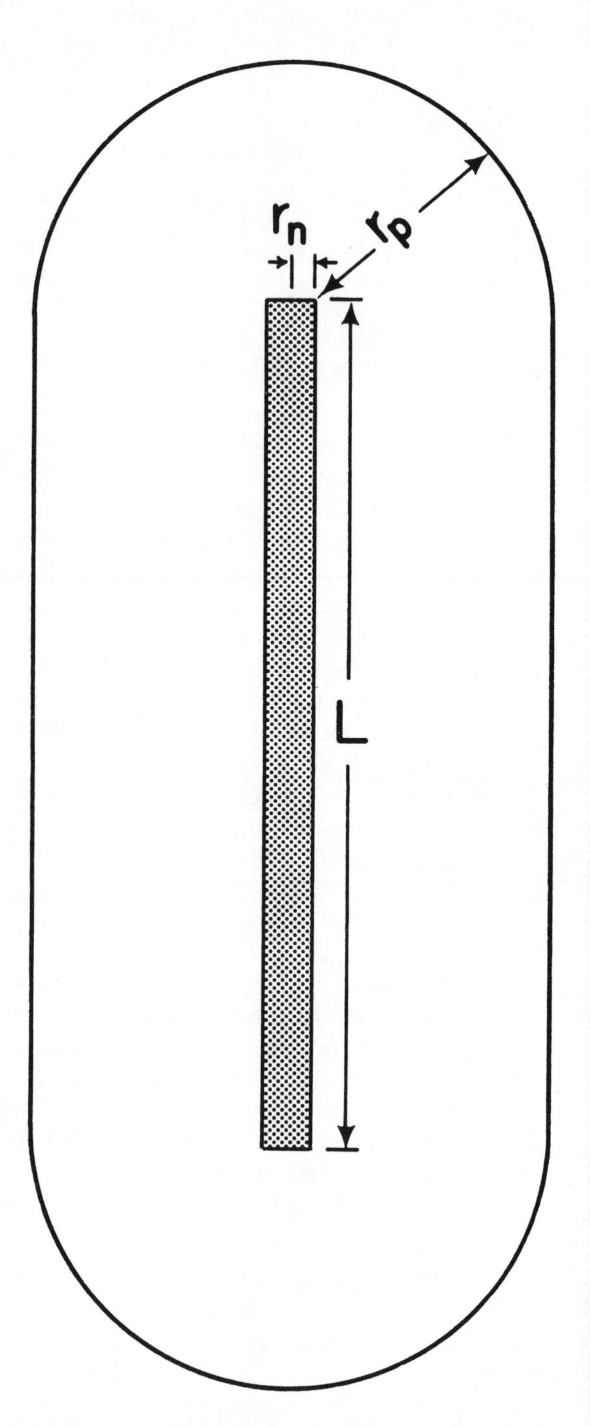

FIG. 2. Diagrammatic illustration of the volume of water effectively sampled by a net of radius r_n; r_p equals the mean patch radius; L equals the tow length. The stippled area represents the cross section of cylindrical volume filtered by the net; the open area, the cross section of the cylindrical volume effectively sampled by the net.

does enlarging the net diameter. Thus, the 25-cm-diam net towed 5,000 m and filtering only about 4% as much water should outperform the 400-cm-diam net towed 500 m, because the volume effectively sampled by the 25-cm net is much larger than that sampled by the 400-cm net (in the case of the 7.5-m patches, the smaller net would effectively sample 236,000 m^3 while the larger would sample only 52,000 m^3).

The evidence presented here suggests that increasing the distance the net is towed is of greater importance in making the sampling effective than increasing net size. It must be noted that this can only be assumed to apply to a volume in which the spatial structure is homogeneous (i.e., the size range of patches and their distribution remains relatively constant). However, in a population structured as in the model a point is reached where further increases in tow length do not provide enough improvement to merit the increase; the estimates derived from 2,000-m tows are substantially better than those from 500 m, but only slightly inferior to estimates from 5,000-m tows. If the spatial structure of zooplankton populations is like that used in the model, which was based on the results of the field study, then the best (in terms of cost in ship time vs. the resulting accuracy and precision) tow length for sampling plankton in the field would be on the order of 2,000 m. A cost function has not been developed but to gain a predicted increase of at least 50% in the accuracy and precision, the increased amount of ship time required to tow a net an additional 1,500 m may be negligible compared to the travel time between stations. If the patches were smaller, or the structure more aggregated, longer tow lengths would be required. The effects of having the patches themselves aggregated has not been studied in the model, and no quantitative information as to how much longer net tows should be extended is available.

The determination of the best (in the sense of cost vs. accuracy and precision) size of net is strongly dependent on the densities of individuals in a patch and their distribution. As densities decrease or the individuals tend toward aggregation in the patches, larger nets are required. The greatest improvement in the estimates of the model population occurred when the net diameter was changed from 25 to 100 cm. A much smaller improvement occurred by increasing the net diameter the same proportion again (to 400 cm). Thus, under the model conditions, the 100-cm net would be the best size to use. Although both the preliminary study and this work have shown that most of the observed differences between replicate tows could be accounted for by the spatial structure of the plankton, the possibility of avoidance of the net cannot be neglected. Barkley (1964), Fleminger and Clutter (1965), and McGowan and Fraundorf (1966), studying the effects of this source of error, have recommended the use of large nets over small ones to reduce avoidance. As the probability of capturing rare species that are randomly distributed is strictly a function of the volume of water filtered, the greater this is, the more confident an investigator can be that the absence of individuals of a rare species from the sample indicates absence from the station area. Increasing the net diameter is the most practical (in terms of the amount of ship time required) way to increase the volume of water filtered. The feasibility of launching and recovering large nets at sea ultimately limits the size that can be used, but nets with diameters up to 200 cm can usually be handled with comparative ease, and the largest possible should be used for routine plankton sampling.

The above statements should be qualified, since criteria for selection involve factors other than a simple striving to increase precision at a minimum cost. Research programs embody sampling questions that can be answered at a given level of precision and this level must be established before meaningful selection can take place. In addition, as samples become larger, the laboratory means of handling them become more complex. The errors associated with laboratory procedures should be considered along with the other factors when a net size and tow length are selected.

The estimates of error for single observations for the field and model experiments are not directly comparable. The method used to calculate the 95% limits in this study does not conform to the procedure introduced by Winsor and Clarke (1940) and in general use since then. This involves using logarithmic transformation to stabilize the variance. Analysis of variance was then applied to the transformed counts. The variance, s^2, of a single observation was obtained; the square root of the variance (the standard deviation, s) was multiplied by the t_{95} limit value (in most cases, 1.96). The antilog of this value and its reciprocal provide 95% limits of a single observation, x_i, as percentages of the observation:

$$\text{upper limit} = x_i \times 100 \times \text{antilog}\,(t_{95} \times s);$$

$$\text{lower limit} = x_i \times 100 \times 1/\text{antilog}\,(t_{95} \times s).$$

Most data sets from my experiments did not require the logarithmic transformation to stabilize the variance. In fact, confidence limits obtained by the Winsor and Clarke method were overestimates of the actual variability in those cases where the transformation was inappropriate. However, to illustrate the contention that the sampling error resulting from patch structure alone in the model is of the same order of magnitude as that observed in field studies, data from the smallest net towed 500 m were logarithmically transformed. The 95% limits, derived as described above, were 425.6 and 23.49% of an observation (i.e., two observations would have to differ by a factor of about 4 to be considered significantly different). As shown by Wiebe and Holland (1968), 95% limits on field samples involve factors of 2 to 5 and, infrequently, 10. Most of the sampling error observed in repeated net tows in the field could thus be due to the patchiness structure of the plankton.

The results of experiment 2 are consistent with those of experiment 1. Increases in net size and tow length did bring about improvements in the estimates of percent similarity, the most significant improvements occurring with extensions of tow length. The differences attributable to net sizes are not as great as in experiment 1 and this is almost certainly due to the greatly increased density of individuals in the patches. As indicated above, the rate of change of the precision estimates is a decreasing function of the numbers captured in a patch and the numbers of individuals captured goes up for all three nets with increases in patch density. Thus differences between them should have become smaller.

The values of percent similarity between observed and known distribution of individuals among species were in general higher when the species were inequitably distributed rather than equitably. This apparently is because most of the index value depends on the few abundant species, which are well sampled. Changes in estimates of the less abundant species change the index only slightly. It is, in fact, possible for a net tow to miss as many as six species and still yield an index value as high as 87%. This, however, only occurred regularly when the patches were smallest and the tow length shortest.

Except for the estimates obtained when the patches were 7.5 m/side, the percent similarities were quite high, usually about 80% (i.e., the "captures" conformed closely with what was known to be present). This raises the possibility that the percent similarity index calculated for adjacent samples using the most abundant forms (5–10 species) may be a more precise indicator of change over an area than estimates of numbers of individuals. Experiments in the ocean are required to substantiate this.

REFERENCES

Barkley, R. A. 1964. The theoretical effectiveness of towed net samplers as related to sampler size and swimming speed of organisms. J. Cons., Cons. Perm. Int. Explor. Mer **29**: 146–157.

Fleminger, A., and R. I. Clutter. 1965. Avoidance of towed nets by zooplankton. Limnol. Oceanogr. **10**: 96–104.

McGowan, J. A., and V. J. Fraundorf. 1966. The relationship between size of net used and estimates of zooplankton diversity. Limnol. Oceanogr. **11**: 456–469.

Taft, B. A. 1960. A statistical study of the estimation of abundance of sardine (*Sardionops caerulea*) eggs. Limnol. Oceanogr. **5**: 245–264.

Whittaker, R. H., and C. W. Fairbanks. 1958. A study of plankton copepod communities in the Columbia Basin, southeastern Washington. Ecology **39**: 46–65.

Wiebe, P. H. 1970. Small-scale spatial distribution in oceanic zooplankton. Limnol. Oceanogr. **15**: 205–217.

———, and W. R. Holland. 1968. Plankton patchiness: effects on repeated net tows. Limnol. Oceanogr. **13**: 315–321.

Winsor, C. P., and G. L. Clarke. 1940. A statistical study of variation in the catch of plankton nets. J. Mar. Res. **3**: 1–34.

9 The Sampling Problem in Benthic Ecology

Alan R. Longhurst

The importance of the role of benthic invertebrates in the bionomics of demersal fisheries is very considerable for it is almost entirely through their agency that the demersal fish are able to utilise as food the organic material in the deposits of the continental shelves (Longhurst, 1958a). Yet quantitative ecology of the benthos lags far behind that of the plankton, in a ratio that seems disparate with the bionomic—and certainly with the economic—importance of the two subjects; it is fairly clear that the reason for this relative neglect lies in the nature of the material and the difficulties inherent in an assessment even of an instantaneous standing crop. The benthic fauna is buried in, or attached to, a substratum of varying consistency into which the gear must make uniform bites; the individual organisms are arranged patchily, both on a large and on a small scale; they may be extremely divergent in size and so sparsely distributed that much time will be expended in obtaining sufficient numbers for statistical treatment. The comparisons between this and a plankton sample with quantitative gear are so obvious as to need no enumeration.

To overcome these difficulties a great variety of grabs and dredges have been designed since Petersen (1911) introduced the first quantitative gear; none appears to be entirely satisfactory, but the later modifications of the Petersen grab, such as those designed by van Veen (Thamdrup, 1933) or Smith and MacIntyre (1954) are probably the best instruments available for quantitative work—in time they will doubtless be replaced by a corer of some sort, but a satisfactory type has yet to be designed and proved at sea. With proper design of the bucket and with adequate weight good results may be had with a grab provided its shortcomings are recognised and the validity of the samples analysed.

Unfortunately, the sampling characteristics of grabs are not yet satisfactorily understood, nor have there been adequate comparative trials at sea between one grab and another, though Ursin (1954), Thamdrup, Smith & MacIntyre, and Birkett (1958) have produced preliminary data; Holme (1953) and Jones (1957) have recorded data on the cumulative curves for recruitment of species to samples using scoop samplers and van Veen grabs respectively. During a recent benthos survey on the West African shelf (Longhurst, 1958b) the sampling characteristics of van Veen and Smith grabs, each covering 0.1m^2, were investigated to determine the number of hauls needed to obtain a valid sample of the population at each benthos station; the results of these trials appear to be of some interest and are discussed here in relation to what is already known of the characteristics of grab samples.

Three statistical stations were worked; these were:—

A. in shallow water of the Sierra Leone estuary, on soft muddy sand, van Veen x 10.

B. off Sierra Leone in six fathoms, on dark grey shelly mud, Smith x 20.

C. off Sierra Leone in 10 fathoms, on shell-sand, Smith x 20.

At the first two stations each sampling unit consisted of a full grab and the instrument bit to its full depth on each occasion; when it did not, the haul was discarded.

Quantitative grab sampling is, in principle, very similar to the quadrat method of plant or soil ecologists, but differs in two important aspects; firstly, complete enumeration of the organisms is scarcely possible, since not all are on the surface of the soil, and secondly, the problems of randomisation of the actual quadrat sites do not obtain at sea where the only difficulty is to keep them close to each other, and in a constant relationship with a marker buoy.

The variation between hauls (sampling units) at each station (sample) has two sources—the patchiness of the fauna on the deposits and the inefficient replication of the sampling technique. Ursin (op. cit.) has shown that the increasing skill of deckhands in working the van Veen sampler is reflected in the rising efficiency of the gear, which depends for its efficiency on

Reprinted from *Proceedings of the New Zealand Ecological Society*, 6:8–12, 1959.

the force with which it is dropped on to the deposits, its alignment in relation to the deposit surface, and the manner in which it is broken free. However, in the shallow water in which the grab was used in this survey this factor is not of importance, for elimination of it in the Smith sampler by the incorporation of twin trigger plates and the helical springs which alone drive the bucket into the deposits did not, in fact, produce greater replication between sampling units. The variation observed in the statistical stations may be considered, then, to be due only to the irregularities in the distribution of the fauna.

There is normally less variation between sampling units for the values of the gross faunal indices — biomass, numbers of species and individuals per unit area—than there is between the numbers of individuals of each species, since the distributional patterns of individual species do not necessarily coincide; this is illustrated by the values for the coefficient of variation $\left(V = \frac{100\ S.D.}{\bar{X}}\right)$ in the following table.

Stations		A	B	C
Sampling Units (N)		10	20	20
Species/sample	— $\bar{X}$	19.70	12.60	17.20
	— V	8.93	19.83	20.43
Individuals/m²	— $\bar{X}$	140.60	32.60	40.60
	— V	34.42	27.85	14.60
Biomass (gm/m²)	— $\bar{X}$	2.71	5.16	2.01
	— V	29.80	39.50	42.64
Mean Value of V for faunal indices	—	24.38	29.06	25.89
Mean Value of V for ind/m² of individual species	—	67.03	155.29	192.30

TABLE 1.—*Coefficient of variation for faunal indices and individual species*

There is an indication in these results that there may be value in the faunal indices as a quick overall measure of the density of the fauna on a particular ground, although it must be recognised that faunas with widely divergent ecology and constituent species may give similar values for biomass, or species and individuals per unit area.

From an analysis of the recruitment of species and individuals to the sample total at each successive sampling unit it is possible to derive some information both about the composition and distribution of the fauna and about the sampling characteristics of the gear being used. From the results of the statistical stations it is possible to take this analysis a little further than those of Holme and Jones.

The population structure at each station was composed—in accordance with expectation — of a very small number of exceedingly abundant species and a very much larger number of increasingly rare ones; the categories containing the greatest number of species were those with one, or at most two, occurrences per station (Fig. 1). The species are clearly arranged according to the principles investigated by Williams (1949) and others, forming a modified logarithmic series in relative abundance. In each case a very few species constitute a very high proportion of the total number of individuals in the sample:—

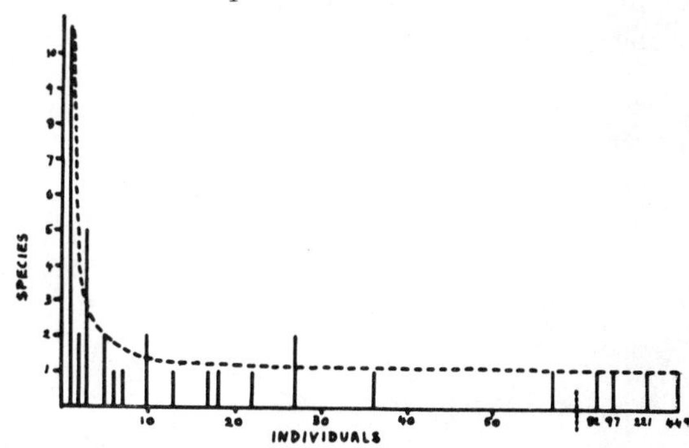

FIGURE 1.—*The relationship between the numbers of species and individuals at Station A.*

A — 4.3% (2) of the species included 64.6% of the individuals.

B — 6.4% (3) of the species included 58.0% of the individuals.

C — 11.0% (8) of the species included 50% of the individuals.

At each station the repetition of sampling units continued to recruit species to the total throughout the sample; this recruitment may be expressed as a cumulative curve which is exponential in form and is similar to those of Holme and Jones; the form of the curve is related in two ways to the total number of species in the sample.

(a) The number of sampling units at which the asymptote, for practical purposes, is reached is positively correlated with the number of species in the sample total.

(b) Similarly, the absolute value of the zero point of the curve is dependent on the mean number of species per sampling unit, and hence on the sample total.

Consequently, the percentage of the sample total which is attained by any given sampling unit is also dependent on the absolute total—

Percentage recruited by haul —	5	10	15	20
A (47 spp. 20 hauls) —	68%	81%	96%	100%
B (73 spp. 20 hauls) —	52%	72%	90%	100%

—and thus the validity of the samples after a given number of hauls depends upon the absolute sample total.

The species structure of the population and the relationship between abundant and rare species leads to two suppositions about the sampling technique. Firstly, that the abundant species may be adequately sampled well before the asymptote of the recruitment curve is reached, and secondly, that the sampling of the rarer species may not give a valid distribution per unit area even when the asymptote *is* reached. In other words, the value of the cumulative recruitment curve as used by Jones and Holme to define the minimal area for the sample is to be doubted.

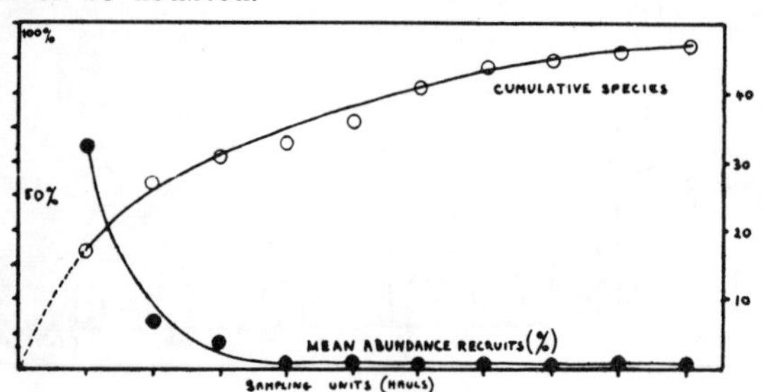

FIGURE 2.—*Decline in relative numbers of recruits during sampling at Station A.*

The first of the above suppositions is supported by the data presented in Fig. 2, which demonstrates for station A that the relative abundance of the recruits at each haul falls very steeply, and that the abundant animals are well represented in the first few hauls. That these are sampled adequately in the first five hauls is indicated by the lack of significant differences between the means (for individuals unit area) for the whole sample and for the first five sampling units; nor is there any trend in the cumulative means for such species plotted over the whole range of the sample. The second supposition is sufficiently obviously true as to need no elaboration.

It should now be possible to assess the validity of a standard 5-haul station, covering 0.5m^2, used in the subsequent faunistic survey; this area was chosen for practical reasons—the sea time of a research vessel being very costly—but proved, in fact, to be a fairly logical and satisfactory choice. From such a sample may be derived adequate information about the density and distribution of the abundant species and an enumeration of about 50–70% of the species which would be taken in a sample four times the size; included among these will be almost all the species which would occur more than once or twice in the larger samples. The faunal indices also appear to be adequately estimated by the 5-haul samples.

The criticism remains that the 5-haul sample, though valid in terms of larger samples with the same gear is not in fact truly representative of the fauna. As Birkett (1958) shows, a grab may miss deeply burrowing organisms entirely and—because of the semi-circular section of its bite—sample differentially organisms which are buried at different depths. Birkett suggests that a critical volume of soil in a sample is necessary before any particular organism can be considered to be sampled satisfactorily, the critical volume corresponding with a bite deep enough to sample effectively the complete stratum in which the organisms are buried. These criticisms were tested, to a limited extent, at stations B and C by taking a long haul in the sampling area with an efficient toothed dredge capable of digging more deeply than the grab; at B only two species occurred which had not previously been taken in the grab, and at C a further six. In the first case, the deposits being soft, the dredge dug deeply and the recruits were burrowing forms—a holothurian and a polychaete; in the second case the ground was harder, the dredge dragged more superficially, and with the exception of a single lamellibranch (*Psammobia*), the recruits were large, semi-mobile forms (*Ethusa, Natica, Strombus, Cymbium, Luidia*). It is evident, then, that a haul with a runner-dredge to sample the large epifaunal organisms at each station, would be of undoubted value in any faunistic survey; this does not, of course, dispose of the differential depth factor in grab sampling which can probably be solved partially by measurement of the soil volume in each haul and the application of factors to correct for the effect of different depth of bite in full or part-full grabs. It is this error which affords the greatest support for the replacement of grabs with some form of coring device to take a parallel-sided bite.

The faunal indices are frequently employed in geographical comparisons of the benthic standing crop (e.g. Sparck, 1935) but there has been little uniformity in the way in which the indices have been derived from the results of grab samples. Both unit area and unit volume have been used, with emphasis on the former; Birkett, in view of the depth-differential effect of grab sampling has recommended the use of unit volume, but this appears to pose at least as many difficulties as it solves. A percentage of the species to be sampled will be found very superficially in the deposits and their numbers/unit volume will apparently decrease with increased depth of bite—the converse of the situation for deeply burrowing organisms which will apparently increase in numbers/unit area with increasing depth of bite; if the emphasis of interest is on an organism which occupies a stratum only just within the sampling power of the grab, as in the case of deeply burrowing fish-food lamellibranchs, then the unit volume method of presentation has much to recommend it, but in general survey work it should be used only with care and in such work the unit area method is still to be preferred.

In order to express densities/unit area from a sample which is generally smaller than the chosen unit the sample results are usually multiplied up to the standard; this appears to be valid except in the case of species/unit area and this index can only with difficulty be standardised. By extending recruitment curves, or plotting them logarithmically (Williams, 1950) it is possible to calculate the number of species in a larger unit area—but this is not possible if the hauls at each station are bulked. The most satisfactory expression of this index is in terms of species per standard $0.5m^2$ sample.

There is a very considerable divergence of opinion as to how the biomass of a sample should be expressed. Some express their results in terms of the total organic nitrogen in the benthos and determine this by direct ashing, or the application of factors to convert alcohol weight to dry weight. Others, less concerned with absolute than comparative values, have expressed biomas as wet, or fresh weight, or as alcohol weight, with or without a factor for the alcohol-soluble solids. Expressions of the standing crop of plankton have proved to be of great utility in regional comparisons; the values given by settled volume or by the Harvey phytoplankton pigment units are no less significant because of the complexity of converting them to absolute values; and it is clear that if such standardised techniques had been introduced early in benthos studies the many geographical surveys now published could usefully be compared with each other. As it is, such comparison is almost valueless, and there is no indication in recently published work in this field of any attempt at standardisation of methods.

It is thought that the alcohol-weight technique adopted for the West African survey could be useful under a diversity of conditions and might afford the basis for a standard method; it has the practical advantage that the specimens are not destroyed and are available for systematic study—an important consideration in an unexplored area; as only comparative figures were required, conversion factors for gut contents, calcareous skeleton, and alcohol solutes were not applied; it seems unlikely that these are of great precision, and it is clear that if the absolute values of organic material are required for productivity studies then a technique which involves ashing, after decalcification if necessary, must be used.

REFERENCES

BIRKETT, L. 1958: A basis for comparing grabs. *J. Conseil.* 23: 202-207.

HOLME, N. A., 1953: Biomass of the bottom fauna of the English Channel off Plymouth. *J. Mar. Biol. Assn. U.K.* 32: 1-48.

JONES, N. S., 1957: Fauna and biomass of a muddy sand deposit off Port Erin. *J. Anim. Ecol.* 25: 217-252.

LONGHURST, A. R., 1958a: The food of the demersal fish of a West African estuary. *J. Anim. Ecol.* 26: 369-387.

LONGHURST, A. R., 1958b: An ecological survey of the West African marine benthos. *Fish. Pubs. London.* 15: 1-102.

PETERSEN, C. G. J., 1911: Valuation of the sea. 1. Animal life of the sea, its food and quantity. *Rep. Dan. Biol. Stn.* 20: 1-79.

SMITH, W. and MACINTYRE, A. D., 1954: A spring-loaded bottom sampler. *J. Mar. Biol. Assn. U.K.* 33: 257-265.

SPARCK, R., 1935: On the importance of quantitative investigations on the bottom fauna in marine biology. *J. Conseil.* 10: 3-19.

THAMDRUP, H. M., 1933: Der van Veen bodengriefer. *J. Conseil.* 13: 206-213.

URSIN, E., 1954: Efficiency of marine bottom samplers of the van Veen and Petersen types. *Medd. Dansk. Fisk.-og Havund.* 1: 1-7.

WILLIAMS, C. B., 1944: Some applications of the logarithmic series and the index of diversity to ecological problems. *J. Anim. Ecol.* 32: 1-44.

WILLIAMS, C. B., 1950: The application of the logarithmic series to the frequency of occurrence of plant species in quadrats. *J. Ecol.* 38: 107-138.

Part 3

POPULATION ECOLOGY

A principal goal of population ecology is to find out how environmental factors and biological characteristics intrinsic to the population interact to determine the number of organisms of each species found in any particular region. This goal is seldom realized in either terristrial or aquatic ecology. The data available are not sufficiently detailed, precise, extensive, or accurate, nor have the theories been sufficiently comprehensive or applicable. The analysis can be attempted in three ways: mathematical studies, experiments, and observations based on field sampling. The most fruitful work employs a combination of these methods to illuminate nature.

In the first, a mathematical model is used in which it is hoped the important biological forces and factors will be expressed not only simply, but also with enough accuracy so that manipulation of the model will permit new conclusions to be drawn. In the second, organisms are grown and investigated in the laboratory under simpler and more controlled conditions than in nature, the experimenter manipulating a small number of the possible sets of conditions to ascertain their effects and relative importance. In the last, either simple observations are made and reported, a method that can be regarded as merely anecdotal, or repeated observations of organisms under both similar and different conditions are made, the observer hoping that the information obtained under different and changing conditions will reveal how the populations are controlled. When all rational analysis fails, a multiple regression can be performed if the observer has no other ways to reduce the mass of data.

When mathematics is applied to a natural science, one attempts to form a mathematical model or picture which states the essential values. These relations can be further examined by manipulations and may respond by giving information and indicating additional relations which were not originally apparent. In exchange, a great portion of the data must be ignored. The art of establishing a successful model lies in choosing the elements truly essential in the relations.

Biological relations are usually so complex that most factors, whether environmental or intrinsic, must be ignored. The interpreter should always remain aware of the simplifying assumptions made, and should make these explicit to others. These assumptions limit the generality of the conclusions, but in return it is possible to examine a few of the presumably important relations when freed from other perturbations. Even the population biology of the simplest organism is complex and variable. The equations derived by Lotka (1925) and Volterra (1928) generated a great deal of research, but it is apparent that the simplifying assumptions have been ignored by many biologists who apply these models to field studies.

The models most often seen in population dynamics—or to employ Lotka's older and more accurate expression, population kinetics—are based on statements about the biological nature of the populations rather than derived from data about a particular organism. They are, therefore, *a priori* models rather than ones established *a posteriori*. Such a model is the one for exponential growth, or for logistic population growth.

A useful informal verbal model used "biotic potential" and "environmental resistance" to explain population growth (Chapman, 1928):

population growth = biotic potential − environmental resistance

In the equation that relates population size (N) of an unchecked population at different times

$$N_t = N_o e^{rt}$$

it is tempting to regard the parameter r as biotic potential, especially since Lotka's use of the term "intrinsic rate of natural increase" implies that r is an internal characteristic of the population unaffected by the environment (Lotka, 1925). This does not bear close examination, however, for r can be measured only in real environments, using data presented in life tables and fecundity tables. Although the experimental procedures minimize environmental effects on survivorship, the l_x column, the effects of different diets or culturing conditions can obviously influence the values in the m_x column. If, following Andrewartha and Birch (1954), r is regarded as the maximum value, part of the difficulty disappears, but the problem of experimentally finding the maximum remains.

In another mathematical simplification, the logistic equation has been used to "explain" the S-shaped growth curve, and Gause (1934) has derived a meaning of environmental resistance from it. As a differential equation, the logistic can be expressed as:

$$\frac{dN}{dt} = rN\,\frac{K-N}{K}$$

where K is the number of organisms at the saturation point. Gause expressed environmental resistance as the quantity $\left[1 - \frac{K-N}{K}\right]$, which may take values between 0 and 1. As measured in his experiments, the estimated values ranged from 0.055 to 0.984. This does not, however, relate how the environment can change r except as crowding acts reflexly on the population, nor are other biological factors in the environment—predators, for example—accounted for. A fuller discussion of some simple population models is found in Smith (1952).

An example of the application of a simple population model to a problem in aquatic ecology is Hall's study of a population of *Daphnia* in a Michigan lake (this collection). A simple equation for population growth is taken as given, the parameters of the equation estimated from experimental data, and conclusions drawn about how environmental factors interact to control the population size. Exploited fish populations have led to more complex models from the large body of data available. Here an unusually efficient predator, man, preys upon a resource, the fishery. The researcher measures the effects of fishing on the fish, with the objective of determining a maximum sustained yield, this maximum often based on economic rather than biological factors. In an unexploited population, all additions to the stock by recruitment and growth are balanced by natural mortality due to disease, predation, and "natural" factors. The new predator, man, causes additional losses, and consequent changes in age composition, recruitment, growth, and mortality. A new steady state is reached in which the amount of exploitation is balanced by positive changes in the other factors. If, however, short-term benefits are pursued too assiduously, the new steady state may not be economically feasible.

LITERATURE CITED

Andrewartha, H. G., and Birch, L. C., 1954. The Distribution and Abundance of Animals. University of Chicago Press, Chicago. 782 pp.

Chapman, R. N. 1928. The quantitative analysis of environmental factors. Ecology, 9:111–122.

Gause, G. F. 1934. The Struggle for Existence. Williams and Wilkins, Baltimore. 163 pp.

Lotka, A. S. 1925. Elements of Physical Biology. Williams and Wilkins, Baltimore. 460 pp. Reprinted as Elements of Mathematical Biology by Dover Publications, New York, 1956.

Smith, F. E. 1952. Experimental methods in population dynamics: a critique. Ecology, 33:441–450.

10

THE BIODEMOGRAPHY OF AN INTERTIDAL SNAIL POPULATION

PETER W. FRANK
University of Oregon, Eugene, Oregon

Abstract. Numbers, deaths and growth rates were observed for three years in a delimited population of *Acmaea digitalis* whose members had been individually marked. The zone of vertical distribution of the species is determined by the behavior of these snails. In the fall and winter they ascend in the intertidal, and descend a smaller distance in the spring. This leads to a differential distribution of size classes. The oldest and largest animals are found highest in the intertidal zone. That the vertical movements are adaptive is suggested by death rates which are highest in the upper portions in summer, but rise in the lowest zone in winter. Probability of survival improves with age. Growth rate was remarkably consistent during the period of observation. It was fastest in the fall and winter months. During the month of July, and in part of June and August growth ceased, at least as measured by a change in shell length. Crowding decreased growth slightly but significantly. Growth rates in different areas suggest that time available for feeding may be a significant variable. Experimental manipulation of density indicates that emigration rates are density-dependent. The mechanism of density regulation is postulated as operating even at low overall densities because of the behavior pattern of the species.

The concept of population balance has a history of controversy (Nicholson 1933; Thompson 1939; Warren 1957). Although semantic difficulties are responsible for part of the problem, something important is at issue: whether and how commonly populations accumulate sufficiently so that they become limited by self-induced shortages in their resources. The wide common concern with the problem perhaps needs no documentation. Its pervasiveness may not be so evident, however, until one realizes that such disparate contributions as Varley's (1947) study of the knapweed gall fly, the comprehensive community analysis outlined by Elton and Miller (1954), MacArthur's (1960) and Preston's (1962) models of community structure and the short theoretical argument of Hairston, Smith and Slobodkin (1960) all bear directly on this point.

Although I was under no illusion that a single specifically oriented research effort could solve the general problem, work with the limpet *Acmaea digitalis* was begun with the rationale that empirical information gathered from natural populations was a requisite for progress towards a solution. This intertidal snail was thought to represent something of a limiting case, since it lives in what is generally conceded an extremely raw environment, yet occurs in seemingly dense populations. The apparent disadvantage that not all stages in the life cycle were accessible, since *Acmaea* has a planktonic larva, was regarded as a potential gain: a density-dependent birth rate was eliminated as a possible regulating factor, and the number of variables was thereby reduced.

At the outset certain limitations inherent in any feasible program were recognized. Except for cases where experimentation is possible, results represent correlations not clearly traceable to their causes. In cases where density can be manipulated, replication is still likely to be insufficient to assure the adequacy of controls. Sampling problems are bound to arise, since the rock surfaces on which the animals live are far from homogeneous. More fundamental than this is the question of the generality of the results: what does a limited though extensive set of observations and measurements, gathered over a relatively short time span and in a small portion of the species' range, signify regarding the performance of this and similar sorts of animals over their total area of distribution? These limitations apply not only to this, but to all work on natural populations conceived on a similar scale. They clearly imply that independent confirmation of significant conclusions is particularly important.

Reprinted from *Ecology*, 46:831–844, 1964.

Animals and Their Habitat

Acmaea digitalis is the commonest limpet of the upper intertidal zone in Oregon, and ranges from the Aleutians to Baja California. Sexes are separate, and spawning occurs at least through winter and spring (Fritchman 1961). The length of planktonic life is unknown. These snails are common on virtually all hard substrates in a vertical zone that extends upwards from median high tide for as much as two to five m, depending on the degree of exposure to spray. They feed on microscopic algae, mainly blue-greens and diatoms, rasped off the substrate. Even casual observations of the sorts of areas where *Acmaea digitalis* is abundant suggest that the limpet lives in an extremely variable and hazardous milieu. At various times the animals are exposed to mechanical stresses from waves and suspended sand, to salinities ranging from that of sea to that of rain water, to large and abrupt temperature fluctuations and to prolonged desiccation.

The only other kinds of animals consistently seen locally with *A. digitalis* are the limpet *A. paradigitalis* (Fritchman 1960) with a narrower vertical range; another snail, *Littorina scutulata,* and the barnacle *Balanus glandula.* At the extreme lower limit of the vertical distribution of *A. digitalis,* a third limpet, *A. pelta,* begins to appear. It is only below this level that animal diversity increases greatly.

The area chosen for our observations lies inside Coos Head along the southern shore and near the entrance of Coos Bay. It is about 0.7 km from the Oregon Institute of Marine Biology, which served as a base of operations. The rocky coast here is partly protected from storms, particularly in summer when major wave trains come from the northwest. In winter, storms sweep up the bay, which opens to the southwest, and there is considerably more effect from rough water. Waves and swells more than a meter high are typical of winter but absent in summer. The difference in seasonal wave action is accentuated by the local situation. Soft sandstone emerges from a sandy beach whose level, during the period of these observations, fluctuated seasonally by about 1.5 m (Fig. 1). In summer the sand level is high, and the force of existing waves is largely broken even at high tide. In winter there is an absence of spray only at the lower tides; when tides are high much of the rock then receives the impact of relatively large waves. During the three years of our study, the beach sand gradually eroded around the base of the rocks in this area in October, and stayed low through March, when gradual accretion lasting into early May began. In 1963, presumably as a result of repair work on the south jetty of Coos Bay, the conditions of previous years were modified, and the winter situation prevailed for that summer.

Although other similar areas were used for subsidiary experiments, the main study site was a single rock that juts out from the main sea cliff (Fig. 1, 2). This soft sandstone rock is roughly conical and has habitable faces to the west, north and east. The cliff of which it is a part abuts to the south. At the two ends of the study area trickles of fresh water run down the rock surface virtually throughout the year. They become sur-

Fig. 1. The experimental rock. Left: typical summer conditions, June 1961. The white area at the upper right is a patch of bleached *Enteromorpha.* Right: typical winter conditions, January 1962. Note the stainless steel fences, which have loosened at this time. The holes in the rock reveal most clearly the differences in sand level.

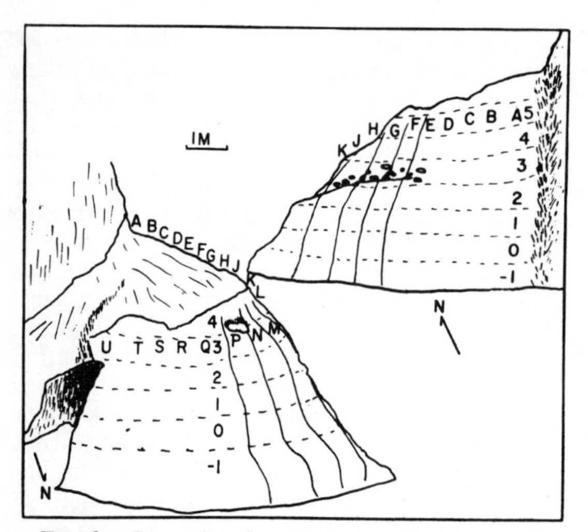

FIG. 2. Composite view of the experimental rock to indicate locations of designated squares, and of experimental fences.

rounded and overgrown by *Enteromorpha*. It was noted early in the preliminary phase of this investigation that limpets do not normally cross such areas, probably because of negative behavior similar to that described for *Patella vulgata* by Arnold (1957). We therefore selected this site and modified water flow by cutting channels into the rock, so that lateral emigration was prevented. The sand formed an effective lower barrier. Those parts of the rock lying within the vertical zone of distribution of *A. digitalis* are generally at an angle of 40 to 80 degrees from horizontal. Of considerable significance are the numerous small depressions and potholes eroded from the sandstone. Many of these are about 8 to 10 cm wide and 3 cm or more deep, and may hold small pools of water. In these areas and near other irregularities in the rock, aggregations of limpets very similar to those observed by Abe (1932) in *A. dorsuosa* often form.

The total area occupied by limpets varies somewhat seasonally, but encompasses something of the order of 40 to 45 m^2 on the experimental site. For convenience in locating animals, a grid of squares, approximately 65 cm along a side, was superimposed on the study area, so that there were 19 columns of such squares arranged in 5 to 7 rows, depending on precise location (Fig. 2). Since the experimental rock is roughly cone-shaped, the lower quadrats are actually a bit larger and the upper ones smaller than the specified size. Corners of these squares were permanently marked with stainless steel screws driven into the rock in 1961. Individual columns of squares are designated by letters A to U (omitting I and O), and levels from low to high are coded from -1 to $+5$. Mean high tide corresponds roughly to the top of row 1. Several gradients, caused by tidal, solar and local features, exist in the area. Aside from the obvious vertical zonation, a gradient of decreased wave exposure exists from column M towards both column A and U. The gradient of increased insolation is generally from U to A, although it is complicated by seasonal variation. The high cliff behind the rock cuts off a certain amount of mid-day sunlight in winter but not summer. Columns S to U receive fresh-water spray from above in the spring, and are generally moister than the rest of the area.

The seasonal moisture régime has a marked effect on algal cover. In summer when the rock remains dry above level 2 most of the time, only traces of algae, particularly bluegreens, can be found in rock scrapings, and the sandstone in the higher regions looks bare to the unaided eye. This is in contrast to a nearby cave area where, at the same level and with roughly comparable limpet densities, diatoms and other small algae form a dense cover. In winter when the experimental rock is continually moist, a heavy diatom blanket develops on it also, and larger algae gain a foothold. These include *Ectocarpus* sp., *Enteromorpha* sp., *Gelidium* sp., *Gigartina papillata, Polysiphonia* sp., *Porphyra perforata, Pterosiphonia robusta, Scytosiphon lomentaria* and *Ulva linza.*

OBSERVATIONAL AND EXPERIMENTAL PROCEDURES

Individual recognition of limpets was basic to many of our observations. Smaller snails could not be handled effectively, but all limpets more than 5 to 6 mm in shell length were marked. First they were carefully removed from the rock and placed on moistened sheets of flexible plastic. In the laboratory the postero-dorsal shell was smoothed with a grinding tool, and an adhesive tape tag with an India ink code number was applied to the shell. Numerals to 1,999 and combinations of numbers and letters were used to designate individual animals. The adhesive tape was then covered with two coats of a quick-drying plastic adhesive (composed of a mixture of Butvar B-76 kindly furnished by Shawinigan Resins, and an acrylic resin 'Dekophane' available from Rona Pearl Corp.). Shell length was measured and recorded before the limpets were replaced, usually at the following low tide or about 12 hours after they were removed. Marked animals received an additional coat of the adhesive *in situ* a year after they were originally labelled. We made no attempt to replace limpets in the exact spot from where they had come, but, except during the first year, replaced them at the same vertical level. The marking procedure leads to losses of limpets from various causes, but these can be kept to about five per cent of those marked. For purposes of census marked limpets were disregarded until they had been recovered on the experimental rock at least once. The immediate losses from tagging were compensated by the addition of a similar number of limpets from near by. From the summer of 1960 on, all limpets 6 mm or more were kept individually tagged through the period of study.

Our marking procedure is related to two major shortcomings in our data: 1) We cannot account in any precise fashion for animals until they have been on the rock for approximately two to three months. Even were it possible to mark younger animals, it is difficult or impossible to identify limpets this small to species. One component of possibly density-dependent mortality is thus removed from precise analysis. 2) Tag losses occur, especially in winter at the lower levels. We can estimate tag losses both from the size of animals and from tag remnants that are normally visible for a period of weeks. However, corrections for tag losses made in the data make conclusions regarding some of the differences in death and emigration rates more tenuous. Over the total period of study, comprising over four years, the total number of limpets of all species marked was about 9,500. Marking was done almost exclusively in June, July and August, when student help was most available. Other observations were also conducted most intensively then. During the rest of the year information was gathered every two weeks, except when occasional severe storms upset this schedule.

Although it may seem simple to estimate density and mortality, the only two parameters that, superficially at least, are significant to the problem posed, one of the major emphases of this report is that, because of heterogeneity of the population and of the habitat, the crude total estimates are almost valueless. The main body of data includes initial numbers and sizes of animals, and periodically repeated measurements of recruitment, disappearance, growth and spatial position. Every two weeks, and oftener in summer, we tried to measure the shell length of a sample of 50 to 200 animals and to census the entire population. To avoid time-consuming duplication during counts, small areas were delimited during census by marking them off with a paste of starch and chalk squeezed from a plastic bottle. Except in summer data were read into a portable tape recorder, since the time available during low tides was rather limited. Efficiency of censuses can be checked, and errors can be corrected by comparisons with subsequent censuses. The number of limpets found on a later census that were missed on a prior one indicates, to a first approximation, the efficiency of the earlier census. The process of approximating the true number of snails present at the time of a particular census can be further corrected for the efficiency of the subsequent census and for the estimated intervening deaths. Usually this correction is of minor dimensions and of uncertain validity.

When conditions were favorable, censuses were about 95% effective. When weather was really inclement efficiency dropped to as low as 70%. Unless they were recovered on a subsequent census, animals that disappeared from our records were assumed dead at the time of their first disappearance. It is highly probable that these animals had died. Reasonably diligent searching of all habitats in the adjoining area—a total of about 200 m of shoreline—revealed only three marked animals outside the experimental area during three years. All of these were on immediately adjoining rocks, where they could have crawled directly from the main area. We repeatedly tried to see what would happen to an animal on sand. The foot invariably became coated with sand grains, and the snail was incapable of moving on its own.

The intended procedure of our study was to gather data for two years under essentially undisturbed conditions to serve as partial controls for density manipulations in the third year. Accordingly, columns F, G, H, M, N and P were fenced off from each other and from the rest of the rock with stainless steel mesh in the summer of 1962 (Fig. 1, 2). Although earlier trials on a nearby rock had indicated that the fences would stay in place and would effectively prevent immigration and emigration, wave action in late fall weakened the experimental fences which had been effective to that time. We therefore removed most of the ineffective remnants in winter (Jan. 7, 1963).

Aside from these observations on the experimental rock proper, other information was gathered from additional areas. Samples of limpets were removed at two-week intervals throughout one year to check on reproductive condition and to determine size-specific biomass. Growth and settlement on a north facing cave wall of about 30 m^2 was assessed by removal of its limpet population and continuing observations of the newly settled individuals. Separate experiments on movements were done on a steep rock face using animals that had been mass marked with various quick-drying paints. Essential details concerning other procedures are best reserved for those sections that deal specifically with them.

TABLE I. Frequency distribution of limpets according to location, July 3, 1961

Row	Column																		
	A	B	C	D	E	F	G	H	J	K	L	M	N	P	Q	R	S	T	U
5	0	0	0	0	0	3	1	0	0	1	0	0	4	1	0	0	0	0	0
4	0	0	13	1	4	11	5	4	2	0	0	2	5	1	1	0	0	2	0
3	0	0	12	1	2	12	11	17	29	25	1	3	2	20	5	20	0	7	3
2	16	0	31	1	10	22	41	58	34	69	31	42	49	69	8	23	2	2	8
1	9	8	13	18	19	15	4	6	24	12	26	92	43	4	7	3	2	2	1
0	0	2	0	4	7	2	12	1	3	3	2	11	9	2	0	0	0	0	0

DISPERSION AND MOVEMENTS

Since this investigation centers about the problem of what limits animals to the zones occupied and the densities achieved, it becomes necessary to describe their spatial distribution. Here and elsewhere the data are too unwieldy to be easily summarized, and I must resort to representative sections of them. The gross pattern of dispersion can be readily illustrated by our quantitative data, as exemplified in Table I, in which numbers of limpets are listed by quadrat. The finer patterning of aggregations of a few to 30 or more limpets in a tight cluster, usually in small depressions or cracks in the rock is difficult to document quantitatively, as is the fact that one rarely finds one limpet on top of another, even where they are closely packed.

From Table I it is evident that the snails center about levels 1 and 2, and are generally more abundant in central than in peripheral columns. On closer inspection the table reveals some of the local differences for which limpet numbers can provide an index. For example, a cleft in column C characteristically contains more limpets than do the surrounding more exposed areas. Most of the heterogeneity illustrated in the table persists over the period of study. It may be worth noting at this point that the columns chosen for eventual density manipulation, G and P, were consistently similar to the adjoining columns that served as their controls. The difference in stratification between column P and columns M and N observable in the table is caused by a short term difference in distribution.

TABLE II. Vertical distribution of *Acmaea digitalis* by season

Level	% of total *Acmaea:*	
	July 3, 1961	Dec. 8, 1962
5	0.9	11.6
4	4.6	17.1
3	15.3	22.6
2	46.4	28.2
1	27.7	17.1
0	5.2	3.4
Total No.	1,113	1,284

Although the main density pattern persists, there are seasonal shifts, as indicated by Table II. Characteristically a greater proportion of animals are found higher in the winter than in summer. Distribution of limpets is not random with respect to size either. As is true of *Patella* (Lewis 1954), the largest *Acmaea digitalis* live highest in the zone occupied by the species (Table III). This is in apparent contradiction to Shotwell's (1950)

TABLE III. Frequency distribution of *Acmaea digitalis*, by size and vertical zone, Aug. 1961

Tidal level	Size class in mm.																					
	5	6	7	8	9	10	11	12	13	14	15	16	17	18	19	20	21	22	23	24	25	26
5	—	—	—	—	—	—	—	—	—	1	1	—	—	3	—	—	—	—	—	—	—	—
4	—	—	—	—	1	—	—	—	—	1	—	3	3	7	3	5	4	1	—	2	1	—
3	—	—	—	1	—	—	—	2	3	8	11	9	21	20	10	8	4	2	1	1	—	—
2	—	—	7	5	10	15	9	14	23	16	37	35	35	33	20	26	6	7	2	—	—	1
1	4	5	26	29	14	16	14	16	8	12	10	19	5	12	7	1	2	—	1	—	1	—
0	—	1	3	1	—	3	1	1	—	1	2	—	6	3	1	—	—	—	—	—	—	—

findings among other members of the genus. A statistically significant regression line may be fitted to the data: Y (tidal level) = 0.143X (size in mm) — 0.22. The relation between size and position is readily explained when it is examined as a function of time. From the censuses, data on successive positions of individual animals are available. By disregarding all except the vertical zone where limpets are found, and restricting our examination to animals one or more years old, it is possible to assemble the information contained in Table IV. Individual animals that have moved upward from the beginning to the end of the period are scored +1, 2 etc., depending on the number of squares moved. Descent is scored as a negative integer. Quantitatively the averages mean very little, since the data reveal a tremendous amount of variation. This only means that not all animals behave alike in any one time interval, and that, at a time when some individuals move up, others may remain at the same level or may even descend. However, with the number of observations available, the reality of the ascent in fall and winter and the lesser downward movement in spring is clearly revealed by the averages, which are five or more times their standard errors. In 1961 to 1962 the upward movement was much less pronounced than in the following fall. Unlike the autumnal rise, which was immediately noticed from the obvious change in position of many limpets, the spring movement in the opposite direction was not detected until the data were analyzed. As a net result of these movements, larger and older animals tend to live higher than do younger ones. Similar observations have been made in littorines by Cranwell and Moore (1938), and in limpets by Abe (1932) and Lewis (1954). Lewis' paper bears a very striking set of resemblances to our observations; the genera *Patella* and *Acmaea* show a number of beautiful convergent adaptations. The general phenomenon of a change in size distribution with position in the intertidal seems relatively common among intertidal animals.

TABLE IV. Average net movement of all animals more than one year old between summer of 1961 and 1962

Time interval	Average movement, No. of squares	No. of *Acmaea*	SD^2	SE
Aug.-Oct.	+0.158	479	0.28	0.019
Oct.-Apr.	+0.242	426	0.41	0.031
May-July	−0.184	255	0.34	0.037

In *Acmaea digitalis* the upward movement is presumably directly related to a behavioral response that anyone who has ever placed these animals in an aquarium will have noted. Within a few hours at most, these limpets will be found above the water surface. Similarly, as winter storms bring increased wave action on parts of the rock receiving only spray in summer, the limpets move upward. The downward movement in spring is not so readily related to a proximate cause. It may perhaps be likened to the presumed behavior of these animals in a hypothetical moisture gradient. At this time of year the upper parts of the rock become significantly drier.

Lateral movements tend to be smaller in extent than vertical ones. Except at the edges of the study site, no obvious barriers to lateral movement exist. Nevertheless the limpets tend to stay in the same or adjacent columns with a high degree of probability for long periods of time (Frank 1964). Especially in the upper parts of their zone of distribution these snails form aggregations of closely clustered individuals. Successive censuses may reveal some changes in relative position of the members of such a group, but suggest that few individuals move more than a few centimeters away for periods of several weeks or months. Strict homing does not occur. In summer those animals located high enough on the rock may be incapable of moving for long periods because the rock may remain dry. Under these conditions the foot of the snail becomes cemented to the underlying rock by dried mucus. Later, when the snail moves away or dies, a noticeable scar remains.

Two other observations are worth recording, since they may be of some interest to students of behavioral physiology. Several times we noted movements of several animals as in a group. The most striking instance was that of 16 animals, all of which were in square H_2 in November 1962. Following removal of the fence, they all appeared in square P_4 in January of the following year. Behavior such as this suggests that at times the limpets may be reacting rather simply to a single environmental stimulus. The second set of observations concerns the reactions of transplanted animals, and parallels more extensive observations of Segal (1956) with *Acmaea limatula*. In summer of 1961 we marked several groups of animals from high in their area of distribution with quick-drying paint to distinguish them from animals from lower in the intertidal, which were labelled with a contrasting color. Subgroups of both classes were repeatedly placed at high and low levels of a selected test area. The limpets that had occupied higher zones moved upwards rapidly as compared with those from lower zones, which tended to remain low. The converse result was not observed, probably because the rock higher up was not wet enough to permit crawling. We did not determine whether the differences are results of acclimation, as in *A. limatula,* or not.

One thus gains the general impression that these limpets move up and down as a result of a number of natural stimuli, but are otherwise conservative with respect to position. Although capable of moving more than a meter in a single tide period, such large scale movements are exceptional. Generally the area grazed by a single individual seems to be restricted to a few square centimeters in any one day. Although there is no return to the same spot, most animals apparently spend their lives, once they have settled, moving in an ambit that is no more than a meter wide. Unfortunately we were consistently unable to observe the limpets at high tide, the time when they feed and move most actively. There is therefore the possibility that effective density is reduced by the animals' dispersing temporarily. This certainly happens immediately around aggregations, but all our indications from records of positions of animals at successive low tides, sometimes for periods up to a month, suggest that grazing in far off areas followed by return to the same small area does not occur.

Numbers, Mortality and Settlement

Figure 3 summarizes census data. It describes total numbers of limpets on the experimental rock during three years and indicates death rates of *A. digitalis.* As may be seen, the greatest overall density existed at the outset. In June of 1960 there were a total of 3,841 limpets large enough to be marked. By June 1961 these had decreased to 1,092. The 1961 year class of recruits consisted of 1,215 newly marked animals, of which, however, many died before the end of summer. The 1962 year class was augmented by 400 limpets that were imported from another area. The graph of total numbers is based on all species of *Acmaea* present on the experimental rock. It includes a few *A. pelta* and the relatively common *A. paradigitalis.* In contrast, the death rates of Figure 3 apply specifically to *Acmaea digitalis.* The ascending portion of the numbers curve of successive year classes is based on rather crude estimates. Despite attempts at total census of small, unmarked limpets, our data on numbers in a new year class are not meaningful until June of the year. Total numbers were undoubtedly somewhat higher in the spring than represented on the graph.

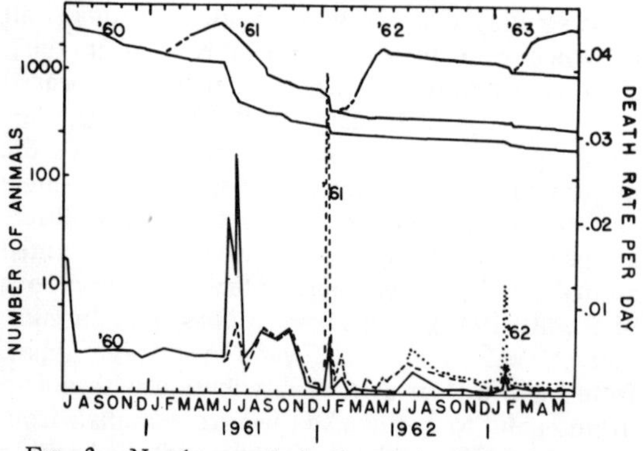

FIG. 3. Numbers and death rates of limpets on the experimental rock. The family of curves at the top indicate numbers of successive year classes as indicated. Note that numbers are graphed on a logarithmic scale. The curves near the bottom indicate instantaneous mortality rates. Those marked '60 apply for animals of all ages marked in 1960. The dashed and dotted lines refer to death rates of the 1961 and 1962 year classes respectively.

Inspection of that part of the figure that deals with mortality is probably more instructive. The indicated death rates were calculated from corrected census data according to the formula

$$d = \frac{\ln N_o - \ln N_t}{t}$$

where d is the instantaneous death rate, N_o the initial and N_t the final number of animals for a time interval t days long. Death rates are plotted separately for the total population of 1960, and for subsequent successive year classes.

For at least some of the deaths general causes can be assigned with some confidence. This applies particularly to the high death rates observed during the summers of 1960 and 1961 and during the winters 1961 and 1962. The two former years were characterized by relatively dry, sunny summers along the coast. As early as the latter part of June of both years, limpets could be seen that adhered to the substrate only by a narrow median piece of the foot, the rest having withdrawn and shrivelled. Some of these animals, when placed in fresh sea water, revived, but the majority did not. By the end of the summer of 1961 *A. digitalis,* especially on south-facing rocks and at the higher levels, had essentially disappeared all along this part of the Oregon coast, and shells of these limpets were washed up on the beaches at each tide.

In both summers a mouse was repeatedly seen feeding on limpets from the experimental site. It destroyed a sizable number, a dozen or so at a time, but removed only limpets so high up that they were already in an advanced stage of desiccation. It is improbable, considering the pull required to remove a healthy limpet, that the latter would be a source of food for mice. Predation by birds, observed only indirectly was only noted at this time. Twice empty shells were left on top of the site under conditions that made it obvious they had been eaten out by birds, probably gulls. The only other predator we have seen feeding on *A. digitalis* does not occur on the experimental rock, although it may be of some significance elsewhere. A polyclad flatworm, probably *Leptoplana,* was repeatedly seen feeding on limpets in nearby places during the late winter months. Often a trail of small limpet shells was left by the worms, some of which had ingested as many as five of these snails. The flatworms are limited to moist, shaded rock, and seem to be abundant among mussel beds. They rarely venture above the equivalent of zone 0 of the experimental area.

It is noteworthy that the deaths during 1961 struck young of the year only half as severely as they did the older year classes. This was probably true the previous summer, but cannot be documented for that period. The explanation lies in the previously described spatial distribution and in its relation to the moisture gradient: the older animals live higher and are thus more subject to desiccation. This is true despite the fact that they tolerate prolonged drying better than do smaller limpets.

The catastrophic mortality of the two winter periods is intimately associated with exceptionally severe frosts during these two years. The frosts seemed to have no direct effect on the limpets *per se*. A census the morning after temperatures fell to —7°C in the vicinity found the snails in good condition, although such temperatures are quite unusual in this area. However, the ice that formed in the top layers of the sandstone caused gradual but extensive exfoliation during the ensuing weeks. Animals attached to areas that exfoliated were carried off. For example, squares M_1 and M_2, which up to this time had had dense populations, lost their total populations in the winters of 1961 and 1962 respectively. Our notes show that in all areas limpet mortality correlates well during this period with the fraction of rock exfoliated. After removal of the top surface, the rock looks noticeably different and has no dense limpet populations for some months.

Aside from these incidents, a general, low level of mortality of about 0.001 to 0.004 per day prevailed. With rates this low, causation is hard to establish, nor is it necessarily particularly significant in regard to the major problem. The age specific differences in mortality are of some interest. Evidently young of the year are at a disadvantage in winter, whereas in summer relations are more variable and probably depend on the severity of drying. That the two freezes affected the young so much more than old animals may be of no general significance. It probably relates only to the nature of the particular rock surface. A survivorship curve based on our data, but disregarding the periods of obvious catastrophes thus would look concave, implying that the probability of survival increases with age at least for those ages for which sufficient data can be gathered. In view of the heterogeneity of the habitat and among the animals, I feel that specific survival probabilities given by such a curve have little meaning at this time. More useful is an analysis of mortality according to location on the experimental rock and to time of year. The data are insufficient to permit useful comparisons by individual quadrats. Too many quadrats lack animals at one time or another. The coarser grouping of Figure 4 illustrates the sort of information that can be gathered. There is a distinct contrast with respect to the time of year and location. The data for July 1961 indicate that very heavy mortality is then typical of the higher portions of the rock. In the fall there is excellent survival in equivalent regions. Little major upward movements had occurred in 1961 by this time, so that these data reflect deaths by vertical zones rather accurately. The bottom half of the figure illustrates in addi-

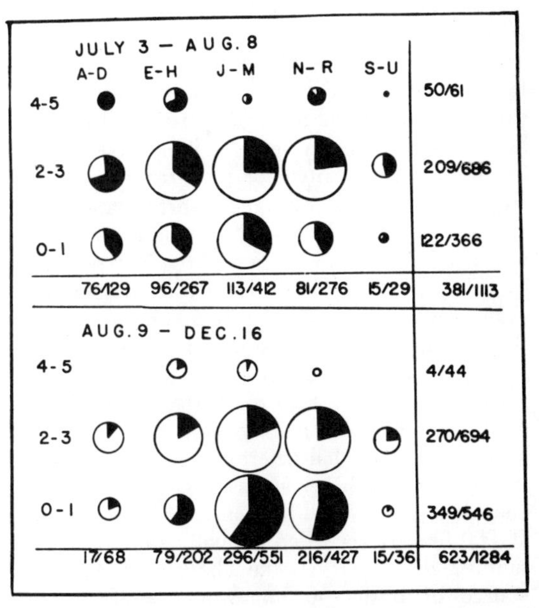

FIG. 4. Numbers and death rates during summer and fall 1961. The sizes of the circles are proportionate to the numbers of limpets in the quadrats indicated on the top and at left. The proportion of the area blackened indicates deaths occurring in the interval. The marginal totals indicate as the numerator the number of deaths out of a total number, given as denominator.

tion that probability of survival then decreases at the lower levels. A significant additional piece of information from a finer analysis is that animals at level 0 consistently, i.e. throughout the year, exhibit a higher death rate than those at level 1.

Except for the period from February 1 to 15 1962, the time of exfoliation, mortality experience similar to that of the lower part of Figure 4 prevails through the following winter and spring. One distinctive element is added: from February to June column group S to U has a higher death rate than does any other group of columns. This increase is correlated with, and undoubtedly caused by, the fresh water spray from the cliff above, which is then most intense.

Data of this sort can be subjected to a Chi square analysis to test additive effects of rows and columns and interaction. Significant differences in survival for rows and columns can usually be found. Moreover, after these effects have been eliminated (by statistical treatment), a statistically significant amount of heterogeneity often remains. One must be somewhat cautious with the data, since they have been corrected for census errors. In several instances, however, these effects are so large that they cannot be considered arithmetic artifacts. The residual heterogeneity was checked for possible correlation with density without significant results. So far these differences are interpretable as resulting from local attributes, but escape more searching analysis.

The view of this limpet population that one forms from these data on mortality and from the earlier section dealing with spatial distribution and movements bears some striking resemblances to the model presented for the grasshopper *Austroicetes cruciatus* by Andrewartha and Birch (1954). Quite evidently the snails are restricted to a relatively narrow region of the intertidal zone by behavior and by high death rates outside their even narrower and shifting optimum zone. Vertical zonation in this species of limpet has thus been "explained." Andrewartha and Birch argue further that, in the case of *Austroicetes* and generally, abundance is merely another aspect of the same phenomena that determine distribution. This is not the case here, as may be seen when one considers what would happen to our scheme were 10,000 limpets recruited annually rather than the number that are. In the absence of additional sources of deaths, no limit would be set to numbers. The population phenomena described for *A. digitalis* to this point give no evidence regarding the maximum densities likely to be achieved by these snails.

A priori, data on settlement are unlikely to be much help. If density of adults has an effect on settling rate of planktonic larvae, it is unlikely to be a density-regulating effect (Nicholson 1954), but rather may be density-disturbing (e.g. Knight-Jones 1951; Wilson 1952). A more likely place to look for negative feedback might be in animals too small to be marked, but already present on the rock. This group of animals may regulate numbers. At least this is a possibility that cannot be eliminated by any data we have. In an attempt to test the question we surveyed a variety of areas containing *Acmaea digitalis* in the summers of 1961 and 1962. Twelve vertical transects 50 cm to 1 m wide were stripped of limpets. Animals less than 6 mm long were separated from the larger ones, and numbers of these classes were then compared per unit area. The correlation coefficient between numbers of adults and young was positive ($r = +0.43$), but not statistically significant at the 95% level of confidence. The positive correlation may imply solely that an area that is suitable for dense populations of limpets one year is also likely to be attractive for settling in another. It certainly lends no credence to the hypothesis that density-dependent mortality of the young snails is an effective population regulator.

Settlement of small limpets could be detected on the rock through much of the year. There was a sharp peak in late April and early May. Spent gonads among animals from nearby rocks were first seen in March of the year. Gonads regressed during the summer months. Otherwise there was no time when a majority of animals in any local population was spawned out. Since it was impossible for us to identify newly settled limpets by species, we cannot be certain that settlement of *A. digitalis* ceased entirely in the fall. Virtually no limpets were recruited to the 6-mm size class during the early months of the year, however.

Growth

Estimates of growth were obtained primarily for the eventual determination of production rate. For present purposes they are suitable both as an index to the quality of environment and as an indicator of age. Since shell shape varies with growth rate (Orton 1933; Moore 1934), it is usually desirable to measure height of the shell. We were restricted to length measurements with vernier calipers, since otherwise the animals would have had to be removed from the rock. We have some evidence from dead animals that shell shape is not a significant variable in the experimental area, although it differs here from shape in some

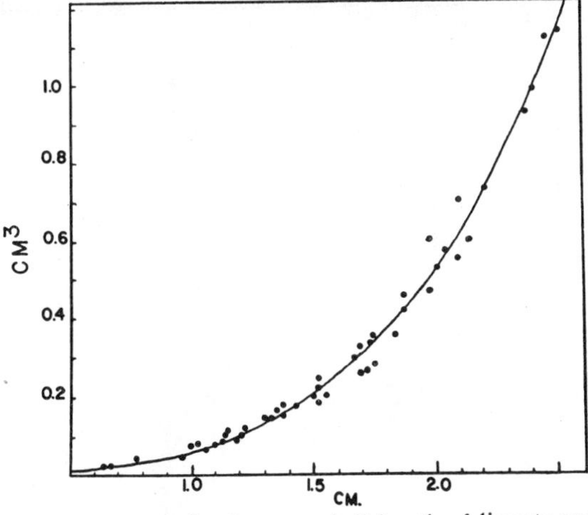

FIG. 5. The relation between shell length of limpets and their volume.

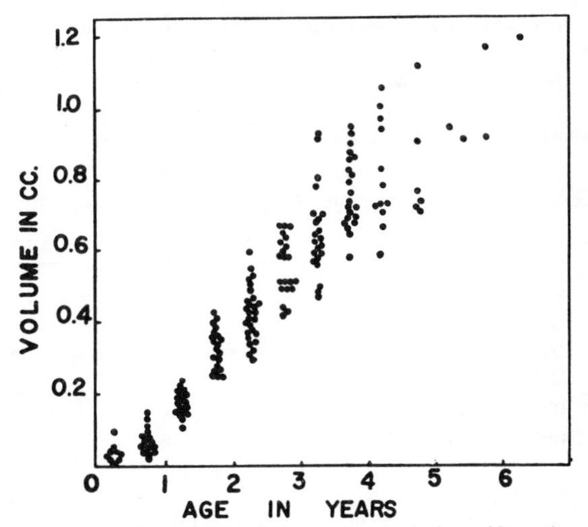

FIG. 6. The relation between age and size. Note that size is not a sufficiently good indicator of age to be used as a predictor. For animals over three years old, age is estimated by extrapolation. For animals less than four years, age and size were determined directly from marked animals.

other nearby areas. To convert shell length to more generally useful measures, we made estimates of the relation between length, volume and dry weight exclusive of the shell. To determine volume, the shells of 47 animals were filled with paraffin, and the casts were weighed. The weights so obtained, when divided by the specific gravity of the paraffin, yield the volumes of Figure 5. Dry weight estimates did not vary significantly between seasons or sexes, so that a simple conversion factor of 0.35 converts the volumes to dry weight.

Table V provides a summary of annual growth rates for June 1961 to 1962. The data are not separable by sex, but neither age specific size distributions nor a small number of growth data for sexed individuals suggest a significant difference in growth of the sexes. Growth rate is, however, quite variable, especially among the larger snails. From these data it is possible to construct the relation of age to size of Figure 6. For animals three years old or less the data are based on growth measurements for the entire period. For older animals an initial estimate of age was made on the basis of initial length.

With the exception of the summer months

TABLE V. A summary of size specific annual growth, 1961 to 1962

Initial size mm	Average yearly increase mm	No.	Coefficient of variation	SE
6	8.7	4	4.8	0.21
6- 8	8.0	46	21.6	0.26
8-10	7.6	44	17.6	0.20
10-12	5.6	7	33.2	0.70
12-14	4.1	18	21.5	0.21
14-16	3.5	29	28.2	0.19
16-18	2.8	34	52.2	0.25
18-20	1.8	24	64.3	0.23
20-22	1.8	12	57.7	0.27
22-24	0.8	4	48.0	0.18
24	0	1	—	—

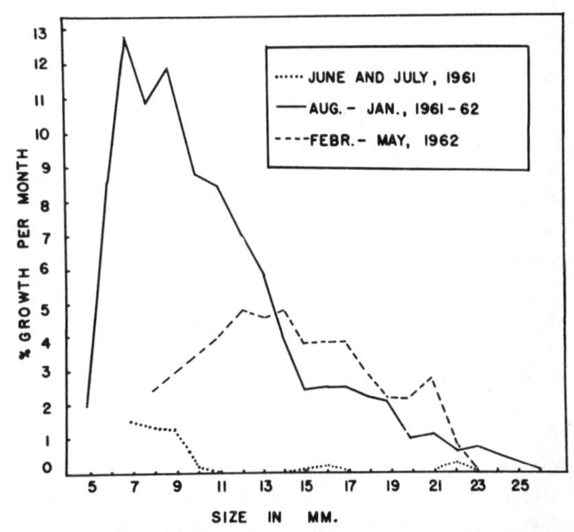

FIG. 7. Growth rate variations between seasons. Values for animals less than 8 mm in August and less than 12 mm in February are probably spurious (see text).

growth rates seem similar throughout the year. This information is conveyed, at least in part, by Figure 7, which is based on growth of 278 animals for the period June to July, 138 for February to May, and 233 for August to January. The graph is somewhat misleading. Snails less than 7 mm long in August or less than 11 mm in Februrary are probably atypical, i.e. are part of a sample selected unintentionally for low growth rate. Groupings by shorter time intervals yields no additional information, since numbers of animals become small, and the relative error of measurement becomes too high. Additional information regarding growth rates is best reserved for the section that deals with density manipulation.

Effects of Experiments

As a result of animals transplanted from one area of the experimental site to another and the addition of animals from elsewhere, columns G and P had significantly augmented numbers by August 10, 1962: column G had 636 limpets, of which 488 were *A. digitalis;* column P contained 475 snails, 393 of them *A. digitalis*. Total volume of limpets (exclusive of shell) in column G is estimated at 274 cc, and for column P at 180 cc. The area actually occupied by the limpets is difficult to estimate in any meaningful terms, but on the basis of where they were at that time, we have assigned an area of 2.8 m^2 to column G and of 2.1 m^2 to column P. This is equivalent to saying that average density per m^2 was 0.98 and 0.86 cc per m^2 respectively. Such an estimate seriously underrates the density in the more crowded middle squares. Although the fences that were to have kept the limpets confined to these columns disappeared or were removed in January, densities in these and in adjoining columns continued to reflect the crowding imposed the previous summer. Thus it is of some interest to determine whether an effect of crowding on growth rate is detectable. Comparison of growth by sub-areas for this year with growth over the area as a whole during the previous two years (Fig. 8) shows most evidently the consistent response of the population in this environment. The data may be subjected to an analysis of covariance, which is simplified by a logarithmic transformation of growth. The analysis supplies proper weighting to the means of the figure. Despite the similarity of the individual curves, the analysis leads to the conclusion that mean growth, adjusted for initial size, differs between them ($F = 9.28$; 99% confidence level is $F_{4,567} = 3.35$). Specifically the heavier line in the figure, which describes growth in columns F, G and H indicates a lower average. The other means do not differ significantly from each other. Since lability of growth rate in limpets is generally recognized and is undoubtedly real, the minor effect on growth was surprising. Fischer-Piette (1948) has supplied good evidence on large differences in growth rates of *Patella* in different environments, and we have observations on growth of *A. digitalis* in a shaded cave that also indicate that much greater variation is potentially possible. Apparently, however, conditions on the experimental site were such that increased crowding did not reduce growth rate significantly in one area (column P), and did so in the other (column G) only slightly. This is true despite the fact that at this time other areas on the rock had densities lower than any experienced since the start of the experimental period.

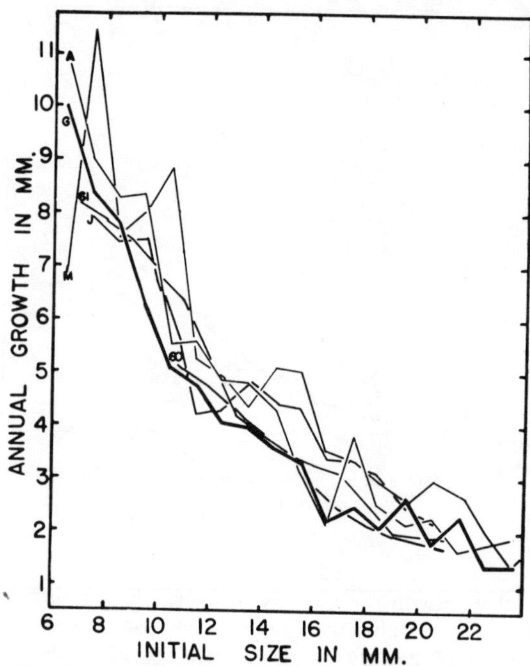

FIG. 8. Growth rate comparisons between years and between groups of columns. (A is growth rate of columns A to E; G applies to F to H; J to J to L; M to M to P; 60 is average growth for all animals measured for 1960-1961; 61 is the same information for the following year).

Because death rates in different columns and rows show so much heterogeneity, it is probably not legitimate to compare mortality except with adjoining squares that did not exhibit significant differences in previous years. Such comparisons must also be restricted to the period when the fences were effective, since after that time it is too difficult to assign death rates to the appropriate column. For column G, 437 of the original 488 limpets survived from August to December, as against 172 of an original 206 for columns F and H. For column P the equivalent values are 339 of 393, as against 282 of 321 for columns M and N combined. The biggest difference is one in the direction of increased survival in column G. However, none of the differences is significant when subjected to the appropriate Chi square test.

During this period there is a striking effect on spatial distribution. It is perhaps most clearly indicated by the fact that limpets for the only time during the period of study moved onto the more or less flat top of the experimental rock in column P (Fig. 2), that is, to an area corresponding to square P_5. The same is true of column G, where the distribution on December 7 is that shown in Table VI. It is quite evident that, either

TABLE VI. Positions of limpets on Dec. 7, 1962 in crowded and in adjacent, uncrowded squares

Vertical zone	Crowded squares G		Uncrowded squares F and H	
	No.	%	No.	%
5	51	11.7	3	1.7
4	88	20.1	14	8.1
3	134	30.7	84	48.8
2	97	22.2	55	32.0
1	49	11.2	11	6.4
0	18	4.1	5	2.9
TotalNo.	437		172	

as a result of the density increase or because of the addition of animals from elsewhere, the crowded quadrats had disproportionately more limpets in the highest and lowest quadrats. The movements thus exhibit a sort of density dependence that, over the year, presumably could effect control. Admittedly, because of the failure of the fences we are unable to prove this point conclusively. Curiously, moreover, tests for differences in rate of lateral dispersion after the fences were removed indicated no significant differences between rates of movement from crowded and from adjacent, less densely populated squares. This may be because density has already been adjusted downward by the vertical movements, and also because the animals may have established home ranges (Frank 1964).

Since these and earlier observations suggested space as important to the snail populations, a number of smaller scale tests of environmental features were made in the summer of 1963. Among these were two attempts to improve the environment: 1. two columns of the experimental site were continually kept moist by a trickle of seawater from a carboy placed on top of the rock and refilled daily; 2. small artificial shelters from insolation were added. These were made of slit inner tubes, anchored to the rock so that they formed a roof but were open at two ends. The effects of the first test were disappointing. Although some increase in algal growth was observed, wave splash this summer was heavier than in previous years. No extensive period of desiccation was thus experienced anywhere, and effects on death rate could not be established. Limpets did not actively aggregate in the moister areas; nor did the artificial shelters, located at levels 3 to 5, ever become significant centers of aggregation, although they usually harbored a few individuals. Apparently such areas do not serve as attractants over any distance, but rather allow limpets to remain when they reach the immediate vicinity. Presumably the snails cease to move in these darker or moister areas rather than actively seeking them out.

DISCUSSION

Our results emphasize the role that behavior plays in determining the characteristics of an animal population. In this regard it complements and parallels studies by Lewis (1954) and by Evans (1951) rather specifically. The vertical level at which *A. digitalis* is found seems determined, at least in large part, by behavioral responses of the animals to their environment. That these responses are correlated with environmental features, i.e. are adaptive, is to be expected. Had we not found increased death rates outside the zone to which the limpets move, their movements would have been puzzling. The ultimate reasons for the upper limit are undoubtedly the physiological possibilities of the animals to tolerate desiccation. The lower limit is not so easily explained. Presumably the behavior of the limpet which determines this lower limit has evolved as a response to the increased predation (Connell 1961) or abrasion at lower levels. Potential competition with other species cannot be ruled out either, nor is there any reason why several factors may not operate together.

Behavior evidently is also involved in regulating abundance, probably in an interaction with space and food. As long as their preferred habitat is not saturated, limpets tend to remain where they are. We have a number of records of limpets that moved no more than a centimeter or two for periods of up to two weeks, when it would have been possible for them to move much larger distances. Rather than leaving, limpets tend to remain in the same small crack or cavity for prolonged periods. Presumably they begin to graze whenever local moisture conditions permit, and cease when they receive a set of signals that suggest that conditions locally are deteriorating. As an area of rock progressively dries, animals tend to be left in the moister spots, where movement finally stops. Since total contact of the foot with the substrate seems to be a necessary condition for normal inactivity, an otherwise favorable area which has a dense aggregation of limpets thus establishes its own upper limit. In a sense, these limpets behave similarly to the titmice studied by Kluyver and Tinbergen (1953), which established constant populations in the optimum habitats and had more variable numbers elsewhere.

Although food in this situation is unlikely to become absolutely limiting, relative scarcity (Andrewartha and Browning 1961) may be a common condition. The concentration of food depends on the production rate and on grazing. The highest areas have limited algal production, and opportunities for grazing there are also reduced. Elsewhere the amount of time available for grazing may determine its intensity. Time of activity as well as limpet numbers will determine how often a particular spot is covered. Limpets will not normally remain in the driest areas, as shown by the downward migration in spring. In other areas, time available for feeding rather than the existing food concentration may be most significant. This is suggested because the great increase in density in columns G and P had no greater effect on growth rate than it did, and because in other areas, e.g. a cave, growth is significantly faster. It is interesting to note that, where similar observations have been made among other limpets (Fischer-Piette 1948), high growth rates are associated with a shortened life span. This seems true in our animals also, and may be of some general importance. The relation of the phenomenon to adaptation and to density regulation is not at all clear, however.

A relative shortage of food resulting from its low concentration may sometimes operate. Castenholz (1961) estimates that a limpet volume of 0.8 cc per dm^2 can keep an area of high productivity free of algal mat. Since his limpet volumes are based on total displacement, they are about 20% higher, for the same biomass, than are ours. In any case algal mats should have been absent from the crowded columns, G and P, no matter which estimate is used. It is true that Castenholz's area was not equivalent, and that his observations stem from the summer and early fall primarily. Nevertheless the fact that during the fall of 1962 there were parts of the crowded quadrats that did have a visible cover of algae seems discordant. There is, however, no real discrepancy between our data and those of Castenholz. Limpets undoubtedly kept the main parts of our quadrats clear of algal mats. However, the habitat is heterogeneous, and as a result densities per m^2 or dm^2 are virtually meaningless, especially since the centers of activity of groups of limpets may shift. Our observations do indicate, however, that areas exist where food concentration is high enough that food is not limiting, even when limpet densities in the general vicinity are high. It is necessary to add that neither our data nor those of Castenholz explain the commonly observed algal mats seen in the upper intertidal in the winter months.

When our limpets were experimentally crowded, the effects on emigration rate were obvious, particularly since lateral movements were restricted. Although we were unable to demonstrate at that time that there were accompanying changes in death rates, these are clearly implied by other parts of our data, and their reality seems assured. Movements therefore can effectively regulate density. If a lower threshold exists below which density plays no part, it is likely to be very low, certainly lower than the existing density in areas normally considered reasonably good habitat for the animals. In a complex and shifting system of gradients, the limpets continually act to acclimate in an appropriate manner. This means that, even when overall densities are low, local crowding may occur. At times of catastrophic mortality the effects of such crowding may be obscured quite effectively. Ultimately, however, density regulation according to this scheme should be regarded as operating at virtually all densities, with increasingly severe, but never spectacular effects as progressively poorer habitats are invaded. This view, complex though it is, is undoubtedly an oversimplification. Different animals will possess different individual optima because of differences in age, history or genome. These individual differences will buffer the system of regulation.

Acknowledgments

Especially during the initial phases of this research, J. H. Connell provided much valuable advice; he and R. W. Morris were kind enough to criticize the manuscript. Discussions with R. W. Castenholz, H. K. Fritchman, A. J. Kohn and J. A. Shotwell were helpful. Several graduate students helped gather the data, and I am happy to be able to thank them here: David C. Coleman, Richard L. Darby and Vernie Fraundorf. The intensive observations and marking regimes pursued during summers would have been impossible without additional help, supplied largely by undergraduates, many of whom were supported by an undergraduate research participation program of the National Science Foundation: Harold Brown, Kenneth Graham, Jon Jacklet, Trudy Kofford, Roberta Langford, Robin Manela, Marjorie McFarland, Stephen Mohler,

John Palmer and Charlotte Patterson. I trust that they profited from the experience as much as they benefited the study. Expenses of the investigation were defrayed by National Science Foundation Grant G 11-3331, for which I am grateful.

Literature Cited

Abe, Noboru. 1932. The colony of the limpet (*Acmaea dorsuosa* Gould). Tohoku Imper. Univ. Sci. Reports, 4th Ser. (Biol.) **7:** 169-187.

Andrewartha, H. G., and L. C. Birch. 1954. The distribution and abundance of animals. University of Chicago Press. p. 782.

———, **and T. O. Browning.** 1961. An analysis of the idea of 'resources' in animal ecology. J. Theoret. Biol. **1:** 83-97.

Arnold, D. C. 1957. The response of the limpet *Patella vulgata* L. to water of different salinities. J. Mar. Biol. Assoc. U.K. **36:** 121-128.

Castenholz, R. W. 1961. The effect of grazing on marine littoral diatom populations. Ecology **42:** 783-794.

Connell, J. H. 1961. The influence of interspecific competition and other factors on the distribution of the barnacle *Chthamalus stellatus.* Ecology **42:** 710-723.

Cranwell, L. M., and L. B. Moore. 1938. Intertidal communities of the Poor Knights Island, New Zealand. Trans. Roy. Soc. New Zealand **67:** 374-407.

Elton, C., and R. S. Miller. 1954. An ecological survey of animal communities: with a practical system of classifying habitats by structural characters. J. Ecol. **42:** 460-496.

Evans, F. G. C. 1951. An analysis of the behaviour of *Lepidochitona cinereus* in response to certain physical factors of the environment. J. Anim. Ecol. **20:** 1-10.

Fischer-Piette, E. 1948. Sur les éléments de prospérité des patelles et sur leur spécificité. J. de Conchyliol. **88:** 45-96.

Frank, P. W. 1964. On home range of limpets. Am. Naturalist **98:** 99-104.

Fritchman, H. K. II. 1960. *Acmaea paradigitalis* sp. nov. (Acmaeidae, Gastropoda) Veliger **2:** 53-57.

———. 1961. A study of the reproductive cycle of California Acmaeidae (Gastropoda). Part III. Veliger **4:** 41-47.

Hairston, N., F. E. Smith and L. B. Slobodkin. 1960. Community structure, population control and competition. Am. Naturalist **94:** 421-425.

Kluyver, H. N., and L. Tinbergen. 1953. Territory and the regulation of density in titmice. Arch. neerland. de Zool. **10:** 265-289.

Knight-Jones, E. W. 1951. Gregariousness and some other aspects of the settling behaviour of *Spirorbis.* J. Mar. Biol. Assoc. U.K. **30:** 201-222.

Lewis, J. R. 1954. Observations on a high-level population of limpets. J. Anim. Ecol. **23:** 85-100.

MacArthur, R. 1960. On the relative abundance of species. Am. Naturalist **94:** 25-36.

Moore, H. B. 1934. The relation of shell growth to environment in *Patella vulgata.* Proc. Malacol. Soc. London **21:** 217-222.

Nicholson, A. J. 1933. The balance of animal populations. J. Anim. Ecol. **2,** Suppl.: 132-178.

———. 1954. An outline of the dynamics of animal populations. Austral. J. Zool. **2:** 9-65.

Orton, J. H. 1933. Studies on the relation between organism and environment. Proc. Liverpool Biol. Soc. **46:** 1-16.

Preston, F. W. 1962. The canonical distribution of commonness and rarity. Ecology **43:** 185-215; 410-432.

Segal, Earl. 1956. Microgeographic variation as thermal acclimation in an intertidal mollusc. Biol. Bull, **111:** 129-152.

Shotwell, J. A. 1950. Distribution of volume and relative linear measurements changes in *Acmaea,* the limpet. Ecology **31:** 51-62.

Skellam, J. G. 1951. Random dispersal in theoretical populations. Biometrika **38:** 196-218.

Thompson, W. R. 1939. Biological control and the theories of the interactions of populations. Parasitology **31:** 299-388.

Varley, G. C. 1947. The natural control of population balance in the knapweed gall-fly (*Urophora jaceana*). J. Anim. Ecol. **16:** 139-187.

Warren, K. B., Editor. 1957. Population studies: Animal ecology and demography. Cold Spring Harbor Symp. Quant. Biol. **22:** P. 437.

Wilson, D. P. 1952. The influence of the nature of the substratum on the metamorphosis of the larvae of marine animals, especially the larvae of *Ophelia bicornis* Savigny. Ann. Inst. Oceanogr. Monaco **27:** 49-156.

11 AN EXPERIMENTAL APPROACH TO THE DYNAMICS OF A NATURAL POPULATION OF *DAPHNIA GALEATA MENDOTAE*

DONALD JAMES HALL

Department of Entomology and Limnology, Cornell University, Ithaca, New York

INTRODUCTION

In general, populations have been studied either in the laboratory under experimental conditions with environmental variables controlled, or in their natural habitat where variables are uncontrolled. In laboratory investigations the analyses are often powerful and yield knowledge of the fundamentals of population growth but are limited to specific conditions seldom found in nature. In field studies, analyses are limited to correlations of population phenomena with environmental variables and frequently involve large errors in estimates of the inferred rates.

The apparent paradox of many field and laboratory population studies can be reconciled in part by manipulating laboratory populations in such a manner that information appropriate to an analysis of the natural population is obtained. To manipulate experimental populations properly, some a priori knowledge of influential variables in the natural population is necessary.

This study is directed to combining an experimental approach with a field description. The purpose of such an analysis is to predict natural population growth of *Daphnia,* and, consequently, to focus attention on the factors which control it.

Because of the difficulties encountered in determining population rate processes and the role of underlying environmental variables, predictive models of population growth have been limited to controlled laboratory populations or unusual natural situations with relatively constant environmental conditions. Predictive schemes applied to natural populations are more likely to succeed the more information they utilize; but such models become hopelessly complicated, requiring vast amounts of empirical information. It is not yet clear whether simple models, requiring relatively little information, can adequately predict population growth in natural situations. However, comparison of a model with the observed population growth focuses attention upon the kinds and amounts of information absolutely essential for prediction. Inappropriate models may prove valuable by emphasizing the effect of disregarding important variables.

Few investigations of this sort have been attempted on plankton populations. Elster (1954) in studying the population dynamics of the copepod, *Eudiaptomus gracilis,* utilized the ratio of eggs-to-females to obtain a reproductive index for the population. By determining the developmental rate of eggs at different temperatures he was able to estimate the increase of the population. Edmondson (1960) applied the experimental-field approach to several rotifer populations, placing great emphasis on the eggs-to-female ratio as a useful tool for determining the birth rates of populations. A model based on birth alone was then used to predict population growth and size.

Although not concerned with zooplankton populations, 2 other studies are pertinent because the approaches are similar. Reynoldson (1961) made a quantitative population study of the triclad, *Dugesia lugubris.* Laboratory experiments allowed Reynoldson to assess the effects of temperature and food on the reproduction of *Dugesia,* and to conclude that population growth was often food-limited. Subsequent field experiments strengthened his conclusion. Morris (1959) in a study of 2 spruce-defoliating insects constructed a predictive model based on a single key factor. The incidence of parasitism in a larval stage of a given generation of defoliators could be used to predict

Reprinted from *Ecology,* 45:94–112, 1964.

the density of the next generation. This approach proved quite successful, in both a predictive and analytic sense.

Daphnia was selected for the present study because much is known of its biology and population attributes (Banta, Wood, Brown, and Ingle 1939; Slobodkin 1954; Edmondson 1955; Frank, Boll, and Kelly 1957). *Daphnia* reproduces parthenogenetically and at frequent intervals, is easily maintained under laboratory conditions, and is usually an important constituent of the zooplankton in lakes and ponds.

The decision on what variables to include was influenced by several sources. Previous investigations of *Daphnia* as well as pilot studies of *Daphnia galeata mendotae* indicate that food and temperature strongly influence population growth rate. Other variables, such as alkalinity, dissolved gases, and light, either tend to remain relatively constant in the zone of the lakes inhabited by *Daphnia* or seem to be of little importance. Predation undoubtedly affects growth rate in many *Daphnia* populations. This variable is intentionally ignored in this study because the necessary labor of estimating predation rates was prohibitive. The difference between observed and predicted population growth, if carefully evaluated, can indicate the extent of predation upon the *Daphnia* population.

Laboratory experiments were performed to determine rate functions of *Daphnia* under controlled conditions and to obtain relevant descriptive population data so that rates could be inferred from similar data on natural populations.

The experiments deal with the effects of food and temperature upon reproduction, development, individual growth, and survival. The choice of experimental temperature conditions was determined by the range of environmental temperatures throughout the year in the natural habitat. Food levels were selected high enough to permit nearly maximal growth, and low enough (it was hoped) to include the range of food levels existing in the lake. The population of *Daphnia galeata mendotae* in Base Line Lake, Michigan, was selected for study because of its dominant position in the zooplankton, and because of the relatively simple morphometric features of the lake.

Materials and Methods

Laboratory Experiments

Daphnia were raised individually from the newborn state in "old-fashion" glasses containing 150 cm³ medium (surface area: 30 cm²; depth: 7 cm) at 25 ± 0.5, 20 ± 1, and 11 ± 1°C.

Three different food levels or concentrations were used at each temperature. Animal responses

TABLE I. Instantaneous rates of population increase (r) of *Daphnia galeata mendotae* with associated standard errors at 3 temperatures and 3 food levels

Temperature (°C)	Food level (Klett units)	r
25	16	0.51±0.006
	1	0.46±0.013
	1/4	0.36±0.015
20	16	0.33±0.016
	1	0.30±0.002
	1/4	0.23±0.002
11	16	0.12±0.006
	1	0.10±0.003
	1/4	0.07±0.005

were so uniform that samples of only a few individually raised *Daphnia* yield a very small sampling error to the population rate of increase data (Table I). With 2 exceptions at 11°C, the sample size for each experimental condition was 10. In addition to these 9 experimental conditions 21 *Daphnia* were grown at 5 ± 1°C.

Before the experiments *Daphnia* were maintained for several generations at the experimental conditions to account for effects of acclimatization and conditioning to food level.

Food consisted of mixed green algae (*Chlorella, Ankistrodesmus,* and other small species) grown at room temperature in large aquaria exposed to natural as well as fluorescent light. No attempt was made to provide pure or uni-algal sources of food. Consequently, some bacteria and detritus were also included, but at all times the algae represented the major food source. The aquaria contained fish and snails. Food rations were passed through a #25 silk bolting cloth in order to remove undesirable large material. At each feeding the optical density of a sample of the algal culture was measured with a Klett-Summerson photometer equipped with a #42 blue filter. The sample was then centrifuged for 15 min at 3,500 rpm in a clinical centrifuge, after which the optical density of the supernatant liquid was measured. The difference in readings was considered the optical density of centrifugable (and, hence, probably filterable) material. Tap water, conditioned for about one week in large aerated aquaria containing various plants and invertebrates, was passed through an HA Millipore filter and used to dilute the stock algae to 3 concentrations representing 16, 1, and ¼ Klett units. Sixteen Klett units equal approximately one million algal cells per cm³. The relationship of optical density of cells to cell count over this range of Klett units appears to be linear. This procedure of quantifying the food resulted in uniform responses as reflected by the uniform brood size of daphnids reared at any given food level in the laboratory during the course of 2 years. This food suspension was then brought to the experimental temperature. The medium was changed daily for the 20 and 25° conditions, bidaily for the 11°, and every 4 days for the 5° conditions.

Daphnia were examined at every change of medium. Molting rate was determined by the frequency of appearance of cast skins. The maximum carapace length of the cast skin was the criterion used to measure size. The duration of egg development and duration of adult instar are nearly identical. *Daphnia* carry eggs in their brood chamber which are released as fully developed young during the molting process, at which time a new batch of eggs enters the brood chamber. Reproduction was measured as the number of newborn present at each change of medium. Calculations of the instantaneous rates of increase (r) were made by combining age-specific survival (l_x) and age-specific birth rates (m_x) following the technique of Evans and Smith (1952). Once data for the rate of increase were complete, experiments were terminated even though not all daphnids were dead.

Field Data

Morphometric information was obtained from the contour map of Base Line Lake published by the Institute for Fisheries Research, Michigan Department of Conservation. Temperature conditions were measured with a Whitney resistance thermometer. Water samples were taken periodically and analyzed for oxygen content according to the PKA modification of the Winkler method. A pigmy Gurley current meter was used to estimate current velocities of the river at the outlet of the lake. Current profiles were then constructed from which flow rates were calculated. Two such estimates were made, one during spring high water and another in midsummer. Stomach analyses were made of fish collected in April 1961 from Base Line Lake.

A #6 mesh, 12-in. diameter plankton net equipped with a Clarke-Bumpus flow meter served as the quantitative sampler. It possessed no closing device, however. Calibration of this instrument indicated that one revolution represents 33.4 liters of water through a range of towing speeds from one to 4 knots. A #6 mesh net retains the smallest *Daphnia galeata mendotae* but allows most of the phytoplankton and much of the zooplankton to pass through, thus facilitating the counting procedure. Collections were made from a rowboat moving at a speed of from one to 2 knots. With few exceptions, samples were not taken on windy days. The sampling procedure consisted of making long oblique tows through each of 3 strata: upper, middle, and deep. This ensured a stratified but extensive sampling of the population. The meter of water immediately overlying the lake bottom was not sampled. The vertical depth of the tows was calculated, using an angle measurer attached to the tow line. Occasionally in homothermous periods a complete oblique tow (from top to bottom of the lake) was made, while during the winter months vertical hauls were taken under the ice.

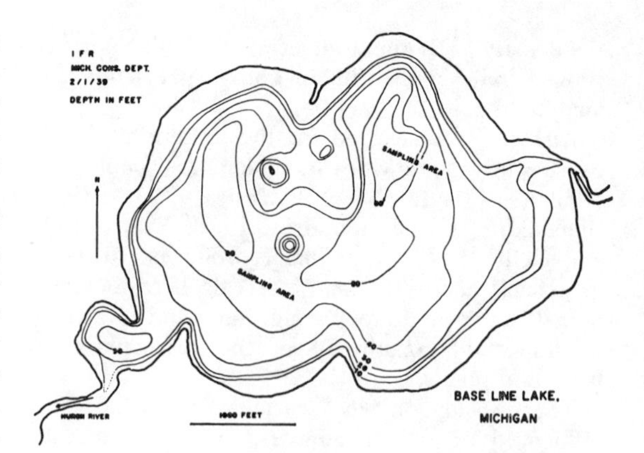

FIG. 1. Contour map of Base Line Lake, Michigan.

The 1960 samples were taken in duplicate at one station. The 1961 samples were taken in duplicate but from opposite sides of the lake (Fig. 1). Samples were also taken at the inlet and outlet of the lake. Each sample represented the contents of 2,000 liters or more of water, excepting those taken in the winter. Thirty-five collections were made during the period from July 1960 to July 1961. *Daphnia* samples were immediately preserved in 95% alcohol. Formalin or less concentrated alcohol caused carapaces to balloon with the consequence that appreciable numbers of eggs fell out of the brood chambers, and it became difficult to assess carapace length.

Samples were too large (about one liter) for complete analysis. After thoroughly mixing the sample, 2 ml subsamples were taken and counted in their entirety until at least 100 *Daphnia* had been counted. Subsamples were placed in a narrow rectangular counting chamber and examined under a dissecting microscope equipped with an ocular micrometer. The maximum carapace length and brood size of each specimen of *Daphnia galeata mendotae* were recorded. Other daphnids were noted if found.

The following equation was used to obtain the density estimate of *Daphnia* in each stratum:

$$\frac{\#\ Daphnia}{100\ \text{liters}} = \frac{\text{average}\ \#\ Daphnia\ \text{in subsample} \times \#\ \text{subsamples in concentrate}}{\#\ \text{revolutions of flow meter} \times 33.4\ \text{liters}} \times 100$$

The density for the total water column was obtained by taking an average of the upper, middle, and lower strata density estimates, weighted according to the thickness of the strata.

Identification and classification of *Daphnia* follow the monograph of Brooks (1957).

Results

Laboratory Studies

Survivorship and fecundity tables were constructed from the age-specific survival (l_x) and reproductive rates (m_x) under 9 experimental conditions. These tables are then combined into an $l_x m_x$ column which is used to solve the following equation for r, the instantaneous rate of population increase:

$$\Sigma\, l_x m_x e^{-rx} = 1.00$$

Birch (1948) gives an excellent discussion of the calculation of r. Since survival through the juvenile and early adult stages is nearly perfect, mortality has little effect on the estimates of r. Consequently, only fecundity tables are presented in Appendix A; l_x values are 1.00 throughout. This treatment permits quantitative evaluation of the effects of food and temperature on the population rate of increase. Population rates of increase (r per day) for each food level are plotted in Figure 2 as a function of temperature. For the chosen conditions, temperature shows a much greater effect upon r than does food level. Obviously, low enough food levels would also have a profound effect upon r. It is clear by inspection of the range of r values ($r = 0.07$ to $r = 0.51$) that any reference to the rate of increase of a population should be accompanied by a statement of the environmental conditions under which the value was determined.

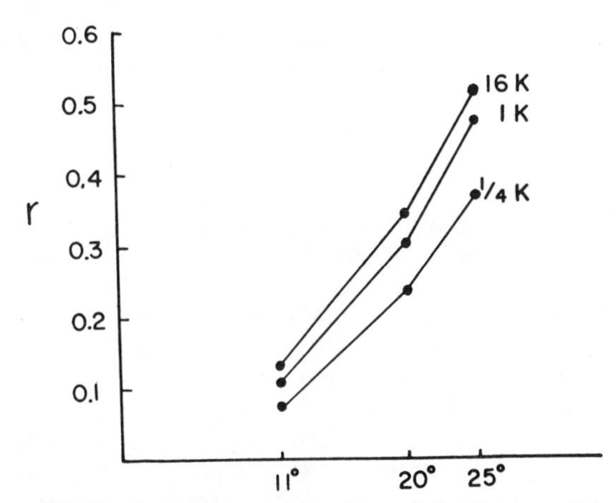

Fig. 2. Instantaneous rates of population increase (r) in relation to temperature (°C) and Klett units of food (K).

Table II. Temperature effects upon duration of instar, egg development, and the ratio of duration of juvenile period to duration of adult instar for *Daphnia galeata mendotae*

Temperature	Duration of instar or duration of egg development	Duration of juvenile period / Duration of adult instar
°C	Days	
25	2.0	3.0 (6/2)
20	2.6	2.9 (7.5/2.6)
15	4.5	
13	6.0	
11	8.0	3.0 (24/8)
9	10.8	
8	12.3	
7	14.2	
5	18.0	
4	20.2	
2	?	

Temperature

Temperature affects the frequency of molting and hence the frequency with which young are produced (Table II). Reproduction occurs every 2.0 days at 25°C; every 2.6 days at 20°C; and every 8.0 days at 11°C. A single adult at 5°C gave birth to young 18 days after producing the eggs.

For each temperature condition the food level had no observable effect upon the frequency of molting and reproduction. Under conditions of near-starvation, however, Banta, Wood, Brown, and Ingle (1939) have demonstrated that the frequency of molting and reproduction is somewhat reduced.

The duration of egg development is identical to the duration of the adult instars at the food levels used. The ratios of the juvenile periods (time from birth to reproductive maturity of instar V) to the duration of adult instars seem temperature-independent (Table II).

Growth is discontinuous in *Daphnia,* for size increases only during and immediately following ecdysis. The amount of growth at each ecdysis does not seem to be dependent upon temperature.

The median lifespan of *Daphnia* is 30 days at 25°C; 60-80 days at 20°C; and about 150 days at 11°C. *Daphnia* grown at 5°C showed no mortality over a 2-month period. These survival estimates suggest that the physiological mortality rate is quite low, probably less than 3% per day

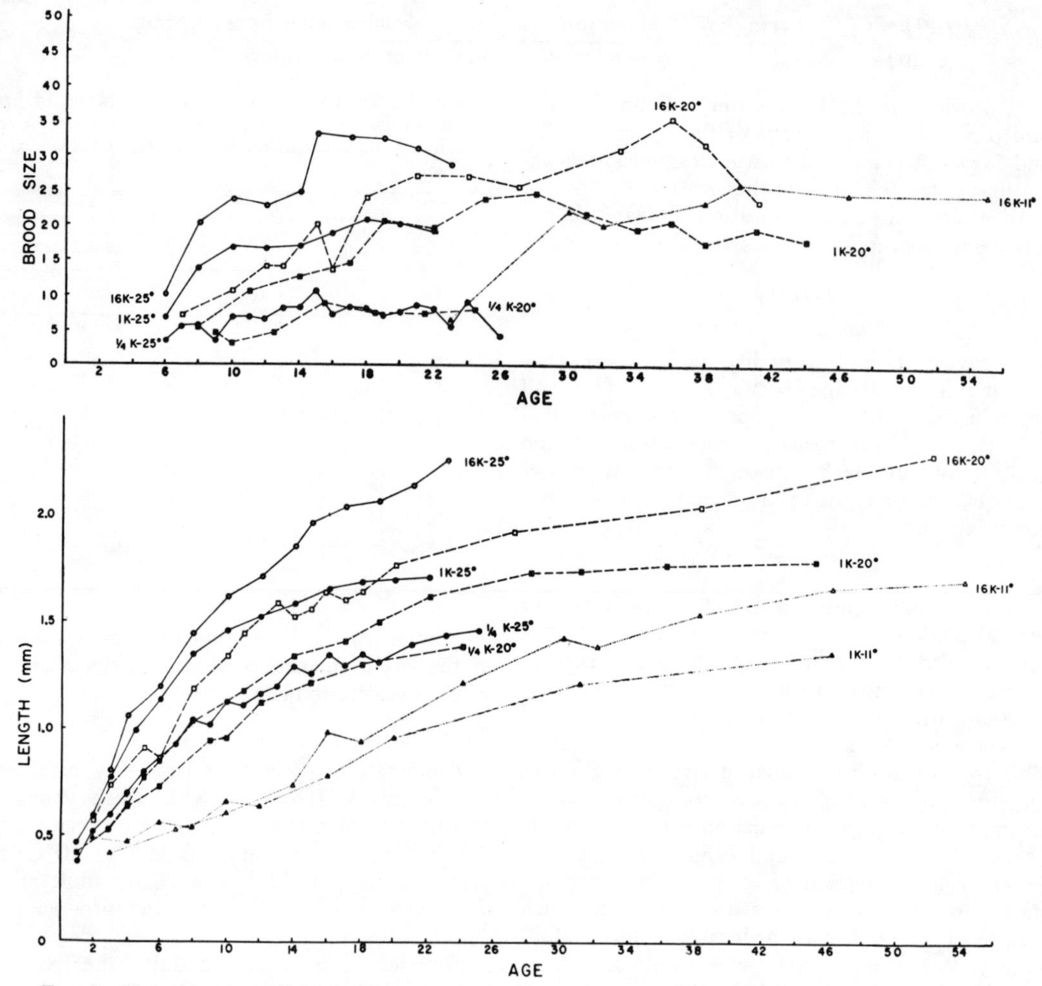

FIG. 3. Upper figure: mean brood size in relation to age (days) under different temperature (°C) and food conditions (Klett units). Lower figure: mean length in relation to age (days) under different temperature (°C) and food conditions (Klett units).

throughout the year. The survival curves were incomplete, but indicate the near-rectangular shape reported for *Daphnia* by Frank, Boll, and Kelly (1957) and Banta, Wood, Brown, and Ingle (1939).

Food

Food level affects reproduction through the number of young per brood (Fig. 3). This effect may appear to be slightly temperature-dependent. When a difference in average brood size does occur for a given food level, the larger size is always associated with a higher temperature. Two features of these experiments might have some bearing on the results. First, at higher temperatures food levels might have been effectively higher than at lower temperatures because the algae seemed to remain suspended better in the warmer water. Second, had the experiments been carried out for the same number of instars, the curves representing brood size at lower temperatures might have reached the same heights as the curves representing brood size at higher temperatures.

Food level also affects the amount of growth per instar (Fig. 3). Adult stages are affected more by different food levels than are juvenile stages. This effect may appear slightly temperature-dependent, but the 2 reservations mentioned in connection with reproduction apply here also.

Since food level strongly affects both brood size and body size, the relation of brood size to body size (Fig. 4) in daphnids is a poor criterion for inferring food level. *Daphnia* from the different food levels show similar trends, although low food

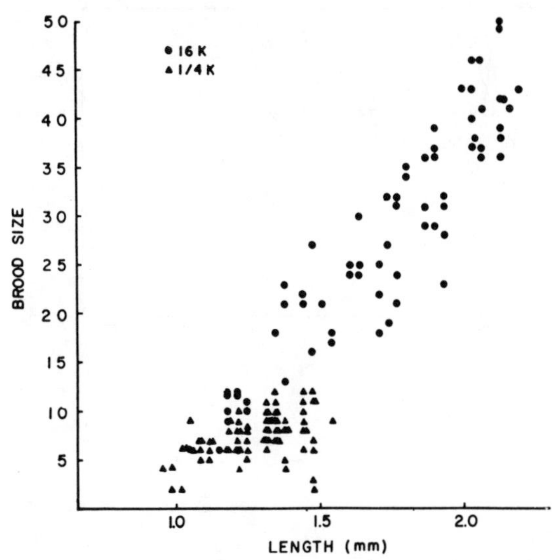

FIG. 4. Brood size in relation to length under high (16 Klett units) and low (¼ Klett unit) food levels at 25°C.

animals remain at the lower end of the distribution.

Although numerous age and size categories are usually present in any natural population, the maximum length and brood size attained by *Daphnia* is apparently determined by the abundance of food. Therefore, an estimate of natural food conditions might be obtained by matching maximum length and brood size from field data to experimentally determined length and brood maxima obtained under specified food levels. Applying this criterion to the collection from Base Line Lake indicates a natural food level of about ¼ Klett unit for much of the year. This technique may prove useful in analyses of other zooplankton populations.

Different food levels did not affect the median

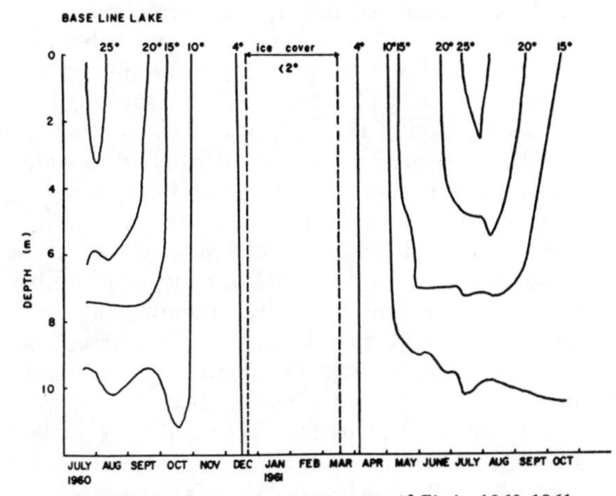

FIG. 5. Base Line Lake isotherms (°C) in 1960-1961.

life span. Banta, Wood, Brown, and Ingle (1939) report that extremely low food levels (near-starvation) increase median longevity.

The separability of the effects of food and temperature greatly facilitates the application of these data to the field collections. Temperature alone may be utilized to predict frequency of molting and reproduction, duration of egg development, and physiological life span. Food level may be inferred from the largest carapace sizes and brood sizes encountered.

Field Studies

Limnological measurements

Base Line Lake in southeastern Michigan (T. 1S.—R. 5E.—Sec. 5, 6; T. 1N.—R. 5E.—Sec. 31, 32) is the last downstream member of a chain of lakes connected by the Huron River. The lake is oval in outline and generally steep-sided, forming a relatively uniform basin ¾ mile long by ½ mile wide (Fig. 1). About 70% of the surface area of the lake overlies water of greater than 8 m depth.

TABLE III. Estimates of volume outflow, lake volume, and flushing rates of Base Line Lake, 1961

VOLUME OUTFLOW	
Spring	1.07×10^6 m³ / day
Midsummer	1.63×10^5 m³ / day
LAKE VOLUME	
Spring	8.73×10^6 m³
Midsummer	8.10×10^6 m³ (total lake)
	5.67×10^6 m³ (epilimnion only)
FLUSHING RATES	
Spring	12.3% / day
Midsummer	3.0% / day (epilimnion only)

Shoal areas are covered with fine marl and are generally devoid of aquatic vegetation. Base Line Lake is a hard-water lake (average surface total alkalinity 155 ppm). Temperature conditions are shown in Figure 5. Stratification occurs from May into October and again from December into March. Oxygen depletion occurred in the hypolimnion in both summers but not during the winter. By June 25, 1961, oxygen concentration was reduced to 1.0 ppm in the hypolimnion; on July 12, 1961, less than 0.10 ppm remained. The effect of summer depletion on *Daphnia* is to reduce the habitable zone of the lake to the epilimnion and thermocline. Prevailing winds are westerly. Circulation patterns within the lake seem to be primarily wind-determined, although the effect of the river flowing through the lake may be considerable during periods of high water.

Table III indicates the volume outflow from the

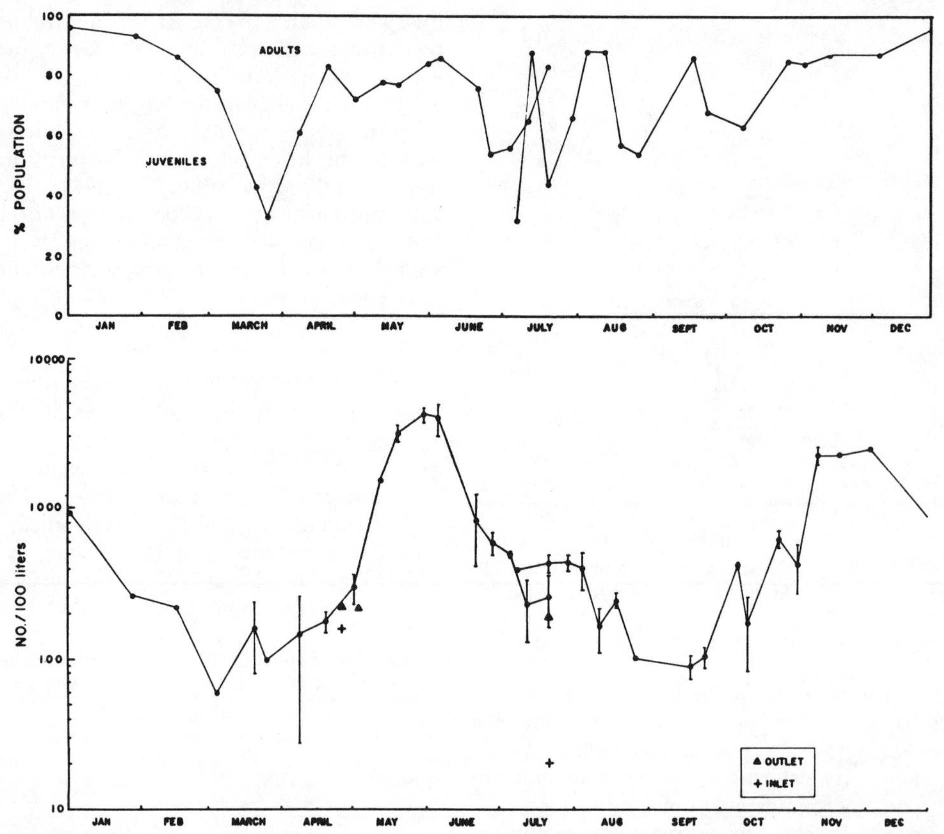

FIG. 6. Upper figure: per cent juveniles and adults in the population. Lower figure: population size (# *Daphnia galeata mendotae* / 100 liters) in 1960-1961. Vertical bars represent standard deviations.

lake at spring high-water level and at summer low-water level. The volume outflow per day is compared with the lake volume yielding a turnover or flushing rate. During May the flushing rate was 12% of the total lake per day. By midsummer the rate had decreased to 3% of the epilimnion per day or 2% of the total lake per day. It is assumed that the outflow rate equals the inflow rate, which is true if the lake level does not fluctuate violently. An annual fluctuation in lake volume of ±3.7%, based on extremes in lake level, lends support to this argument.

Population Attributes

Population size

Population size of *Daphnia galeata mendotae* throughout the course of a year is indicated in Figure 6. The values given are densities (numbers/100 liters) rather than estimates of total population size. Since the densities were determined as an average for all depths, the only assumption involved in converting densities to total numbers is that the densities apply to all parts of the lake. Samples were taken in duplicate, which should be considered the minimum allowable for population studies. However, samples were very large when compared to most sampling procedures utilizing plankton traps or Clarke-Bumpus samplers. The average coefficient of variability of the population for the pairs of samples is 17.3%, which is relatively low as far as field studies go. Subsamples represented a very small fraction of the total sample (1/200 or less). If any subsample has an equal (but very small) probability of including a given daphnid from the concentrate, then the distribution of the daphnids in the concentrate may be considered a Poisson distribution in which the variance is equal to the mean. Comparing the variance to the mean from samples taken in spring, summer, and fall yields a chi-square value with a probability of 0.10-0.250. The conclusion is that the method of mixing the concentrate ensured randomness, and that the subsampling error was random. The average coefficient of variability associated with subsampling is 18.5%.

Two population maxima occurred; one in late spring, the 2d in late fall. The maximum densities were 4331/100 liters and 2556/100 liters, respectively. During the summer months population density gradually decreased from about 400/100 liters in July to 100/100 liters in September. A large number of *Daphnia galeata mendotae* were present throughout the winter. The minimum density of 60/100 liters occurred in early March just as reproduction was commencing.

Exceedingly few individuals were observed to carry ephippia. All ephippia obtained from bottom samples proved to be *Daphnia pulex*. A few males were observed in fall and spring. *Daphnia galeata mendotae* apparently overwinters largely in the free-swimming stage in Base Line Lake. No reproduction occurred between January 1 and March 4. The water temperature was less than 2°C during the period. Although reproduction occurred in the laboratory at 5°C, it is possible that at 2°C reproduction is prevented. Such a low temperature may also alter the mortality rate.

The vertical distribution of the population seemed dependent upon the temperature stratification of the lake. Under homothermous conditions nearly equal densities of *Daphnia* occurred in the upper, middle, and deep strata. As soon as the lake began to stratify, however, the majority of the daphnids were found in the upper and middle strata. Densities during summer were 10 times greater in the upper stratum than they were in the deep stratum. Lack of oxygen precludes the permanent existence of *Daphnia* in the hypolimnion, and thus few, if any, *Daphnia* would be expected from the deep stratum. Their occurrence in samples from this stratum may be a sampling artifact, since the sampler lacked a closing device. This would bias the density estimates, but the bias would be quite small because the time spent in ascending through the upper and middle strata is very short compared with the time spent in the deep stratum. Another explanation may be that the deep samples frequently included the lower limits of the thermocline where daphnids could live.

Diurnal vertical migration of *Daphnia galeata mendotae* is probably confined to the epilimnion and the thermocline. Attempts to demonstrate vertical migration showed no pattern whatsoever. If vertical migration does occur, it probably does not markedly change the daily temperature environment of the population as a whole. McNaught and Hasler (1961) believe that this species migrates vertically through one m or less of water in Lake Mendota, Wisconsin.

Reproduction

The average brood size (ratio of eggs to mature *Daphnia*) served as a population reproductive index. The criterion of reproductive maturity is

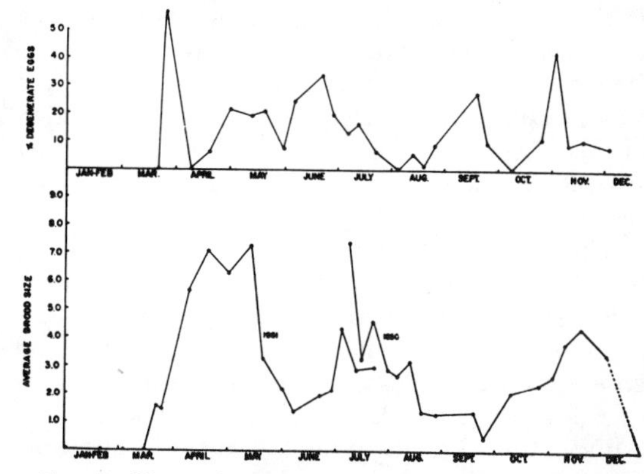

FIG. 7. Upper figure: per cent degenerate eggs in the population. Lower figure: average brood size (average number of eggs per adult).

somewhat arbitrary, since no morphological distinction can be made between mature and immature *Daphnia* unless the adults are reproductive. Carapace length at the onset of reproduction (instar V) was determined under various food and temperature conditions. The range of such lengths was from 0.97 mm to 1.25 mm. All shorter lengths occurred at low food levels. Since it appears that the natural population also experienced a low food level, the size at maturity was arbitrarily set at 0.97 mm. At a high food level this would include many of the last juvenile instars, but would nevertheless yield a very high average brood size.

Average brood sizes are plotted in Figure 7. Three distinct peaks occur. The spring and fall brood size maxima immediately precede the two population density maxima. The summer increase in brood size is not accompanied by an increase in daphnids. The average brood size ranged from 2.0 to 4.0 eggs throughout most of the reproductive season. The largest average brood size was 7.5 eggs.

The per cent reproductive adult females was usually high, especially during peak egg production, but dropped to 50% or less after the spring and summer reproductive peaks and immediately after the onset of reproduction in March. If degenerate broods were also included the per cent would be higher.

Degenerate eggs were frequently observed in the brood chambers of *Daphnia* taken from Base Line Lake. These eggs were a dark gray-brown color and often appeared to be disintegrating. Normal eggs appeared blue-green or yellow-green in color and were of a firm, less fragile texture. Degenerate eggs were excluded from the calculation of average brood size. The percentage degenerate eggs in relation to the total eggs present is shown in Figure 7. There is no suggestion of any relationship with normal brood size. The frequency of entire broods appearing degenerate was greater than the frequency of only a fraction of the brood appearing degenerate, whereas under laboratory conditions the few degenerate eggs observed occurred singly.

The stage of development of the eggs in the brood chamber was also noted. Three stages were defined: the egg stage (no segmentation), first embryo stage (segmentation, but no eyespot), and 2d embryo stage (possession of an eyespot). This classification is similar to Edmondson's (1955). Although the 3 stages are not of identical duration, the relative frequency of their occurrence may still be used to evaluate the question of reproductive periodicity. All 3 stages were present in roughly equal numbers in all samples, indicating that no diurnal periodicity in reproduction occurs under natural conditions. This observation permits reproduction to be treated as a continuous function of the population.

Size structure

Knowledge of the age and size structure of metazoan populations is necessary to analyze properly population growth (Slobodkin 1954). *Daphnia* exhibit no age-specific characters, making it difficult to describe the age structure of the population. A method of estimating the age structure of a population would be to determine, under laboratory conditions, the length of *Daphnia* grown under a variety of experimental conditions, and then to apply the appropriate length-age relationship to the observed field size distribution. The validity of such an estimate depends upon how well the laboratory-determined length-age relationship can be applied.

In this study an attempt was made to estimate the instar structure of the population (later, using temperature data from the laboratory, ages of immature instars will be estimated). The ranges of lengths of instars grown in the laboratory were used to construct a series of size categories, which, when applied in the field samples, probably represent distinct instars at least for the immature period. Once maturity is reached the individual growth rate (increases in length per molt) decreases and becomes more strongly dependent upon food level. Thus, it becomes increasingly difficult to assign an individual to a specific instar. The size structure is therefore based upon 4 immature instars and 3 mature "categories." Figure 6 represents the frequency distribution of adults and juveniles throughout one year. The increase in the proportion of adults from January through March was caused by the absence of reproduction coupled with the slow maturation of the existing immature daphnids. However, as reproduction commenced in late March the population size structure shifted rapidly until by mid-April less than 20% of the *Daphnia* were mature. For 2 months the proportion of mature daphnids remained at this level with only minor fluctuations. This period coincides with the spring period of rapid population growth. By mid-June the proportion of adults began to increase, reaching 46% of the population before again decreasing to the 20% level. A series of 3 rather violent fluctuations in all size categories occurred during the summer and early fall. Each increase in percentage of adults was followed by a decrease to the 10 to 20% level. In October a 4th such decrease occurred, but this time the proportion of adults reached the 15% level and remained there for more than a month. This last period of stable age structure coincides with the fall period of rapid population growth.

Population size structure in relation to depth showed little variation among the upper, middle, and lower strata. The juveniles were slightly more frequent in the upper and middle strata in the summer. Because most of the population was located in the epilimnion (upper and middle strata), the small differences of size frequency categories in relation to depth are considered unimportant. The population size structure was similar at all depths in spring and fall. The size structure of samples taken at the outlet at spring high water was very similar to the size structure of the population in the lake. The size structure of outlet samples taken in midsummer was composed entirely of juveniles and did not correspond to the size structure of the lake population.

Population losses

Observations on the abundance and feeding habits of several predators indicate that *Leptodora* (a cladoceran), *Chaoborus* (phantom midge

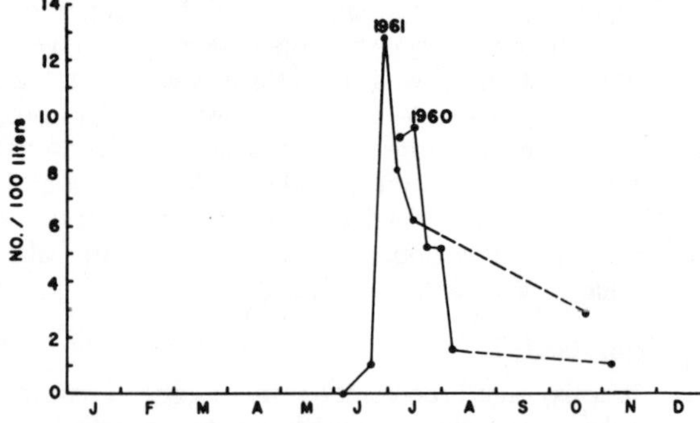

FIG. 8. Population size of *Leptodora* (# / 100 liters) in 1960-1961.

larva), and fish prey upon *Daphnia*. Population density was estimated for *Leptodora* only. *Leptodora* appears in June and reaches peak numbers in July (Fig. 8). Andrews (1949) reports a similar seasonal abundance of *Leptodora* in Lake Erie, but with a 2d minor peak of males in October. *Chaoborus* appear in the upper waters only at night, spending the day in or just above the bottom of the lake. Stomach analyses of adult cisco (*Leucichthys artedi*) and crappies (*Pomoxis nigromaculatus*) taken from Base Line Lake indicate that these fish feed extensively on large *Daphnia pulex* in early spring, but not upon the smaller adults of *Daphnia galeata mendotae*.

Another population loss is caused by the Huron River. This loss may be considered the net effect of passive immigration and emigration of *Daphnia*. Once carried out of the lake, *Daphnia* are lost to the system and soon die (Chandler 1939). The net loss of the population was determined by subtracting the density of *Daphnia* in the river immediately below the lake from the density in the river immediately above the lake. During late April and early May the high flushing rate of the lakes was accompanied by considerable but nearly identical immigration and emigration (Fig. 6). The net loss was low (perhaps as high as 3% of the population per day). During midsummer the low flushing rate of the lake was accompanied by lower immigration and emigration rates. The net loss was one per cent of population per day. These estimates indicate that the river flowing through Base Line Lake exerts a relatively constant and small source of mortality.

Other zooplankton populations

Four other species of *Daphnia* were found in Base Line Lake: *Daphnia pulex, D. retrocurva, D. ambigua,* and *D. longiremis*. Only the first 2 species ever reached appreciable numbers. *D. pulex* reaches a peak abundance in spring after which numbers become sparse. At no time was *D. pulex* more abundant than *D. galeata mendotae*. During summer *D. pulex* is found only in the colder regions of the lake, i.e., the thermocline, where it is the dominant daphnid. The fact that this species does not appear in the warmer epilimnion, and often possesses considerable haemoglobin which is an adaptation to enable it to withstand low oxygen concentration (Fox 1948), would indicate that this species occupies a restricted habitat during much of the year. *D. retrocurva* appeared in the plankton only after the lake had warmed to 20°C. Population size of this species increased during July and early August, and, by late August, this species became the most abundant cladoceran. On September 24, 1960, *D. retrocurva* occurred at a density of 730/100 liters in the epilimnion compared to a density of about 175/100 liters of *D. galeata mendotae*. By late October *D. galeata mendotae* again dominated other cladocera. Indeed, during spring and fall maxima, this species was far more abundant than any species of copepod. *Cyclops* is the predominant copepod. *Diaptomus* is present also. *Bosmina* appears as an important member of the zooplankton during the colder months.

PREDICTIONS

A first-order prediction of *Daphnia* population growth may now be attempted utilizing much of the information obtained from the field and laboratory observations. Such a prediction will indicate if important variables have been overlooked in the population analysis, and may prove a valuable means of gaining insight into this biological system.

Two related kinds of predictions are made: predicted rates of increase and predicted population size or density.

Population growth rate

The rate of increase of the population may be estimated if the birth, death, emigration and immigration rates are known. Although no animal reproduces continuously, many do reproduce frequently enough so that the instantaneous growth equation,

$$N_t = N_o e^{rt} \qquad (1)$$

is a close approximation to natural population growth. N_o and N_t denote initial population size and size t time units later; e is the base of natural logarithms; t represents time units; and r denotes the instantaneous rate of increase and is determined by the 4 rates mentioned above.

If the effects of death, emigration, and immigration are ignored, the growth equation becomes,

$$N_t = N_o e^{bt}$$

in which b is the instantaneous rate. Although the symbols r, b, and d (instantaneous death rate) have been defined under stable age conditions (see Lotka 1925), they may be applied to populations without a stable age distribution as well. The only distinction is that under stable age conditions r, b, and d remain constant; whereas under a continually changing age distribution the instantaneous rates will change accordingly. The instantaneous birth rate, b, may be estimated if the finite birth rate, B is known:

$$b = \ln (1 + B)$$

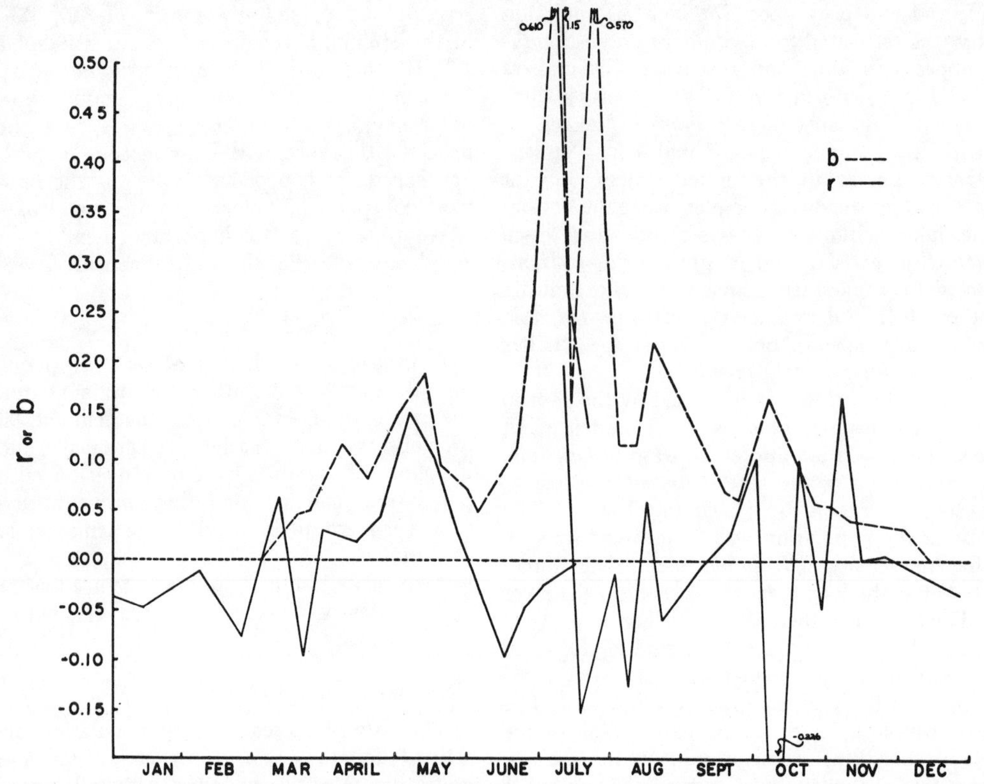

FIG. 9. Observed (r) and estimated (b) instantaneous rates of population increase of *Daphnia galeata mendotae* in Base Line Lake.

The finite birth rate is defined as the number of newborn occurring during an interval of time divided by the number of animals already existing:

$$B = \frac{\text{number of newborn (during interval t to t + 1)}}{\text{population size at t}} \quad (2)$$

The problem then is to predict the number of newborn during any interval of time. Four kinds of information are used for this prediction: the population size (N_o); the number of adults (N_A); the average brood size of adults ($\bar{E}$); and the rate of development of the eggs expressed as the fraction of development accomplished per day ($1/D$). The finite birth rate or number of newborn per individual per day is given by the equation:

$$B = \frac{N_A \cdot \bar{E} \cdot 1/D}{N_o} \quad (3)$$

The values of N_A, $\bar{E}$, and N_o are estimated from a sample of the natural population. In order to determine $1/D$, the temperature at which the eggs develop must be known. Table II indicates values of D for various temperatures. The lengths of time required for development (D) at 25, 20, 11, and 5°C were determined under constant laboratory conditions. The remainder of the values were obtained by curvilinear interpolation, but not

TABLE IV. Average temperatures (°C) during different periods of the year in the upper, middle, and lower strata of Base Line Lake

Periods	Strata		
	Upper	Middle	Deep
1960			
July-Aug.	25	20	15
Sept.	20	20	15
Oct.	13	13	11
Nov.	7	7	7
1961			
Jan.-March	2	2	2
1/2 March	4	4	4
April	6	6	6
1/2 May	11	11	8
1/2 May	15	15	9
June	20	15	9
July	25	15	11

by logarithmic interpolation, since Q_{10} changed considerably throughout this temperature range. Table IV indicates the estimated average temperatures for the 3 strata of Base Line Lake for various months of the year. These average temperatures were determined from the annual temperature profile of Base Line Lake (Fig. 5). During periods of rapid temperature changes in Base Line Lake the averages were based on shorter periods. When the 3 strata differed in average temperatures, 3 separate values of b were calculated from which a weighted average of the 3 was determined. The calculated values of b for the 28 dates are plotted in Figure 9.

Successive pairs of points from Figure 6 yield values of N_o and N_t, from which an average r for each time period can be calculated.

From equation (1):

$$r = \frac{1_n N_t - 1_n N_o}{t} \qquad (4)$$

The 32 points of r in Figure 9 were calculated in this manner.

The properties of b include a lower limit of zero (whenever reproduction ceases). Rapid shifts in the age structure of the population as well as sampling errors produce large changes in the value of b. The lower limit of r is—∞. Values of b should always exceed values of r, since $r = b - d$. Population rate of increase based on b alone is, except for 2 periods of time, always higher than the observed rate of increase.

Despite fluctuations in b and r definite trends in the 2 sets of points are apparent. During spring and fall months r and b values are both positive and nearly equal. Maximum values of b are attained in the summer months, whereas r is negative during this period indicating a population decrease. During the winter b is zero, and r is negative.

Obviously, the curve of r is related to the population numbers from which it was derived. The maximal densities of spring and fall are preceded by maximal r values whereas the minimal summer and winter densities are preceded by minimal r values. The seasonal curve for b is very different. Values of b reach a single large peak during the summer months rather than 2 peaks in spring and fall.

Comparison of birth rates and rates of population change show that in spring and fall, predation, natural death, and net loss to the Huron River are not important variables affecting population growth. Average total population losses of 3.9 and 5.6% per day for April-May and October-November respectively were determined by subtracting r from b during these periods. During

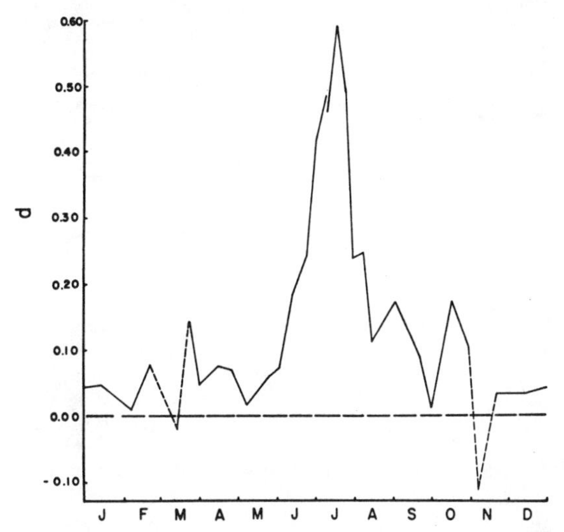

FIG. 10. Estimated instantaneous death rate (d) of *Daphnia galeata mendotae* in Base Line Lake obtained by subtracting r from b in Fig. 9.

the winter a 4.4% loss per day occurs, which probably reflects natural death and loss to the river alone. From mid-June through mid-September the very large difference between r and b indicates that the total population loss is very high (Fig. 10). An average total loss of 28.5% per day was determined by subtracting r from b during July and early August.

Since the net loss to the population due to the river system is 3% per day or less in the spring and 1% in the summer (0.03 and 0.01 daphnids lost per daphnid per day respectively), the effect of animals entering and leaving the population in this manner is small and reasonably constant. Physiological mortality as discussed above is probably low also. Predation then would seem to be the cause of the negative population growth rate throughout most of the summer.

A 2d method of estimating the loss rate is to construct survivorship curves for the population from the size distributions for various periods of the year. A survivorship curve yields the rate of survival (or conversely the death rate) if plotted as a semilogarithmic graph. Knowledge of the age structure and growth rate of the population is necessary to construct a survivorship curve from field data. Laboratory data on the length-age relationship under different food and temperature conditions were here employed to estimate the age and instar of *Daphnia* taken from the natural population. If population size remains unchanged for a length of time greater than one generation period, then the survival curve will become identical or nearly identical to the stable age distribution curve. To obtain a survival curve when the

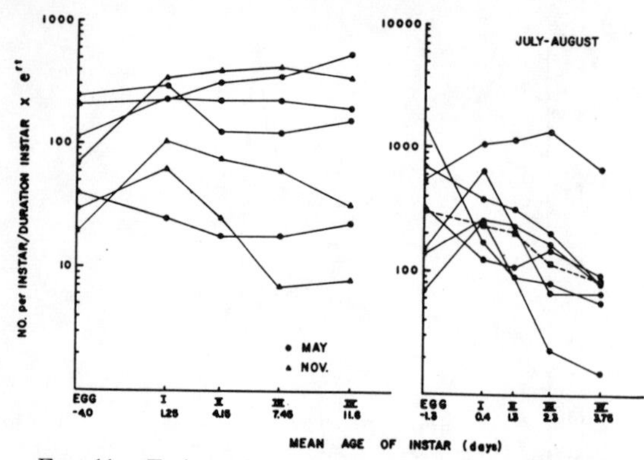

FIG. 11. Estimated survivorship curves relating numbers (# per instar / duration of instar X e^{rt}) to age (days) for May, November, and July-August. Average survivorship in July-August is represented by the dotted line.

population is changing size, the number of animals in each age category of the age distribution must be multiplied by a correction factor which is based upon the population rate of change and the age of the animal. This method should work providing the rate of population change is known and remains relatively constant over a length of time greater than the generation period.

Figure 11 represents survival curves estimated by the above technique for May, July-August, and November. These periods were chosen because population rate of change was relatively constant (Fig. 6). On the abscissa are plotted the mean ages of the eggs and estimated instars I-IV. The numbers of daphnids within each of these categories was determined by (1) dividing the number of individuals existing as eggs, instar I, etc., by the duration of the category (to adjust for different instar durations), and (2) multiplying this by the finite rate of population change over the time period between birth ($t = o$) and mean age of category. In the case of eggs, negative time units were employed since the eggs had not entered the population yet. This procedure adjusts the numbers in each category so that they represent the estimated survival in time of a population of the size that exists at $t = o$. The slope of the line indicates the survival rate since a logarithmic scale was used to plot numbers. A horizontal line indicates perfect survival. It is interesting to note that during May survival appears to be complete; in other words, the death rate is negligible. The survivorship curves for November indicate much the same thing, although not so clearly. In both cases sampling error may produce an apparent increase in numbers with age, which is unrealistic, but the average horizontal trend is obvious. During July and August the curves are no longer horizontal. Their negative slopes reflect a high death rate. The average summer survivorship curve indicates a death rate of 0.259 daphnids per daphnid per day. Since net immigration-emigration effects indicate a loss of 0.01 daphnids per daphnid per day and the physiological death rate amounts to no more than 3% per day, the conclusion is again reached that predation is the most important factor controlling population size in the summer.

The average summer survivorship curve indicates a reasonably constant loss rate for all age categories. This suggests that predation viewed for the entire period is not age-specific.

Population size

Population size was predicted for 4 intervals; May 1-May 19, June 5-June 28, July 8-August 4, and September 24-November 10. The rate of increase, r, was estimated for one generation period by estimating the number of *Daphnia* that would be produced during the period that it took for a newborn *Daphnia* present at $t = o$ to mature. Thus mature daphnids present at the beginning of each interval were assumed to produce young throughout the entire interval at a rate dependent upon the average brood size and developmental rate 1/D. Daphnids in the last immature instar at the onset of the period were treated in the same manner except the number of "reproductive days" in the period for these animals was shortened by the time it took to reach the mature instar. All subsequent immature instars were treated in the same manner with the number of reproductive days decreased by how many days short of maturity they were. The estimated duration of immature instars at various temperatures was determined from the laboratory experiments. The total number of *Daphnia* born during this generation period was added to the initial number to obtain a predicted population size for the end of the generation period. The instantaneous rate of increase was then determined by solving equation (4), in which t equals the number of days from birth to maturity at the given temperature. The 4 values of r were then utilized to solve for predicted population size at the end of each of the 4 intervals of prediction.

If but one generation occurred during the interval (owing to a low temperature or the shortness of the interval) then the estimated population size would be a good estimate of the expected population size based upon birth rate alone. However, if several generations occurred during the interval, then the r associated with the initial population size structure would be insensitive to changes in size structure occurring during the in-

TABLE V. Observed population size compared to estimated population size of *Daphnia galeata mendotae* at the end of 4 intervals

Time interval	Population size (/100 liters)	
	Observed	Estimated
May 1-May 19	3,225	2,661
June 5-June 28	607	114,723
July 8-August 4	413	8,745,000
Sept. 24-Nov. 10	2,337	3,663

terval. It is impossible to assume that the initial size structure would not change during some of these intervals (Fig. 6); therefore, this method must be criticized for this shortcoming. However, population size estimated in this manner (Table V) indicates much the same thing as the estimate of population growth did when compared to the observed population growth.

The estimated population size of 2,661 *Daphnia*/100 liters is similar to the observed population size of 3,225 *Daphnia*/100 liters present at the end of the interval May 1-May 19. The estimated population size of 3,663 *Daphnia*/100 liters is also similar to the observed population size of 2,337 *Daphnia*/100 liters present at the end of the interval from September 24-November 10. However, the 144,723 *Daphnia*/100 liters estimated from June 5-June 28 is far greater than the observed population of 607 *Daphnia*/100 liters. An even greater discrepancy exists at the end of the period of July 8-August 4 when the estimated population is 8,745,000 *Daphnia*/100 liters compared to observed 413 *Daphnia*/100 liters.

DISCUSSION

The population growth model based on birth rate alone is an adequate expression of population growth of *Daphnia galeata mendotae* in Base Line Lake during spring and fall, but is inadequate during the summer. Predation seems to be a key variable affecting the population throughout the summer.

Population losses during the summer, estimated by determining the average difference between the expected and observed population growth (b-r) for this period, amount to 28.5% of the population per day. This estimate is remarkably similar to the loss rate of 25.9% of the population per day obtained from the estimated survivorship curves for the same period.

Although both estimates make use of the parameter r, the effect of r in correcting the numbers in each age category in the construction of the survivorship curves is small. The 2 estimates of population loss may be considered virtually independent, since they utilize mainly different information in different ways. Confidence in the reality of such a high population loss rate is increased by the agreement of the estimates.

An average predation rate of just over 25% of the population per day when the population is decreasing very slowly implies that the population turnover time is about every 4 days during July and early August. The turnover rate applies to biomass as well as numbers since the slope of the average summer survival curve is nearly constant (Fig. 11). This appears to be an exceptionally high turnover rate for a microcrustacean population. Stross *et al.* (1961) estimated the turnover time of *Daphnia longispina* in a lime-treated lake to be 2.1 weeks, whereas the estimated turnover time of a *Daphnia pulex* population in the untreated control lake was 4.6 weeks. The slightly lower average temperature of these lakes in comparison to Base Line Lake would, in itself, increase the population turnover time.

Under laboratory conditions *Daphnia galeata mendotae* exhibits a maximal rate of increase of 0.51 at 25°C. If we assume that predation rates would be proportional to the age frequency classes (i.e., predation does not change the age distribution), the theoretical minimum turnover time for *Daphnia* would be slightly less than 1.5 days. This is the length of time necessary to double population size calculated from the exponential growth equation (1).

Temperatures frequently reach 25°C in the surface waters of lakes but not for long periods of time. Food levels in natural habitats rarely, if ever, reach the levels associated with a maximal rate of increase. Even at the lowest laboratory food level, which is approximately the same as that of the lake, the maximum rate of increase is 0.36. Thus, a higher predation rate than that estimated (0.25) is possible, although the daphnids might risk extinction.

The turnover rate of once every 4 days observed in *Daphnia galeata mendotae* indicates that practically all the biomass produced by this population is passed on rapidly to the next trophic level. Since this species is a major component of the zooplankton, it must play a major role in the food chain of the lake during the summer. The role of *Daphnia galeata mendotae* as food would be much less important in spring and fall, since predation appears to be negligible at these times.

The bimodal curve representing population size of *Daphnia galeata mendotae* throughout a period of one year (Fig. 6) is typical of many zooplankton populations. This pattern of population growth is generally interpreted as a reflection of a food-limited system, in which phytoplankton become abundant as the result of increased nutrients following spring and fall mixing of the lake. The zooplankton respond to this increased food level by an increase in numbers until the food supply is essentially exhausted. A decline in numbers results from the shortage of food. In summer the phytoplankton (mainly blue-green and filamentous green algae) may be abundant but are considered unavailable as food for the zooplankton, resulting in low zooplankton densities. In winter phytoplankton growth is limited so that food is again a limiting factor in the growth of zooplankton populations.

The bimodal population curve of *Daphnia galeata mendotae* in this study is not simply the result of a food-limited system. This conclusion stems from considerations of reproductive potential and age-size structure of the *Daphnia* population.

The carrying capacity of the environment in terms of food is not reached during the summer because considerable energy is still available for the production of eggs. A very low average brood size is expected in an equilibrium population (which has reached the carrying capacity of the environment in terms of food supply), for at equilibrium only one egg, on the average, will be produced by each daphnid, and yet the life expectancy of daphnids includes several adult instars. Slobodkin (1954) has observed that equilibrium laboratory populations of *Daphnia* exhibit very little egg production. The average brood size in his equilibrium populations was less than 0.5 eggs/adult.

On one date only (September 24, 1960) did the average brood size fall below 1.0 egg/adult in Base Line Lake. During July and early August the average brood size ranged between 2 and 5 eggs per adult, which is much too large a brood size to be associated with a completely food-limited population.

The population growth rate in spring and fall is initially slowed by a decrease in reproduction indicating lower food levels, but only low enough to reduce the rate of increase and not to cause the rapid decline in population size that was observed in late May and early June.

Thus population size in summer is probably regulated more by predation pressure than by the scarcity of food. The predators are removing the center portion out of what otherwise would be a unimodal population curve. Without predation *Daphnia* would become food limited, but at a considerably higher population density.

Other zooplankton populations may show the same bimodal population curve as a result of similar heavy predation pressures in the summer months. Nelson and Edmondson (1955) report that following the fertilization of Bare Lake, Alaska, the phytoplankton increased rapidly in density. This was followed by increased reproduction of the zooplankton (especially rotifers), but the size of the zooplankton population increased very little. The interpretation is that most of the increased growth of zooplankton populations was immediately consumed by the next trophic level (Nelson 1958).

In a study of *Daphnia retrocurva* in Bantam Lake, Connecticut, Brooks (1946) reports a population maximum of 3,650 *Daphnia*/100 liters in late spring. The density rapidly decreased thereafter, reaching 340 *Daphnia*/100 liters by July 9, and remained at this level for more than one month. The average brood size (mean number of eggs/adult) increased from less than 1.0 in mid-June to 2.5 on July 21. The mean water temperature during this period was 20°C. The similarity of his observations to those of the present study is striking. Again, the interpretation is that the rapid population growth in the spring is initially slowed by a food shortage, but the population size is drastically reduced by predation during the summer. Brooks did not estimate predation rates, however.

Stross *et al.* (1961) report a rapid increase in population size of *Daphnia longispina* in July. Average brood size exceeded 1.0 egg/adult during most of the period of population increase. However, before the population maximum was reached the brood size dropped to about 0.2 eggs/adult and remained at this size for 2 weeks while the population density decreased by a factor of 10. In this case food scarcity may have played a larger role in the control of population size than in the other studies discussed, although predation would still have to play a major role to account for the observed rate of decrease of population size.

A population with constant age-specific survival and reproductive rates will eventually reach a stable age distribution (Lotka 1931). Slobodkin (1954) found that a stable age-size population structure is not achieved in laboratory populations of *Daphnia* because of time lags inherent in the populations. In natural populations a stable age distribution is unlikely because the age-specific survival and reproductive rates change in association with changing environmental conditions.

The high, relatively constant proportions of juveniles in the spring and fall are certainly reflective of a population growing under relatively stable conditions. However, these periods are followed by a rapid shift toward an adult-dominated population indicative of lowered reproduction and /or selection predation upon the juveniles. These interpretations are borne out by observation of the brood size and estimates of population loss (Figs. 7 and 9).

The role of fungal diseases and other parasites conceivably could contribute greatly to the high summer loss rate. However, extensive examinations of field collections during the summer indicated no visibly parasitized *Daphnia*.

Applying developmental rates obtained from but a few clones of *Daphnia* to the entire natural population assumes that no significant genetic variability of developmental rate exists. The validity of this assumption has not been examined.

The frequent occurrence of degenerate eggs in the natural population (Fig. 7) was unexpected from the results of the laboratory studies. Brooks (1946) reports a similar finding in a natural population of *Daphnia retocurva*. He suggests that egg degeneration is caused by inadequate nutrition. He also postulates that brood size is indicative of nutritional level. However, a comparison of brood size and frequency of degenerate eggs in both his and the present investigations indicates that these 2 criteria of nutritional level are inconsistent. Laboratory studies show that brood size is a reflection of the amount of food or energy available. Degenerate eggs may reflect a specific nutritional deficiency, a change in food level, temporary anoxia, or other conditions.

Causal analyses of natural population growth have frequently been attempted using the correlation approach. Although correlations are valuable in indicating the degree of relationship between variables, their biological significance is often difficult to evaluate. Hazelwood and Parker (1961) examined the effect of various environmental factors upon population size of *Daphnia* and *Diaptomus* (Copepoda). Their approach was based entirely upon correlations of several factors with population size. Accurate population estimates are essential if correlations are to be meaningful. Their analyses are based upon single samples at one depth at one station, and, may thus confound distributional variation with population density. Considerable distributional variation was, in fact, borne out by a single series of samples taken August 13, 1959, from different parts of the lake. Little, if any, experimental evidence was obtained about the effects of variables upon population parameters which would have permitted a stronger analysis of the system.

To perform adequate analyses of natural populations, experimental approaches must be utilized. In the present study the effects of different food and temperature conditions upon various attributes of *Daphnia* were found to be important. The range of laboratory temperatures tested was probably adequate to discern the effects on *Daphnia* of the ambient temperature conditions throughout the year. The laboratory food levels extended to much higher levels than those occurring in Base Line Lake. Testing the effects of food levels lower than ¼ Klett unit would have improved the present study.

Reduction of the complexity of natural population systems to a minimal number of variables is needed before any generalities can be made about the operation of such systems. The birth rate model is a beginning in this direction. Its value does not lie so much in the actual prediction of population growth (in the sense of forecasting), for the model requires empirical information about the population during the interval of prediction. Instead, the model is valuable in testing the effects upon population growth of omitting all but a few variables.

Where the model fails to predict growth rates adequately, attention is then focused upon those variables which seem most influential. An experimental analysis of these suspect variables in both the laboratory and the field will provide further insight into the dynamics of the system. Because reproduction appears to be so labile in most zooplankton populations, the birth rate model may prove of widespread use.

Summary

The dynamics of the population of *Daphnia galeata mendotae* in Base Line Lake, Michigan, were analyzed by combining field and laboratory studies. Life-table and fecundity-table experiments were performed at 25, 20, and 11°C. Three food levels were tested at each temperature. Daily or bidaily observations enabled determination of the frequency of molting, duration of egg development, number of newborn per day, individual growth rate, and median life span.

Calculated population rates of increase, r, ranged from 0.07 to 0.51. Temperature influences r more strongly than food level does under the conditions examined. The effects of food and temperature are separable, thereby facilitating the application of experimental data to the field collection. Frequency of molting and reproduction, duration of egg development, and physiological life span are influenced principally by temperature. Growth per instar, maximum carapace length, and brood size are influenced principally by food.

The population in Base Line Lake was sampled for one year. All samples were examined permitting the estimation of population density, size structure, and average brood size. Population density shows a bimodal annual curve with maximal number in late spring and fall. Average brood size shows a trimodal annual curve. Two peaks in brood size precede and apparently cause the density maxima, whereas the 3d peak occurs in midsummer and is not followed by an increased population size. Average brood size ranged between 2 and 4 eggs per brood during most of the reproductive season. Population loss rates caused by the Huron River are relatively constant and small (3% and 1% population loss/day in spring and summer).

Population rates of increase based on birth rate alone, b, are estimated and compared with observed rates of increase, r. In spring, fall, and winter the small differences between b and r (3.9, 5.6, and 4.4% per day respectively) are accounted for by physiological mortality and river loss. Average summer population loss rate of 28% per day seems primarily due to predation. Summer survival curves estimated from field data indicate an average loss rate of 25% per day.

The summer population turnover rate of both numbers and biomass is estimated to be once every 4 days. *Daphnia galeata mendotae* plays an important role in the food chain of the lake during the summer. Population size does not appear to be solely determined by food. Predation pressure prevents the population from being large in the summer.

The birth rate model focuses attention upon the effect of ignoring important variables and should prove valuable in analyzing other zooplankton populations.

Acknowledgments

The author wishes to acknowledge the advice and encouragement of Dr. Frederick E. Smith, Department of Zoology, University of Michigan. Also, appreciation goes to the Horace H. Rackham, School of Graduate Studies, University of Michigan, which awarded the author a scholarship and fellowships during this study.

Literature Cited

Andrews, T. F. 1949. The life history, distribution, growth and abundance of *Leptodora kindtii* (Focke) in western Lake Erie. Abstracts of Doctoral Dissertations, 57. Ohio State University Press.

Banta, A. M., T. R. Wood, L. A. Brown, and L. Ingle. 1939. Studies on the physiology, genetics, and evolution of some Cladocera. Carnegie Inst. Wash., Dept. Genetics, Paper no. 39. 285 pp.

Birch, L. C. 1948. The intrinsic rate of natural increase of an insect population. Jour. Anim. Ecol. **17:** 15-26.

Brooks, J. L. 1946. Cyclomorphosis in *Daphnia*. I. An analysis of *D. retrocurva* and *D. galeata*. Ecol. Monog. **16:** 409-447.

———. 1957. The systematics of North American *Daphnia*. Mem. Conn. Acad. Arts & Sci. **13.** 180 pp.

Chandler, D. C. 1939. Plankton entering the Huron River from Portage and Base Line lakes, Michigan. Trans. Am. Micro. Soc. **58:** 24-41.

Edmondson, W. T. 1955. The seasonal life history of *Daphnia* in an arctic lake. Ecology **36:** 439-455.

———. 1960. Reproductive rates of rotifers in natural populations. Mem. Inst. Ital. Idrobiol. **12:** 21-77.

Elster, H. J. 1954. Über die Populationsdynamik von *Eudiaptomus gracilis* Sara und *Heterocope borealis* Fischer im Bodensee-Obersee. Arch. für Hydrobiol., Suppl. 20: 546-614.

Evans, F. C., and F. E. Smith. 1952. The intrinsic rate of natural increase for the human louse, *Pediculus humanus* L. Amer. Nat. **86:** 299-310.

Fox, H. M. 1948. *Daphnia* hemoglobin. Roy. Soc London Proc. Ser. B., **135:** 195-212.

Frank, P. W. 1960. Prediction of population growth form in *Daphnia pulex* cultures. Amer. Nat. **94:** 357-372.

———, **C. D. Boll, and R. W. Kelly.** 1957. Vital statistics of laboratory cultures of *Daphnia pulex* De Geer as related to density. Physiol. Zool. **30:** 287-305.

Hazelwood, D. H., and R. A. Parker. 1961. Population dynamics of some freshwater zooplankton. Ecology **42:** 266-273.

Lotka, A. J. 1925. Elements of physical biology. Baltimore: Williams and Wilkins. 460 pp.

———. 1931. The structure of a growing population. Human Biol. **3:** 459-493.

McNaught, D. C., and A. D. Hasler. 1961. Surface schooling and feeding behavior in the white bass, *Roccus chrysops* (Rafinesque), in Lake Mendota. Limnol. and Oceanogr. **6:** 53-60.

Morris, R. F. 1959. Single-factor analysis in population dynamics. Ecology **40:** 580-588.

Nelson, P. R. 1958. Relationship between rate of photosynthesis and growth of juvenile red salmon. Science **128:** 205-206.

Nelson, P. R., and W. T. Edmondson. 1955. Limnological effects of fertilizing Bare Lake, Alaska. U.S. Fish and Wildlife Service Fishery Bull. no. 102, 56: 415-436.

Reynoldson, T. B. 1961. A quantitative study of the population biology of *Dugesia lugubris* (O. Schmidt) (Turbellaria, Tricladida). Oikos **12:** 111-125.

Slobodkin, L. B. 1954. Population dynamics in *Daphnia obtusa* Kurz. Ecol. Monog. **24:** 69-88.

Stross, R. G., J. C. Neess, and A. D. Hasler. 1961. Turnover time and production of the planktonic crustacea in limed and reference portion of a bog lake. Ecology **42:** 237-244.

APPENDIX A

Age-specific reproduction of *Daphnia galeata mendotae* under 9 different conditions of food and temperature expressed as total number of offspring per day. The values of N remain unchanged with respect to age in these tables.

Age (days)	Food levels (Klett units)		
	1/4 (N=12)	1 (N=10)	16 (N=10)
TEMPERATURE: 25°C			
6.0	31	75	100
7.0	33	0	0
8.0	27	138	202
9.0	18	0	0
10.0	46	168	237
11.0	48	0	0
12.0	33	166	227
13.0	41	0	0
14.0	58	153	248
15.0	42	0	229
16.0	57	170	36
17.0	33	0	260
18.0	40	185	0
19.0	28	0	258
20.0	38	161	0
21.0	0	0	248
22.0		152	
TEMPERATURE: 20°C			
	(N=8)	(N=9)	(N=10)
7.5	0	0	28
8.5	0	36	6
9.5	8	0	8
10.5	13	12	86
11.5	3	56	42
12.5	22	12	98
13.5	14	7	24
14.5	1	66	80
15.5	45	14	64
16.5	25	8	25
17.5	0	82	171
18.5	45	15	43
19.5	13	117	33
20.5	0	18	193
21.5	21	36	18
22.5	8	98	48
23.5	5	31	141
24.5	14	18	21
25.5	8	92	46
26.5	21	39	129
27.5	0		
28.5	0		
29.5	9		
30.5			
TEMPERATURE: 11°C			
	(N=2)	(N=5)	(N=9)
22	0	5	13
24	0	7	44
26	0	7	4
28	0	7	0
30	8	11	88
32	0	25	80
34	0	0	0
36	0	20	0
38	10	11	140
40	0	21	52
42	0	0	16
44	0	21	0
46	11	12	156
48	0	20	41
50			0
52			0
54			182

APPENDIX B

Daphnia galeata mendotae, Base Line Lake, 1960-1961

Date	(1)[a] Stratum	(2) N_o	(3) N_A	(4) $\bar{E}$	(5) 1/D	(6) b	(7) r observed
8 July 1960	U	667	395	7.3	0.50	1.15	
	M	341	290	7.8	0.39	1.27	
	D	107	88	6.8	0.22	0.81	
(8) Weighted average		N_o^a:401		$\bar{E}$:7.4		b:1.15	0.20
14 July 1960	U	1492	146	3.0	0.50	0.14	
	M	2255	333	3.5	0.39	0.18	
	D	194	14	2.8	0.22	0.04	
	(8)	N_o:1313		$\bar{E}$:3.3		b:0.16	0.15
21 July 1960	U	497	95	4.6	0.50	0.37	
	M	750	571	4.7	0.39	0.87	
	D	94	76	4.0	0.22	0.54	
	(8)	N_o:447		$\bar{E}$:4.6		b:0.66	0.00
29 July 1960	U	778	52	0.5	0.50	0.02	
	M	722	459	3.9	0.39	0.67	
	D	77	31	3.1	0.22	0.25	
	(8)	N_o:450		$\bar{E}$:2.9		b:0.33	0.01
4 Aug. 1960	U	458	0	0	0.50	0.00	
	M	693	142	2.6	0.39	0.19	
	D	89	19	3.3	0.22	0.15	
	(8)	N_o:413		$\bar{E}$:2.7		b:0.12	0.13
11 Aug. 1960	U	245	5	2.0	0.50	0.02	
	M	217	35	3.8	0.39	0.21	
	D	39	16	2.3	0.22	0.19	
	(8)	N_o:167		$\bar{E}$:3.2		b:0.12	0.06
18 Aug. 1960	U	265	118	1.8	0.50	0.34	
	M	400	154	1.1	0.39	0.15	
	D	85	57	1.3	0.22	0.18	
	(8)	N_o:250		$\bar{E}$:1.4		b:0.22	0.06
26 Aug. 1960	(8)	N_o:103					0.01
18 Sept. 1960		128	17	1.4	0.39	0.07	
	(8)	N_o: 91					0.03
24 Sept. 1960		143	45	0.5	0.39	0.06	
	(8)	N_o:106					0.10
8 Oct. 1960	U	827	283	2.3	0.22	0.16	
	M	322	162	2.2	0.22	0.22	
	D	155	49	1.3	0.17	0.07	
	(8)	N_o:434		$\bar{E}$:2.1		b:0.16	0.23
25 Oct. 1960		647	97	2.4	0.17	0.06	
	(8)	N_o:647					0.05
2 Nov. 1960		438	70	2.7	0.13	0.05	
	(8)	N_o:438					0.16
10 Nov. 1960		2337	304	3.4	0.09	0.04	
	(8)	N_o:2337					0.00
19 Nov. 1960	(8)	N_o:2348					0.01

APPENDIX B (Continued)

Date	(1)[a] Stratum	(2) N_o	(3) N_A	(4) $\overline{E}$	(5) 1/D	(6) b	(7) r observed
3 Dec. 1960		2256	332	3.5	0.07	0.03	
	(8)	N_o:2256					0.03
1 Jan. 1961							
	(8)	N_o:941					0.05
28 Jan. 1961							
	(8)	N_o:264					0.01
15 Feb. 1961							
	(8)	N_o:222					0.08
4 Mar. 1961							
	(8)	N_o: 60					0.06
20 Mar. 1961		162	92	1.6	0.05	0.05	
	(8)	N_o:162					0.10
25 Mar. 1961		99	66	1.5	0.05	0.05	
	(8)	N_o: 99					0.03
8 Apr. 1961		152	58	5.7	0.05	0.11	
	(8)	N_o:148					0.02
19 Apr. 1961		180	31	7.1	0.07	0.08	
	(8)	N_o:180					0.04
1 May 1961		307	84	6.2	0.09	0.15	
	(8)	N_o:307					0.15
12 May 1961		1576	342	7.6	0.13	0.19	
	(8)	N_o:1576					0.10
19 May 1961		3225	744	3.4	0.13	0.10	
	(8)	N_o:3225					0.03

Date	(1)[a] Stratum	(2) No	(3) N_A	(4) F	(5) 1/D	(6) b	(7) rob-served
30 May 1961	U&M	11597	1939	2.2	0.22	0.08	
	D	1728	242	2.0	0.09	0.03	
	(8)	N_o:4331				b:0.07	0.01
5 June 1961	U	9177	1379	0.9	0.39	0.05	
	M	2478	426	1.4	0.22	0.06	
	D	667	80	1.5	0.09	0.02	
	(8)	N_o:4124				b:0.05	0.10
21 June 1961	U	1664	306	2.0	0.39	0.13	
	M	776	213	1.8	0.22	0.10	
	D	198	44	1.5	0.09	0.03	
	(8)	N_o:854				b:0.11	0.05
28 June 1961	U	1204	511	2.3	0.39	0.32	
	M	524	237	2.4	0.22	0.22	
	D	170	77	2.7	0.09	0.11	
	(8)	N_o:607				b:0.27	0.03
5 July 1961	U	1275	562	4.8	0.50	0.72	
	M	229	106	3.4	0.22	0.30	
	D	150	46	3.9	0.13	0.14	
	(8)	N_o:510				b:0.61	
12 July 1961	U	709	166	2.1	0.50	.022	
	M	184	141	3.8	0.22	0.50	
	(8)	N_o:238				b:0.28	0.04
21 July 1961	U	831	81	2.5	0.50	0.12	
	M	114	45	2.5	0.22	0.20	
	D	129	59	3.8	0.13	0.20	
	(8)	N_o:268				b:0.13	

[a] (1) Upper, middle, and lower strata of lake; (2) total population size (total/100 liters); (3) total adults (adults/100 liters); (4) average brood size; (5) developmental rate of eggs which is the reciprocal of duration of egg development from Table II; (6) instantaneous birth rate b = ln (1+B); (7) observed instantaneous rate of population increase from equation (1); (8) weighted average of N_o, $\overline{E}$, and b for the entire water column.

Predation, Body Size, and Composition of Plankton

The effect of a marine planktivore on lake plankton illustrates theory of size, competition, and predation.

John Langdon Brooks and Stanley I. Dodson

During an examination of the distribution of the cladoceran *Daphnia* in the lakes of southern New England, it was noted that large *Daphnia*, although present in most of the lakes, could not be found among the plankton of several lakes near the eastern half of the Connecticut coast. The characteristic limnetic calanoid copepods of this region, *Epischura nordenskioldi* and *Diaptomus minutus*, and the cyclopoid *Mesocyclops edax* were also absent. Small zooplankters were abundant, especially the cladoceran *Bosmina longirostris* and the copepods *Cyclops bicuspidatus thomasi* and the small *Tropocyclops prasinus* (*1*).

All of these lakes lacking large zooplankters have sizable "landlocked" populations of the herring-like *Alosa pseudoharengus* (Wilson) = *Pomolobus pseudoharengus* (Fig. 1), known by several common names including "alewife" and "grayback" (*2*). This is originally an anadromous marine fish, breeding populations of which have become established in various bodies of fresh water, including Lake Cayuga, New York, and the Great Lakes (*3*). The marine populations live in the coastal waters of the western Atlantic, from the Gulf of St. Lawrence to North Carolina, and ascend rivers and streams to spawn in springtime. The young return to the sea in summer and autumn (*4*). The seven Connecticut lakes (Fig. 2) with self-perpetuating populations of alewives are within about 40 kilometers of the present coastline, and each is drained directly, by a small stream or river, or indirectly, through the estuaries of the larger Connecticut or Thames rivers, into Long Island Sound (Fig. 3). As such streams and rivers are normally ascended by marine alewives, it is assumed that the establishment of these self-sustaining populations in the lakes is natural.

The authors are affiliated with the Biology Department, Yale University, New Haven, Connecticut, where Dr. Brooks is associate professor of biology.

The "alewife lakes" are diverse in area and depth. Although we have not examined the food of the alewives in these Connecticut lakes (alewives are difficult to catch), studies in other lakes have revealed that planktonic copepods and Cladocera are the primary food. The indifference of alewives to nonfloating food is not surprising in view of the adaptation of the parent stock to feeding on zooplankton in the open waters of the sea (*5*).

The dominant crustaceans in the plankton of all the alewife lakes are the same small-sized species, *Bosmina longirostris* (or *Ceriodaphnia lacustris*) being most numerous; *Cyclops bicuspidatus thomasi* and *Tropocyclops prasinus*, present in varying ratios, are also numerous. By contrast, in the nonalewife lakes *Diaptomus* spp. and *Daphnia* spp. are always dominants, usually accompanied by the larger cyclopoids *Mesocyclops edax* and *Cyclops bicuspidatus thomasi*. The absence of all but one of these last-named larger zooplankters from the lakes inhabited by the planktivorous alewife may be due to differential predation by the alewives. The elimination of these pelagic zooplankters allows the primarily littoral species, such as *Bosmina longirostris*, to spread into the pelagic zone, from which, we conclude, they would otherwise be excluded by their larger competitors.

Changes in Crystal Lake Plankton

An opportunity to test this hypothesis was provided by the introduction into a lake in northern Connecticut of *Alosa aestivalis* (Mitchell), "glut herring," a species closely related, and very similar, to *Alosa pseudoharengus* (*6*). The plankton of Crystal Lake had been quantitatively sampled by one of us in 1942 before *Alosa* was introduced. At that time the zooplankton was dominated by the large forms (*Daphnia, Diaptomus, Mesocyclops*) expected in a lake of its size (see Table 1). Resampling 10 years after *Alosa* had become abundant should reveal plankton similar in composition to that common in the lakes with natural populations of *Alosa pseudoharengus* and unlike that characteristic of Crystal Lake before *Alosa* became abundant.

The plankton of the entire water column of Crystal Lake was sampled quantitatively on 30 June 1964 (*7*). All the crustacean zooplankters caught in the 1942 and 1964 collections (a total of 6623 specimens) were

Reprinted from *Science*, Vol. 150, No. 3692, pp. 28–35, October 1, 1965. Copyright 1965 by the American Association for the Advancement of Science.

identified, and those belonging to each species were enumerated. As the nauplii of all species were lumped and merely enumerated as "nauplii," the figures for copepods given in Table 1 indicate only the percentages of adults and copepodids. The copepodid stages of some coexisting cyclopoids were so similar that differentiation of species was sometimes uncertain. For these species a lumped total is given in the table. The plankton of four lakes with natural populations of *Alosa* was sampled for comparison with that of Crystal Lake in 1964, and the plankton of four lakes, without *Alosa*, in the "alewife" region of southern Connecticut was sampled for comparison with the Crystal Lake plankton of 1942. All these lakes were sampled between 5 June and 7 July 1964 (*8*). The relative frequencies of those crustacean zooplankters which comprise 5 percent or more of the total are given in Table 1. The relatively large predaceous rotifer *Asplanchna priodonta*, the only noncrustacean recorded in Table 1, was remarkably numerous in some of the alewife lakes. The plankton of Crystal Lake in 1964, when *Alosa aestivalis* was abundant, was quite like that of the natural alewife lakes, and not at all like the plankton of Crystal Lake before *Alosa* was a significant element in the open-water community. Crystal Lake in 1942 resembled the lakes without alewives in that its plankton was dominated by *Diaptomus* and *Daphnia*. It might be added that resampling (*9*) of a majority of the other Connecticut lakes after the same 20-year interval has not revealed such a major change in the composition of the zooplankton anywhere else.

In order to examine more carefully the differences in body size between the dominants of alewife and those of non-alewife lakes, the size range of each species was determined. Body size was measured as body length, exclusive of terminal spines or setae. (The posterior limit of measurement for each genus is shown on the drawings of Fig. 4.) A summation of the numbers of each dominant that fell within each size interval yields a size-frequency diagram for the crustacean zooplankters. Although such diagrams were prepared for each lake, only those for Crystal Lake in 1942 and 1964 are presented here (Fig. 4). As size intervals represented by less than 1 percent of the population sample were left blank, the histogram does not indicate the presence of large but relatively rare forms. In Crystal Lake in 1942, specimens of the genera depicted occurred up to a length of 1.8 millimeters, and the predaceous *Leptodora kindtii* was represented by a few specimens between 5 and 10 millimeters. In 1964, by contrast, no zooplankters over 1 millimeter long could be found, although in other *Alosa* lakes there were occasional specimens up to 1.25 millimeters. The nauplii and metanauplii were counted but not measured, so that the totals of measured specimens are decreased by these amounts (see Table 1). The histograms (Fig. 4) show that the majority of the zooplankton are less than about 0.6 millimeter in length when *Alosa* is abundant, whereas the majority of specimens of the dominant species in the same lake before *Alosa* became abundant were over 0.5 millimeter long. The modal size in the presence of *Alosa* was 0.285 millimeter, whereas the modal size in the absence of *Alosa* was 0.785 millimeter. This seems clear evidence that predation by *Alosa* falls more heavily upon the larger plankters, eliminating those plankters more than about 1 millimeter in length.

Effects of Predation by Alosa

Whether or not a species will be eliminated (or reduced to extreme rarity) by *Alosa* predation will in good part depend upon the average size of the smallest instar of egg-producing females; a sufficient number of females must survive long enough to produce another generation. To assess the significance of this critical size, specimens approximating the average size of the smallest mature instar of the dominant species of both years were drawn to scale and appropriately placed on the size-frequency histograms. The smallest mature females of *Daphnia catawba*, *Mesocyclops edax*, *Epischura nordenskioldi*, and certainly *Leptodora kindtii* are too large to have a reasonable chance of surviving long enough to produce sufficient young. However, the elimination of *Diaptomus minutus* (depicted in Fig. 4) but not of *Cyclops bicuspidatus thomasi*, indicates that, for species maturing at a length between 0.6 and 1.0 millimeter, factors other than size (such as escape movements, spatial distribution) are also of significance. Aside from *Cyclops bicuspidatus thomasi*, all the characteristic zooplank-

Table 1. The relative frequency of planktonic Crustacea in lakes with and without *Alosa*. C, Cedar Pond; BA, Bashan Lake; BE, Beach Pond; G, Lake Gaillard; L, Linsley Pond; A, Amos Lake; Q, Lake Quonnipaug; R, Rogers Lake. Relative frequencies expressed in percentages.

Organism	Lakes without alewives				Crystal Lake		Natural alewife lakes			
	C	BA	BE	G	1942 (No *Alosa*)	1964 (*Alosa*)	L	A	Q	R
Cladocera										
Leptodora kindtii		*			*					
Holopedium spp.		*	6			*				5
Diaphanosoma spp.		*		*	5		*		*	
Daphnia galeata	59									
Daphnia catawba		*	7	11	14					
Ceriodaphnia lacustris						*	*	*	43	*
Bosmina coregoni			*	13						
Bosmina tubicens			*							
Bosmina longirostris	*					34	39	44	16	5 10
Copepoda										
Epischura nordenskioldi		*	*	*	*					
Diaptomus minutus		84	76		52					
Diaptomus pygmaeus	*			50	*		*	6	*	5
Mesocyclops edax	21			}12						
Cyclops bicuspidatus	*					}29	34			35
Tropocyclops prasinus							*	11	9	20
Orthocyclops modestus									16	
Nauplii	11	13	10	12	11	28	7	30	8	18
Rotifera										
Asplanchna priodonta						*	17	6	*	
Dimensions										
Area (hectares)	9	112	158	440	80		9	42	45	107
Maximum depth (m)	5	14	19	20	14		14	14	14	20
Mean depth (m)	3	5	6		7		7	6	4	6

* Present, but comprising less than 4.5 percent.

ters of *Alosa* lakes mature at lengths of less than 0.6 millimeter.

Since *Alosa*, except during spawning, avoids the shores, its predation falls more heavily upon, and may eliminate, "lake species" of zooplankton that tend to avoid the shore, allowing littoral or "pond species" to thrive. Among the Cladocera, for example, a large *Daphnia* (*D. catawba, D. galeata*) usually occurs as a dominant in the zooplankton of lakes over 5 meters deep, while *Bosmina longirostris* is a common dominant in shallower bodies of water. When *Alosa* is present, any population of large *Daphnia* is eliminated or severely reduced, and *Bosmina longirostris*, relatively spared by its smaller size and the more littoral habits of a large part of its population, replaces *Daphnia* as the open-water dominant. The differential predation due to the tendency of *Alosa* to feed in open water away from the bottom (*5*) can also be seen among the medium-sized copepods. Of the species usually dominant in lake plankton, *Diaptomus minutus*, the smallest, is eliminated by *Alosa* in small lakes, and in those lakes the larger *Cyclops bicuspidatus thomasi* (see Fig. 4) achieves a dominance that it seldom enjoys otherwise. This differential predation is probably due to the fact that adult *Diaptomus minutus* are primarily epilimnetic in summer, while the adult *Cylops bicuspidatus thomasi* tend to be heavily concentrated in the inshore and bottom waters, only the immature being found in the open water. *Diaptomus pygmaeus*, intermediate between the above two species both in size at the onset of maturity and in spatial distribution of adults, often survives in *Alosa* lakes as a minor component of the plankton.

The alewife lakes of Connecticut are small. It is, therefore, of interest to examine the plankton of larger lakes into which *Alosa* has been introduced. *Alosa* has long been abundant in Lake Cayuga, the largest of the Finger Lakes of New York, with an area of 172 square kilometers and a maximum depth of 132 meters. The plankton of Cayuga is dominated by *Bosmina longirostris* (or *Ceriodaphnia* sp.). Large Cladocera, such as *Daphnia, Leptodora*, and *Polyphemus*, are never common, and *Diaptomus* is scarce. The calanoid *Senecella calanoides* (2.7 mm) is present only at depths below 80 meters. This numerical ascendancy of the smaller zooplankters in the upper waters is consistent with the concept of size-dependent predation by *Alosa*. The persistence of large zooplankters in the Laurentian Great Lakes indicates that *Alosa* has had a less dramatic effect on the plankton of these immense lakes (*10*).

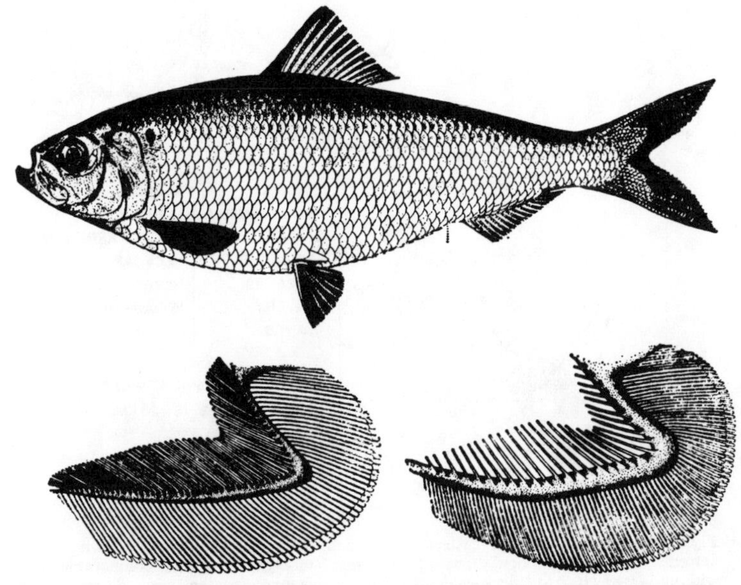

Fig. 1. *Alosa* (= *Pomolobus*) *pseudoharengus* (Wilson). Top, mature specimen, 300 mm long. Note that mouth opens obliquely. Bottom left, first branchial arch, with closely spaced gill rakers that act as a plankton sieve. Compare with (bottom right) the widely spaced gill rakers of *A. mediocris*, a species that feeds primarily upon small fish. [After Hildebrand (*4*), with the permission of the Sears Foundation for Marine Research, Yale University]

Size and Food Selectivity

To assess the significance of *Alosa* predation it is necessary to consider the importance of the size of food organisms throughout the open-water community. To simplify the discussion we shall consider the open-water community of a lake to comprise four trophic levels. Level four, the piscivores, consists chiefly of fish, even though their fry are planktivorous and thus belong to the third trophic level, planktivores. On the third level, also, fish are quantitatively most important (*11*). Level two consists of the herbivorous zooplankters. (Some zooplankters are predators and therefore on level three, but are quantitatively usually negligible.) The planktonic herbivores feed upon the microphytes—the larger algae (net phytoplankton) and the small algae (nannoplankton)—that comprise level one in the open waters, together with bacteria and a variety of nonliving organic particles. The nannoplankton, together with all the other particles that can pass through a 50-micron sieve, will be called nannoseston.

Animals choose their food on the basis of its size, abundance, and edibility, and the ease with which it is caught. However, there is a fundamental difference between food selection by herbivorous zooplankton and that by the predators of higher levels. For planktivores and piscivores, other things being equal, the least outlay of energy in relation to the reward is required if a smaller number of large prey, rather than a larger number of small ones, are taken (*12*). When the environment provides a choice, therefore, natural selection will tend to favor the predator that most consistently chooses the largest food morsel available. At the highest trophic levels, where the number of available prey is often low, a predator must often take either a small morsel or none at all. The variety (*13*) and abundance of the zooplankton, however, usually provide the planktivore with an array of sizes from which to choose. One would expect, therefore, that planktivores would prey upon larger organisms more consistently than do the piscivores. It should further be noted that visual discrimination is indispensable in the feeding of piscine planktivores, even of those such as the alewife, whose gill-rakers serve as plankton strainers (Fig. 1) (*14*).

In the selection of food by herbivorous zooplankters, on the other hand, visual discrimination plays a negligible role; indeed, the absolute dimensions of both herbivore and food particle may well determine the restricted role of vision (*15*). The lower limit of food-particle size for planktonic herbivores is determined by the mechanism that removes these particles from a water current flowing over a part of the body near the mouth. Studies of feeding in rotifers, cladocerans, and calanoid copepods have demonstrated that all can secure particles in the 1- to 15-micron range. This represents the entire range that can be taken by most herbivorous rotifers, while the upper size limit of particles that can also be taken by large cladocerans and calanoids probably accords roughly with the body size of the zooplankter, and commonly includes particles up to 50 microns (*16*). Among food particles of usable size, both rotifers and calanoids can exercise selection by rejecting individual particles, apparently on the basis of chemical or surface qualities, but the chief control that cladocerans exert is by varying the rate of the feeding movements (*17*).

Size-Efficiency Hypothesis

To differentiate between these two types of feeding, the planktivores and piscivores can be called "food selectors," because they continuously make choices, in large part on the basis of size. The herbivorous zooplankters, on the other hand, can be called "food collectors," because the size range of their food is more or less automatically determined. The ecological implications of size-dependent predation upon the array of planktonic food collectors are outlined in what we shall call the "size-efficiency hypothesis":

1) Planktonic herbivores all compete for the fine particulate matter (1 to 15 μ) of the open waters;

2) Larger zooplankters do so more efficiently and can also take larger particles;

3) Therefore, when predation is of low intensity the small planktonic herbivores will be competitively eliminated by large forms (dominance of large Cladocera and calanoid copepods).

4) But when predation is intense, size-dependent predation will eliminate the large forms, allowing the small zooplankters (rotifers, small Cladocera), that escape predation to become the dominants.

5) When predation is of moderate intensity, it will, by falling more heavily upon the larger species, keep the populations of these more effective herbivores sufficiently low so that slightly smaller competitors are not eliminated.

The data supporting this hypothesis are summarized below.

The view that the small particles present in open waters are the most important food element for all planktonic herbivores is supported by the following: Rotifers and large Cladocera (*Daphnia*) and calanoids (for example, *Eudiaptomus*) are all able to collect particles of the 1- to 15-micron range, as noted above. Particles in this range are heterogeneous (algae, bacteria, organic detritus, organic aggregates) and therefore constitute a relatively constant and demonstrably adequate source of food. Also, they are more digestible than many of the net phytoplankton (such as diatoms) which have a covering that impedes digestion and assimilation (*18*).

The competitive success of the larger planktonic herbivores is probably due to (i) greater effectiveness of food-collection; and (ii) relatively reduced metabolic demands per unit mass, permitting more assimilation to go into egg production.

The greater effectiveness of the larger zooplankters in collecting the nannoseston appears to be largely responsible for the replacement of small by larger species in nature, whenever circumstances permit (*19*). The probable basis for this greater effectiveness is the fact that in related species (with essentially identical food-collecting ap-

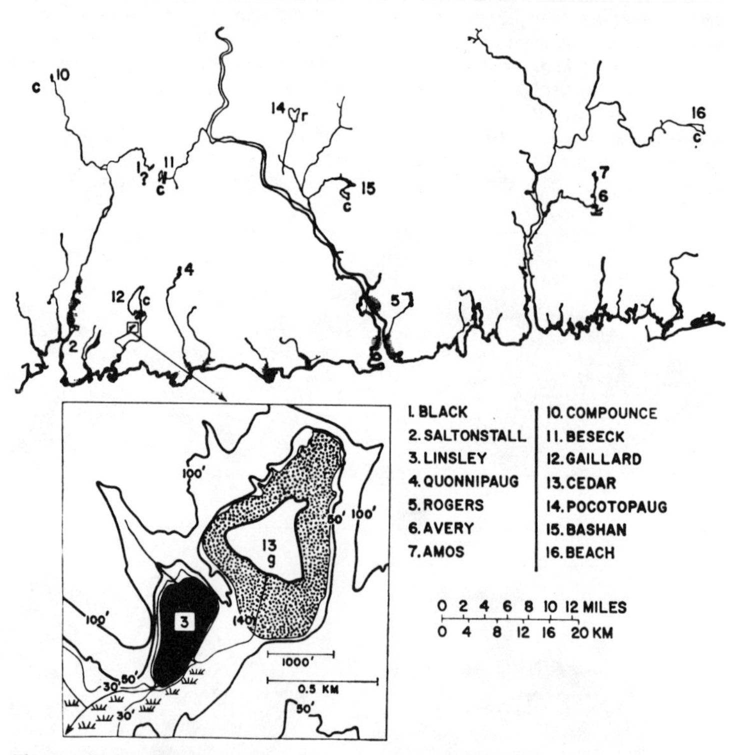

Fig. 2. The coastal strip of eastern Connecticut with the lakes (1–7) known to have natural "landlocked" populations of *A. pseudoharengus*. For the comparable lakes without *Alosa* (10–16), the species of openwater *Daphnia* present are indicated by the following symbols: *c*, *D. catawba; g, D. galeata; r, D. retrocurva*. Large *Daphnia* are missing in all the "alewife" lakes. The bars at the outlets of lakes 11 and 12 indicate that they have been dammed by man. Major intertidal marshes are cross-hatched. The query next to Black Pond (1) indicates that the plankton has not been studied. Inset shows details of Linsley and Cedar ponds. Stippled area around Cedar Pond is bog forest (see Fig. 3).

Fig. 3. Aerial view of Cedar and Linsley ponds (Branford, Connecticut). Cedar, in the foreground, lacks alewives, although they are common in Linsley, into which the outflow from Cedar drains through the surrounding bog forest (lighter in hue). Linsley in turn drains through a short meandering stream into Long Island Sound (Branford Harbor) which can be seen in the upper left corner (see Fig. 2). [Photograph by Truman Sherk]

paratus) the food-collecting surfaces are proportional to the square of some characteristic linear dimension, such as body length. In Crystal Lake, for example, the body length of *Daphnia catawba* is about four times that of *Bosmina longirostris*, so that the filtering area of the *Daphnia* will be about 16 times larger than that of *Bosmina*. Studies by Sushtchenia have shown that the relative rates at which *Daphnia* and *Bosmina* filter *Chlorella* are, indeed, proportional to the squares of their respective body lengths, suggesting that in Cladocera the area of the filtering surfaces is a major determinant of filtration rate. In addition to this greater ability to collect particles in the 1- to 15-micron range, the larger species can also exploit larger particles not available to smaller species; this appears to be especially significant in the greater competitive success of large calanoid copepods (*20*).

There is some indication that both basal metabolic rate and at least a part of the ordinary locomotor activity may be lower per unit mass in the larger than in the smaller of related species of zooplankters, although the depression of the basal metabolic rate may be slight (*21*). Locomotor activity is difficult to relate to body size, but it is likely that for herbivores a considerable proportion of such work is done in overcoming sinking. The rate of passive sinking of zooplankton up to the size of *Daphnia galeata*, with a carapace length of 1.5 millimeters, is proportional to the square of the body length. However, in *D. pulex*, which are about 1 millimeter longer, the rate is almost proportional to the body length itself (*22*). Therefore, locomotor activity probably increases with no more than the square of the length, and in larger forms shows an even lower rate of increase.

Thus both greater efficiency of food collecting and somewhat greater metabolic economy explain the demonstrably greater reproductive success of the larger of related species. This, together with the fact that generation time is but little greater in large cladocerans than in small ones, undoubtedly underlies the rapidity with which dominance can shift in this group (*23*).

The size-efficiency hypothesis predicts that whenever predation by planktivores is intense, the standing crop of small algae will be high because of relatively inefficient utilization by small panktonic herbivores, and that of large algae will also be high because these cannot be eaten by the small herbivores. Whenever the intensity of predation is diminished and large zooplankters predominate, the standing crops of both large (net phytoplankton) and small algae (nannoseston) should be relatively low, because of the greater efficiency of utilization of nannoseston and because some of the net phytoplankton can also be eaten.

When this prediction is tested, it is important that the biomass of the standing crops of large and small zooplankters be more or less equal, and the only data that meet these requirements are those obtained by Hrbáček *et al.* from Bohemian fish ponds with low and high fish stocks (*19*). In this excellent study, the authors made a qualitative and quantitative comparison between the zooplankton, net phytoplankton, and nannoseston of Poltruba Pond in 1957, when the fish stock was low, and the situation in the same pond in 1955, when the fish stock was high. In 1957, Poltruba Pond was also compared with a pond of roughly similar size (Procházka) with a large fish stock. Poltruba drains through a screened outlet. The biomasses (measured as organic nitrogen) of the zooplankton in the three situations were roughly equivalent, but in Poltruba in 1957, a large *Daphnia* comprised 80 percent of the zooplankton, whereas in the other two situations *Bosmina longirostris* was dominant and rotifers and ciliates were common. Where fish stocks were high and *Bosmina* was dominant, the net phytoplankton (especially diatoms and *Dinobryon*) was much more abundant than when *Daphnia* was dominant. Moreover, in both situations with *Bosmina* dominant the standing crop of nannoseston was two to three times greater than in the presence of *Daphnia*. This was true for both its organic nitrogen and its chlorophyll content. (Photosynthesis of the nannoseston in the presence of *Bosmina* was about five times greater than in the presence of *Daphnia*, a clear indication that the increase in the nannoseston in the presence of *Bosmina* was not merely an increase in the amount of slowly dying algae or detritus.) This result is precisely what the size-efficiency hypothesis predicts.

Size of Coexisting Congeners

In both aquatic and terrestrial habitats, pairs of closely related species of food selectors living in the same community and exploiting the same food source often apportion the available size array of food bits, in rough accord with their own divergent body sizes; that is, the larger species takes the larger bits and the smaller one the smaller bits. This is such a common and well-known phenomenon among congeneric birds (coexisting species of which may differ principally in body size, beak size, and size of food taken) that specific instances thereof and its evolutionary significance do not require discussion here (*24*). We wish only to emphasize that food apportioning according to body size is a path to stable coexistence seldom available to planktonic food collectors. Only in rare circumstances could it be advantageous for a species of large planktonic food collectors to abandon the 1- to 15-micron size range in favor of large par-

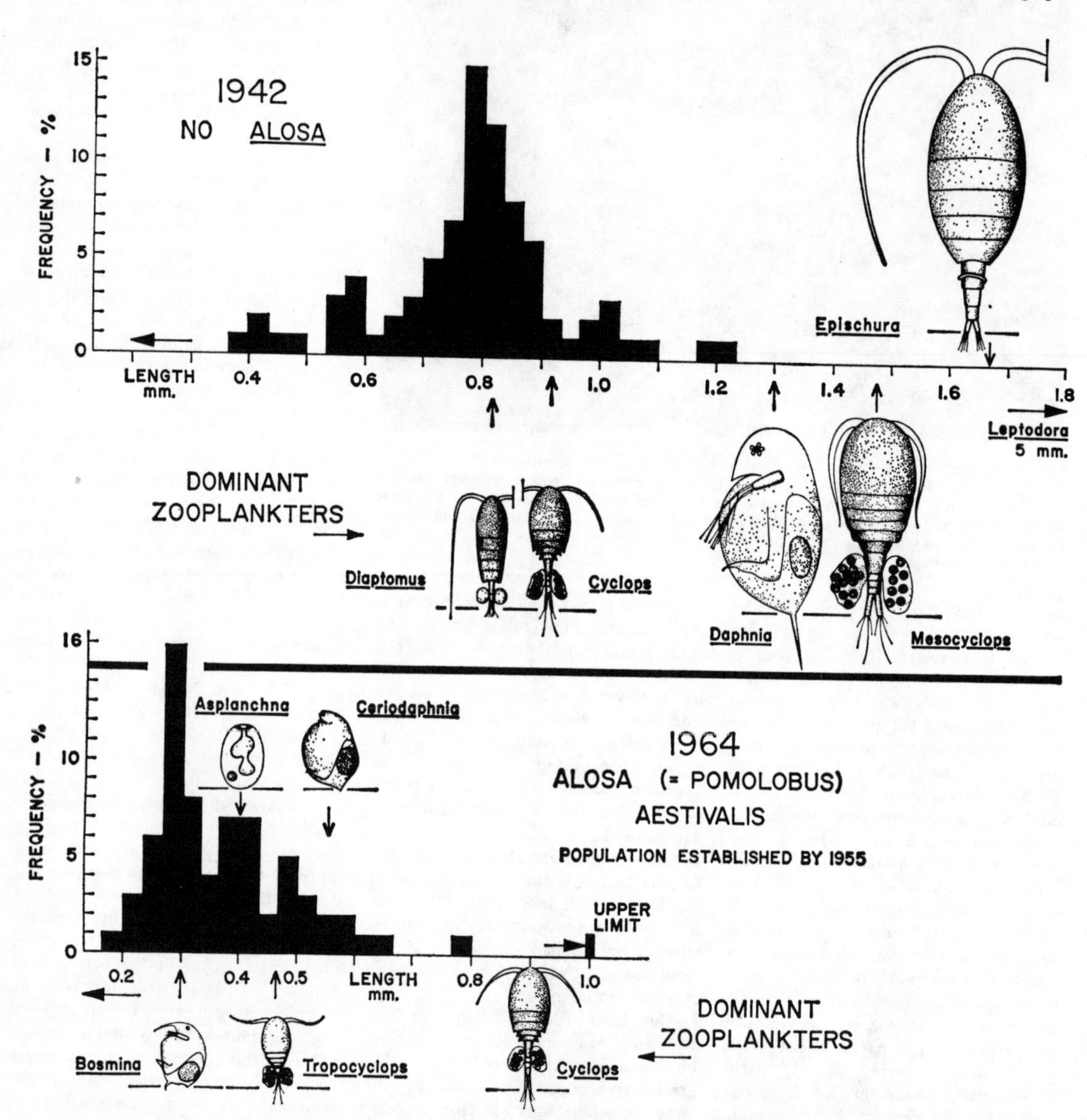

Fig. 4. The composition of the crustacean zooplankton of Crystal Lake (Stafford Springs, Connecticut) before (1942) and after (1964) a population of *Alosa aestivalis* had become well established. Each square of the histogram indicates that 1 percent of the total sample counted was within that size range. The larger zooplankters are not represented in the histograms because of the relative scarcity of mature specimens. The specimens depicted represent the mean size (length from posterior base lines to the anterior end) of the smallest mature instar. The arrows indicate the position of the smallest mature instar of each dominant species in relation to the histograms. The predaceous rotifer, *Asplanchna priodonta*, is the only noncrustacean species included; other rotifers were present but not included in this study.

ticles alone, leaving these small particles to small-sized congeneric competitors (*25*).

On the contrary, many coexisting congeneric zooplankters are of roughly similar size, and presumably—according to the size-efficiency hypothesis—of similar efficiency in food collecting (*26*). This tendency towards similarity in body size can be illustrated by the association in European lakes of *Daphnia galeata* and *D. cucullata*, the latter almost certainly derived from the former. This pair is well suited to our purpose because the various populations of *D. cucullata* exhibit a range of body size unusually large for *Daphnia*. During midsummer, some pond populations of *D. cucullata* mature when the carapace is only about 500 microns long, whereas in lake populations the body size at the onset of maturity may be as much as 900 microns. There is, of course, a complete array of intermediate body sizes in other populations. At the onset of maturity in midsummer, females of *Daphnia galeata* are slightly larger than the largest *D. cucullata*, usually at least 1 millimeter. As all these populations can be passively disseminated, clones of large, intermediate, and small forms of *D. cucullata* have almost certainly been introduced many times into each of the lakes in which *D. galeata* lives. But, as Wagler (*27*) has pointed out after examining 87 European populations of *D. cucullata*, it is only the clones with the largest body size that are found coexisting with *D. galeata*. The dwarf *D. cucullata* would be competitively excluded by *D. galeata* from lakes, just as Hrbáček observed it to be eliminated from ponds by larger species of *Daphnia*, whenever decreased predation allows the larger species to exist. In fact, clones of dwarf forms (carapace length in midsummer less than 550 μ) were found only in ponds where they were associated with *Bosmina longirostris*. It is also clear from Hrbáček's studies of fish ponds that dwarf *D. cucullata*, like *Bosmina longirostris*, can dominate only when predation by planktivores is intense (*28*).

Summary

In the predation by the normally marine clupeoid *Alosa pseudoharengus* ("alewife") upon lake zooplankton, the usual large-sized crustacean dominants (spp. of *Daphnia, Diaptomus*) are eliminated, and replaced by small-sized, basically littoral, species, especially *Bosmina longirostris*. The significance of size in food selection by planktivores as opposed to planktonic herbivores is examined, and it is proposed that all planktonic herbivores utilize small organic particles (1 to 15 μ). The large species, more efficient in collecting these small particles and capable of collecting larger particles as well, will competitively exclude their smaller relatives whenever size-dependent predation is of low intensity. Intense predation will eliminate the large species, and the relatively immune small species will predominate. These antagonistic demands of competition and predation are considered to determine the body size of the dominant herbivorous zooplankters.

References and Notes

1. For classification and distribution of North American *Daphnia*, see J. L. Brooks, *Mem. Conn. Acad. Arts Sci.* **13**, 1 (1957). Identifications and nomenclature of Cladocera, Calanoida, and Cyclopoida in this article accord with accounts by J. L. Brooks, M. S. Wilson, and H. C. Yeatman in *Fresh Water Biology*, W. T. Edmondson, Ed. (Wiley, New York, ed. 2, 1959).
2. C. W. Wilde, *A Fisheries Survey of the Lakes and Ponds of Connecticut* (Connecticut State Board of Fisheries and Game, Hartford, 1959). Although the term "land-locked" is commonly used to refer to self-sustaining freshwater populations of typically marine fish, not all of these populations are barred from returning to the sea, although low dams on most rivers probably isolate them from the anadromous marine stock.
   Opinions appear about equally divided between *Alosa* and *Pomolobus* as the generic name for the alewife. For example, *Pomolobus* is favored by S. F. Hildebrand, in "Fishes of the Western North Atlantic, Part III," *Mem. Sears Found. Marine Res.* **1**, 111 (1963); "*Alosa*" by G. A. Moore in *Vertebrates of the United States* (McGraw-Hill, New York, 1957). Opinion is similarly divided on the use of English names. Hildebrand favors "grayback" for *A.* (= *P.*) *pseudoharengus*, whereas the American Fisheries Society suggests "alewife." J. Hay [*The Run* (Doubleday, New York, ed. 2, 1965)] discusses the ecology of anadromous alewives.
3. Anadromous marine alewives are common, at least during the summer, in many coastal lakes in the Cape Cod area of Massachusetts and along the central portion of the Maine coast. [For references to pertinent surveys see J. L. Brooks and E. S. Deevey, Jr., in *Limnology in North America*, D. G. Frey, Ed. (Univ. of Wisconsin Press, Madison, 1963), pp. 117–62.] Permanent natural populations of *A. pseudoharengus* are known from several freshwaters of northeastern U.S. in addition to the Connecticut lakes. There are resident breeding populations of both this species and *A. aestivalis* (Mitchell) in the lower Mohawk River, New York, and in a few small lakes, different for each species, tributary thereto [*N.Y. State Dept. Conserv. Biol. Surv. Suppl. Ann. Rept.* (1935)]. There is a difference of opinion as to whether the *A. pseudoharengus* population of Lake Ontario is natural or has been artificially established [see R. R. Miller, *Trans. Am. Fisheries Soc.* **86**, 97 (1956)]. Regardless of its origin in Lake Ontario, it has spread into the other Great Lakes through man-made waterways. By the latter part of the 19th century, it had spread from a tributary of Lake Ontario via canals into Lake Cayuga and thence into the two adjoined Finger Lakes, Seneca and Keuka.
4. S. F. Hildebrand, "Fishes of the Western North Atlantic, Part III," *Mem. Sears Found. Marine Res.* **1**, 111 (1963).
5. For food habits in freshwater populations, see T. T. Odell, *Trans. Am. Fisheries Soc.* **64**, 118 (1934); A. L. Pritchard, *Univ. Toronto Biol. Ser. 38* (1929), p. 390. For a summary of food in sea, see S. F. Hildebrand (*4*).
6. *Alosa* (= *Pomolobus*) *aestivalis* (Mitchell) is referred to by several common names, among them "glut herring" which we shall use, and "blueback herring" or "blueback" which are preferred by Hildebrand (*4*). In a previous fisheries survey *Alosa* had not been found in Crystal Lake. See L. M. Thorpe, *Conn. State Geol. Nat. Hist. Surv. Bull. 63* (1942). For the 1955 survey that found *Alosa*, see C. W. Wilde (*2*). The Connecticut State Board of Fisheries and Game suspects that the glut herring was inadvertently introduced into Crystal Lake by the dumping of live bait, with which a few glut herring could have been intermixed, from the Connecticut River, some 15 miles away.
7. Sampled with Forest-Juday plankton trap of 10-liter capacity. In 1942 one sample was taken at each meter depth interval from the surface to just above the bottom. In 1964, two samples (2 meters apart horizontally) were taken at each depth interval.
8. Beach and Bashan lakes were sampled by the method used in sampling Crystal Lake in 1964 (*7*); the others were sampled by vertical tow-nettings, bottom to surface. All specimens taken in each trap series were counted. Subsamples from net collections were counted until a total of 350 to 600 specimens was reached for each lake.
9. J. L. Brooks, unpublished data.
10. E. B. Henson, A. S. Bradshaw, D. C. Chandler, *Mem. Cornell Univ. Agri. Exp. Sta. 378* (1961). Studies of Cayuga plankton: W. C. Muencher, *N.Y. State Dept. Conserv. Biol. Surv. Suppl. Ann. Rept.* (1928), p. 140; A. S. Bradshaw, *Proc. Intern. Assoc. Theor. Appl. Limnol.* **15**, 700 (1964). Bradshaw also gives data on the plankton of Lake Erie. For other references to plankton of the Laurentian Great Lakes, as well as a general summary, see A. M. Beeton and D. C. Chandler in *Limnology in North America*, D. G. Frey, Ed. (Univ. of Wisconsin Press, Madison, 1963), p. 535.
11. This is not to deny that predation by various invertebrates, such as phantom midge larvae (*Chaoborus* spp.) and true plankton predators (*Leptodora*, polyphemids) can be significant in some lakes. However, as predators they are almost certainly not as important as fish.
12. For the significance of size and number in biotic communities, see C. S. Elton, *Animal Ecology* (Sidgwick and Jackson, London, 1927). Analyses of the factors influencing food selection by various freshwater fish are given by V. S. Ivlev, *Experimental Ecology of the Feeding of Fishes*, D. Scott, Transl. (Yale Univ. Press, New Haven, 1961).
13. The ease with which zooplankters are passively dispersed makes it probable that most species present in any continental area will be introduced into a given lake within a reasonably short time (10 to 25 years).
14. M. Kozhoff, *Lake Baikal and Its Life* (Junk, The Hague, 1963); V. S. Ivlev (see *12*).
15. The aberrant cladoceran *Polyphemus pediculus*, although rare, is the smallest widespread member of the open-water community to have an eye sufficiently complex to form a distinct image. This species uses its relatively large, movable, many-lensed eye to locate prey that are 100 to 300 μ in length (or larger). While medium-sized herbivorous zooplankters are the same size as *Polyphemus*, the particulate food they collect is one-tenth the size of the food particles seen and seized by *Polyphemus*.
16. Information on food size in planktonic rotifers is summarized by W. T. Edmondson, *Ecol. Monographs* **35**, 61 (1965). Although *Polyarthra* (150 μ long) can take particles up to 35 μ, this does not vitiate the general statement that 15 μ is the characteristic upper size limit for the food of planktonic herbivorous rotifers.
   Sources of information on the size of food of Cladocera are too numerous to list here. Large *Daphnia*, among the largest of the cladoceran herbivores, are cultured on bac-

teria or algae (*Chlorella*, *Chlamydomonas* spp.) less than 10 μ in length. Most analyses reveal that most of the gut content is unidentifiable small particles, several microns in diameter. Over half the gut content of two large cladoceran species, *Daphnia galeata* and *D. catawba*, from a Maine lake was unidentifiable fine particulate material [D. W. Tappa, *Ecol. Monographs* **35**, 395 (1956)]. The identifiable algae present in the guts of both species were quantitatively similar and represented the forms, up to about 75 μ, that occurred in the water. The only times during the two summers included in the study by Tappa when this fine particulate material constituted less than half of the gut contents was during a *Dinobryon* bloom in July 1961, when the guts of both species were full of (the tests of) *Dinobryon*. *Dinobryon* sp. (single or in small clusters up to about 25 μ) was the most frequent identifiable alga in the guts of both species for the two years, and the diatom *Stephanodiscus* sp. (about 50 μ) was the second most abundant, being the dominant alga about one quarter as frequently as *Dinobryon*. Aside from the short period when the guts were full of *Dinobryon*, there were never more than 20 identifiable algal cells per gut; the mean for each species for all 26 dates during the two summers is less than 10 algal cells per *Daphnia*.

L. M. Sushtchenia [*Nauchn. Dokl. Vysshei Shkoly Biol. Nauki* **4**, 21 (1959)] notes that food particles about 3 μ are filtered more readily by *Bosmina*, *Diaphanosoma*, *Simocephalus* and *Daphnia* spp. than those 15 to 20 μ.

The great differences in the methods of food gathering by various species of cyclopoids makes their inclusion here impossible, but see G. Fryer, *Proc. Zool. Soc. London* **129**, 1 (1957); *J. Animal Ecol.* **26** (1957). The nutrition of *Diaptomus (Eudiaptomus) gracilis* Sars, a common Eurasian calanoid species of medium size (adults about 1.2 mm long), consists of small algae (nannoplankton), and G. Fryer [*Schweiz. Z. Hydrol.* **16**, 64 (1954)] also reported considerable amounts of fine detritus, apparently of vegetable origin, in the guts of this species in Lake Windemere. E. Nauwerck [*Arch. Hydrobiol.* **25**, 393 (1962)] demonstrated that this species in culture could ingest and derive nourishment from particles of animal detritus between 0.1 and 10 μ in diameter. We will consider this well-studied species to be characteristic of most freshwater calanoids, although, of course, larger species are expected to ingest larger particles as well.

17. For observations of selectivity in rotifers on the basis of qualities other than size see works of W. T. Edmondson, especially *Ecol. Monographs* **35**, 61 (1965). See also L. A. Erman, *Zool. Zh.* **41**, 34 (1962).

    Selectivity by *Diaptomus gracilis* in laboratory culture (see *16*) has been noted by A. G. Lowndes [*Proc. Zool. Soc. London* **1935**, 687 (1935)] and for cyclopoid copepods by G. Fryer (see *16*).

    L. M. Sushtchenia, (see *16*), provides data on the rates at which four species of planktonic Cladocera filter suspensions of various algae. The rate of food ingestion in *Daphnia magna* is controlled at various food concentrations by the rate of food collection and at high concentrations by rejection of the excess collected food from the food groove leading to the mouth [J. W. McMahon and F. H. Rigler, *Can. J. Zool.* **41**, 321 (1963); see biblography of that paper for other references to feeding rates].
18. See J. W. G. Lund, *Proc. Intern. Assoc. Theor. Appl. Limnol.* **14**, 147 (1961), for discussion of seasonal occurrence of microalgae in some English lakes. The quantitative importance of bacteria and detritus as food for planktonic herbivores has been mentioned too frequently for listing here. Detritus alone (although probably with some bacteria admixed) has been noted as food, but of poor quality, for *Diaptomus gracilis* (see Nauwerck, *16*) and for *Daphnia* [A. G. Rodina, *Zool. Zh.* **25**, 237 (1946)]. Although organic aggregates have been most thoroughly investigated to date in the sea [see G. A. Riley, *Limnology and Oceanography* **8**, 372 (1963)], their size range in the sea, 1 to 50 μ, is probably also characteristic of fresh waters where they certainly occur (G. A. Riley and P. Wangersky, private communications).

    The relative digestibility of flagellates, small green algae, and some bacteria as compared to *Scenedesmus*, *Raphidium*, and *Pediastrum* was noted by M. A. Kastal'skaia-Karzinkina [*Zool. Zh.* **21**, 153 (1942)]. M. Lefevre [*Bull. Biol. France Belg.* **76**, 250 (1942)], using egg production as a measure, assesses the nutritive value of 21 species of algae and shows that it is small size and fragile cell walls which make algae the most suitable as food.
19. For changes consequent upon alterations in fish stock in Bohemian ponds, see J. Hrbáček, M. Dvořakova, V. Kořínek, L. Procházkóva, *Proc. Intern. Assoc. Theor. Appl. Limnol.* **14**, 195 (1961); J. Hrbáček, *Cesk. Akad. Ved, Rada. Mat. Prirod. Ved* **72**, 1 (1962). W. Pennington [*J. Ecol.* **5**, 29 (1941)] details rapid replacement of rotifers by *Daphnia* in tub cultures of microalgae.
20. Sushtchenia (see *17*), in her Table 3, provides data on the filtration rates of *Bosmina longirostris*, *Diaphanosoma brachyurum*, *Simocephalus vetulus*, and *Daphnia magna*, at several temperatures and concentrations of *Chlorella*. When these results are reduced to the approximate values of 19°C with a standard concentration of *Chlorella* cells, the relative filtration rates of these four species are 1 : 1 : 10 : 18. The relative body lengths (from her Table 2) are 1 : 1.5 : 3 : 4, while the squares of the body lengths are as 1 : 2 : 9 : 16. Thus the filtration rates of these four cladoceran species are approximately proportional to the squares of their respective body lengths.

    Calanoid copepods may be quite unlike the Cladocera in the relationship between body length and rate of food collection, because of a compensatory decrease in the rate at which the head appendages beat. R. Schröder, *Arch. Hydrobiol. Suppl.* **25**, 348 (1961) has shown that in a *Diaptomus* species 2.4 mm long the rate of beating was one-half that in a species 1.2 mm long.
21. A general treatment of the realtionship between size and metabolism is given in E. P. Odum, *Fundamentals of Ecology*, (Saunders, Philadelphia, ed. 2, 1959), pp. 56–59. For zooplankton see G. C. Vinberg, *Zh. Obshch. Biol.* **11**, 367 (1950). Vinberg's "rule" states that decreasing the weight of a plankter by one decimal place increases its O_2 consumption per unit weight by 155 percent (transl. J. Hrbáček, see *19*). For summary of Crustacea in general, see H. P. Wolvekamp and T. H. Waterman in *Physiology of Crustacea*, T. H. Waterman, Ed. (Academic Press, New York, 1960), vol. 1.
22. J. L. Brooks and G. E. Hutchinson, *Proc. Nat. Acad. Sci. U.S.* **36**, 272 (1950).
23. At 20°C the eggs of large limnetic *Daphnia*, such as *D. galeata*, develop into neonates in 2.5 days and become egg layers (in nonturbulent culture) in a week; see D. J. Hall, *Ecology* **45**, 94 (1964). J. Green [*Proc. Zool. Soc. London* **126**, 173 (1956)] gives data relating body size to egg size and clutch size in *Daphnia* and other Cladocera. The longer generation time and the complexity of the life cycle of copepods may obscure the effects of body size upon competition.
24. T. W. Schoener [*Evolution* **19**, 189 (1965)] considers the ecological and evolutionary implications of this phenomenon in birds.
25. G. E. Hutchinson [*Ecology* **32**, 571 (1951)] based his suggestion of food-size selectivity in relation to body size in calanoid copepods on Lowndes's observation (see *17*) of selectivity in the feeding of *Diaptomus gracilis*. Although selectivity is clearly demonstrated by this copepod, it seems likely that attributes other than size of the algae were the basis of this selection. G. Fryer (see *16*) adduced his observation of differential food habits of *Diaptomus gracilis* and *Diaptomus laticeps* in support of such a food-partitioning hypothesis. During February, March, and April, when diatoms are abundant in Lake Windemere, *D. laticeps* (1.6 mm) had their guts full of *Melosira italica*, whereas *D. gracilis* had apparently been feeding primarily on nannoplankton. That it is the larger copepod that takes a somewhat larger food particle is not surprising, but it does not necessarily mean that *D. laticeps* is incapable of feeding on nannoseston under other conditions. Comparison with the food habits of the *Daphnia* in Aziscoos Lake, Maine (Tappa, see *16*), may be useful. One might assume that *Daphnia galeata* is a selective feeder on the basis of the July 1961 samples at which time the guts were crammed with *Dinobryon*. However, at all other times during the open water season of 1961 and all of 1962, over half of the gut was filled with ingested nannoseston. While copepods can be highly selective feeders, their discriminations do not appear to be based primarily upon size. To summarize, at certain times in certain ecosystems a population of a large-sized calanoid may ingest large food particles not utilizable by a coexisting, small-sized, congener. But it does not necessarily follow that the larger species always feeds exclusively on such large particles.
26. Such a stable association of *Daphnia galeata* and *Daphnia catawba* is not uncommon in the lakes of central Connecticut (J. L. Brooks, unpublished). The nature of the association between these two species in a Maine lake has been carefully examined by D. W. Tappa (see *16*). K. Patalas [*Roczniki Nauk Rolniczych Ser. B* **82**, 209 (1963)] presents data on the seasonal changes in the crustacean plankton of several Polish lakes and considers the relation of fish predation to the population size of the various competing zooplankters.
27. E. Wagler, *Intern. Rev. Ges. Hydrobiol. Hydrog.* **11**, 41, 262 (1923).
28. For a possible relationship between the seasonal incidence of predation and seasonal changes of size and form in *Daphnia* (cyclomorphosis), see J. L. Brooks, *Proc. Nat. Acad. Sci. U.S.* **53**, 119 (1965).
29. We thank the following for their assistance: D. W. Tappa for help with field work; Dr. S. Jacobson, New Haven Water Co., for permission to sample Lake Gaillard; J. Atz, American Museum of Natural History, for sharing his extensive knowledge of fish and their literature; G. E. Hutchinson and W. T. Edmondson for making translations of Erman and Sushtchenia available; and C. Goulden, G. E. Hutchinson, and G. A. Riley for critical perusal of the manuscript. S.I.D. was supported in part by the NSF Undergraduate Science Education Program at Yale University. The research of J.L.B. has been supported by grant GB1207 from the National Science Foundation.

Stock and Recruitment[1]

By W. E. Ricker

Pacific Biological Station, Nanaimo, B.C.

ABSTRACT

Plotting net reproduction (reproductive potential of the *adults* obtained) against the density of stock which produced them, for a number of fish and invertebrate populations, gives a domed curve whose apex lies above the line representing replacement reproduction. At stock densities beyond the apex, reproduction declines either gradually or abruptly. This decline gives a population a tendency to oscillate in numbers; however, the oscillations are damped, not permanent, unless reproduction decreases quite rapidly *and* there is not too much mixing of generations in the breeding population. Removal of part of the adult stock reduces the amplitude of oscillations that may be in progress and, up to a point, *increases* reproduction.

INTRODUCTION

General

There exists today a considerable body of knowledge which goes by the name of "the theory of fishing" or "the modern theory of fishing"—the work of a succession of the most distinguished fishery biologists of our time. It is concerned mainly with predicting what catch can be obtained from a given number of young fish recruited to a fishery, if their initial size and the growth and natural mortality rates prevailing are known. That is, methods have been developed for computing the effects of different rates of exploitation, of changes in rate of exploitation from year to year, of different minimum size limits, etc., upon the yield obtained. Not only that, but much progress has been made in developing methods of determining the actual magnitudes of the population statistics required to make these calculations.

Valuable as the above contributions have been, they comprise only half of the biological information needed to assess the effects of fishing and an optimum level of exploitation. Fishing changes the absolute and the relative abundance of mature fish in a stock, and the effect of this upon the number of recruits in future years has often been considered only in the most general manner. The points of view encountered usually range from an assumption of direct proportion between size of adult stock and number of recruits, to the proposition that number of recruits is, for practical purposes, independent of the size of the adult stock. The possibility of a *decrease* in recruitment at higher stock densities has less often been considered.

The scarcity of information on this subject is quite explicable, since it usually requires many years of continuous observation to establish a relation between size of stock and the number of recruits which it produces. However, it has become an urgent problem to have a scientific description of the regulation of abundance of fish stocks, in order to complete the basis for predicting optimum levels of exploitation. This paper attempts to summarize some of the theoretical and factual information available, both from fish populations and from other animals, and to provide a stimulus to studies which will eventually put the subject on a solid foundation.

Reprinted from *Journal of the Fisheries Research Board of Canada*, 11:559–586; 609–623, 1954.

Theory of Population Regulation

Basic in any stock-recruitment relationship is the fact that a fish population, even when not fished, is limited in size; that is, it is held at some more or less fluctuating level by natural controls. Ideas concerning the nature of such controls were first clarified and systematized by the Australian entomologist Nicholson (1933). He showed that, while the level of abundance attained by an animal can be affected by any element of the physical or biological environment, the immediate mechanism of control must always involve competition, using that word in a broad sense to include any factor of mortality whose effectiveness increases with stock density[2]. The term *density-dependent* mortality was used for the same concept by Smith (1935). More strictly, density-dependent causes of mortality should include both those which become more effective as density increases and those which become less so. The former are the ones which provide control of population size; and they have been called *concurrent* (Solomon, 1949), *compensatory* (Neave, 1953) or *negative* (Haldane, 1953). The opposed terms are *inverse*, *depensatory* and *positive*, all referring to density-dependent factors which become *less* effective as density increases.

No sharp line can be drawn between the kinds of mortality which are compensatory and those which are not, although Nicholson, Smith and others have felt that as a rule biological factors tend to predominate among the former, and physical agents among the latter, for insects at least. Among fishes, extremes of water temperature, drought and floods, are physical agents which may often cause mortality whose effectiveness is independent of stock density; whereas deaths from such biological causes as disease, parasitism, malnutrition and predation will usually become relatively more frequent as stock density increases. Yet exceptions to the rule above are sufficiently numerous to make the rule itself of doubtful applicability to fishes. For example, the biological factor of predation may have a uniform effectiveness over a considerable range of prey abundance, or at times may even become more effective at lower prey densities; and most if not all physical causes of mortality are compensatory when stock becomes dense enough that some of its members are forced to live in exposed or unsuitable environments. In addition, it is of course often difficult to ascribe a death to any single cause.

There is no necessary relation between the relative magnitudes of the causes of mortality existing at a given time, as measured by the fraction of the stock which each kills, and their relative contribution to compensation. An important and deadly agent of mortality may be strongly density-dependent, or weakly so, or not at all; and different agents may have their maximum compensatory effect over quite different ranges of density.

Age Incidence of Compensatory Mortality

Density-dependent causes of mortality could affect the abundance of either the existing adult stock or the young which it produces. In this paper we will

[2]This almost axiomatic proposition is implied in the writing of various earlier authors back as far as Malthus, but Nicholson was the first to formulate it explicitly and to emphasize its importance: Haldane (1953) calls his inspiration "a blinding glimpse of the obvious". The theorem and its diverse consequences were elaborated mathematically by Nicholson and Bailey (1935). Subsequent writers have developed it with varying emphasis, but nothing very substantial seems to have been added. Solomon (1949) gives a useful review of this literature, and Varley's (1947) quantitative assessment of various agents controlling the abundance of a trypetid fly population is outstanding.

consider mainly the effects of density dependence in the mortality which strikes the younger members of a population—among fishes, the eggs, larvae, fry and fingerlings. That is, the relative abundance of a brood will be considered to be determined by the time the first of its female members begin to mature: subsequent mortality is assumed to be non-compensatory.

This distinction between immature and mature stages of the life history, and the restriction of compensatory mortality to the former, is the principal difference between the thesis of this paper and that of earlier treatments of effects of density upon reproduction (e.g., Hutchinson, 1948; Haldane, 1953; Fujita and Utida, 1953; and many others). These have usually taken the Verhulst logistic equation as a point of departure:

$$\frac{dN_t}{dt} = \frac{bN_t(K - N_t)}{K}$$

(N_t is abundance at time t; K is equilibrium abundance; b is the instantaneous rate of increase of the population at densities approaching zero.) This expression implies a continuous tendency for the population to adjust itself toward the equilibrium size K, whether it is currently less than or greater than K. The adjustment would evidently have to involve compensatory mortality among the adult stock when its density is greater than K. Though such mortality is not impossible, there is in fish populations, at least, little indication of it; and as a matter of fact this aspect of the logistic equation has never been applied to any concrete biological situation.

Our assumption that *no* compensatory mortality occurs among the mature stock is unlikely to be strictly true of any species, and may not be even approximately true of some. Nevertheless it has seemed worth while to follow out the consequences of making this distinction between mature and immature, with fishes particularly in mind. The opportunities for compensatory effects are so much greater during the small, vulnerable early stages of a fish's life, that restriction of compensation to those stages seems likely to have wide applicability as a useful approximation.

To further simplify the initial approach to the problem, fishing is considered to attack only mature individuals, so that the *recruitment* produced by a given density of mature stock means both the number of commercial-sized fish, and the number of maturing fish, which result from its reproductive activity. The term *reproduction* will be used in a similar but slightly more general sense, to mean the number of young surviving to any specified age after compensation is practically complete; it will *not* mean the initial number of eggs or newborn young.

The above conditions and definitions will be assumed to apply unless exception is made specifically.

Kinds of Population Control Mechanisms

Compensatory types of mortality can be of various kinds. Some of the more likely possibilities are as follows:

1. Prevention of breeding by some members of large populations because all breeding sites are occupied. Note that territorial behaviour may restrict the number of sites to a number less than what is physically possible.

2. Limitation of *good* breeding areas, so that with denser populations more eggs and young are exposed to extremes of environmental conditions, or to predators.

3. Competition for living space among larvae or fry, so that some individuals must live in exposed situations. This too is often aggravated by territoriality—that is, the preemption of a certain amount of space by an individual, sometimes more than is needed to supply necessary food.

4. Death from starvation or indirectly from debility due to insufficient food, among the younger stages of large broods, because of severe competition for food.

5. Greater losses from predation among large broods because of slower growth caused by greater competition for food. It can be taken as a general rule that the smaller an animal is, the more vulnerable it is to predators, and hence any slowing up of growth makes for greater predation losses. Since abundant year-classes of fishes have often been found to consist of smaller-than-average individuals (Hile, 1936), this may well be a very common compensatory mechanism among fishes (cf. Ricker and Foerster, 1948; Johnson and Hasler, 1954).

6. Cannibalism: destruction of eggs or young by older individuals of the same species. This can operate in the same manner as predation by other species, but it has the additional feature that when eggs or fry are abundant the adults which produced them tend to be abundant also, so that percentage destruction of the (initially) denser broods of young automatically goes up—provided the predation situation approaches the type in which kills are made at a constant fraction of random encounters (cf. page 609, below).

7. Larger broods may be more affected by macroscopic parasites or micro-organisms, because of more frequent opportunity for the parasites to find hosts and complete their life cycle.

8. In limited aquatic environments there may be a "conditioning" of the medium by accumulation of waste materials that have a depressing effect upon reproduction, increasingly as population size increases.

Not all of the above compensatory effects need exist, or be important, in any given population, or in the same population every year. To a considerable extent they are likely to be complementary, so that if, for example, exceptionally favourable conditions permitted a good hatch of even a large spawning of eggs, a reduced growth rate of the fry would permit increased predation and so reduce survival in that way.

TYPES OF REPRODUCTION CURVES

Whatever the various kinds of compensatory and non-compensatory mortality acting on a brood may be, the average resultant of their action, over the existing range of environmental conditions, is represented by the average size of maturing brood, or recruitment, which each stock density produces. A graph of this relationship between an existing stock, and the future stock which the existing stock produces, will be called a "reproduction curve". It is most convenient to label the axes in terms of the *eggs* in present and future generations, respectively[3]. The abscissa represents the mature eggs produced by the current year's stock. The ordinate represents the total of mature eggs produced by the progeny re-

[3]The argument is developed here in terms of populations of oviparous fishes which spawn once a year, but it can readily be modified to apply to other kinds of animals; an example for a viviparous animal is given on page 596. Among mammals, choice of the most suitable census age for plotting on reproduction curves may require care. If newborn young are used, effects of stock density upon frequency of conception and uterine mortality may be overlooked. If number of mature females is used, it should preferably be adjusted to take care of age variation in litter size or frequency, as the age structure of the population changes.

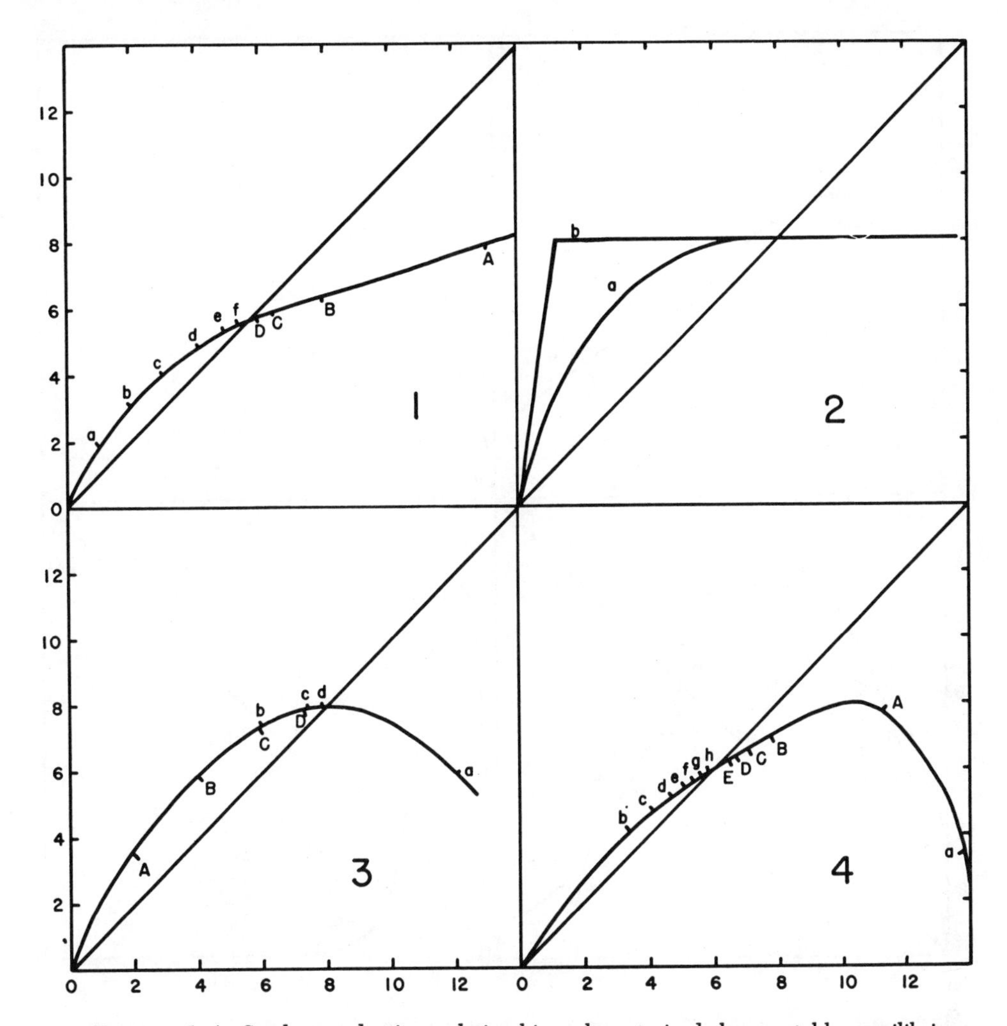

FIGURES 1–4. Stock-reproduction relationships characterized by a stable equilibrium. *Abscissa*— number of eggs produced by parent stock in a given year; *ordinate*—number of eggs produced by the progeny of that year.

sulting from the current year's reproduction (obtained by summing over such period of time as the current year's hatch is a component of future years' stocks). In a state of nature, or with a stable fishery, the average size of parental and filial egg production, defined as above, tends to be equal over any long period of years, although striking changes may occur between individual years, or generations.

Figures 1–8 show a number of possible types of reproduction curve. In each of them the straight diagonal constitutes a useful boundary of reference which will be called the "45-degree line". Any curve lying wholly above this line describes a stock which is increasing without limit, hence such a curve cannot exist in practice. Similarly a curve below the 45-degree line describes a stock

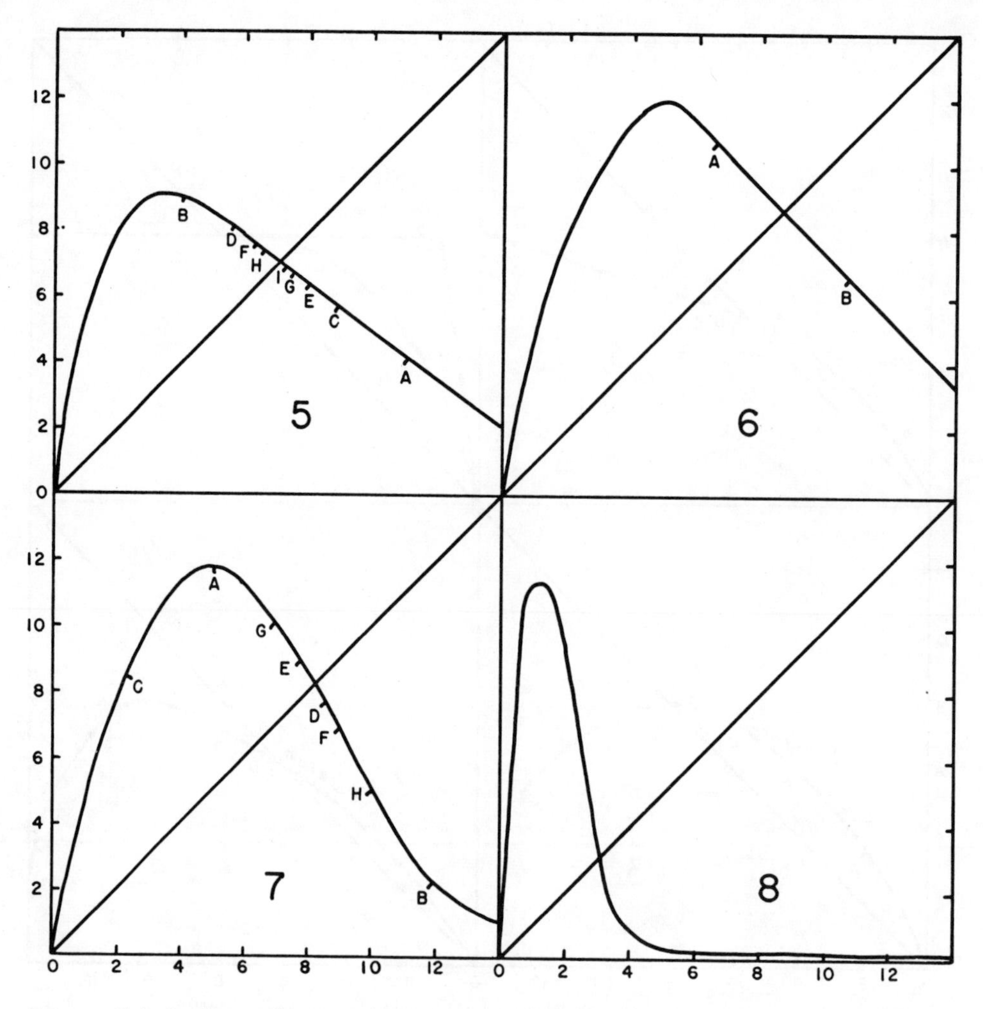

Figures 5–8. Stock-reproduction relationships in which there is an oscillating equilibrium, or (in 5) an oscillating approach to stable equilibrium. Axes as in Figures 1–4.

that will decrease to zero in a few generations[4]. The 45-degree line itself would describe a stock in which density dependence is absent, the filial generation tending always to be equal to the parental, except as factors independent of density deflect it. Such a stock would have no mechanism for the regulation of

[4]This statement is likely to be literally true only if the reproduction curve is plotted and fitted on logarithmic axes. On arithmetic axes a curve lying wholly below the 45-degree line *could* describe a stable population (having more than one age in the breeding stock) provided there were moderate to large random deviation from the average relationship. This point is discussed further on page 578, where it is evident that it would be advantageous to fit reproduction curves on logarithmic axes so that deviations above and below the curve would be in better balance. However such plotting has the serious disadvantage that statements of parent-progeny relationships in terms of the slopes of the two limbs of the curve become too complex to be practical. The best way out of this difficulty would be to fit the curve on a logarithmic plot, then transform it to arithmetic axes for interpretation.

its numbers: if only density-independent causes of mortality exist, the stock can vary without limit, and must eventually by chance decrease to zero. Thus the first qualification of a reproduction curve is that it must cut the 45-degree line at least once—usually only once—and must end below and to the right of it.

Figures 1–8 indicate some of the types of recruitment curve that might exist in actual populations. All are characterized by having a region above the 45-degree line in which reproduction is more than adequate to replace the existing stock, and a region below the 45-degree line in which reproduction is inadequate to replace existing stock.

In the population of Figure 1, rate of reproduction (ratio of filial to parental eggs) decreases continuously as size of stock increases, although the slope of the curve becomes stabilized soon after the 45-degree line is crossed, and the actual number of young produced continues to rise indefinitely. Beyond the 45-degree line, however, this number is inadequate to fully replenish the stock.

In the population of Figure 2a, rate of reproduction also declines continuously; the actual number of recruits produced reaches an asymptotic level and thereafter does not change. Curve 2b differs from 2a in that it rises more steeply and it is initially straight; i.e., at low stock densities rate of reproduction is large, and constant. The range of densities over which there is constant recruitment is quite broad in b, simulating the condition loosely described by the statement "there are always as many recruits as the grounds can support".

Figures 3–8 show reproduction curves in which numbers of recruits begin to decrease after stock reaches some large magnitude. Such curves have an ascending left limb, a dome and a descending right limb. The descending right limb provides of course a more severe control of stock size at the higher densities. The differences in position of the dome and in the slope of the two limbs are of importance in determining changes in stock abundance, as described below.

All these curves are meant to represent the net effect of the sum total of density-dependent mortality factors acting upon the population. The reproduction of any actual year is affected also by density-independent factors, so that the actual number of young produced will deviate from the number indicated by the curve. In a later section the effect of such density-independent factors will be examined. The first task will be to consider how a population behaves, under the strict control of a reproduction curve, when it is given some initial randomly chosen position.

This subject, and others later, will be considered for two situations separately. The first is where the currently hatched brood will constitute the whole of the breeding stock of a subsequent generation; that is, there is no mixing of ages in the spawning stock. This situation is fairly common among insects; for example, many mayflies, stoneflies, caddis flies, etc., have a single brood per year, usually with a one-year life cycle. Among vertebrates this condition is exceptional, but it exists in some fishes. There the length of life is commonly more than one year, so that two or more separate populations or "lines" exist concurrently.

The other situation to be considered is of course that where two or more broods contribute to the spawning stock at any given moment. Among vertebrates at least, this is the usual state of affairs.

REPRODUCTION IN THE ABSENCE OF DENSITY-INDEPENDENT VARIABILITY

SINGLE-AGE SPAWNING STOCKS

In all of the relationships shown in Figures 1–8, the stock is in equilibrium at the density at which the reproduction curve cuts the 45-degree line—that is, the stock is then producing enough progeny, and only enough, to replace its current numbers. In Figure 1, an initial deflection of abundance to either side of the equilibrium point is compensated by a gradual, asymptotic return to equilibrium, for example by paths A–D or a–f. In Figure 2a things work the same way to the left of the equilibrium point; to the right of it any deflection, no matter how great, is returned to equilibrium in a single generation. In Figure 2b deviations to the left, as far as 1.5 units of stock, are also returned immediately to the equilibrium level.

In Figure 3 the equilibrium point is at the top of the dome of the reproduction curve. The curve resembles 1 as regards deflections to the left of the equilibrium point; but a displacement to the right is followed by an immediate return across the 45-degree line to the ascending limb, after which it "climbs" this limb to the equilibrium point (a–d). The curve of Figure 4 is similar, but the dome lies to the right of the 45-degree line; A–E and a–h are possible paths to the equilibrium point.

In Figure 5, a stock deflected from equilibrium to any position along the descending limb will oscillate back and forth about the equilibrium point with decreasing amplitude, for example by the route A–I. If the deflection is great enough, the stock may have to climb the left limb before it is swung over to the right limb and begins the oscillating phase.

In Figure 5 the right limb has a downward (negative) slope numerically less than —1. In Figure 6 this slope is exactly —1 after the dome is passed. Here any moderate deflection along the straight part of the right limb results in a swing back across the 45-degree line of exactly the same magnitude, so that the deflection tends to be perpetuated indefinitely; for example, A and B are two such conjugate points. The intersection of the 45-degree line is a point of "indifferent" equilibrium; it is itself stable, but there is no inherent tendency for the stock to return to that level.

Finally, when the slope of the right limb of a reproduction curve lies between —1 and $-\infty$, equilibrium at the 45-degree line is not merely indifferent, it is unstable. That is, any deflection from equilibrium, no matter how small, initiates a series of oscillations along the right limb whose amplitude increases until the dome of the curve is reached or surpassed. The latter event usually sends the stock back to the right limb and the cycle begins again. No matter where they begin, all such cycles eventually reach the dome of the curve, and a stable oscillation series is established for which the dome is a convenient starting point. The cycle in Figure 7 is A, B, C, D, E, F, G, H, A, etc. (cf. Fig. 11D, below); the number of stages in this cycle depends upon the exact shape of the curve, chiefly upon whether or not one stage lands close to the 45-degree line. Figure 8 represents a more extreme situation, in which substantial reproduction is obtained over only a narrow range of stock densities considerably below the equilibrium level, and the stock would be subject to violent oscillations.

TABLE I. An illustration of how to obtain the definitive distribution of population abundance for given conditions and a given reproduction curve, in this case Figure 7. Four years are assumed to elapse between a brood's existence as fertilized eggs and its first year of contribution to reproduction, while its total egg production is divided among the 4 successive years of its mature existence in the ratio 2 : 3 : 3 : 2. An arbitrary stock density (measured in terms of its egg production) is taken as a starting point, in this case 2.0 units (column 5). The reproduction corresponding to this abscissal value is read off from the ordinate of Figure 7, namely 7.9 egg units (column 6). These eggs are divided among the 4 years in which they are actually laid, beginning 4 years after the year in which the fish that carry them were hatched. For example, the 10.2 units of "recruitment" in line 5 are divided among line 9 (2.0 units), line 10 (3.1 units), line 11 (3.1 units), and line 12 (2.0 units). Breeding potential for each year (column 5) is obtained by horizontal addition of the contributions of the 4 age-groups in columns 1–4. In this example the definitive 11-year period of oscillation is achieved in the first peak-to-peak interval, while the average definitive amplitude, approximately 3.6 to 11.3, is apparent in the second peak-to-peak interval. Adjustment to the conditions imposed by the reproduction curve is not always as rapid as this.

Eggs produced at successive ages					
IV	V	VI	VII	Stock	Recruitment
0.4	0.6	0.6	0.4	2.0	7.9
0.4	0.6	0.6	0.4	2.0	7.9
0.4	0.6	0.6	0.4	2.0	7.9
0.4	0.6	0.6	0.4	2.0	7.9
1.6	0.6	0.6	0.4	3.2	10.2
1.6	2.4	0.6	0.4	5.0	11.9
1.6	2.4	2.4	0.4	6.8	10.4
1.6	2.4	2.4	1.6	8.0	8.7
2.0	2.4	2.4	1.6	8.4	8.0
2.4	3.1	2.4	1.6	9.5	6.1
2.1	3.6	3.1	1.6	10.4	4.6
1.7	3.1	3.6	2.0	10.4	4.6
1.6	2.6	3.1	2.4	9.7	5.8
1.2	2.4	2.6	2.1	8.3	8.2
0.9	1.8	2.4	1.7	6.8	10.4
0.9	1.4	1.8	1.6	5.7	11.5
1.2	1.4	1.2	1.2	5.2	11.9
1.6	1.7	1.4	0.9	5.6	11.7
2.1	2.5	1.7	0.9	7.2	9.9
2.3	3.1	2.5	1.2	9.1	6.8
2.4	3.5	3.1	1.6	10.6	4.2
2.3	3.6	3.5	2.1	11.5	2.8
2.0	3.5	3.6	2.3	11.4	3.0
1.4	3.0	3.5	2.4	10.3	4.8
0.8	2.0	3.0	2.3	8.1	8.5
0.6	1.3	2.0	2.0	5.9	11.4
0.6	0.8	1.3	1.4	4.1	11.5
1.0	0.9	0.8	0.8	3.5	10.7
1.7	1.4	0.9	0.6	4.6	11.8
2.3	2.6	1.4	0.6	6.9	10.3
2.3	3.4	2.6	1.0	9.3	6.4
2.1	3.4	3.4	1.7	10.6	4.2
2.4	3.2	3.4	2.3	11.3	3.1
2.1	3.5	3.2	2.3	11.1	3.5
1.3	3.1	3.5	2.1	10.0	5.3
0.8	1.9	3.1	2.4	8.2	8.3
0.6	1.3	1.9	2.1	5.9	11.4
0.7	0.9	1.3	1.3	4.2	11.6
1.1	1.0	0.9	0.8	3.8	11.1
1.7	1.6	1.0	0.6	4.9	11.9
2.1	2.5	1.6	0.7	6.9	10.3
2.3	3.1	2.5	1.1	9.0	7.0
2.2	3.5	3.1	1.7	10.5	4.4
2.4	3.3	3.5	2.1	11.3	3.1
2.1	3.6	3.3	2.3	11.3	3.1
1.4	3.1	3.6	2.2	10.3	4.8
0.9	2.1	3.1	2.4	8.5	7.8
0.6	1.3	2.1	2.1	6.1	11.2
0.6	0.9	1.3	1.4	4.2	11.6
1.0	0.9	0.9	0.9	3.7	11.0

Multiple-Age Spawning Stocks

When a spawning population consists of two or more age-groups, the young produced in a given year contribute to the stock of more than one future year, and the results of a deflection from equilibrium abundance are much modified. A fairly plausible example is where each brood contributes to the spawning stocks of four future years, in the ratio 2:3:3:2, and first spawning occurs 4 years after a brood was produced (Table I). Figure 9 shows the result of an initial deflection of such a stock to an abundance of 12, on some of the reproduction curves of Figures 1–8. The course of events for the most part reflects what was learned in the single-spawning situation. Those based on Figures 1–6 all end up at the stable equilibrium level; this being reached by direct approach for 1, 2 and 3, with one hesitation for 4, and by a series of damped oscillations for 5 and 6. From Figures 7 and 8 series of undamped oscillations are obtained, that is, permanent cycles of abundance.

It may seem surprising that curve 6 too does not generate permanent oscillations, but the mixing of year-classes gradually brings the stock to a steady level, in the absence of any tendency toward divergence.

The permanent oscillations of the type produced by curves 7 and 8 will repay more extended discussion. Figure 10 depicts series of oscillations, based on Figure 7, but with each brood contributing to the spawning stock for only two years. Time of first maturity is successively delayed one, two and more years, so that the average contribution to reproduction is made 1.5, 2.5, 3.5, etc., years after the brood in question existed as mature eggs. In every case stable cyclical fluctuations exist, just as in Figure 9F. Their *period* is always double the mean interval from egg to egg, that is, 3, 5, 7, etc., years.

The *amplitude* of the cycle varies. If the fish spawn first in the year after their appearance, amplitude is very small, but it quickly increases if maturity is delayed. More generally, appreciable amplitude in cycles of this type depends upon a preponderance of the reproduction of a brood occurring after one or more spawning periods have elapsed from the time of its birth.

With the longer intervals of lag between hatching and first spawning the cycles of Figure 10 become less regular; minor peaks appear, and the pattern is duplicated more closely at 2-cycle intervals, a feature which can be detected also in Figure 9F. However, these tendencies are much less apparent when more than two ages occur in the spawning stock, and only the dominant peaks would be detectable with any combination of average age of spawners and average age at first maturity that is apt to occur in nature.

Endless examples of reproduction-curve cycles can be constructed. Any desired combination of period and amplitude can be obtained, in more than one way, by selecting appropriate combinations of reproduction curve and age distribution of breeding. The general characteristics of such cycles can be summarized as follows:

1. Cycles occur when the outer part of the reproduction curve slopes downward, provided this slope begins at some point above the 45-degree line.
2. Cycles are damped and eventually disappear when the slope of the outer limb of the reproduction curve lies between 0 and -1. They are permanent when the slope is numerically somewhat more than -1, the exact critical limit de-

pending upon the amount of mixing of generations in the spawning stock and the interval to first spawning.

3. Period of oscillation is determined by the mean length of time from parental egg to filial eggs, being twice that interval or close to it. It is independent of the exact shape of the reproduction curve, and also independent of the number of generations in the spawning stock, provided there is more than one.

4. Amplitude of oscillation depends partly on the exact shape of the reproduction curve.

5. Amplitude of oscillation tends to decrease with increase in the number of generations comprising the spawning stock.

6. Amplitude of oscillation increases rapidly with increase in number of generations between parental egg and the first production of filial eggs, up to a

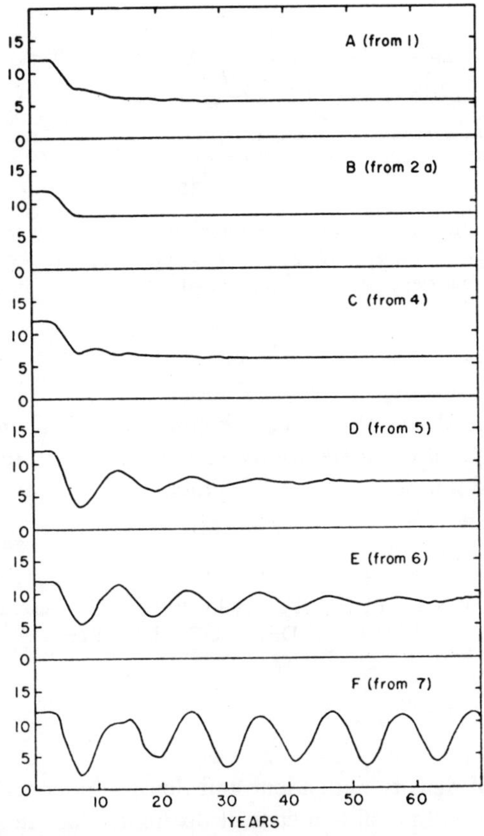

FIGURE 9. Change in abundance of the stocks of Figures 1, 2a and 4–7, following an initial sustained deflection to an abundance of 12 units, when the spawning stock is composed of 4 year-classes and first spawning is in the fourth year after hatching. *Abscissa*—years (generations); *ordinate*—relative abundance (egg production) of the mature stock.

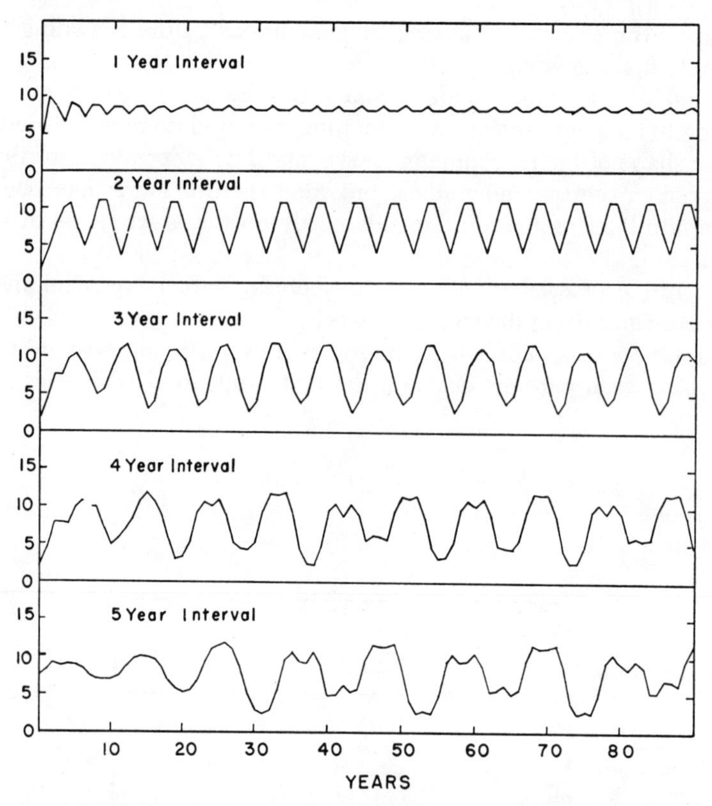

FIGURE 10. Population oscillations determined by the reproduction curve of Figure 7, when there are two ages in the spawning stock and spawning first occurs after 1, 2, 3, 4 and 5 years, respectively, from deposition of the parental eggs. Axes as in Figure 9.

limit imposed by the shape of the reproduction curve. When reproduction by a brood begins strongly in the generation following its birth, the oscillations are so weak that they could not be recognized in practice.

VARIATIONS IN REPRODUCTION PRODUCED BY FACTORS INDEPENDENT OF DENSITY

GENERAL

A comparison of density-dependent and density-independent reproduction is desirable in order to find possible means of distinguishing the two by their effects on population abundance, particularly since it has been suggested that some of the apparently periodic variations in animal numbers may reflect random variability alone (Hutchinson, 1948; Palmgren, 1949; Cole, 1951, 1954).

As an introduction, it is known that quantitative events selected completely at random tend to have an average peak-to-peak interval of exactly 3, provided they are classified finely enough that like values do not occur in adjacent positions. Cole (1951) demonstrates this mathematically and found approximately this period in a selection from Tippett's table of random numbers.

Cole uses a sequence of Tippett's numbers as a model with which to compare cycles of animal abundance in nature. Such a model, however, is not appropriate for our present purpose, because it must be interpreted as a reflection of random variation in the capacity of the environment to sustain the animal in question, rather than random variation in number of mature progeny produced per female. This is true for the following reason. When using Tippett's table or any similar assemblage, the number of numbers available is finite. The smallest number can as easily be followed by the largest as by any other; the largest cannot be followed by anything larger. The biological characteristics of a corresponding population are that it must have a very large potential rate of increase (since the greatest possible abundance can follow directly on the least), and it must have severe compensation at the higher stock densities (since they are close to a ceiling of abundance which cannot be exceeded). Few if any populations could meet both these conditions in the course of a single reproductive period, so that Cole's model, to have verisimilitude, must be applied only to situations where a census is taken at intervals of several generations. Thus it is not appropriate, for example, to fish populations when censused yearly, since these usually spawn only once a year.

In any event, a simple series of random numbers is not a suitable model of random variation *in success of reproduction*. What then *is* a suitable model? Since extremes of environmental conditions are less common than conditions approaching normal, our first postulate will be that a normal frequency distribution provides a fairly realistic picture of the relative frequencies of the resultants of the various independent factors making for success or failure in reproduction; from which resultants a random selection is to be made. Table 8.6 of Snedecor (1946) was used for this purpose, selection of t-values being made corresponding to the figure in the ".00" column closest to each of a sequence of selections from Tippett's random table. The series obtained begins as follows: 1.8, 0.5, 0.2, 0.1, 0.2, 0.0, 1.0, 0.1, etc. Tippett's table was again used to divide these into items representing reproduction above and below the replacement level, respectively; this gave the series below, decreases being followed by *d*.

1.8*d*, 0.5, 0.2, 0.1, 0.2*d*, 0.0, 1.0*d*, 0.1, etc.

The next question is to decide how these are to be applied to an existing reproductive potential. An illustration will be of assistance. Suppose that a breeding population matures 10 billion (10^9) eggs; and let us define "average" environmental conditions here as those which will permit the maintenance level of reproduction, that is, 10 billion future mature eggs produced by the progeny of the current year's sexual activity. Variations in environmental conditions add to or delete from this average production. The most severe conditions imaginable can reduce reproduction to zero. The most favourable conditions possible will permit the survival of all the eggs and thus, assuming equality of the sexes, produce future reproductive units to the number of 5 billion times the total expectation of egg production of a female fish just maturing—which expectation might easily be 2,000,000 in the case of the cod, for example. The mean and the extreme limits of reproduction would thus be:

	Mature eggs	*No. of fish (at first maturity)*
Lower limit:	0	0
Average:	10,000,000,000	10,000
Upper limit:	10,000,000,000,000,000	10,000,000,000

This shows that in a fairly typical instance the stock produced by average environmental conditions is located very asymmetrically with respect to the two extreme limits, on this arithmetic scale. On a logarithmic scale the average production becomes much more nearly central, if a minimum of 1 fish is assumed:

	Log No. of fish
Lower limit:	0
Average:	4
Upper limit:	10

From this and other considerations it appears likely that our symmetrically distributed environmental variations will tend to act in relative rather than absolute fashion; so that, for example, if a given negative deviation produces a decrease in reproduction to half of the average level, the same positive deviation will increase reproduction to twice the average level. In other words, the figures representing environmentally caused deviations from the reproductive norm must be multiplicative rather than additive. To put them into this form, unity is added to each, so that deviations indicating below-average conditions (the *d*-items above) become divisors, those indicating above-average conditions become multipliers, and zero deviation is multiplication or division by unity. The result is shown in Table II. The absolute magnitude of the items shown is of course arbitrary, and it can be varied at will by multiplying the original random series (before unity was added) by any desired integer or fraction.

Single-Age Spawners

Since the figures of Table II constitute a model of variation in success of reproduction under the influence of density-independent factors alone, a picture of corresponding population fluctuations is obtained by multiplying some initial stock density by each item in succession. This is done in the lower line of

TABLE II. Series of numbers selected randomly from a population having the positive half of a normal frequency distribution with a standard deviation of unity, each number augmented by unity, and the series divided randomly into multipliers and divisors (the latter indicated by the suffix "*d*").

2.8*d*	1.1*d*	3.0*d*	2.2	1.9*d*	1.3*d*
1.5	1.1*d*	1.9	1.1*d*	1.0	1.6*d*
1.2	1.9*d*	1.4	1.0	2.0*d*	1.2
1.1	2.0*d*	1.1	1.4	1.1*d*	2.2*d*
1.2*d*	2.4	2.4*d*	1.7*d*	2.8	1.3*d*
1.0	1.5*d*	1.7*d*	1.7*d*	2.7	2.3*d*
2.0*d*	2.2	1.3	1.6	2.8*d*	1.2
1.1	1.3*d*	2.7*d*	1.8	3.5*d*	1.3
2.3*d*	1.2	1.5*d*	1.4*d*	1.2*d*	3.0
1.8*d*	1.9	1.9	1.0	1.2*d*	1.3
1.1*d*	2.0*d*	1.9*d*	1.8	1.7	2.2*d*
1.7	2.1*d*	2.1*d*	1.0	1.9*d*	3.2*d*
1.8*d*	1.5	1.8*d*	2.2	2.6	1.4*d*
2.5	1.3*d*	1.8	1.0	2.4*d*	1.5
2.0	1.6*d*	2.6*d*	1.5*d*	2.5	1.9*d*
2.0*d*	1.5*d*	2.3	2.2*d*	2.0*d*	1.1*d*
2.0	1.0*d*	1.7*d*	1.6*d*	2.2*d*	2.6
2.2*d*	1.9*d*	1.1	1.9	2.1*d*	2.5
2.2	1.9*d*	2.8	1.7	1.9	2.1*d*
1.3	1.3	1.3	2.6	1.8*d*	2.6*d*
1.4	2.1*d*	1.1*d*	2.4*d*	2.8	2.1*d*
1.2*d*	2.1	1.6	1.5	1.3	1.8
2.0*d*	1.2	1.2*d*	2.0*d*	1.7*d*	2.6*d*
1.7*d*	1.8	1.0*d*	1.7*d*	1.6	1.3
1.1*d*	2.0	1.7	1.0	1.7*d*	2.5*d*
1.5	1.1*d*	1.4	1.2*d*	1.8	2.7*d*
2.2	2.3	1.1	1.5	1.0	1.1
1.6	1.2	2.1*d*	2.0	2.0*d*	1.3
1.4*d*	2.0*d*	2.1	1.3*d*	2.0	1.7*d*
2.0*d*	1.6*d*	1.9*d*	1.8*d*	1.6	1.1*d*
1.3*d*	2.4*d*	1.2*d*	2.3	1.0*d*	2.0
3.4	1.1*d*	1.1*d*	1.6	1.3*d*	1.6
1.3	1.7*d*	1.6	1.3*d*	3.7*d*	3.3*d*
3.2*d*	1.9*d*	1.5*d*	1.7*d*	1.3	2.3*d*
1.3	2.7*d*	1.3	1.3*d*	1.0	1.5*d*
2.2*d*	1.5	2.2*d*	1.7	1.3	1.6*d*

Figure 11A; the upper line is from a different random series. The lines fluctuate a good deal and at times diverge considerably from the original abundance: at the end of the 70 generations shown in Figure 11 the Table II line has changed by a factor of 260; while if the same is continued through the whole 216 generations of the Table the net change is 5.5×10^{-7}, representing a relative decrease

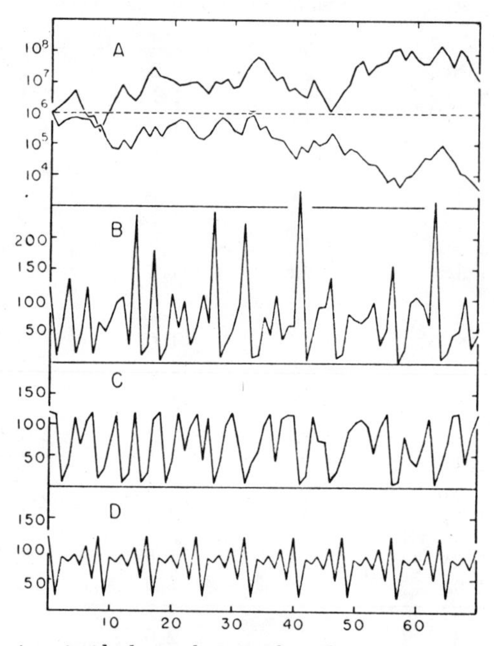

FIGURE 11. Fluctuations in ideal populations of single-age spawners. *Abscissa*—generations; *ordinate*—number of units of stock.

A. Two series of changes in a population from an initial density of 1,000,000, when it varies under the influence of density-independent factors only. The lower line corresponds to the first 70 entries of Table II, multiplied in sequence.

B. Changes in a population from an initial abundance of 119, as determined by the random factors of Table II and the reproduction curve of Figure 7, when the action of the latter precedes the former.

C. As in B, but with the random factors acting before the compensatory mortality shown by the reproduction curve.

D. Population cycle generated by Figure 7 alone.

from 55,000,000 to 1. This large change occurs in spite of the fact that Table II is constructed in such a way that, if it were continued over any really long period, the population should be above its initial level about as often and as much as it is below it[5]. The net change results partly from an excess of *d*-items in Table II, and partly from the fact that the *d*'s happen to be larger numbers, on the whole.

[5]A series of 1,000 "generations" constructed in a different but analogous fashion by Hutchinson and Deevey (1949, fig. 4) shows similar major long-period trends.

By chance the change indicated by Table II is considerably greater than would be expected to occur often in a series of only 216 generations. However it is easy to show that as the number of generations increases, the most probable divergence of population size from the original level also increases continuously. When the number of generations approaches anything appropriate to a geological time scale the likely change is very great indeed. For example, after 10,000 generations the standard deviation of the number of increases (or decreases), from the most probable number 5,000, is $\sqrt{10{,}000 \times \frac{1}{2} \times \frac{1}{2}} = 50$. This means that there is about 1 chance in 3 that the excess of increases or decreases will be 50 or more at that time, while the chances of the excess having been 50 or more at some time *during* the 10,000 generations are much greater. With an average change factor of 1.67 from one generation to the next (as in Table II), this means that the population is fairly certain to have been increased or decreased by a factor of about 1.67^{50}, or 140,000,000,000, somewhere along the line. Without wasting thought on the impossibility of a population of any sizable organism *increasing* this much, such a *decrease* would obviously bring even the most numerous fish stock, for example, to extinction. Thus no population of single-age spawners can survive if its reproduction is completely at the mercy of density-independent influences in the environment. In the same way a gambler will in the long run lose everything if his opponent has equal skill and unlimited resources. (In practice, of course, the effects of random environmental factors would usually vary as between different parts of the animal's range, and local extinction could be followed by recolonization from adjacent areas.)

The average interval between peaks of Table II should be close to the 3 which is characteristic of a random series; in actuality it was 3.09 when ties were randomly divided into higher and lower values. However this is not what determines the period of a series such as Figure 11A. In the latter a peak occurs whenever a multiplier is followed by a divisor, and the average interval between peaks is nearly 4 years, again adjusting for ties. The theoretical or long-term average period can be determined as follows: since a peak or trough occurs whenever there is a change from multiplier to divisor or divisor to multiplier, the expectation that the first such change will occur between adjacent cells is one-half, the expectation that it will occur between a cell and the next but one is one-fourth, that it will occur at the third cell is one-eighth, and so on. The average of these intervals 1, 2, 3, etc., weighted as to frequency, is exactly 2, which represents the average peak-to-trough interval. Since peaks and troughs are equally common, the average peak-to-peak or trough-to-trough interval is therefore 4.

However it must be emphasized that this 4 represents the average interval between *all* peaks, regardless of size. Casual inspection of Figure 11A might give a different impression. The eye tends to ignore the smaller humps, and to impose a certain regularity among the rest by magnifying those of intermediate size when they happen to fall into a sequence with large ones, and diminishing them when they do not. In this manner the lower line could "suggest" a cycle of 11 or 12 years, with 4 peaks and 4 troughs actually showing.

Multiple-Age Spawners

To obtain a model of the consequences of random fluctuations upon reproduction of multiple-age stocks, the random values of Table II have been applied to a "population" constructed on the same basis as Figure 9—that in the spawning stock there are 4 ages, that each year-class produces eggs in the ratio 2:3:3:2 in the 4 years of its reproductive activity, and that each fish spawns for the first time 4 years after it was itself a fertilized egg. The resulting curve is shown in Figure 12A (it starts from an age distribution characteristic of the reproduction-curve cycle of 12D).

The most distinctive characteristic of the line of Figure 12A is that it tends to rise. This increase occurs in spite of the fact that divisors happen to be in excess in Table II, and if the series is continued a little farther much higher values are encountered. The fact is that, if continued, the population will increase without limit. The reason is that the contributions of the several year-classes to a given year's spawning must be added arithmetically, while expectation of

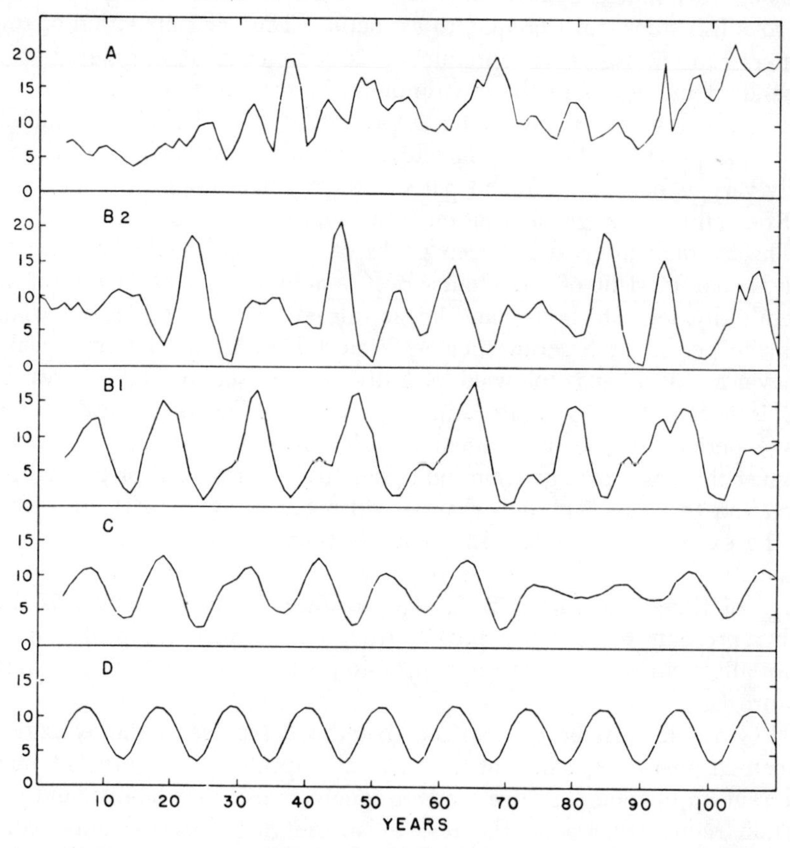

Figure 12. Fluctuations in a population of multiple-age spawners, as determined by the random series of Table I (curve A), by the reproduction curve of Figure 7 (curve D), and by combinations of these (curves B and C). Axes as in Figure 9.

survival of an egg is proportional to random environmental effects having a symmetrical distribution so constituted that any increase is greater, absolutely though not relatively, than the corresponding decrease. It would be difficult, of course, to prove that environmental effects upon reproduction are actually distributed in just this manner. On the other hand it was shown above that they are very unlikely to be merely additive; and any intermediate condition will result in a slower but still unlimited upward trend to a graph such as Figure 12A.

It thus appears that any likely random sequence of environmental changes, even one which for single-age spawners causes a catastrophic decline, can be made to produce instead a large increase in abundance of stock by the simple device of dividing the spawning of each year-class among several calendar years. Even though wholly non-compensatory reproduction cannot occur in nature, this circumstance may have considerable advantage for any population in which random variability is large and/or the left limb of the reproduction curve not very steep (e.g., as in Figs. 1, 2a, 3 or 4). Indeed it is easy to show that in the presence of random variability a multiple-age spawning stock can maintain itself even when its reproduction curve lies wholly below the 45-degree line—though of course it cannot be *too far* below. Thus there is a sound ecological basis for the customary occurrence, in nature, of reproductive assemblages consisting of more than one age-group.

As regards period, we find again that the peaks and troughs of Figure 12A have a certain apparent regularity in spite of their random origin. Discounting the first 15 years as being still somewhat under the influence of the reproduction-curve cycle, and counting all peaks no matter how small, there are 17 peak-to-peak intervals in the figure, of which no less than 10 are of 5 years' duration—the average interval is 5.2 and the range 2 to 7. The fact that the average interval is longer than the 3 which is characteristic of a simple random series (and longer than in the series for single-age spawners) is explained by the method used to obtain it. This is quite similar to taking a "running average" of four items at a time, and Cole (1951, fig. 1) has shown that applying such a procedure to a random series increases the mean peak-to-peak interval. "Artificial" trends of really long period are not as apparent in Figure 12A as in 11A, but a longer series might have confirmed them.

In summary, the characteristics of population changes which would result from density-independent factors acting alone (if that were possible) are as follows:

1. A population of single-age spawners would vary widely above and below its initial abundance; eventually, after a few thousand generations at most, it would either become extinct or, more likely, be fragmented temporarily into small independent units.

2. Populations of multiple-age spawners would increase in abundance indefinitely, though not without ups and downs.

3. The average peak-to-peak period for lines of single-age spawners would be 4.

4. The average peak-to-peak period for multiple-age spawners would be

more than 3, and would be the greater, the greater was the number of ages in the spawning stock; in the example used it was about 5 years.

5. For single-age spawners, and probably for multiple-age spawners as well, "cycles" having longer periods than these would tend to be apparent in graphs of abundance, because of conscious or unconscious mental suppression of small peaks and troughs, and regularization of large ones.

COMBINATIONS OF COMPENSATORY AND NON-COMPENSATORY MORTALITY

In natural populations non-compensatory mortality is superimposed upon the reproduction expected from density-dependent factors, and the curve of population fluctuation is the resultant of both. The random effect can be introduced either before or after the amount of reproduction indicated by the curve is written down, according as the random factors are thought to act before, or after, the compensatory ones. The latter procedure gives greater influence to the random element, and may correspond better to events in nature, though the two types of mortality may of course act concurrently, and to a considerable extent probably do.

SINGLE-AGE SPAWNERS

Combinations of Table II or similar series can be made with the various kinds of reproduction curve shown in Figures 1–8. It seems unnecessary to reproduce examples of most of these. With a flat-topped curve such as Figure 2b, and the scale of random variability shown in Table II, the resultant reproduction is practically always equal to the product of the equilibrium abundance and the multiplier or divisor for the year in question. Only an improbably large deviation could shift the population over to the ascending part of the curve for a year. Relatively slight modifications of this situation are obtained when broad-domed curves having a stable equilibrium point are used (e.g., Figs. 3 or 4); the principal difference being that, on the whole, reproduction is somewhat less.

Fluctuations produced by a combination of the reproduction curve of Figure 7 with the random series of Table II have some of the characteristics of either component (Fig. 11). The regular series A–H of Figure 7, repeating at 8-year intervals, is replaced by an irregular one. The average peak-to-peak period is about 3.5–slightly less than that of the random series (which was close to 4) and greater than the 2 of the reproduction-curve sequence. The average amplitude of the "cycles" (ratio of peak to preceding trough) is greater than what either component series exhibits, but there are of course no extreme trends in abundance such as resulted from random causes alone. There is a pronounced tendency for each peak to be followed immediately by a trough, which is a peculiarity also of the random-curve series.

The above description applies whether the random influence operates before the reproduction-curve (Fig. 11C) or after it (Fig. 11B). The two kinds of series are quite similar, but the latter of course has the greater amplitude of changes.

Multiple-Age Spawners

Combination of compensatory and non-compensatory reproduction in multiple-age stocks will be considered only for the case where the former precedes the latter, and will be illustrated by means of a population in which contributions to reproduction are spread over four years in the ratio 2:3:3:2, as in Figure 9.

In general, the kind of fluctuation which results from any combination of compensatory and non-compensatory mortality factors depends upon the relative magnitudes of the two components. The action of the same series of random factors is shown in Figures 12B and 12C at two levels of intensity which are in the ratio of 5:1; Table II is used for B, and values one-fifth as large for C, i.e., 1.36*d*, 1.10, 1.04, etc. In each case they are combined with the reproduction indicated by Figure 7. The same initial age distribution is used for all series of Figure 12.

By itself, Figure 7 yields the steady oscillation of Figure 12D. At the lower intensity of random effects (line C) the population cycle determined by the reproduction curve is not too seriously altered: there is variation in amplitude, and the peaks move out of phase by as much as a fourth at times, yet the prevailing periodicity can be determined fairly accurately from even a short series. At the higher intensity of random effects, however, much greater disturbance is evident; it is followed through in two lines of Figure 12, beginning with B1 and continuing in B2. As regards *period*, the 11-year cycle is first increased to 15–17 years, then reduced to 7 or 8 (counting major peaks only). Though the average over a long period may thus tend to 11 years, peaks and troughs move out of and back into phase with the reproduction-curve cycle and also with the random "cycle", which are its determiners. More serious is the fact that the smaller peaks and troughs of what appears to be the main series are in some cases impossible to distinguish from the "artifacts" resulting from random fluctuation: hence the average 11-year period above could not be discovered in practice, even with a very long record. As regards *amplitude*, the maximum ratio of peak to adjacent trough is much greater than is found in either the reproduction-curve cycle or the random series.

When the random element is given still greater relative importance, the reproduction-curve element becomes unidentifiable as such. However the latter continues to make an important contribution to the resulting population changes. It makes peaks and troughs much less numerous than they would otherwise be, and it provides a control of limits of abundance—that is, the progressive upward tendency of the random curve is effectively curbed.

The relative importance of random and compensatory factors in determining population abundance also depends somewhat upon the number of ages represented in the spawning stock. When this is large, random factors are less effective in disturbing reproduction-curve cycles.

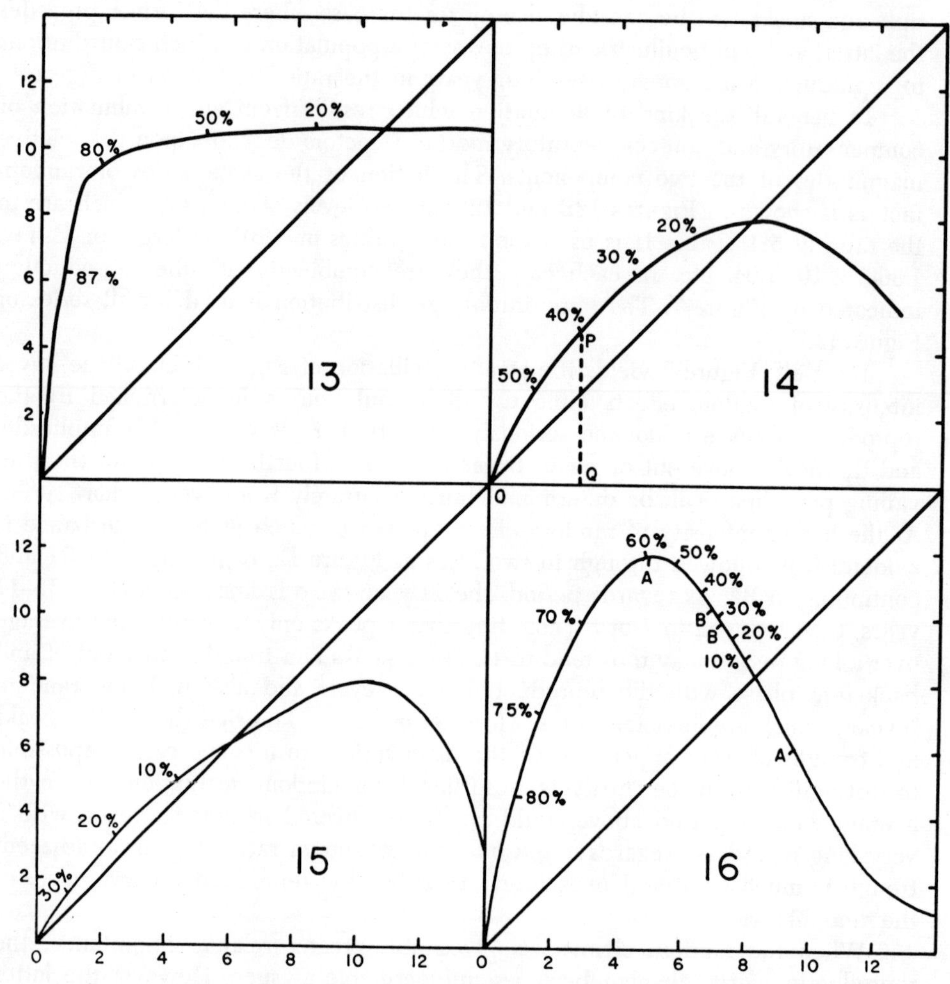

FIGURES 13–16. Equilibrium densities of stock for the various rates of exploitation indicated by percentages on the reproduction curve. Stock density before fishing is the ordinate value corresponding to the point on the curve opposite the percentage exploitation in question. Axes as in Figures 1–4.

EFFECTS OF REMOVAL OF MATURE STOCK

Single-Age Spawning Stocks

Many natural populations have a part of their members removed by man, either because they are useful to him or, in the case of harmful species, because he hopes to reduce their abundance. The case where only *mature* stock is taken by the fishery is considered in examples to follow. Many fisheries take a portion of the immature stock also, but this appears not to add any new principle to what is considered below, provided the region of compensatory mortality is not invaded.

The effect upon the various population parameters of removing a constant percentage of the mature stock, before reproduction, is shown for four types of single-age-group populations in Figures 13–16. In Figure 14, for example, a spawning stock OQ produces filial spawners equal to PQ. However, after the fishery has removed 40 per cent of the mature stock the number remaining is equal to OQ, which again produces PQ, and so on. Thus P is the equilibrium position for 40 per cent exploitation. The fraction of the mature stock which must be removed prior to spawning, in order to maintain equilibrium at any point on the reproduction curve, is equal to the complement of the reciprocal of the slope of a line joining that point to the origin.

The general effect of exploitation is to move the point of equilibrium abundance to the left on the reproduction curve. In Figures 13–15 this means that under exploitation the equilibrium population is always smaller than it is under natural conditions. In Figure 16 (which is the same as 7), removal of part of the spawning stock at first results in a *larger* equilibrium level of stock, and the same is true of Figures 5, 6 and 8. When rate of exploitation becomes large enough, however, the stock is again reduced. Equilibrium points for various rates of exploitation are indicated on Figures 13–16.

Oscillating equilibria are much changed by exploitation. Their amplitude is reduced, and the number of positions through which they swing is decreased, or oscillation may be eliminated entirely. However, stable oscillations persist when exploitation is light to moderate. In Figure 16, for example, with 10 per cent removal the stock traces the cycle 11.9, 4.0, 10.9, 5.5, 11.9, etc.; with 20 per cent removal it alternates between the levels of 6.1 and 11.9 units (A–A); with 30 per cent removal the oscillation is reduced to between 9.9 and 10.3 (B–B), so that it would not be distinguishable in practice; and with 60 per cent removal it disappears completely[6].

The point of maximum sustained yield can easily be computed from graphs like Figures 13–16. For example, in Figure 16 it is at 65 per cent exploitation;

[6]When equilibrium is unstable the *average* size of the populations of the cycle tends to be less than the equilibrium level. In Figure 16 the average is 7.7 for no exploitation, 7.8 for 10 per cent, and 9.0 for 20 per cent, as compared with the equilibrium values of 8.2, 8.6 and 9.4. For 30 per cent or greater exploitation, equilibrium stock and average stock are the same.

in Figure 13 it is close to 80 per cent; in 15 only about 18 per cent. In general, the greater the area between the reproduction curve and the 45-degree line above the latter, the greater is the optimum rate of exploitation.

Beyond the level of maximum yield lies a level of maximum permissible exploitation above which the stock is progressively reduced to zero. This limit is a function of the maximum angle made by any line joining a point on the reproduction curve to the origin; and as noted above, the maximum rate of removal which can be sustained is equal to the complement of the cotangent of that angle.

If a man's interest in a single-age population, of a noxious insect for example, is to reduce it to as low a level as possible by increasingly intensive destruction, he can be sure of making *direct* progress toward the desired goal only if the reproduction curve is one of the types shown in Figures 1 and 2. If it is like Figures 3 or 4, and stock happened to be on the outer limb, some moderate rate of removal would dampen a crash which otherwise was imminent. With curves of types 5–8 a paradoxical situation develops. Moderate destruction of adults will in general tend to stabilize the population at or about some magnitude *greater* than its primitive average abundance. In Figure 16 maximum abundance is consistently achieved when as much as 60 per cent of the spawning stock is destroyed each year. To reduce the population below the primitive average, more than 73 per cent must be destroyed. Thus although sufficiently intensive effort will be successful, moderate destruction of the mature population is worse than no action at all[7]. Furthermore, if an intensive campaign of continuing control of a variable species is decided upon, it is most efficient to begin it at a time that the population is at a low point of its cycle, i.e., when the pest may be doing no particular damage.

Multiple-Age Spawning Stocks

The effect of a fishery upon stocks containing more than one age-group of spawners parallels what has just been found for single-age-group stocks. If the stock is one for which a stable equilibrium exists (Figs. 1–5), adding a fishery does not disturb the stability, but the abundance of the population is changed (in 2b the change begins only after the inflection point is reached). With the curves of Figures 1–4 the change is in the direction of a decrease in abundance, but with Figure 5 exploitation at a moderate rate *increases* the adult stock, until the dome of the reproduction curve is reached.

[7]A similar conclusion is reached in Nicholson and Bailey's (1935) detailed analysis of host-parasite interaction in insects. Varley (1947) concluded that a trypetid fly population was maintained at about 10 times the density it would otherwise have had, because of the presence of non-compensatory causes of mortality which killed host and parasite equally.

Stocks which perform regular oscillations in the absence of a fishery (Figs. 7 and 8) are changed in three respects when exploitation begins: (1) their average equilibrium abundance is at first increased, but later decreases again if exploitation becomes sufficiently intensive; (2) the amplitude of oscillation decreases; and (3) the period of oscillation tends to decrease slightly.

1. The first-named effect can be estimated rather easily, since the average abundance of multiple-age stocks proves to be practically the same as the equilibrium abundance. The latter has been calculated for Figure 16 using various rates of exploitation, as shown on that Figure and in Table III. The maximum abundance is at the dome of the curve, and maximum catch is obtained slightly to the left of it, at 65 per cent exploitation.

2. The advent of fishing not only affects the total abundance of a stock, but also gives it a younger age composition, because survival rate from year to year is reduced. In the example of Table IV, based upon Figure 16, the fish are assumed to mature first at age III. Fishing takes place just prior to spawning and rate of fishing is the same for all ages. Natural mortality occurs between successive fishing-and-spawning seasons, and is 20 per cent from age III to age IV, 20 per cent from IV to V, 30 per cent from V to VI, and 100 per cent after the spawning at age VI. The average weight of a fish at time of fishing-and-spawning is 2 units at age III, 4.28 units at IV, 7.14 at V and 9.52 at VI; and egg production is proportional to weight. Under these conditions the *equilibrium* distribution of the contributions of the several age-groups to egg production, at different rates of exploitation, is shown in Table IV.

TABLE III. Average or equilibrium abundance of stock, by weight, in a stock having more than one age at maturity, and the catches at various rates of exploitation; from Figure 16.

Rate of exploitation	Stock before fishing	Catch
%		
0	8.2	0
10	8.8	0.9
20	9.5	1.9
30	10.1	3.0
40	10.8	4.3
50	11.5	5.8
60	11.9	7.1
65	11.4	7.4
70	9.7	6.8
75	7.2	5.4
80	5.0	4.0

TABLE IV. Relative weights and contributions to reproduction of the age-groups in the stock described on page 173.

Rate of exploitation	Fraction of eggs contributed by age			
	III	IV	V	VI
%				
0	0.200	0.300	0.300	0.200
10	0.233	0.314	0.283	0.170
20	0.273	0.327	0.261	0.139
30	0.320	0.336	0.235	0.109
40	0.377	0.339	0.204	0.081
50	0.446	0.333	0.165	0.055
60	0.526	0.315	0.127	0.032
70	0.619	0.278	0.083	0.020
80	0.737	0.221	0.042	0.000

The first line of Table IV was arranged to have the same distribution of contributions to spawning as used for Figure 9F. In Figure 17 are shown cycles obtained from the 10 per cent and 20 per cent lines of Table IV, starting from the equilibrium distribution characteristic of no fishery[8]. At 10 per cent exploitation the oscillation of the population is maintained, though at reduced amplitude and about a higher mean level. At 20 per cent exploitation oscillation gradually decreases to an inconsiderable amplitude, and at higher rates of exploitation it disappears completely. However even at 20 to 40 per cent exploitation the oscillation persists through several cycles.

3. The length of a cycle was found earlier to be twice the mean length of a generation (interval from egg to egg). In the first line of Table IV this is 4½ years, and the cycle is 9 years. With the higher exploitations of Table IV a shift of age distribution toward younger fish occurs, and we accordingly expect a shorter cycle to appear. In this example, however, the cycles disappear before much decrease in period can take place. At 10 per cent exploitation average age of contribution to reproduction is reduced to only 4.39, and at 20 per cent it is 4.27. The corresponding predicted cycle lengths are 8.8 and 8.5 years, and the peaks shown in Figure 17 do in fact become progressively a little closer together. In a somewhat more realistic example there would have been more than four age-groups in the stock under conditions of light exploitation or none, and in that event the reduction in period of oscillation with increased fishing would be

[8]This procedure is the only practicable one, but it ignores the fact that the age distributions of Table IV would themselves be completely realized only after a period of 4 years from the beginning of fishing. Consequently the equilibrium situations in Figure 17 would actually be approached more gradually than the figure indicates.

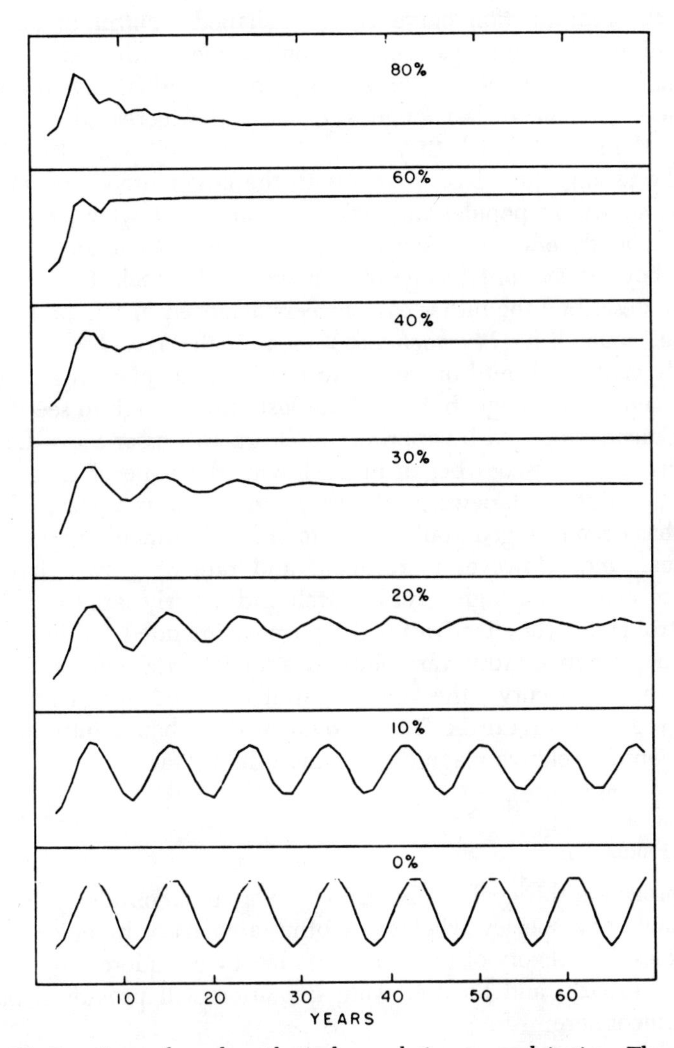

FIGURE 17. Reactions of an hypothetical population to exploitation. The population is one regulated by the reproduction curve of Figure 7, with ages III, IV, V and VI contributing to egg production in the ratio 2:3:3:2 when there is no exploitation. The stable cycle characteristic of no exploitation is shown by the 0% line. The other lines show the effect of annual removal of the indicated percentage of the mature stock from a population which initially had the age distribution characteristic of the start of the ascending phase of the no-exploitation cycle. (See also footnote 8.)

greater. However it is doubtful whether in nature the reduction would ever be great enough to be identifiable with certainty, before increasing exploitation removed oscillation of this type entirely.

When the reproduction curve is dome-shaped, *control* of an undesirable population of multiple-age spawners presents some of the same difficulties as described earlier for single-age stocks. However the initial favourable reduction in abundance will tend to last longer—i.e., until the increased broods of young can grow to the size at which they cause damage. Only when this first phase is over, and the fishing effort has to cope with the larger broods which come from the reduced spawning populations, will it be apparent whether or not control can be permanently effective. Few recorded attempts to reduce nuisance fishes have lasted beyond the initial stage of removal of old stock. Foerster and Ricker (1941) have described the rather easy success achieved in this preliminary phase of control of squawfish (*Ptychocheilus*) in a small lake. The experiment was subsequently continued until one or more large broods of young began to grow into the damaging size range, but it did not last long enough to see if these could be kept sufficiently reduced by appropriate effort. In Lesser Slave Lake, Alberta, unlimited fishing for ciscoes began in 1941 with the hope of thus reducing the lake's population of the tapeworm *Triaenophorus*, and as a result the spawning population has been changed gradually from one 5–7 years old to one now (1951) mostly 2 years old. However recruitment and rate of growth have increased sufficiently to maintain a high level of catch and a fairly large population mass, so that control is not yet effective for the purpose intended (Miller, 1950, 1952).

When non-compensatory variability is added to the effects of exploitation and the reproduction curve, the kind of result obtained differs little from what was discovered from Figure 11. The amount of disturbance introduced depends of course upon the relative magnitude of the random factors.

Theory of Predation

An approach to this goal can be made by way of a consideration of predation upon the young of a species, whether by other animals or by older individuals of its own species. The theory of predation was briefly considered by the writer in a recent paper (1952), and the following quotation will provide a basis for the further argument here:

> It is convenient to distinguish three types of numerical relationship between predators and a species of prey which they attack.
>
> A. Predators of any given abundance take a fixed number of the prey species during the time they are in contact, enough to satiate them. The surplus prey escapes.
>
> B. Predators at any given abundance take a fixed fraction of prey species present, as though there were captures at random encounters.
>
> C. Predators take all the individuals of the prey species that are present, in excess of a certain minimum number. This minimum may be determined in different ways: (1) There may be only a limited number of secure habitable places in the environment, so that some prey are forced to live in exposed situations where capture is inevitable.

The number of such secure niches may be partly governed by territorial behaviour of the prey. (2) The maximum "safe" density of prey may be the one at which predators no longer find it sufficiently rewarding to forage for them, and move to other feeding grounds.

The three situations above tend to intergrade, of course, but it is useful to keep their differences in mind.

SITUATION A

This is likely to occur when a prey species is temporarily massed in unusual numbers, for example, adult herring in spawning schools, or newly-emerged fry of pink and chum salmon going downstream. The main characteristic of such situations is that the number of prey eaten depends on the abundance of predators, but not on the abundance of prey. Hence such situations cannot last long, and the predators cannot make the prey in question their principal yearly food; otherwise they would almost surely increase in abundance and the situation would change to type B or type C. If a type A situation persisted for long, it would come to an abrupt end with the extermination of the prey.

SITUATION B

Here the number of the prey species eaten is proportional to the abundance of predators and to the abundance of prey. Unlike A, this type of predation can easily occur over long portions of the year; and the prey species may comprise the larger portion of the predators' annual ration. This situation was observed at Cultus Lake, British Columbia, for predation of squawfish, char, coho and trout upon fingerling sockeye, over a wide range of abundance of the latter (Foerster and Ricker, 1941).

SITUATION C

The classical example of this situation was described by Errington for bob-white in Iowa, where a given range would winter safely a fixed number of birds, practically independently of the number which were present in autumn, the surplus being taken by predators. Studies conducted from the Atlantic Biological Station of the Fisheries Research Board of Canada, St. Andrews, N.B., suggest that in some rivers predation upon Atlantic salmon parr tends toward type C, because a marked increase in the number of young fish planted was followed by only a relatively small increase in number of surviving smolts (Elson, 1950). In type C situtions the predators must tend to have ample alternative foods, and they are often mobile so that they can leave an area where the food supply has been "cleaned up" for the season.

Consider a prey species which is subject to type B predation over some part of its life history. Assume first that predation causes the whole of the mortality that the prey is subject to during that period. Then the prey species decreases in abundance according to the well-known exponential formula

$$N/N_0 = e^{-it}, \tag{1}$$

where: N_0 is initial abundance, N is abundance at time t, $e = 2.718 \ldots$, and i is a statistic representing the fraction of the prey which would be eaten in a unit of time if its abundance were held constant for that long; i is often called the *instantaneous* mortality rate.

Under the conditions postulated, instantaneous mortality rate is directly proportional to the abundance of predators. This can be illustrated by considering a situation where predators attack a prey population of 1,000,000 individuals under type B conditions, and they inflict losses corresponding to $i = 0.8$,. where the unit of time, t, is the whole season that predator and prey are in contact. Equation (1) indicates that in, for example, 1/1000 of the season, the predators will eat $i/1000 = 0.0008$ of the prey present, or 800 fish. During the next thousandth of the season, the predators eat $i/1000$ of the surviving prey, or $999{,}200 \times i/1000 = 799$ fish. The following interval they eat $998{,}401 \times i/1000 = 799$; then $997{,}602 \times i/1000 = 798$. This continues until all the thousand time-intervals have elapsed, at the end of which there are:

$$1{,}000{,}000(1 - 0.0008)^{1000} = 472{,}400$$

survivors, or 47.2 per cent. What happens if the number of predators, is doubled? In that event, during the first thousandth of the season twice as many fish will be eaten, i.e., 1,600, leaving 998,400. In the next thousandth the fraction eaten is likewise double, namely 0.0016; multiplied by the number of survivors this gives 1,597; and so on. At the end of the season $1{,}000{,}000(1 - 0.0016)^{1000}$ survive, which is 201,900, or 20.2 per cent.

Thus doubling the number of predators doubles the instantaneous mortality rate (which is true generally), but it increases actual mortality by only 27.0/52.8 = 51 per cent (which is true of only this particular example).

The general relation between predator abundance and actual mortality, or survival, is most conveniently expressed as:

$$\frac{p_2}{p_1} = \frac{\log s_2}{\log s_1}, \qquad (2)$$

where p_2 and p_1 represent two levels of predator abundance, and s_2 and s_1 are the corresponding survival rates for the prey. This expression can easily be derived from (1), since p is proportional to i, and $s = N/N_0$ when $t = 1$.

CANNIBALISM

Of all the methods of population regulation listed earlier, cannibalism is the one in which the abundance of the control agent is most closely and inseparably allied to that of the population controlled. That is, an increase in mature stock not only increases the number of eggs laid or young born in a given reproductive season, but it also decreases the rate of survival of those young. What is the combined effect of these two opposed influences?

Let the size of a stock be measured by the number of eggs it lays, and let this size be proportional to its *instantaneous* efficiency in cannibalism (the fraction of young eaten by the adults in a short interval of time). Consider two situations characterized by different sizes of stock and let the ratio of the second to the first be w, corresponding to the p_2/p_1 of equation (2) of the quotation above. Finally, let all other sources of mortality, whether they occur before, after or during the period of the cannibalism, be density-independent, their total effect being to reduce survival rate of eggs and young to the fraction k of what it would otherwise be. The number of eggs laid, in situation 1, is E_1; in situation 2 it is therefore $E_2 = wE_1$. The reproduction (absolute number of recruits produced) in situation 1 is:

$$R_1 = E_1 k s_1; \qquad (3)$$

and in situation 2 it is:

$$R_2 = E_2 k s_2 = w E_1 k s_2. \qquad (4)$$

From equation (2) of the quotation above,

$$w = \frac{\log s_2}{\log s_1}; \quad \text{hence } s_2 = s_1^{w}. \qquad (5)$$

From (4) and (5), the reproduction in situation (2) becomes:

$$R_2 = w E_1 k s_1^{w}. \qquad (6)$$

Expressed as a fraction of the reproduction in situation 1, this is:

$$ws_1^{w-1}. \tag{7}$$

Plotting numerical values of (7) against w, for various values of s_1, yields a family of curves each of which has an origin at zero, a dome, and an extended right limb which approaches the abscissa asymptotically. No matter what survival rate is chosen for s_1, each curve describes the whole range of possible survival values as w is varied; hence only one curve of the family is needed. The convenient one to choose is that for which (7) is a maximum when $w = 1$. To locate this maximum, (7) is differentiated with respect to w and equated to zero:

$$ws_1^{w-1} \log_e s_1 + s_1^{w-1} = 0; \tag{8}$$

whence,

$$-\log_e s_1 = 1/w; \quad \text{or, } s_1 = e^{-1/w} \tag{9}$$

The value of s which makes (7) a maximum is $e^{-1/w}$, and if this is taken as the initial value of s (that for which $w = 1$), s becomes equal to $1/e$ or 0.3679. Substituting $1/e$ for s_1 in (7), we thus finally obtain the expression

$$we^{1-w}. \tag{10}$$

This shows the actual level of reproduction as a fraction of the maximum, when w represents the ratio of the actual density of mature stock to the density which gives maximum reproduction. Values of (10) are plotted as curve B of Figure 33.

From the equations and the curve the following conclusions are evident:

1. Since the curve approximates to the abscissa as w is increased, then *if the breeding stock is made sufficiently large, cannibalism reduces reproduction practically to zero*, in spite of the greatly augmented egg deposition. In theory this is true no matter how small the original rate of cannibalism; in practice, a type B predation situation between adults and young could not be maintained over *too* wide a range of stock densities.

2. Since, as w approaches 0, e^{1-w} approaches e, it follows that at minimal densities the survival rate from cannibalism is e times the survival characteristic of maximum reproduction. In other words, the instantaneous mortality rate at maximum reproduction is greater by unity than it is at vanishingly small stock densities (since $-\log_e (1/e) = 1$; cf. expression (1) of the quotation above).

Curve B of Figure 33 can be changed into a reproduction curve by changing the scale of each axis to represent actual numbers of parents and progeny measured in comparable units—for example the egg production unit described in an earlier section. Equation (10) gives no information concerning the steepness of this actual reproduction curve, which will depend chiefly on the magnitude of k, the survival rate from factors other than cannibalism. On Figure 33 a number of possible curves have been drawn, all based on equation (10) but using different ratios of abscissal to ordinate scales. Referring each to the 45-degree line indicated, shapes reminiscent of several of the arbitrarily drawn types of Figures 3–8 can be identified, as well as most of the curves observed in actual populations. Curve E, of course, could not describe a stable population because it lies well below the 45-degree line, while even D would be rather precarious.

In addition to cannibalism, the argument of this section can be extended to

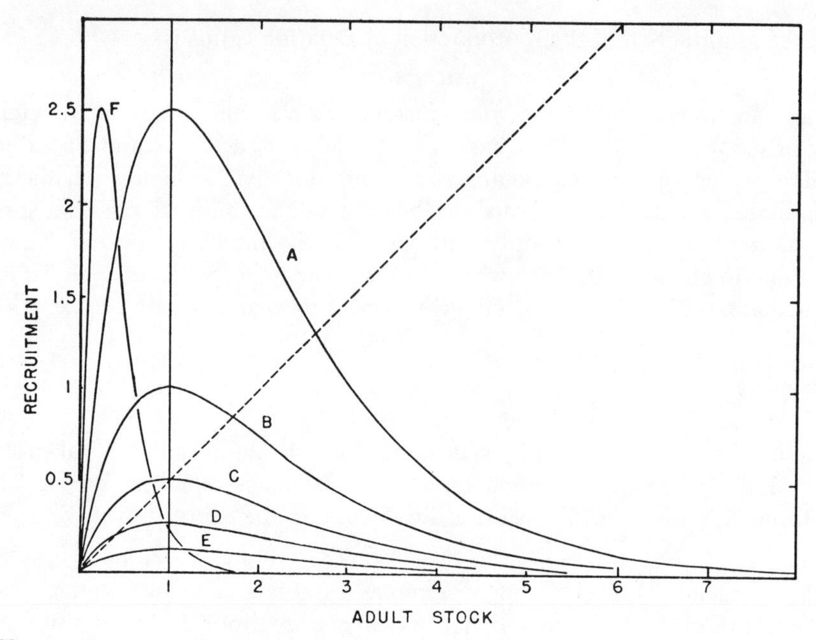

FIGURE 33. Graphs of we^{1-w} ("recruitment") plotted against w ("adult stock"). Curve B corresponds to the axes as labelled; the other curves are obtained by varying the ordinate and (for F) abscissal scales.

the analogous situation where strife between mature and young animals occurs and results in frequent killing of the latter, but without necessarily any eating of one by the other; this has been described among muskrats, for example (Errington, 1954). In so far as these contacts are of type B—with the older animals killing a fixed fraction of the young they encounter—they would have the same population status as described above.

It is not easy to assess the probable importance, in nature, of cannibalism and allied effects. In most of the fish populations whose reproduction curves were examined earlier there is little or no evidence of cannibalism or of other direct injury to young by their parents, and among *Oncorhynchus* this is impossible because the parents die. However, in assessing the possible role of cannibalism (or any other agent of predation) in population regulation, it is not possible to weight it directly on the basis of the number of units which it kills. The latter will depend mainly upon the stage at which it is active. For example, suppose that cannibalism by mature trout fell upon their fingerling progeny following an average density-independent egg mortality of 50 per cent and fry mortality of 90 per cent (95 per cent in all); and that the adult trout ate 63 per cent of the 5 per cent remaining, or 3 per cent of the original brood. In that event cannibalism would seem to be quite insignificant either from the point of view of the percentage of the total young which it killed, or from the point of view of the frequency of occurrence of small trout in stomachs of larger trout. In spite of this, under the conditions just postulated cannibalism would be the sole mechanism regulating the abundance of the population.

Predation by Other Organisms

Even granting that lethal contact between adults and young is the most direct and least fallible population-regulating mechanism, and that it is quite apt to exist undetected, the writer's present opinion is that it will probably not prove to be important in more than a minority of populations. In its absence, other compensatory agents must take over the role of regulators which determine the shape of the reproduction curve.

TYPE B SITUATIONS. A predator can qualify as an agent of population control only if it can increase in density or effectiveness as the abundance of prey increases, and vice versa. One way this can occur is by migration of additional predator units to the region of predation when prey abundance is large. For example, greater-than-average abundance of herring spawn might attract birds from a wider area than usual. It is possible (not necessarily probable) that, after gathering together, such aggregations would remain in the vicinity for a few days longer than the supply of eggs justified it, with the result that an initially superior spawning would yield a less-than-average number of eggs hatched.

A predator might also actually increase its total abundance with that of the prey in question. Consider, for example, a plankton predator or assemblage of predators which feeds on the pelagic eggs and young of mackerel. Suppose that in years when mackerel eggs approach zero abundance these predators (which have additional foods) are numerous enough to consume the fraction $(1 - s_1)$ of the eggs. Suppose also that for each unit increase in egg numbers the average predator population increases by one-tenth of its "basic" number. (This average would be taken over the whole time that predators and mackerel are in contact because, just before the eggs are available, initial predator abundance is assumed to be the same in all years.) During the later stages of the predator-prey contact the abundance of predators would have "overshot" that of the prey, with the result again that mackerel broods initially larger than average would end up smaller than average. From expression (2) or (7) it is easy to calculate that relative reproduction is equal to

$$Zs_1^{1+Z/10}, \qquad (11)$$

where Z is the density of mackerel eggs immediately after spawning in a given year, in the unit mentioned above, and the predator density corresponding to no eggs is taken as unity. Values of (11) are calculated in Table VI, which shows the relative number of mackerel surviving the pelagic stage, for a series of initial prey densities and for two values of s_1. Sette (1943) estimated total survival of Atlantic mackerel eggs and pelagic larvae, in 1932, to lie between 0.000001 and 0.00001, so that the $s_1 = 0.0001$ column of Table VI would not be unrealistic.

TABLE VI. Survival rate and absolute number of survivors when average predator abundance varies with initial prey density in the manner indicated by columns 1 and 2. Pairs of survival rates and survivor numbers are shown for initial survival rates of 0.1 and 0.0001. The last two rows show the computation of the maximum number of survivors for each situation, which occurs when survival is 0.3679 ($1/e$) of its initial value.

1 Prey abundance	2 Predator abundance	3 Survival rate of prey	4 Relative number of survivors	5 Survival rate of prey	6 Relative number of survivors
			(1) × (3) × 10^3		(1) × (5) × 10^7
0	1.00	0.1000	0	0.0001000	0
0.2	1.02	0.0955	19	0.0000832	166
0.4	1.04	0.0912	37	0.0000692	277
0.6	1.06	0.0871	52	0.0000575	345
0.8	1.08	0.0832	67	0.0000479	383
1.0	1.10	0.0794	79	0.0000398	398
1.2	1.12	0.0759	91	0.0000331	397
1.4	1.14	0.0724	101	0.0000275	385
1.6	1.16	0.0692	111	0.0000229	367
1.8	1.18	0.0661	119	0.0000191	343
2.0	1.20	0.0631	126	0.0000158	317
2.5	1.25	0.0562	141	0.00001000	250
3.0	1.30	0.0501	150	0.00000631	189
4	1.4	0.0398	159	0.00000251	100
5	1.5	0.0316	158	0.000001000	50
6	1.6	0.0251	151	0.000000398	24
8	1.8	0.0158	127	0.000000063	5
10	2.0	0.0100	100	0.000000010	1
15	2.5	0.0032	47	...	...
20	3.0	0.0010	20	...	...
30	4.0	0.0001	3	...	...
1.086	1.1086	...	...	0.00003679	399
4.343	1.4343	0.03679	160	...	...

The distributions of columns 4 and 6 of Table VI are of course the same as those of Figure 33. Expression (11), like (7), is a maximum when the rate of survival from the predator in question is $1/e$ of what obtains when egg abundance is close to zero (cf. Table VI). Thus we again conclude that *the instantaneous mortality rate at maximum reproduction is greater by unity than the rate characteristic of a very small population density.* For example, if $s_1 = 0.01$, the instantaneous mortality rate at minimum stock density, from the action of the controlling predator, is equal to $-\log_e 0.01 = 4.61$. At the density of maximum reproduction, mortality from this predation becomes 5.61. Mortality from other factors ($= -\log_e k$) must be added to this 5.61 to give the total average egg-to-egg instantaneous mortality rate at maximum reproduction. Note however that at the *replacement* density of stock the compensatory mortality is greater than 1, its exact value depending on the steepness of the reproduction curve (relative to the 45-degree line).

The form of expression (11) indicates that the magnitude of the arbitrary factor 10, relating initial egg density to mean predator density, does not affect the general shape of the recruitment curve, though it of course affects its steepness: the larger this factor, the broader is the range of initial egg densities which afford substantial reproduction. Furthermore the rule italicized above holds even if the relation between prey and predator is quite irregular.

Similarly, the magnitude of s_1 affects the shape of the reproduction curve only in that the smaller s_1 is, the greater is the percentage change in reproduction produced by a given change in predator abundance (Table VI).

SITUATIONS OF TYPES A AND C. So far the theory presupposes type B predation situations. Type A situations need not be discussed in detail: for reasons given earlier, they tend to be restricted in space or time, and they lack the qualifications of a population control mechanism. Their effect on reproduction curves would be to introduce irregularities such as are shown in Figures 30 and 31.

Type C situations however *can* regulate a population, and in a pure form they produce reproduction curves like Figure 2b. The horizontal part of the curve corresponds to the limit of surviving young which the habitat will sustain. Type C situations may grade into type B, in which case it would in practice be difficult to distinguish a broad-domed type B curve from the mixed type.

It is also possible for type B and C situations to follow each other, and this is examined in Figure 34. If the type B situation comes first, its typical reproduction curve is truncated by the limit of reproduction imposed later by Type C (curve A of Fig. 34). If the type C situation comes first, it will at a certain point limit the brood to a density which may be less than, equal to, or more than what gives maximum reproduction in the subsequent type B situation; these three possibilities are illustrated by curves B–D of Figure 34. Note that while the type C situation can *reduce* the difference in instantaneous mortality rate between minimal and maximal reproduction, it cannot *increase* this difference.

The various curves of Figure 34 should be looked for in nature when the biology of the animal in question points in their direction. Among the examples given earlier, a limit of environmental capacity is strongly suspected for coho salmon; actually Figure 21 could readily be fitted with a truncated curve like 34A, but there is only one point for the outer decreasing phase.

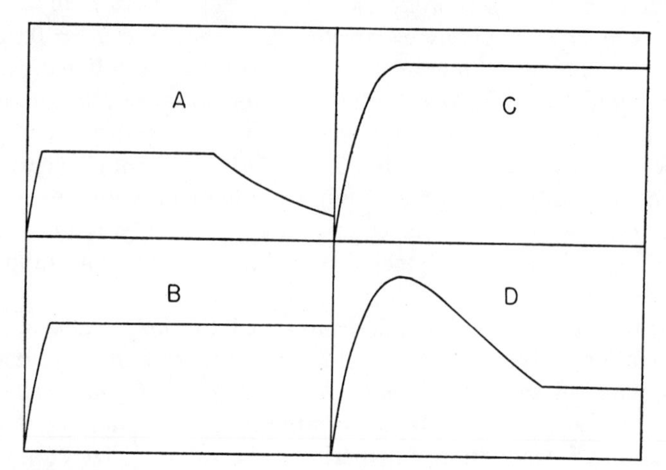

FIGURE 34. Reproduction curves when predation situations of types B and C occur in succession. *Curve A*—when type B precedes type C; *curves B–D*—when type C precedes type B. The type B reproduction curve is the same for all four situations.

SIZE OF PREDATOR AND SIZE OF PREY. It has sometimes been said that large predators (e.g., foxes) cannot control or regulate the abundance of a smaller prey species (e.g., mice), because the rate of reproduction of the prey is so much greater than that of the predator. This argument would be valid only if the predation situation were of type A, with the foxes always able to kill as many mice as interested them—a situation which could scarcely last indefinitely. However if the fox-mouse predation situation were of type B, calculations from (7) show that any increase at all in the number of foxes reduces the survival rate of mice more than proportionately, provided the foxes initially are the cause of an instantaneous mortality rate of 1 or more in the mouse broods; and the greater the increase of the foxes, the less this initial mortality need be to accomplish the result stated. To put it another way, the effectiveness of foxes in reducing the survival of mice will increase more rapidly than does the actual number of foxes, whenever they are already reducing mouse survival to 37 per cent or less of what it would be in the absence of foxes. This effect gives the foxes an advantage which tends to counterbalance the greater rate of reproduction of the mice.

More generally, there is evidently no reason why a large predator, or assemblage of predators, might not regulate the abundance of a smaller and more prolific prey species, provided their abundance can change with that of the prey to *some* extent. Indeed, since a given change in predator abundance has a greater effect upon survival, the smaller the average survival rate of the prey (and hence the greater its fecundity), it could be argued that great fecundity should make a species more—not less—easily controlled by small changes in predator abundance. Against this must be laid the fact that great fecundity makes the prey more likely to occasionally "slip out from under" by becoming so numerous that the type B predation situation cannot be maintained.

Other Compensatory Agents

We have seen that both cannibalism and "ordinary" predation lead to the Figure 33 type of reproduction curve under type B conditions. For other compensatory agents only a qualitative examination can be attempted here, but they must be mentioned briefly.

The special type of predation situation in which insect "parasites" eat other insects regularly leads to low host reproduction at high host densities, as shown in Nicholson and Bailey's (1935) detailed analysis.

"True" parasites (those which do not regularly and consistently destroy their hosts) and disease organisms can also exhibit non-linear effectiveness which can seriously reduce reproduction at high stock densities—though their activity in this respect must usually be very irregular, if the human situation is a guide.

In the *Daphnia* cultures described earlier no predation or cannibalism existed, nor apparently did disease. The most plausible suggestion which has been made is that some metabolic product acted as a depressor of reproduction. If, for example, a metabolic waste killed a fixed fraction of *Daphnia* embryos per unit concentration and per unit time throughout prenatal life, it would have exactly the same effect upon survival as random cannibalism, and lead to the reproduction curve of Figure 33. The points of Figure 28 can be fitted by expression (10) with good apparent agreement.

Again, consider the compensatory situation where a brood, initially more numerous than average, grows more slowly than average because of competition for limited food; and because of this smaller individual size it is more vulnerable to predation than a faster growing brood at corresponding ages. It is fairly clear that here too there *could* be a lag effect whereby the brood would lose by increased predation more than the numerical advantage which it originally had; though it would be difficult to show that this should (or should not) occur as a general rule.

Finally, consider what happens when spawning facilities are limiting, as for some salmon or trout populations, for example. Natural selection will probably see to it that the population uses the best spawning facilities first. At increasing densities there is both a spread to less suitable gravel in other parts of the stream and a crowding and superimposition of eggs on the favourable grounds. With extremely heavy seeding, fungus may cover and kill almost the whole of the eggs in even the best redds; something of the sort probably decimated the Karluk pinks in the disastrous year 1924 (Barnaby, 1944). The sum of the above effects would usually add up to a domed reproduction curve with an extended right limb. Though an *exact* fit to expression (10) would not be likely, the latter might serve as a picture of the normal expectation.

FITTING THE CURVE $z = we^{1-w}$

A first approximation to a fit of expression (10) to any body of data can be obtained very quickly, because the distribution is completely determined by the position of the maximum and the latter can be selected by eye. The work consists merely of dividing each observed estimate of reproduction (Z) by the estimated maximum Z, and each observed parental stock density W by the density which produced the estimated maximum W; this gives w and z, respectively. From (10),

$$\log_e \hat{z} = \log_e w + (1 - w). \qquad (12)$$

A sample calculation is shown in the last line of Table VII.

To find the *best*-fitting curve of this type it would be necessary to calculate the residuals (differences between observed and calculated values) for a series of trial positions of the maximum point. The latter can be varied in two dimensions, so the process might be fairly protracted. The accepted criterion of best fit would be that for which the sum of the squares of the residuals is least, but an easier

TABLE VII. Computation of expected reproduction ($\hat{Z}$) and residuals (R) for a set of stock-reproduction (W–Z) observations, using $z = we^{1-w}$. W = 13, $Z_{max.}$ = 38, is the trial position of the dome of the curve. The last line is a computation of $\hat{Z}$ for an arbitrary W. (See text for further details.)

1 W	2 Z	3 z	4 w	5 $\log_e z$	6 $\log_e w$	7 Unity	8 R	9 R^2	10 $\log_e \hat{z}$	11 $\hat{z}$	12 $\hat{Z}$
5	18	0.47	0.38	−0.75	−0.95	1.00	−0.42	0.18	−0.33	0.72	27
8	78	2.05	0.61	+0.72	−0.49	1.00	+0.82	0.67	−0.10	0.90	34
16	20	0.53	1.23	−0.64	+0.21	1.00	−0.62	0.38	−0.02	0.98	37
28	5	0.13	2.15	−2.02	+0.77	1.00	−1.64	2.69	−0.38	0.68	26
33	8	0.21	2.54	−1.56	+0.93	1.00	−0.95	0.90	−0.61	0.54	21
78	28	0.74	6.00	−0.30	+1.79	1.00	+2.91	8.47	−3.21	0.04	1.5
Sum	..	..	..	..	..	..	+0.10	13.29	..	..	..
50	..	..	3.85	..	+1.35	1.00	..	..	−1.50	0.22	8

criterion—that the sum of the residuals should approximate to zero—would be useful in the earlier stages. Table VII shows a computation of residuals (R) for the pink salmon data of Figure 19. The dome of the curve is estimated to be at a parental abundance of W = 13 and filial abundance of $Z_{max.}$ = 38. Dividing the observed values of columns 1 and 2 by these values gives z and w in columns 3 and 4. The residual (R) is the observed value $\log_e z$ less the estimated $\log_e \hat{z}$ from (12), which is equivalent to

$$R = \log_e z + w - \log_e w - 1. \qquad (13)$$

The four right-hand terms of (13) are in the order of columns 4–7 of Table VII; adding columns 4 and 5 and subtracting 6 and 7 gives R in column 8, and R^2 is shown in 9. The sum of the R's is close to zero, showing that the trial position of the dome of the curve cannot be too far from the best one; however since more than half of the R^2 total is contributed by the last observation, a somewhat better fit could probably be obtained by moving the dome a little to the right or upward.

Expected reproduction values, $\hat{Z}$, are calculated in the last three columns of Table VII. Column 10 is equal to 8 less 5; or it can be obtained directly by adding columns 6 and 7 and subtracting 4 (cf. formula 12). The latter method is used to calculate $\acute{Z}$ for arbitrary values of W, as shown in the last line of the Table.

Base-10 logarithms can be used in Table VII if columns 4 and 7 are multiplied by 0.4343, but the procedure shown is more convenient when a comprehensive table of natural logarithms is at hand.

Fitted curves have not been used on the observed reproduction figures above (Figs. 18–28). Apart from the work involved, the theory of expression (10) needs further examination, and uncritical use of it now might conceal variant or alternative situations, such as that suggested for Figure 21. All the curves of Figures 18–28 were drawn freehand before expression (10) had been developed, and it is interesting that most of these observations suggested a concave and tapering right limb. Even the original hypothetical curves of Figures 7 and 8 were drawn this way, because of a feeling that reproduction couldn't really decline right to zero as spawners became increasingly numerous.

Returning to the question introduced at the start of this division of the paper, no universally applicable theory of reproduction has been discovered, or is likely to be. However several possible reasons are apparent why recruitment can decline, and usually will decline, at higher stock densities. Also, from reasonable assumptions a simple mathematical expression has been developed which could be used to represent most of the observed reproduction data over the range of densities which they cover.

SUMMARY

1. The general theory of reproduction indicates that density-dependent causes of mortality set a limit to the size which a population achieves. The "reproduction curve" for a fish species is defined as a graph of the average number of eggs produced by a filial generation against the number produced by its parental spawning assemblage, under the existing frequency distribution of environmental conditions for survival (Figs. 1–8).

2. The level of adult stock at which a *maximum number* of recruits (on the average for existing variations in environmental conditions) is obtained is not necessarily the level at which the *replacement number* is obtained (on the average). Maximum recruitment may occur either at the replacement level of adult stock (Figs. 2, 3), or at a higher level (Figs. 1, 4), or at a lower one (Figs. 5–8).

3. When, with increasing stock density, the maximum of recruits exceeds and precedes the replacement number, a population tends to oscillate in abundance. These oscillations are stable if the reproduction curve crosses the 45-degree line (representing the replacement level of reproduction) with a slope between -1 and $-\infty$ (Figs. 7, 8); they are damped if the curve crosses at any numerically lesser slope (Fig. 5).

4. Under the conditions of Figures 7 and 8 a population of single-age spawners has an irregular but permanent cycle of abundance, such as shown in Figure 11D for example, if enivronmental conditions are stable.

5. Under the same conditions a population of multiple-age spawners tends to have a more regular cycle, whose peak-to-peak period is close to twice the median length of time from oviposition by the parent generation to oviposition by all its progeny (Figs. 9F, 10).

6. For such populations to have cycles of appreciable amplitude, it is necessary that reproduction be absent or light during the first year or two of the animal's life (assuming that reproduction occurs once a year).

7. Random fluctuation in reproductive success, *by itself*, leads to very large changes in abundance and eventual extinction or fragmentation, for populations of single-age spawners (Fig. 11A), and to unlimited increase for multiple-age spawners (Fig. 12A).

8. Combinations of random fluctuation in reproduction with reproduction curves produce populations which neither increase indefinitely nor decline to extinction. The cyclical changes produced by steep reproduction curves are maintained when random fluctuation is small or moderate, but they become variable in period and eventually unrecognizable as random effects become more important.

9. Under the combined influence of a steep reproduction curve and variable non-compensatory mortality, populations fluctuate more widely than when controlled by either of these factors alone.

10. When the dome of a population's reproduction curve lies above the 45-degree line, the result of light or moderate exploitation is to *increase* the abundance of the stock in subsequent generations; more intensive exploitation will decrease it (Fig. 16).

11. Another result of exploitation is to reduce the amplitude and complexity of any reproduction-curve oscillations that may be in progress; sufficiently intensive exploitation eliminates such oscillation entirely (Fig. 17).

12. Among multiple-age spawners exploitation tends also to reduce the *period* of oscillation somewhat, because the spawning stock gradually becomes younger.

13. Reproduction curves, or approximations to them, are plotted for fish populations and four invertebrate populations in Figures 18–28. No example was discovered where the maximum of recruitment is at or to the right of the 45-degree line (as in Figs. 1, 3, 4). The left limb was usually steep. The most probable position for the right limb was always sloping downward, sometimes only slightly, sometimes quite steeply; the most typical situation has a slope about as in Figure 5.

14. Cyclic population changes that are apparently the direct result of a steep reproduction curve are those of Herrington's haddock (1912–29) and the *Daphnia* cultures of Pratt. Most other well-known cycles seem not to be of a *simple* reproduction-curve type, but the shape of this curve must profoundly influence the course of population abundance in any animal.

15. More complex reproduction curves have been suggested by indirect evidence (Figs. 30, 31), and a depression of percentage reproduction at extremely low densities of stock may be fairly common.

16. An asymmetrical dome-shaped reproduction curve can be developed from simple assumptions involving random cannibalism or compensatory preda-

tion: if w represents adult stock density as a fraction or multiple of the density which provides maximum reproduction, we^{1-w} represents the actual reproduction at density w, as a fraction of the maximum reproduction. In this event the survival rate at maximum reproduction is always $1/e$ or 37 per cent of what it is at densities near zero; or in other words, at maximum reproduction the instantaneous mortality rate from compensatory activity is 1.

17. Curves computed from the above formula, shown in Figure 33, include most of the types actually observed. However if "environmental capacity" sets an upper limit to number of survivors at some stage of the pre-adult life history, this "typical" curve could either be levelled off at any point of its course, or else truncated (Fig. 34).

ACKNOWLEDGMENTS

A number of individuals have read the manuscript, and numerous changes, deletions and additions have resulted from their suggestions. Those who co-operated in this manner include Y. M. M. Bishop, L. M. Dickie, J. R. Dymond, R. E. Foerster, S. D. Gerking, J. L. Hart, D. J. Milne, F. Neave, A. L. Pritchard, W. B. Scott, W. M. Sprules, J. C. Stevenson, F. H. C. Taylor, F. C. Withler and D. E. Wohlschlag. I am indebted to Messrs. C. E. Atkinson, C. J. Burner and J. C. Stevenson for permission to peruse, and in some cases use, unpublished or incompletely published data. Mrs. D. Gailus has had the difficult task of maintaining the legibility and accuracy of the manuscript through a long series of revisions.

REFERENCES

ALLEE, W. C. 1931. Animal aggregations. Univ. Chicago Press, 431 pp.

BABCOCK, J. P. 1914. Annual Report of the British Columbia Commissioner of Fisheries for 1913, Victoria, B.C.

BARANOV, F. I. 1918. [On the question of the biological basis of fisheries.] *N.-i. Ikhtiologicheskii Inst., Izvestiia*, **1**(1): 81–288.

BARNABY, J. T. 1944. Fluctuations in abundance of red salmon, *Oncorhynchus nerka* (Walbaum), of the Karluk River, Alaska. *U. S. Fish and Wildlife Serv., Fish Bull.*, **50**: 237–295.

BURKENROAD, M. D. 1946. Fluctuations in abundance of marine animals. *Science*, **103**(2684): 684–686.

CHIANG, H. C., AND A. C. HODSON. 1950. An analytical study of population growth in *Drosophila melanogaster*. *Ecological Monogr.*, **20**: 173–206.

CLEMENS, W. A. 1952. On the cyclic abundance of animal populations. *Canadian Field-Naturalist*, **66**: 121–123.

COLE, LAMONT C. 1951. Population cycles and random oscillations. *J. Wildlife Management*, **15**: 233–252.

1954. Some features of random population cycles. *Ibid.*, **18**:2–24.

ELSON, P. F. 1950. Increasing salmon stocks by control of mergansers and kingfishers. *Fish. Res. Bd. Canada, Atlantic Prog. Repts.*, No. 51, pp. 12–15.

ERRINGTON, P. L. 1954. On the hazards of overemphasizing numerical fluctuations in studies of "cyclic" phenomena in muskrat populations. *J. Wildlife Management*, **18**: 66–90.

FOERSTER, R. E., AND W. E. RICKER. 1941. The effect of reduction of predaceous fish on survival of young sockeye salmon at Cultus Lake. *J. Fish. Res. Bd. Canada*, **5**: 315–336.

FUJITA, H., AND S. UTIDA. 1953. The effect of population density on the growth of an animal population. *Ecology*, **34**: 488–498.

GALLAGHER, H. R., A. G. HUNTSMAN, J. VAN OOSTEN AND D. J. TAYLOR. 1944. Report of the International Board of Enquiry for the Great Lakes Fisheries, pp. 1–24, Govt. Printing Office, Washington, D.C.

GALTSOFF, P. S., AND V. L. LOOSANOFF. 1939. Natural history and method of controlling the starfish (*Asterias forbesi* Desor). *Bull. U. S. Bur. Fish.*, **49**(31): 75–132.

HALDANE, J. B. S. 1953. Animal populations and their regulation. *New Biology*, **15**: 9–24. Penguin Books, London.

HERRINGTON, W. C. 1941. A crisis in the haddock fishery. *U. S. Fish and Wildlife Serv., Fish. Circ.*, No. 4, 14 pp.

1944. Factors controlling population size. *Trans. 9th North Am. Wildlife Conf.*, pp. 250–263.

1947. The role of intraspecific competition and other factors in determining the population level of a marine species. *Ecol. Monogr.*, **17**: 317–323.

1948. Limiting factors for fish populations. Some theories and an example. *Bull. Bingham Oceanogr. Coll.*, **9**(4): 229–279.

HILE, R. 1936. Age and growth of the cisco, *Leucichthys artedi* LeSueur, in the lakes of the northeastern highlands, Wisconsin. *Bull. U. S. Bur. Fish.*, **48**: 211–317.

HUNTSMAN, A. G. 1952. How Passamaquoddy produces sardines. *Fundy Fisherman*, **24**(24): 5. St. John, N.B.

1953. Movements and decline of large Quoddy herring. *J. Fish. Res. Bd. Canada*, **10**: 1–50.

HUTCHINSON, G. E. 1948. Circular causal systems in ecology. *Annals New York Acad. Sci.*, **50**: 221–246.

HUTCHINSON, G. E., AND E. S. DEEVEY, JR. 1949. Ecological studies on populations. *Survey of Biol. Progress*, **1**:325–359. Academic Press, New York.

JOHNSON, W. E., AND A. D. HASLER. 1954. Rainbow trout production in dystrophic lakes. *J. Wildlife Management*, **18**: 113–134.

KESTEVEN, G. L. 1947. Population studies in fisheries biology. *Nature*, **159**: 10–13.

LOTKA, A. J. 1925. Elements of physical biology. Williams and Wilkins, Baltimore, 460 pp.

MILLER, R. B. 1950. Observations on mortality rates in fished and unfished cisco populations. *Trans. Am. Fish. Soc. for 1949*, **79**: 180–186.

1952. The role of research in fisheries management in the Prairie Provinces. *Canadian Fish Culturist*, No. 12, pp. 13–19.

NEAVE, F. 1952. "Even-year" and "odd-year" pink salmon populations. *Trans. Royal Soc. Canada* (V), Ser. 3, **46**: 55–70.

1953. Principles affecting the size of pink and chum salmon populations in British Columbia. *J. Fish. Res. Bd. Canada*, **9**: 450–491.

NICHOLSON, A. J. 1933. The balance of animal populations. *J. Animal Ecol.*, **2**: 132–178.

NICHOLSON, A. J., AND V. A. BAILEY. 1935. The balance of animal populations. Part I. *Proc. Zool. Soc. London* for 1935, pp. 551–598.

PALMGREN, P. 1949. Some remarks on the short-term fluctuations in the numbers of northern birds and mammals. *Oikos*, **1**: 114–121.

PRATT, D. M. 1943. Analysis of population development in *Daphnia* at different temperatures. *Biol. Bull.*, **85**: 116–140.

PRITCHARD, A. L. 1948a. Efficiency of natural propagation of the pink salmon (*Oncorhynchus gorbuscha*) in McClinton Creek, Masset Inlet, B.C. *J. Fish. Res. Bd. Canada*, **7**: 224–236.

1948b. A discussion of the mortality in pink salmon (*Oncorhynchus gorbuscha*) during their period of marine life. *Trans. Royal Soc. Canada* (V), Ser. 3, **42**: 125–133.

RICKER, W. E. 1952. Numerical relations between abundance of predators and survival of prey. *Canadian Fish Culturist*, No. 13, pp. 5–9.

1954. Effects of compensatory mortality upon population abundance. *J. Wildlife Management*, **18**: 45–51.

Ricker, W. E., and R. E. Foerster. 1948. Computation of fish production. *Bull. Bingham Oceanogr. Coll.*, **11**(4): 173–211.

Royal, L. A. 1953. The effects of regulatory selectivity on the productivity of Fraser River sockeye. *Canadian Fish Culturist*, No. 14, pp. 1–12.

Scott, W. B. 1951. Fluctuations in abundance of the Lake Erie cisco (*Leucichthys artedii*) population. *Contr. Royal Ontario Mus. Zool.*, No. 32, 41 pp.

Smith, F. E. 1952. Experimental method in population dynamics: a critique. *Ecology*, **33**: 441–450.

Smith, H. S. 1935. The role of biotic factors in the determination of population densities. *J. Econ. Entomol.*, **28**: 873–898.

Smoker, W. A. 1954. A preliminary review of salmon fishing trends on Inner Puget Sound. *Washington Dept. Fish., Res. Bull.*, No. 2, 55 pp.

Snedecor, G. W. 1946. Statistical methods. 4th ed., Iowa State College Press, Ames, Iowa.

Solomon, M. E. 1949. The natural control of animal populations. *J. Animal Ecol.*, **18**: 1–35.

Tester, A. L. 1948. The efficacy of catch limitations regulating the British Columbia herring fishery. *Trans. Royal Soc. Canada* (V), Ser. 3, **42**: 135–163.

Thompson, D. H. 1941. The fish production of inland streams and lakes. *In* "A Symposium on Hydrobiology", pp. 206–217. Univ. Wisconsin Press, Madison.

Tippett, L. H. C. 1927. Random sampling numbers. *Tracts for Computers*, No. 15.

Varley, G. C. 1947. The natural control of population balance in the knapweed gall-fly (*Urophora jaceana*). *J. Animal Ecol.*, **16**: 139–187.

Vevers, H. G. 1949. The biology of *Asterias rubens* L.: Growth and reproduction. *J. Marine Biol. Assn. U. K.*, **28**: 165–187.

Part 4

COMMUNITY AND ECOSYSTEM ECOLOGY

There exist two principal opposing views of the nature of biological communities. One, which seems to follow naturally from such early studies of aquatic systems as that by Möbius (1877) or Forbes (1887), suggests that communities are natural, organized systems of plants and animals having parallel structures of species presence and abundance and concomitant functional similarities. These systems have evolved and result from the consequent mutual interdependence of the organisms present. The opposing viewpoint is that what we examine are not communities in the former sense, but rather randomly collected assemblages of organisms whose physiological tolerances permit them to exist together, and that such groupings are artificial. This viewpoint appears more commonly among plant ecologists (see, for example, Gleason, 1926). Perhaps more simply, a group of organisms which do in fact occur together in time and space must coexist and function together, and therefore should be studied together.

Presumably, a community of organisms could be studied additively—that is, by investigating all the species separately and treating the community as the sum. Such a course will recommend itself to no one with practical experience, even to an ecologist who has the aid of a capacious and sophisticated computer. This simplistic view of how to study an assemblage is not only practically impossible, it is also theoretically inadequate, for such an assemblage has properties not shared by the species taken separately. There is a structure based on the kinds and proportions of species present, an organization based on the flow of energy and matter, and a succession of the populations themselves. More realistically, therefore, concepts and assumptions must be simplified to permit the community ecologist to understand his subject matter. Lindeman (1942) suggested a history of viewpoints guiding synecological theory; the principal stages he suggests are (1) the static species-distributional, (2) the dynamic species-distributional, and (3) the trophic-dynamic viewpoint.

The first should be understood to include not only the studies which give species lists—for example, lists of fish or oligotrophic and eutrophic lakes as given by Forbes—but also the attempts to ascertain statistical relations among populations which will permit them to be treated as something other than mere random assemblages. An early attempt at stating this problem and suggesting a solution was Forbes' study on the local distribution of Illinois fishes for which he developed a mathematical "coefficient of association" (Forbes, 1907). Solutions to similar problems continue to be suggested, usually based either on chi-square contingency tables (see, for example, Nichols in this collection) or on an analysis of recurrent groups (Fager, 1957).

Once a community—which is here presumed to be, following Englemann (1961), "an assemblage of populations coexisting in time and space, mutually regulative and interdependent, and depending ultimately upon some common energy source"—is identified, a principal function of the ecologist is to try to make sense of the data. Ideally, one expects the ecologist to move now from the static to the dynamic aspect of distribution, where the causes are treated, but this expectation is seldom met. Among the attempts are the several models suggested by MacArthur (1960), the use of information theory following the suggestions of Margalef (1957), and the theories relating stability to diversity. These suggestions have induced a great deal of research, but only the last one is a realistic attempt to relate structure and function.

More successful in making functional sense out of masses of data are the attempts to sort the organisms into functional groups—the producers, herbivores, carnivores, and reducers—and then to study the flow of energy. Lindeman's study of Cedar Bog Lake gave this simplifying and integrating procedure a strong stimulus in aquatic biology. His paper is readily available in reprints, and in this collection the work of Riley and Wright is closely related to that of Lindeman.

In these studies the object to be studied by ecologists is not treated as a mere community of plants and animals, but rather such communities are regarded as artificially set off from the more natural unit, the ecosystem. This concept was proposed by Tansley (1935), who said, "But the more fundamental conception is, as it seems to me, the whole *system* (in the sense of physics), including not only the organism-complex, but also the whole complex of physical factors forming what we call the environment of the biome. . . . These ecosystems, as we may call them, are of the most various kinds and sizes. They form one category of the multitudinous physical systems of the universe, which range from the universe as a whole down to the atom."

Lindeman's study was a model flow through an ecosystem, the fundamental equation being given as

$$\frac{d\Lambda_n}{dt} = \lambda_n + \lambda'_n$$

where Λ_n is any trophic level in a food chain, (producers, herbivores, or carnivores), λ_n is positive and is the rate of assimilation of energy from the next lower trophic level (the $n-1$ level), and λ'_n is negative and represents the sum of the rates of energy dissipation resulting from decomposition, respiration, and feeding by the next higher level (the $n+1$ level). λ_n is defined as the gross productivity of the level n. Net productivity is $\lambda_n + \Lambda'_n$, or $d\Lambda_n/dt$. In words, this equation says that

the rate of change of energy content of any trophic level	=	the rate of energy inflow from the next lower level	−	sum of the rates of energy dissipation

Efficiency of energy utilization at a trophic level can also be measured, the measurement preferred by Lindeman being λ_n/λ_{n-1}. Lindeman's estimates for productivities and efficiencies in a lake food chain are given in Table 1. Productivities decrease at each succeeding trophic level; this is necessarily true, because each level dissipates some of its energy through respiration.

To arrive at these values, Lindeman first estimated the annual changes in standing crop (the mass of protoplasm that can be harvested) of all the trophic levels present in Cedar Bog Lake. The lowest level in the food chain contained the phytoplankton and the rooted aquatic plants; the next level contained the herbivorous animals, both planktonic and bottom-dwelling; and the highest level contained the carnivores in the lake. Decomposers were not studied separately, nor were higher levels of carnivores represented, since game fish (predators on other carnivores) are not usually found in a bog lake. These changes in standing crop are given in the first data column of Table 1. From laboratory data and from reports of other investigators Lindeman obtained average values for respiration rates of plants, herbivores, and carnivores (column 2 of Table 1). Values for decomposition rates were also obtained from laboratory estimates and reports of earlier investigations (column 4). The sum of the change in standing crop, respiration, and decomposition gives the total caloric value taken in from the preceding trophic level or, for producers, made available by photosynthesis. The total intake for carnivores, 3.1 gcal per cm^2 per year in Table 1, is entered in the table in the line above it as predation loss by the herbivores; there is no other source of energy for carnivores. Likewise, the total herbivore intake is the same as the loss at the producer level due to predation. In this way a complete energy budget can be worked out.

Although Lindeman's values for uncorrected productivity (change in standing crop) are adequate to show energy relationships, they are certainly too high, and his values for decomposition are too low. For an ecosystem in approximately steady state, the net productivity values should be close to zero. That is, the mass of organisms in each trophic level should be constant, as is true, for example, in a forest or in the sea. Generally, in most ecosystems the amount of organic

Table 1. Productivity Values for the Cedar Bog Lake Food Chain (gcal/cm^2/year)*

Trophic level	*Changes in energy content; uncorrected net productivity*	*Respiration*	*Predation*	*Decomposition*	*Corrected Net productivity*	*Efficiency (%)*
Producers	70.4	23.4	14.8	2.8	111.3	.1
Primary consumers	7.0	4.4	3.1	.3	14.8	13.3
Secondary consumers	1.3	1.8	0	0	3.1	22.3

*After Lindeman, 1942

material produced by green plants by photosynthesis is entirely metabolized by animals and decomposers.

The meaning of productivity as the rate of contribution of energy from the next lower level is not in accord with the usual meaning of the term. It is preferable to consider net increase in energy content $\frac{d\Lambda}{dt}$ as productivity, especially since only the latter has a clear meaning at trophic levels higher than producers. This definition of efficiency is also special, and is often replaced by a definition based on transforming input into stored chemical energy.

A complete understanding of ecology cannot be based on reductionist principles. The formal separation between individuals, populations, and communities is useful, for each level has its own special characteristics and appropriate methods of study, but each level must be understood with at least a minimum of reference to those nearest it in the hierarchy of biological organizations. Studies of animal behavior and of birth rates of populations are part of an analysis of a community. In a reciprocal fashion, the community provides a framework within which studies of lower organizations make sense.

LITERATURE CITED

Engelmann, M. D. 1961. The role of soil arthropods in the energetics of an old field community. Ecol. Monog., 31:221–238.

Fager, E. W. 1957. Determination and analysis of recurrent groups. Ecology, 38:586–595.

Forbes, S. A. 1887. The lake as a microcosm. Bull. Peoria Sci. Assoc., 77–87.

Forbes, S. A. 1907. On the local distribution of certain Illinois fishes: an essay in statistical ecology. Bull. Ill. Lab. Nat. Hist., 7:1–19.

Gleason, H. A. 1926. The individualistic concept of the plant association. Bull. Torrey Bot. Club, 53:7–26.

Lindeman, R. L. 1942. The trophic-dynamic aspect of ecology. Ecology, 23:399–418.

MacArthur, R. 1960. On the relative abundance of species. Amer. Nat., 94:25–36.

Margalef, R. 1957. La teoría de la informacíon en ecología. Mem. Real. Acad. Ciencias y Artes de Barcelona, 32:373–449.

Möbius, Karl 1877. Die Auster und die Austernwirtschaft. Berlin. [Translated 1880, The Oyster and Oyster Culture. Rept. U.S. Fish Comm., 1880:683–751.]

Tansley, A. G. 1935. The use and abuse of vegetational concepts and terms. Ecology, 16:284–307.

Benthic polychaete assemblages and their relationship to the sediment in Port Madison, Washington

14

F. H. NICHOLS

Department of Oceanography, University of Washington; Seattle, Washington, USA

Abstract

Two indices of community association and a 3 dimensional ordination of stations were used to elucidate the relationship between small-scale changes in species, composition of polychaete assemblages and changes in the physical character of the sediment, at 9 stations along a 1.5 km subtidal transect. There were no faunal or physical discontinuities that might have been used to delineate boundaries of polychaete communities, except at a very shallow station where the effects of extreme fluctuations of temperature and salinity were evident. The degree of similarity between assemblages was related to the similarity of the sediments at the different stations. In particular, the changes in species composition appeared to correspond most clearly to differences in the clay content of the sediment.

Introduction

Terrestrial plant communities have long been defined with the aid of mathematical analyses, permitting objective examination of the structure, and variation of vegetation in terms of environmental factors and species' distributions, as well as evaluation of the nature of community boundaries (GREIG-SMITH, 1964; WHITTAKER, 1967). In contrast, until very recently, marine benthic communities have been defined in terms of the few largest, most commonly distributed, or numerically important species in the manner of PETERSEN (1913). As a result, there is much confusion as to the definition and delineation of a benthic community. Because of the essentially 2-dimensional nature of both terrestrial plant and benthic infaunal communities, an approach to the study of benthic communities similar to those used by terrestrial plant ecologists should offer meaningful results in relating community structure and variation to the physical and chemical environment, as suggested by THORSON (1957).

Of the physical environmental parameters that affect the distribution of benthic fauna, the most significant seem to be temperature, salinity, and the nature of the substrate. That faunal assemblages are related to sediment types has been well documented (JONES, 1950). SANDERS (1956, 1958) has related faunal assemblages to specific sediment types in Long Island Sound and Buzzard's Bay, suggesting that the clay content of the sediment determined the relative amounts of biomass, and played a role in determining the dominant feeding type. However, he did not observe small-scale changes in both sediment and fauna between closely arranged stations; rather, his stations covered large areas of sea bottom and encompassed a wide range of sediment types.

As a result of a 2 year study (1963 to 1964) of the benthic infauna of Puget Sound Washington, USA, LIE (1968) established that the nature of the substrate is the most important factor governing differences between faunal assemblages in this area, since temperature and salinity do not vary greatly subtidally. However, LIE's stations were geographically isolated and encompassed a wide range of sediment types. The present study was designed to permit examination of the small-scale changes in species' composition and substrate character between 2 of LIE's stations.

Only the distribution of polychaetes was examined. Among the organisms retained by a 1.0 mm screen, polychaetes comprise about $^1/_2$ of the species, and more than $^1/_2$ of the biomass at the 2 reference stations (LIE, 1968), and they include all major feeding types. Thus, their patterns of distribution might be considered representative of the total macrofaunal community.

Research area and stations

Port Madison (Fig. 1) is located on the west side of Puget Sound, Washington, USA. A mixed tide, with a range of approximately 3.5 m, creates currents in Port Madison reaching a maximum strength of nearly 2 m/sec through Agate Passage. The prevailing winds are from the south, and large wind waves occur during the winter. Eroded beach cliffs on the northern shore, a large spit in Miller Bay, and a small delta at the entrance to Miller Bay, indicate that an east-to-west longshore current is an active agent in the sedimentary processes of Port Madison. This current is probably related to both tidal activity and wind waves. Recent sediments are derived from the streams that enter Puget Sound, and from cliff and submarine erosion. A small stream entering Miller Bay probably contributes to the sedimentary processes in the research area.

Reprinted from *Marine Biology*, 6:48–57, 1970.

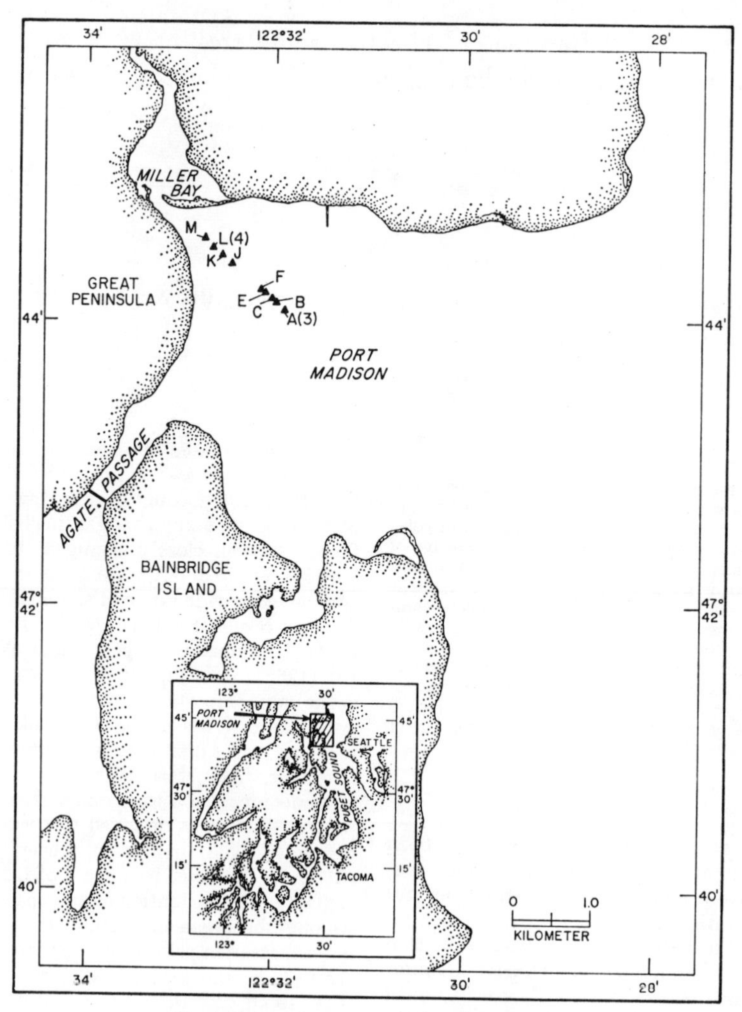

Fig. 1. Research area, showing sampling sites

Although seasonal hydrographic data are not available for Port Madison, monthly hydrographic samples were taken in 1964 and 1965 south-east of Port Madison (G. C. ANDERSON, unpublished data) and in 1965 and 1966 in Port Orchard to the south-west (T. S. ENGLISH, unpublished data) by the Department of Oceanography, University of Washington. These data indicate that surface temperatures in Port Madison vary seasonally about 10 C° (7° to 17 C°), and those at 35 m (the depth of the deepest station considered here) vary about 5 C° (8° to 13 C°). Salinities at the surface vary from 25.3 to 30.5‰, and at 35 m from 28.8 to 30.5‰. At neither station, does the dissolved oxygen concentration fall below 4.0 ml/l in the upper 35 m. Dissolved oxygen, therefore, is not considered an important factor in this study and is not treated further.

The study area lies along a transect (Fig. 2), including LIE's stations 3 and 4 (LIE, 1968). Station 3 is located on the top of a small seaknoll about 1.85 km southeast of the entrance to Miller Bay, at a depth of 23 m. Station 4 is located about 1.25 km to the northwest of station 3, near the entrance to Miller Bay, at a depth of 10 m. The depth between these 2 stations increases to 34 m near the base of the seaknoll. Samples were collected on 21 and 22 October 1965 at 12 stations, 11 of which were selected at approximately 5 m depth intervals. The 12th station was located approximately 0.25 km shoreward of the shallow reference station, at a depth of 2 m. Nine of the stations, chosen to represent different depths and sediment types, are considered in this study.

Material and methods

Three replicate samples collected at each station with a 0.1 m van Veen grab were used for species' identification and sediment analysis. A small portion of the sediment (approximately 150 ml) was removed from each sample with a 47 mm diameter core liner, for sediment analysis. The remainder of the sample was washed in a 1 mm screen to separate the organisms

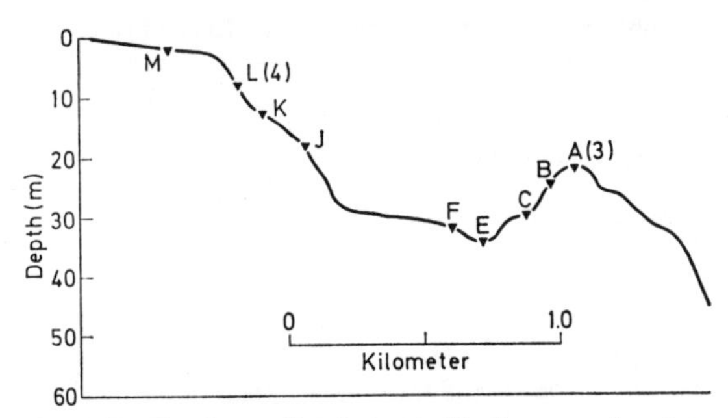

Fig. 2. Profile of sampling transect. Depths are referred to mean lower low water. Stations A and L coincide with LIE's stations 3 and 4 (1968) as indicated

and large debris from the sediment. The material remaining on the screen was preserved in a 5% formalin solution (buffered with sodium borate) for sorting and identification in the laboratory.

The sediment size distribution was ascertained for each sample using the standard dry-sieve and pipette technique (KRUMBEIN and PETTIJOHN, 1938) based on $^1/_2$ phi-size intervals. The size-fraction data in phi notation, in which phi is defined as the negative logarithm to the base two of the ratio of grain diameter in millimeters to the standard graim diameter of 1.00 mm (McMANUS, 1963), were programmed for computer analysis and processed utilizing the computer program by CREAGER et al. (1962). The parameters considered in this study were mean phi (the measure of mean grain size), the sand-to-mud ratio (a ratio between the material coarser than 0.0625 mm to that finer than 0.0625 mm), clay content (the percentage of the clay fraction), and sorting (the standard deviation of mean phi in phi units). The modal or most common grain size paralleled the mean grain size in this study and has not been considered separately.

The animals were sorted into major groups, using a 1.5 × sorting lens. All polychaetes, including fragments, were identified to species (with the exception of the family Capitellidae), those genera that were represented by poorly preserved specimens or fragments of unidentifiable specimens, those genera for which species' identifications are as yet uncertain, and undescribed species. With the exception of those species within the family Capitellidae and those genera which contained 2 or more unidentified species, all unidentified species are referred to by their generic names and are considered single taxonomic categories. As such, they are included in the analyses described below. Identification was based primarily on the keys of BERKELEY and BERKELEY (1948, 1952), and keys for local polychaetes (K. D. HOBSON and K. BANSE, unpublished). Additional references were USHAKOV (1955) and O. HARTMAN's reports of the Allan Hancock Pacific Expeditions (HARTMAN, 1961). Species names are used, except where noted, as in BANSE et al. (1968), which includes references to the original literature.

The species' compositions of individual stations were compared using KENDALL's coefficient of association T (see LOOMAN and CAMPBELL, 1960), BRAY and CURTIS' (1957) index of similarity, and a mechanical ordination technique to determine the degree of similarity of the assemblages at these stations. A ranking of stations, derived from the degree of similarity of pairs of stations, was compared with rankings of stations arranged in order of increasing or decreasing values of different environmental parameters, to ascertain the presence of causal relationships between the physical environment and species' distributions.

Results and discussion

Sediment characteristics

A summary of those abiotic parameters that would seem to have most biological meaning appears in Table 1. Sample means are used in the comparison of stations.

The nearshore station M consists predominantly of fine-to-very fine, moderately sorted sand, with a low clay content, and a high sand-to-mud ratio. Farther from shore, in deeper water, the sediments have a finer mean grain size (larger mean phi), greater clay content, and poorer sorting. This is most marked at the deep stations E and F. At the seaknoll end of the transect, the sediments, containing shell fragments and glacial debris, are more coarse than those found at the other stations. They have a mean grain size in the fine-sand range, a relatively high sand-to-mud ratio, intermediate clay content, and moderate-to-poor sorting.

In general, there exists a sediment gradient from the very fine, moderately sorted, clean sand of the shallow, inshore area, through the poorly sorted, silty sand of the deep area, to the moderately-to-poorly sorted fine sand with shell fragments of the seaknoll.

Polychaete distribution

The exact number of species in the collection is uncertain (between 130 and 135; NICHOLS, 1968) because certain genera, for example *Nereis*, *Magelona*, and *Polycirrus*, and those in the family Capitellidae, appear represented by more than one unidentified species. These genera were not included in the analyses. This omission is unfortunate in that, at the 3 deep

Table 1. *Summary of sediment size analysis, sample means only. Mean grain diameters in millimeters are shown in parentheses. See text for definition of phi notation*

Station	Depth (m)	Mean phi	Sand/mud ratio	Clay (%)	Sorting (phi units)
A	23	2.78 (0.144)	6.34	3.76	1.22
B	25	2.53 (0.172)	9.90	3.04	0.97
C	30	2.94 (0.129)	5.03	4.05	1.30
E	34	3.57 (0.082)	1.68	5.90	1.46
F	32	3.91 (0.066)	1.05	6.45	1.39
J	18	3.26 (0.104)	3.21	3.44	1.23
K	13	3.43 (0.092)	3.30	3.31	1.01
L	8	3.30 (0.100)	4.06	2.31	0.89
M	2	3.07 (0.119)	9.12	1.29	0.65

stations C, E and F, the capitellids and the *Polycirrus* species, respectively, comprised 16, 22 and 20% of the total number of specimens, with an average of 12.4% at all stations. However, by considering the lumped species as single taxonomic categories, they would have been given undue weight in the analyses. The *Nereis* and *Magelona* species contributed less than 2% of the total at all stations.

It must be pointed out that there was a large number of rare species; 1/3 to 1/4 of the species at each station were represented by a single specimen. As discussed below, these species, as well as the conspicuously abundant species, may contribute to the degree of association between stations.

With the exception of the differences between the shallowest stations L and M, faunal changes from station to station are gradual. At the seaknoll stations A and B, there are many species (77 and 74, respectively), none of which is dominant. *Chaetozone setosa*, *Nephtys ferruginea*, and several other species are equally important, although not found in large numbers. There are fewer species (61, 55 and 59) at the deeper stations C, E and F, and the deposit-feeding species, primarily *Pectinaria californiensis* and *Polycirrus* spp., predominate. In the finer sediments of stations E and F, *P. californiensis* is overwhelmingly dominant. Station J is dominated by *P. californiensis*, but *Owenia fusiformis* and *Prionospio malmgreni* are also well represented. Station K is dominated by these latter 2 species, while the number of *P. californiensis* decreases markedly. Station L is somewhat different from station K, primarily because of the appearance of *Platynereis bicanaliculata*, which accounts for nearly 40% of the specimens found at that station. In addition, *P. malmgreni* and *Glycinde picta* are important members of this assemblage.

Thus, the species' composition of polychaete assemblages in Port Madison changes gradually from one station to another between the 2 reference stations A and L (LIE's stations 3 and 4; LIE, 1968), as in the plant ecologists' continuum, and corresponds to the sediment gradient as LIE (1968) suggested. There are no abrupt physical barriers in the environment preventing species from inhabiting substrates of differing sediment size characteristics although, for any particular species, the greater the departure from the apparent optimal habitat type, the fewer the specimens found.

An important species change occurs, however, between stations L and M. At station M the number of species (21) is less than 1/2 of that found at any other station. *Owenia fusiformis* is the dominant species, followed closely in numerical importance by *Platynereis bicanaliculata* and *Glycinde picta*. Many species that live in the similar sediment of station L are missing. Because it is located in shallow water, this station is subjected to relatively wide fluctuations of temperature and salinity, due not only to normal sea-surface fluctuations, but to the influx of brackish surface water from Miller Bay. Although station M represents extremes in clay content and sorting, the significant change in fauna between stations L and M seems to be related rather to an ability to withstand the rigors of a shallow estuarine environment. In addition, the active sedimentary processes taking place at this location, on the delta near the mouth of Miller Bay, may add to the instability of the physical environment. The species found at station M may be those that are able to contend with high rates of sedimentation and shifting of sand in response to tidal activity.

Measures of association

KENDALL's coefficient of association *T*. As used by LOOMAN and CAMPBELL (1960), KENDALL's *T*,

$$T_{ab} = \frac{Cw - xy}{\sqrt{(Au)\,(Bv)}},$$

is a measure of the association between 2 stations, based on the number of species common to both stations, in which A and B are the numbers of species at stations a and b, C is the number of species common to both a and b, u = S − A, v = S − B (S = number of species at all stations), x = A − C, y = B − C, and w = u − y = v − x.

Utilizing a 2 × 2 contingency table, coefficients were computed for all possible combinations of station pairs using 132 species, including fragments (Table 2). (Coefficients were also computed using a reduced species' list as described below.) Stations with a coefficient of association of less than 0.171 show a lack of association at the 5% level (X^2 test with 1 df). Station A on the seaknoll lacks association with stations E and F, the deep stations, and with station J, the next station shoreward from station F. Station M at the shoreward end of the transect lacks association with stations A, C, E and F, and is barely similar to station J. All other stations show a significant degree of association. The relatively high coefficients of association between adjacent stations suggest that the faunal transition is smooth from one assemblage to the next.

Because of ease of computation, lack of emphasis on extreme values, and independence of the distribution of the data, rank-correlation techniques were employed to determine whether a relationship exists between the degree of faunal association and the similarity of sediment characteristics.

Table 3. *Rank correlation: ranking of stations by faunal association* (KENDALL's T) *compared with ranking by sediment parameter and depth. (The degree of correspondence between rankings is measured with* KENDALL's τ. *Significance is indicated by the probability of occurrence by chance* p *as shown)*

	Ranking									τ	p
Faunal association	M	L	A	B	K	C	J	F	E		
Mean grain size	B	A	C	M	J	L	K	E	F	+0.39	0.090
Sand/mud	B	M	A	C	L	K	J	E	F	+0.61	0.012
Clay content	M	L	B	K	J	A	C	E	F	+0.72	0.003
Sorting	M	L	B	K	A	J	C	F	E	+0.83	<0.001
Depth	M	L	K	J	A	B	C	F	E	+0.72	0.003

Table 2. *Matrix of* KENDALL's *coefficient of faunal association* T. *Shown are coefficients based on a comparison of total species lists as well as a comparison of lists comprising only those species included in the list of 48 numerically most important species (latter values in parentheses; see text). Boldface indicates lack of association at the 5% level*

	A	B	C	E	F	J	K	L	M
A		0.481 (0.435)	0.301 **(0.148)**	**0.115** **(0.022)**	**0.151** **(−0.031)**	0.175 **(0.092)**	0.310 **(0.246)**	0.242 **(0.027)**	**0.141** **(0.173)**
B			0.451 (0.402)	0.232 **(0.122)**	0.301 **(0.174)**	0.385 (0.348)	0.489 (0.374)	0.295 **(0.051)**	0.237 **(0.184)**
C				0.463 (0.540)	0.450 (0.533)	0.665 (0.770)	0.540 (0.575)	0.357 (0.087)	0.185 (0.197)
E					0.739 (0.856)	0.530 (0.545)	0.423 (0.381)	0.277 **(0.068)**	**0.109** **(0.107)**
F						0.545 (0.556)	0.476 **(0.258)**	0.250 **(0.000)**	0.158 **(0.135)**
J							0.543 (0.487)	0.363 (0.195)	0.214 (0.244)
K								0.466 (0.344)	0.298 **(0.221)**
L									0.464 (0.420)

It is possible to arrange the stations sequentially, based on the degree of association between pairs of stations. By placing those stations with high degrees of association closest together, and those with lower degrees of association farther apart, beginning with the station showing the least association with all others (station M), the sequence becomes as shown in the first row of Table 3. This ranking can be compared with rankings of the stations according to decreasing mean grain size, decreasing sand-to-mud ratio, increasing clay content, increasing standard deviation (poorer sorting), and increasing depth, using a simple rank-correlation coefficient, KENDALL's τ (tau) (KENDALL, 1962),

$$\tau = \frac{2S}{n(n-1)}$$

in which S is the sum of the "scores" of the ranking of individual pairs and n is the number of pairs being ranked (Table 3). The degree of association between assemblages appears to be strongly related to the clay content and the sorting of the sediment, and to some degree to the sand-to-mud ratio. There is no correlation with mean grain size at the 5% level. The arrangement of stations is strongly correlated with depth. At the same time, depth is strongly correlated with clay content and sorting of the sediment (with an r correlation coefficient of + 0.77 in both cases). These interrelationships are discussed below.

When comparing faunal lists from different localities, it is often most convenient in terms of time spent in species' identification and statistical computation to use lists comprising only those species which are conspicuous by their size or numerical abundance. This procedure is undertaken with the assumption that small rare species (those species represented by 1 or 2 specimens) probably do not contribute significantly to the characterization of the faunal assemblage at those localities. There are undoubtedly certain species that, although occurring in small numbers, are recurring members of the assemblage and might contribute to the degree of association between stations. However, as their importance is not appreciated in a few 0.1 m^2 samples, they cannot be identified as such and hence are omitted with the other rare species.

To test the above assumption, coefficients of association were computed in the manner described above, but based on the number of numerically important species common to individual pairs of stations. These species, defined arbitrarily as those contributing the first 80% of the specimens at each station when the species were listed by descending numerical importance, constituted a composite list of 48 species. As shown in Table 2, the comparison of stations based solely on numerically important species (5% significance level at 0.283) tends to exaggerate the degree of association between stations or lack of such association: stations C, E, F and J demonstrate a higher degree of mutual association than before, while stations A, L and M are significantly associated only with their respective neighbors. However, with the exception of the transposition of stations L and M and stations E and F, the subjective arrangement of stations based on the new coefficients of association is similar to the previous arrangement. Consequently, the correlations between this new ranking of stations and the ranking by sediment parameters are little changed.

BRAY and CURTIS index of similarity. The KENDALL coefficient, because it is based on a convenient test statistic (X^2) and is relatively easy to compute for a limited number of sampling sites, is a useful means for quantifying the similarity between different sampling sites. A drawback, however, is that

Table 4. *Matrix of the* Bray *and* Curtis *(1957) index of similarity*

	A	B	C	E	F	J	K	L	M
A		68.08	55.38	36.05	35.13	43.65	37.89	27.42	9.37
B			64.05	39.74	42.32	50.44	45.67	31.02	17.31
C				56.09	53.43	61.06	54.58	29.48	16.25
E					70.91	52.61	46.01	22.90	16.10
F						55.25	49.46	27.52	15.22
J							60.05	36.85	19.43
K								45.33	29.26
L									33.35

it is based solely on the presence or absence of species. In a small area where the species' composition of adjacent stations does not vary greatly, a method that incorporates abundance data might offer a more informative comparison. Such a technique is described by Bray and Curtis (1957). They use a modified coefficient of community, or "index of similarity", 100 (2 w/a + b), in which a and b are the numbers of species at 2 stations, and w the number of species in common at the 2 stations. In order to reflect the differences in abundance of the various species, the measure of species at each station is expressed as a percentage (score) of the maximum number of specimens occurring at any station, and thus is an expression of the success of a species at the different stations, relative to an "optimum" condition at the station with the highest number of specimens. This new measure is further refined to account for the differences in total scores between stations: the score for each species within a station is expressed as a percentage of the total score for that station. The terms a and b of the equation expressing the "index of similarity" are, therefore, each equal to 100%, and the equation reduces to 200 w/(100 + 100) or w. The term w is, by definition (Bray and Curtis, 1957), the sum of the lesser scores for those species that have a score above zero at the 2 stations being compared.

In light of the amount of work involved in computing this index of similarity and the uncertainty of the usefulness of comparing entire species lists, the list of numerically important species mentioned above was used here. The measures of association derived from the Bray and Curtis index shown in Table 4 indicate, again, that adjacent stations have a high degree of similarity, although between stations L and M there is less similarity than is seen in the matrix of Kendall's T. The obvious differences between the species' lists for these 2 stations (48 versus 21 total species), as well as the large differences in number of specimens of the species found in greatest abundance, may explain this lower degree of association.

When the stations are ranked according to the Bray and Curtis index (1957) values and compared with ranks of environmental parameters, the results (Table 5) are essentially the same as those obtained with Kendall's T (Table 3). The only noticeable differences are that the sand-to-mud ratio is no longer significant at the 5% level, and the mean grain size is clearly non-significant.

The major difficulty with the Bray and Curtis measure of association is that it does not have a statistical basis, and thus cannot be tested for significance. Nevertheless, it permits a precise differentiation between stations within an area of medium to high similarity (Bray and Curtis, 1957).

A preference for one of the 2 indices of association is made only by weighing the advantage of any additional information provided by abundance data

Table 5. *Rank correlation: ranking of stations by faunal association* [Bray *and* Curtis *(1957) index*] *compared with ranking by sediment parameters and depth. (See also Table 3)*

	Ranking	τ	p
Faunal association	L M K B A J C F E		
Mean grain size	B A C M J L K E F	+0.22	0.238
Sand/mud	B M A C L K J E F	+0.44	0.060
Clay content	M L B K J A C E F	+0.78	0.001
Sorting	M L B K A J C F E	+0.89	<0.001
Depth	M L K J A B C F E	+0.78	0.001

against the ability to test, statistically, the significance of the association. When comparing stations based on the presence or absence of numerically important species, there is a tendency toward a decrease in the degree of association between stations that otherwise are similar. It is evident that rare species significantly contribute to the degree of similarity when only presence and absence are considered, as their status is the same as that of the numerically important species. On the other hand, when numerical abundance is included in computing an index such as that described by Bray and Curtis (1957), the effect of rare species is greatly reduced. In the present study, the omission of the rare species from the computation of the Bray and Curtis index resulted in associations between stations very similar to those achieved by comparison of stations based on the presence or absence of all species. Thus, the choice of association index seems to rest on the importance attached to the role of the rare species in the overall study to which the data are to be applied. Because of the limited geographical area covered in this study, and the concomitant limited range of environmental parameters and high degree of similarity between species' lists, the results of the present comparison of faunal indices are inconclusive. It is intuitively felt, however, that if a reduced species' list is to be used as the basis for a comparison of stations, abundance data should be used, especially if any degree of similarity between remotely spaced stations is to be expected.

Ordination

When examining the composition of biotic assemblages in relation to the physical environment, it is useful to attempt to interpret the cumulative effect of all environmental parameters as well as to examine the relative importance of individual parameters. This is most easily accomplished through the process of ordination, a method used by plant ecologists who assume the existence of a continuum, in which the sampling sites are arranged in a single or multidimensional framework of composition gradients based on the degree of biotic intercorrespondence among the sites. The position of the sites along each axis of this framework can then be correlated with the environmental factors, and the positions in multidimensional space can be used to interpret the interactions of these factors. The advantage of ordination is in separating less obvious environmental factors affecting the distribution of organisms from the dominant factors, and thus approximating factor analysis (GREIG-SMITH, 1964). FAGER (1963) suggests that ordination is applicable to the study of benthic communities in which multiple samples have been taken in regions where physical and chemical characteristics change over relatively short distances.

Table 6. *Station coordinates in the* BRAY *and* CURTIS *(1957) ordination*

Station	x axis	y axis	z axis
A	0.0	34.7	−4.6
B	8.5	34.2	0.0
C	10.8	44.5	12.3
E	20.1	55.2	27.8
F	19.7	50.4	22.6
J	18.6	37.2	28.2
K	29.5	28.0	26.4
L	39.5	0.0	22.1
M	70.6	11.5	17.8

An ordination following BRAY and CURTIS (1957) was used in this study. It requires that the indices of similarity be converted to hypothetical inter-station distances by subtraction from the maximum possible index value. The distance between stations is thus inversely proportional to the degree of their similarity.

The stations were plotted along 3 perpendicular axes (Table 6) as described by BRAY and CURTIS (1957). Because of the large variability among the samples within stations, the small number of stations used, and the few environmental parameters measured at those stations, the 3-dimensional framework was constructed only with some lack of precision. Stations plotted on the y and z axes failed to maintain exact positional relationships with the x axis as prescribed by BRAY and CURTIS. This problem was encountered by these authors when plotting their z axis, and their method for plotting that axis was used for plotting the y and z axes in this study.

Significance of the correlations between the arrangement of stations on the 3 axes with the environmental data was tested with the correlation coefficient r (SNEDECOR, 1956) (Table 7). Except for the transposition of stations J and E, the stations are aligned on the x axis in their actual geographic positions, undoubtedly reflecting the importance of the interrelationships of adjacent stations. The significant correlations (at the 5% level) of depth, clay content, and sorting with this axis, reflect the generally smooth transition from the shallow, nearshore clean sand to the deeper, offshore silty sands. The alignment of stations on the y and z axes appears to be more directly related to sediment gradients: highly significant correlations (at the 1% level) exist between clay content, sorting, depth and the y axis, and between mean grain size and the z axis, and a significant correlation (at the 5% level) between the sand-to-mud ratio and the z axis. The station alignment on the y axis is exactly the same as that resulting from the aforementioned subjective arrangement of stations using the BRAY and CURTIS index of similarity (Table 5).

Table 7. *The* r *correlation coefficient values: depth and sediment parameters versus station positions on ordination axes.* $(r = \pm 0.66$ *at the 5% significance level)*

	x axis	y axis	z axis
Depth	−0.77	0.92	−0.13
Mean phi	0.22	0.24	0.81
Sand/mud ratio	0.19	−0.43	−0.70
Clay content	−0.73	0.86	0.20
Sorting	−0.72	0.89	0.13

BRAY and CURTIS (1957) imply that the environmental factors associated with the x axis are more clearly correlated with compositional similarity than those of the y and z axes. If this is the case, the juxtaposition of stations appears to contribute most to the degree of faunal similarity; the polychaete assemblages overlap, despite the changing physical environment, because of their close proximity. The high degree of association between pairs of adjacent stations, seen also in the primary diagonal of the 2 matrices (Tables 2 and 4), may result from the influx of stray specimens from neighboring locations (LIE, 1968) but it is difficult to assess this without seasonal or annual data. However, because the results of the analysis based on all species (using KENDALL's T) were very similar to the results of the analysis based on dominant species alone, the effect of stray species appears to be of limited importance.

The highly significant correlation between the arrangement of stations on the y axis, and the sediment parameters of clay content and sorting, reinforces the earlier suggestion that these parameters have a measurable effect on the composition of faunal assemblages in this area. On the other hand, the often-used parameter, mean grain size, is clearly shown to be of limited importance here.

Using the coordinates of Table 6, a 3-dimensional model was constructed to evaluate the spatial arrangement of stations, an arrangement based on compositional gradients and established, at least in part, by the cumulative effect of all environmental factors. As expected, the deep stations E and F are very closely spaced, with stations C, J and K not far removed. Stations A and B, and to a greater extent, stations L and M, are well separated in space from the first group. In addition to the associations based solely on the juxtaposition of adjacent stations, the arrangement seems to reflect the sedimentary environment of the area. The 2 shallowest stations L and M, are subjected to the near-surface conditions associated with tides and surface currents resulting in a predominance of coarse sandy sediment. The deeper stations represent a more quiet environment, characterized by the deposition of silt and clay. The sedimentary environment of the seaknoll stations A and B, located on the rim of a submarine valley extending from Agate Passage to the main basin of Puget Sound, represents a third situation undoubtedly influenced by the strong tidal flow through the valley. The presence of gravel and shell fragments may be evidence of a low-deposition environment.

Significance of sediment parameters

Mean grain size has some influence on the degree of similarity between polychaete assemblages as demonstrated by the ordination, but its use as a biologically meaningful sediment characteristic can be questioned. While it does indicate the average grain size, it does not reflect the shape of the distribution curve as in the cases of bimodal or skewed distributions.

The sand-to-mud ratio as a measure of the hardness of the sediment does not have a pronounced effect on the degree of similarity between stations. Because this ratio does not vary through a wide range, its significance is not strongly felt in the distribution of polychaetes of Port Madison.

All analyses indicate that the degree of similarity between polychaete assemblages in Port Madison is, at least partly, dependent on the clay content of the sediments. SANDERS (1956, 1958) concluded that clay content is the most valid grain-size parameter for determining faunal distributions; clay particles tend to bind organic matter in greater quantities owing to their high surface-to-volume ratio, making it more readily accessible to deposit feeders. NEWELL (1965) had found a greater abundance of microorganisms in finer grades of marine sediments, and a corresponding greater abundance of 2 deposit-feeding molluscs (*Hydrobia ulvae* and *Macoma balthica*). He suggests that the abundance of microorganisms is related to the greater surface area of fine sediments and reflects the amount of food (nitrogen) available to deposit feeders. The presence of *Pectinaria californiensis* in large numbers in the silty sediment of the deep stations E and F in the present study, is perhaps evidence for this relationship.

That depth was shown to be correlated with the distribution of several of the important species and the degree of similarity between stations (Tables 3 and 5) appears to be a reflection of the nature of the sedimentary processes associated with depth: the increased deposition of silts and clays in the quieter, deeper water. Depth itself is undoubtedly only indirectly related to the distribution of polychaete species in this area, especially in light of the limited depth range considered here.

The biological significance of sorting is not well understood. In the geological sense, sorting is the measure of the spread of grain sizes around the average size. Since well-sorted sediments are characterized by the predominance of one grain size, they can support, in terms of niche specificity, only those species whose requirements are met by that single grain size, in addition, of course, to those that are indifferent to sediment particle size. On the other hand, the poorly sorted sediments offer a wider range of grain sizes, capable of meeting a wider variety of needs.

It must be noted, however, that in this area, the sediments with low clay content have the best sorting, and the sediments with the highest clay content have the poorest sorting (Table 1). Perhaps the sorting is here a manifestation of the clay content and, for this reason, does not represent a separate environmental factor with which faunal distributions can be correlated. Because there is little evidence suggesting an explanation for the importance of sorting, it is assumed that the clay content is the dominant environmental factor determining faunal distributions.

Summary

1. Polychaete assemblages were sampled at 9 stations located along a 0.8 mile transect in the subtidal environment of Puget Sound, Washington, USA. Depth ranged from 2 to 34 m, and sediment ranged from fine, clean sand to very fine, silty sand. Adjacent assemblages show a high degree of similarity without noticeable discontinuities between them, as determined by indices of association and a 3 dimensional ordination, with the exception of a faunal discontinuity between a very shallow station located at the end of the transect and the adjacent deeper station. This discontinuity was apparently caused by the combined effect of the substrate, wide temperature and salinity fluctuations, and active sedimentary processes.

2. As determined by rank correlation techniques, the degrees of association between polychaete assemblages along this transect are related to differences in the nature of the sediment. Clay content and sorting

are highly correlated with station similarity, while mean grain size and the sand-to-mud ratio are less important. Clay content appears to have biological significance, while sorting may only be a manifestation of the clay content.

3. The KENDALL coefficient of association (T), based on the test statistic Chi-square, is a convenient method for determining the significance of differences between sampling sites. However, its use is limited when the assemblages are very similar and the major differences are in numbers of specimens per species.

4. The BRAY and CURTIS index of similarity, $2\,w/(a + b)$, is useful for comparing similar stations where the abundance of species is as important as the presence or absence of species. However, this index does not have a statistical basis, and limits cannot be placed on its values.

5. The BRAY and CURTIS 3 dimensional ordination has application to the marine benthic environment, although in this study its use was limited by too few stations and measured environmental parameters.

Acknowledgements. Contribution No. 524 of the Department of Oceanography, University of Washington, Seattle, USA. Based on a thesis submitted in partial fulfillment of the requirements for the degree of Master of Science, University of Washington, 1968. Special thanks are due Drs. K. BANSE and U. LIE for their advice and assistance during the course of the research and for their critical review of this manuscript. This research was supported, in part, by Atomic Energy Commission Research Grant AT(45-1)-1725 (ref: RLO-1725-139), and National Science Foundation Ecology Training Grant GB-6518X.

Literature cited

BANSE, K., K. D. HOBSON and F. H. NICHOLS: Annotated list of polychaetes, pp 521—556. *In:* U. LIE: A quantitative study of benthic infauna in Puget Sound, Washington, U.S.A., in 1963—1964. FiskDir. Skr. (Ser. Havunders.) **14**, 229—556 (1968).

BERKELEY, E. and C. BERKELEY: Annelida, Polychaeta Errantia. Can. Pacif. Fauna **9b** (1), 1—100 (1948).

— — Annelida, Polychaeta Sedentaria. Can. Pacif. Fauna **9b** (2), 1—139 (1952).

BRAY, J. R. and J. T. CURTIS: An ordination of the upland forest communities of southern Wisconsin. Ecol. Monogr. **27**, 325—349 (1957).

CREAGER, J. S., D. A. MCMANUS and E. E. COLLIAS: Electronic data processing in sedimentary size analyses. J. sedim. Petrl. **32**, 833—839 (1962).

FAGER, E. W.: Communities of organisms. *In:* The sea, Vol. 2, pp 415—437. Ed. by M. N. HILL. New York: Interscience Publishers 1963.

GREIG-SMITH, P.: Quantitative plant ecology, 256 pp. London: Butterworths 1964.

HARTMAN, O.: Polychaetous annelids from California. Allan Hancock Pacif. Exped. **25**, 1—226 (1961).

JONES, N. S.: Marine bottom communities. Biol. Rev. **25**, 283—313 (1950).

KENDALL, M. G.: Rank correlation methods. 3rd ed., 199 pp. New York: Hafner Publishing Co. 1962.

KRUMBEIN, W. C. and F. J. PETTIJOHN: Manual of sedimentary petrography, 549 pp. New York: Appleton-Century-Crofts, Inc. 1938.

LIE, U.: A quantitative study of benthic infauna in Puget Sound, Washington, U.S.A., in 1963—1964. FiskDir. Skr. (Ser. Havunders.) **14**, 229—556 (1968).

LOOMAN, J. and J. B. CAMPBELL: Adaptation of SØRENSON's K (1948) for estimating unit affinities in prairie vegetation. Ecology **41**, 409—416 (1960).

MCMANUS, D. A.: A criticism of certain usage of the phinotation. J. sedim. Petrol. **33**, 670—674 (1963).

NEWELL, R.: The role of detritus in the nutrition of two marine deposit feeders, the prosobranch *Hydrobia ulvae* and the bivalve *Macoma balthica*. Proc. zool. Soc. Lond. **144**, 25—45 (1965).

NICHOLS, F. H.: A quantitative study of benthic polychaete assemblages in Port Madison, Washington, 78 pp. Seattle: M. S. Thesis, Univ. Washington 1968.

PETERSEN, C. G. J.: Valuation of the sea. 2. The animal communities of the sea-bottom and their importance for marine zoogeography. Rep. Dan. biol. Stn **21**, 1—44 (1913).

SANDERS, H. L.: Oceanography of Long Island Sound, 1952—1954. 10. The biology of marine bottom communities. Bull. Bingham oceanogr. Coll. **15**, 345—414 (1956).

— Benthic studies in Buzzards Bay. 1. Animal-sediment relationships. Limnol. Oceanogr. **3**, 245—258 (1958).

SNEDECOR, G. W.: Statistical methods. 5th ed., 534 pp. Ames: Iowa State Univ. Press 1956.

THORSON, G.: Bottom communities (sublittoral or shallow shelf). Mem. geol. Soc. Am. **67**, 461—534 (1957).

USHAKOV, P. V.: Polychaeta of the far eastern seas of the USSR. [In Russ.]. Opredeliteli po faune SSSR, Akad. nauk SSSR **56**, 1—445 (1955). (Transl. from Russian. pp 1—419. Jerusalem: Israel Prog. Sci. Transl. 1965).

WHITTAKER, R. H.: Gradient analysis of vegetation. Biol. Rev. **42**, 207—264 (1967).

Author's address: Mr. F. H. NICHOLS
Department of Oceanography
University of Washington
Seattle, Washington 98105, USA

15 A SAND–BOTTOM EPIFAUNAL COMMUNITY OF INVERTEBRATES IN SHALLOW WATER

Edward W. Fager
Scripps Institution of Oceanography, La Jolla, California 92037

ABSTRACT

A community of nine species of epifaunal invertebrates living on sand in shallow water was censused for six years. The community consisted of three coelenterates, three gastropods, two echinoderms and one decapod. Seven of the species had aggregated distributions, but individuals of the first and fourth most abundant ones were randomly distributed. The relationship of distribution to settlement and mortality is discussed. The importance of each species population in relation to its demands on the environment, and its effects on it, was assessed in terms of frequency, density, biomass, cover, and motility. The results suggest that it is often misleading to label some species in a community "important" and others "unimportant," especially if this means that the latter are ignored in studies of community structure and dynamics. Populations of the nine species comprising the community remained constant (within sampling variability) over the six years. This community of few species, thus, appears to be a steady-state system. This was unexpected in view of the short species list, the lack of indication of substantial interspecific interactions, and the rigorous conditions in the environment.

INTRODUCTION

This paper is concerned with the community of invertebrates living on the sand plain between the submarine canyons in La Jolla Bight, under water of 5–10-m depth. It is an attempt to describe the spatial and temporal structure of a relatively simple community in a uniform environment and from this to derive some understanding of community ecology.

The 5- to 10-m depth range was originally chosen because it allowed 50–60 min working time underwater per dive and avoided the complications of surf in shallower water and the problems of decompression introduced by prolonged work at greater depths. It has turned out that this region supports a characteristic invertebrate community, clearly separated from those animals that live in the surf zone and with relatively few overlaps with the fauna found at depths greater than 10 m.

It will be evident from later sections that the topography and sediment are essentially uniform over a large area and that while water movement, temperature and, to a lesser extent, salinity vary temporally, the variations are widespread over the area studied. The sand plain is sufficiently large to reduce edge effects and within it solid substrates occupy only a small fraction of the total area. The barriers provided by the deep water of the submarine canyons and the limited amounts of shallow water at their heads make this effectively a closed system for the adult invertebrates considered in this study. It is, of course, not closed for any planktonic species nor for the larvae of the resident species if they have extended pelagic stages.

This study owes a great deal to the work of R. J. Ghelardi and A. O. Flechsig who made most of the dives with me. The nitrogen determinations were done by Miss T. Schultze. The anemones were determined by Dr. Cadet Hand, the gastropods by Dr. R. Stohler, and the fish by Mr. A. O. Flechsig. Financial support came from National Science Foundation Grants G-7141 and GB-5800 and from the Marine Life Research Program, the Scripps Institution's component of the California Cooperative Oceanic Fisheries Investigations, sponsored by the Marine Research Committee of the State of California.

BOTTOM CHARACTERISTICS

La Jolla Bight is a rather shallow indentation in the coastline. It is somewhat

Reprinted from *Limnology and Oceanography*, 13:448–464, 1968.

protected to the north by the general westerly trend of the California coast and to the south by Point La Jolla and Point Loma but receives long-period swell approaching from the southwest to northwest (70–80° window). It is primarily this swell that gives rise to the wave surge that affects the bottom at depths of 5–10 m. Short-period waves have little effect at these depths.

Two submarine canyons come into La Jolla Bight (Fig. 1). One of them, Scripps Canyon, is about 960 m north of the Scripps Institution of Oceanography (SIO) pier; the other, La Jolla Canyon, is off the La Jolla Beach and Tennis Club, about 1,325 m south. They bring relatively deep water close to shore.

North of Scripps Canyon, the bottom sediment is fine sand, although at the head of the canyon there is a small intertidal rocky reef, and at times rock is exposed subtidally on the southern edge of the canyon and attached algae grow there. Directly south of La Jolla Canyon there is a small area of coarse sand and then an extensive area of rock from the intertidal out to depths over 20 m. Attached algae and surf grass, *Phyllospadix scouleri*, are abundant on the rocks.

The region between the canyons where this work was done is roughly triangular, with the base of the triangle along the shore and its apex at the confluence of the inner edges of the canyons. Except along the edges of the canyons whose walls are very steep, the sand plain has a nearly constant slope of 2–3% seaward from the surf zone out to a depth of about 30 m. The area between the depths of 5 and 10 m below MLLW is about 3.15×10^5 m^2.

The only solid substrates in the plain are the SIO pier extending out to a depth of about 5 m, a few large concrete blocks and oil drums put down for experimental purposes in the past, 20–30 brass stakes and four hose buoys inserted for this study, dense patches of the sand dollar *Dendraster excentricus*, and isolated individuals and beds of two species of tube-building polychaetes. In total these stable substrates do not represent more than a small fraction of 1% of the area. Various intertidal organisms that attach to solid substrates extend out to sea on the pier pilings and some small attached algae grow on the pilings, on the concrete blocks and oil

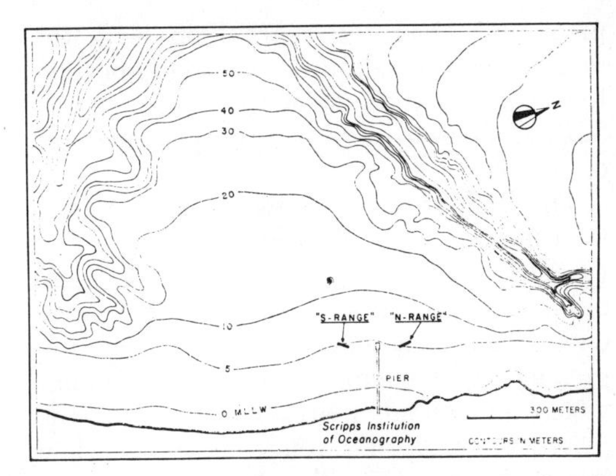

Fig. 1. Map of area of study. "*S-range*" and "*N-range*" mark the locations of the two series of permanent stations.

drums, on the hose buoys, and on the polychaete tubes. There are no macroscopic plants on the sand, but there is a sparse diatom flora on the sand grains (F. Round, personal communication).

The sand is fine and well sorted. The median grain diameter decreases gradually from 0.20 mm on the beach to 0.09 mm at a depth of about 30 m. At the depth of this study it is about 0.12 mm; 90% is in the size range, 0.08 to 0.19 mm. Most of the sand is quartz, mixed with about 5% heavy minerals, 3% micaceous materials, and less than 3% silt. There are only minor seasonal and local variations in the character of the sediment (Inman 1953).

An annual sediment cycle moves sand off the beach during autumn and winter storms and puts it back during spring and summer. This beach cut and fill is reflected in a concurrent fill and cut at depths out to about 10 m. Based on 15 sets of 10 measurements each, made at intervals of 1–9 weeks over a period of 17 months, the median net change in sand level at water depths of 6–8 m was 0.5 cm/week. Only five measurements showed zero change in sand level over a week. The maximum change measured was a fill of over 25 cm in less than five days. It was about two months before the sand returned to its former level and the measuring rods could be located.

WATER CHARACTERISTICS

The area receives little runoff from the land except during the uncommon winter storms. Although these provide a surface layer of clay-filled low salinity water, the salinity at 5-m depth is not affected appreciably. Over the years 1956–1962, the range of daily salinities recorded for bottom water at the end of the SIO pier was 32.07 to 34.04‰, with a seven-year average of 33.63‰.

The highest monthly mean bottom temperatures are recorded in July–September (around 18C) and the lowest in December–February (around 15C). The mean temperatures, however, give a false impression of the temperature regime under which the animals live, because during the period from April through October when the water is thermally stratified, the inshore edge of the thermocline is often found at depths of 5–10 m. Internal waves can then produce rapid changes in temperature. With a thermometer laid on the bottom we have observed changes of over 5C in a matter of minutes and rapid changes of up to 8C have been recorded by thermistors placed near the bottom just off the SIO pier (C. L. Hubbs, personal communication).

At the depths considered in this study, the wave surge resulting from the long-period swell keeps the top few millimeters of sand almost constantly in suspension and moves it inshore–offshore so that animals living on the sand surface are usually in a miniature sandstorm. The water movement is sometimes so violent that a diver, even though heavily weighted, can not stay in position, and the bottom can be seen through a cloud of sand only at rare intervals. The movement scours away sand around solid objects that project above the surface. This puts a stress on the animals that is met either by some sort of anchor imbedded in the sediment or by the ability to burrow rapidly. All of the species appear to be able to adjust to rapid changes in sand level.

In addition to the inshore–offshore movement, there is a net water movement shoreward along the bottom and often a longshore current. The latter most frequently tends toward the north (Shepard 1950). It may be important in the transport of plant detritus and associated animals from the rocky area south of La Jolla Canyon.

Rip currents occur all along the beach, most frequently at certain locations. The seaward heads of these currents are generally at depths of 5–6 m. They may have an effect on the distribution of some of the animals—directly during settlement (Fager 1964) and indirectly by affecting the availability of detrital material used as food.

METHODS

Most of the work was done in the years 1957–1959, with just enough to check for changes in 1960–1963. The field observations and sampling were all done using SCUBA. With this tool it has been possible to carry out repeated, detailed studies of the distribution patterns of the animals without disrupting the habitat or depleting the populations. It has also provided some observations on behavior, especially species interactions. More than 500 man hours were spent underwater, and 402 quantitative counts at "random" locations and 270 at "permanent stations" were obtained. The apparatus and methods used in obtaining these data have been described by Fager et al. (1966) and only briefly will be discussed here.

The larger animals living on the sand surface or extending through it were sampled by dropping a brass circle (area, $m^2/4$) in such a way that the diver could not see the exact location where it would fall. Two divers then recorded the animals in the two halves of the circle. Repetition of counts by experienced divers indicated little or no variation between divers. If carried out carefully, the approach of the diver and the dropping of the ring had no apparent effect on the animals.

Practical considerations made this method of positioning the samples the one of choice, but it did result in a haphazard, though presumably unbiased, pattern of samples rather than a strictly random one. A few sets of samples were placed randomly in the study area by dropping markers from a skiff run at relatively constant speed across it. The timing of dropping the markers was determined by sets of numbers drawn from a random number table. Analysis of the resulting counts did not indicate any significant differences between these random samples and those placed haphazardly.

It was initially thought that the circle counts would be reliable for animals down to a size of 5 mm or a little less if they lived on the sand surface or moved about just beneath it. To check this, a comparison was made between population density estimates obtained from $m^2/4$ circle counts and from sand cores (5 cm deep, 35-cm^2 area) taken during the same months and in the same areas (394 circles/230 cores). Three species of small gastropods (*Balcis* sp., *Olivella baetica, Turbonilla attrita*) and one species of isopod (*Ancinus* sp.) that were easily visible when moving on the sand surface were used in the comparison. In all cases the ratios, mean number per m^2 (cores)/mean number per m^2 (circles), were significantly greater than 1.0 (Table 1). Thus, many of the individuals must have been buried in the sand, probably near the surface but quiescent and, in the case of the gastropods, with the siphon retracted. This would mean, for example, that on the average only one individual in 27 (95% limits, 13–42) of the *Olivella baetica* population was active at any one time.

The circle counts seemed to be more satisfactory for the larger animals. However, because locations where any of these animals were evident were avoided when taking cores, the only species that was taken frequently enough in the cores to allow a comparison was *Amphiodia occidentalis*. The ratio, mean number per m^2 (cores)/mean number per m^2 (circles), for this species was 1.8 ± 1.2.

Two sets of five permanent stations each were established, one north and one south of the pier (Fig. 1). Each station was marked by two 0.6-mm-diameter brass stakes driven into the sand 1.5 m apart to position a template that fitted onto them. We could, thus, return repeatedly to the same place on the bottom and determine the position of an animal. Tests showed that the determination of position was good to ±1–2 cm, the error arising mostly from the diver's difficulty in reaching and maintaining a position directly above the animal. When the template was not in place, there was little or no disturbance of either the bottom sediment or the water movements in the area where counts were made.

TABLE 1. *Comparison of core and circle density estimates. Values are mean numbers per m^2, ratios of these and 95% confidence limits of the means and ratios, based on 230 cores and 394 circle counts*

	Cores*	Circles†	Cores/circles
Balcis sp.	18.7 ± 9.0	0.28 ± 0.10	66.8 ± 41.2
Olivella baetica	29.8 ± 14.5	1.10 ± 0.30	27.1 ± 14.6
Turbonilla attrita	6.2 ± 3.6	0.37 ± 0.10	16.8 ± 10.7
Ancinus sp.	6.2 ± 5.0	0.43 ± 0.18	14.4 ± 13.1

* Cores—5 cm deep, 35 cm^2.
† Circles—m^2/4, only animals visible on the surface counted.

The species studied at the permanent stations were mostly coelenterates. As these were not easily marked, it was assumed that if an individual of a species was seen repeatedly within ±2 cm of the location of previous sightings of the species, it was the same individual. Given the relatively low densities of the species, the probability of having the same locations occupied repeatedly would be small if their occupation depended on random movements of different individuals.

All determinations of biomass, size, and motility of a species were made on a series of individuals selected to represent the size structure of the field population. Measurements of coelenterates and distance between arms of brittle stars were made in the field on animals that were as little disturbed as practicable. Sizes for the other species are a combination of field and laboratory determinations. For the motile species, rates of movement were also estimated in the field for undisturbed animals. Organic nitrogen was determined on oven-dried (100C, 15–18 hr) material by standard micro-Kjeldahl technique. Wet and dry weights were considered unsatisfactory measures of biomass because of large differences between the species in the amounts of structural material, mostly inorganic. As Paine (1964) has pointed out, ashing may also give unsatisfactory results when comparisons are made between species differing widely in inorganic content. Organic nitrogen as a measure of biomass avoids these difficulties and, as most of it can be equated with protein, is a good measure of the living tissue present.

CHARACTERISTICS OF THE EPIFAUNA

At these depths, 39 species were seen resting on, crawling over, or extending through the sand (Table 2). Eight of the species were fish and, as discussed later, seemed to be well separated ecologically from the abundant epifaunal invertebrates. Nineteen of the invertebrates were seen on less than half of the dives and occurred less than five times in the quantitative samples. They are not considered further

TABLE 2. *List of epifaunal species recorded during the study*

Anthozoa
- *Harenactis attenuata* Torrey
- *Renilla köllikeri* Pfeffer
- *Stylatula elongata* (Gabb)
- *Zaolutus actius* Hand

Polychaeta
- *Diopatra splendidissima* Kinberg
- *Owenia fusiformis* Delle Chiaje

Malacostraca
- *Ancinus* sp.
- *Blepharipoda occidentalis* Randall
- *Cancer gracilis* Dana
- *Crangon nigromaculatus* (Lockington)
- *Holopagurus pilosus* Holmes
- *Heterocrypta occidentalis* (Dana)
- *Inachoides tuberculatus* Lockington
- *Lepidopa myops* Stimpson
- *Portunus xantusii* (Stimpson)

Gastropoda
- *Acteon punctocaelatus* (Carpenter)
- *Balcis* sp. (= *micans* Carpenter ?)
- *Epitonium tinctum* Carpenter
- *Nassarius fossatus* (Gould)
- *Nassarius perpinguis* (Hinds)
- *Olivella baetica* Carpenter
- *Olivella biplicata* (Sowerby)
- *Pleurophyllidia californica* Cooper
- *Polinices recluzianus* (Deshayes)
- *Turbonilla attrita* Dall and Bartsch
- *Turbonilla tridentata* Carpenter

Asteroidea
- *Astropecten armatus* Gray
- *Astropecten californicus* Fisher

Ophiuroidea
- *Amphiodia occidentalis* (Lyman)
- *Amphiodia urtica* (Lyman)

Echinoidea
- *Dendraster excentricus* (Eschscholtz)

Holothurioidea
- *Molpadia arenicola* (Stimpson)

Pisces
- *Citharichthys stigmaeus* Jordan and Evermann
- *Hypsopsetta guttulata* (Girard)
- *Paralichthys californicus* (Ayres)
- *Platyrhinoides triseriata* (Jordan and Gilbert)
- *Pleuronichthys ritteri* Starks and Morris
- *Pleuronichthys verticalis* Jordan and Gilbert
- *Rhinobatos productus* (Ayres)
- *Urolophus halleri* Cooper

because of lack of information. The remaining 12 species were present in every month, were recorded on more than half of the dives and in more than five quantitative samples. One of these species, *Dendraster excentricus*, generally occurs in dense beds at depths greater than 10 m. A large proportion of the individuals observed in the depth range here considered had injured tests, often with barnacles growing on them. It appears reasonable to consider the *Dendraster* as displaced from the normal habitat and not part of this community of animals, although, as shown later, they are a source of food. Another species, *Balcis* sp., is an ectoparasite on the *Dendraster* and has the same status as the latter. A third species, *Olivella baetica*, has also been left out of consideration because the sampling techniques used for this study gave gross underestimates of its population density (*see* discussion in methods section). The remaining nine species are considered to constitute the characteristic epifaunal assemblage in this habitat—the species that were always seen and that dominated the habitat, both numerically and in terms of their demand and impact on it. Table 3 presents information on habit, type of distribution, reproduction, food, and predators. Table 4 lists the frequency of occurrence and estimates of numbers, organic nitrogen, area covered per square meter, and motility.

The burrowing anemone *Harenactis attenuata* was the most frequent and abundant invertebrate on the sand at these depths. These anemones live with the column buried in the sand and only the sand-colored disc and tentacles showing at the surface. They are alternately partially covered and uncovered by sand as it is moved by the wave surge. The animals are not attached to any solid substrate but hold their position by means of an anchoring bulb. This easily breaks off so that uninjured animals are difficult to obtain for laboratory studies. Observations at the permanent stations have shown that the anemones move up and down in the sand to adjust for moderate change in sand level (2 cm or less per week) and do so without appreciable lateral movement (Table 5). Even following an unusually large and rapid fill (over 25 cm in less than five days), individuals were found at the same positions after the sand surface had returned to its usual level two months later.

Tentacle spread of what appeared to be

mature individuals was 2.5 to 3.0 cm. The animals have turned out to be unexpectedly long-lived in the field. At the permanent stations, 11 of 45 individuals that were first seen in July 1958 were still present over five years later in October 1963 (*see* Table 5 for examples) and the median period of persistence was over two years. During this time there was no noticeable change in size although this must be interpreted with caution because coelenterates are notoriously difficult to characterize by measurements of disc width or tentacle spread, and these were the only measurements available without uprooting the animals. There was little recruitment during the five years; only seven young were recorded as having settled in the 2.5 m^2 covered by the permanent stations. These aspects of the life history are reflected in the lack of appreciable change in estimates of population density over a six-year period (Fig. 2); the 95% confidence intervals of the estimates of mean numbers per $m^2/4$ show broad overlaps and 22 of the 26 intervals include the overall mean value of 1.62 individuals per $m^2/4$.

The food items most commonly seen in the grasp of the tentacles were small *Dendraster*. Short strands of *Phyllospadix* were also frequently caught, presumably for the animals on the surf grass; in two cases isopods of the genus *Idothea* were found partially digested by the anemone but still clinging to a surf grass fragment. Other observed food items were a badly broken up polychaete and a damaged *Holopagurus*. These observations suggest that the food of this species is composed largely of animals displaced from their normal habitat and rolled slowly along the bottom by the net inshore movement of the water. *Harenactis* was never observed to capture active prey; nothing has been seen to eat *Harenactis*.

This species is very rare at depths less than 5 m and greater than 10 m. Within this depth range the density did not change over the region between the submarine canyons except for a small increase in density just off the end of the SIO pier, possibly due to the presence of a more or less persistent rip current at this position. The pattern of distribution was examined in four different ways: by fitting the distribution of numbers of individuals per $m^2/4$ sample to a Poisson distribution; by examining the change in the ratio of variance to mean as successively larger (up to 3,600 cm^2) samples were formed by combining adjacent samples in a line of 144 contiguous samples of 225 cm^2 each; by counting runs of like signs along lines, placed parallel, perpendicular, or at 45° angles to the shore, where a plus was recorded when at least one *Harenactis* was present within an interval of 10 cm along the line and a minus when none was present; and by looking at the distribution of distances to nearest neighbor and the percentage of reflexives (Clark and Evans 1954, 1955) in randomly selected plots of 3.24 m^2 each. None of these gave significant evidence of nonrandom distribution of individuals. This is in direct contrast to the distribution patterns of the other abundant organisms, except *Amphiodia*, all of which have provided evidence of patchiness. The implication is that, for *Harenactis* within this depth range, the probabilities of settling and of survival after settling are independent of physical or biological differences between locations, including the presence or absence of other members of the same species. The data from the permanent stations indicate that, once settled, individuals stay in place (Table 5) and live a long time.

The anemone, *Zaolutus actius*, was the second species in abundance and third in frequency in the sand (Table 4). Its distribution will here be considered only outside of the *Owenia* bed that was described earlier (Fager 1964). It is smaller than *Harenactis*; individuals that appeared to be mature were 1 to 1.5 cm across the tentacles. The disc and tentacles were generally held 0.5 to 1.0 cm above the sand surface. Also in contrast to *Harenactis*, this species was always attached to some solid substrate, usually to the tubes of polychaetes but sometimes to small pebbles. In the laboratory, individuals moved laterally a few centimeters without emerging from the sand. This would mean that individuals living in dense beds of tube-building polychaetes could change their positions. The median period of persistence at the permanent stations was four months, measured by resightings of individuals; the maximum was 11 months. If

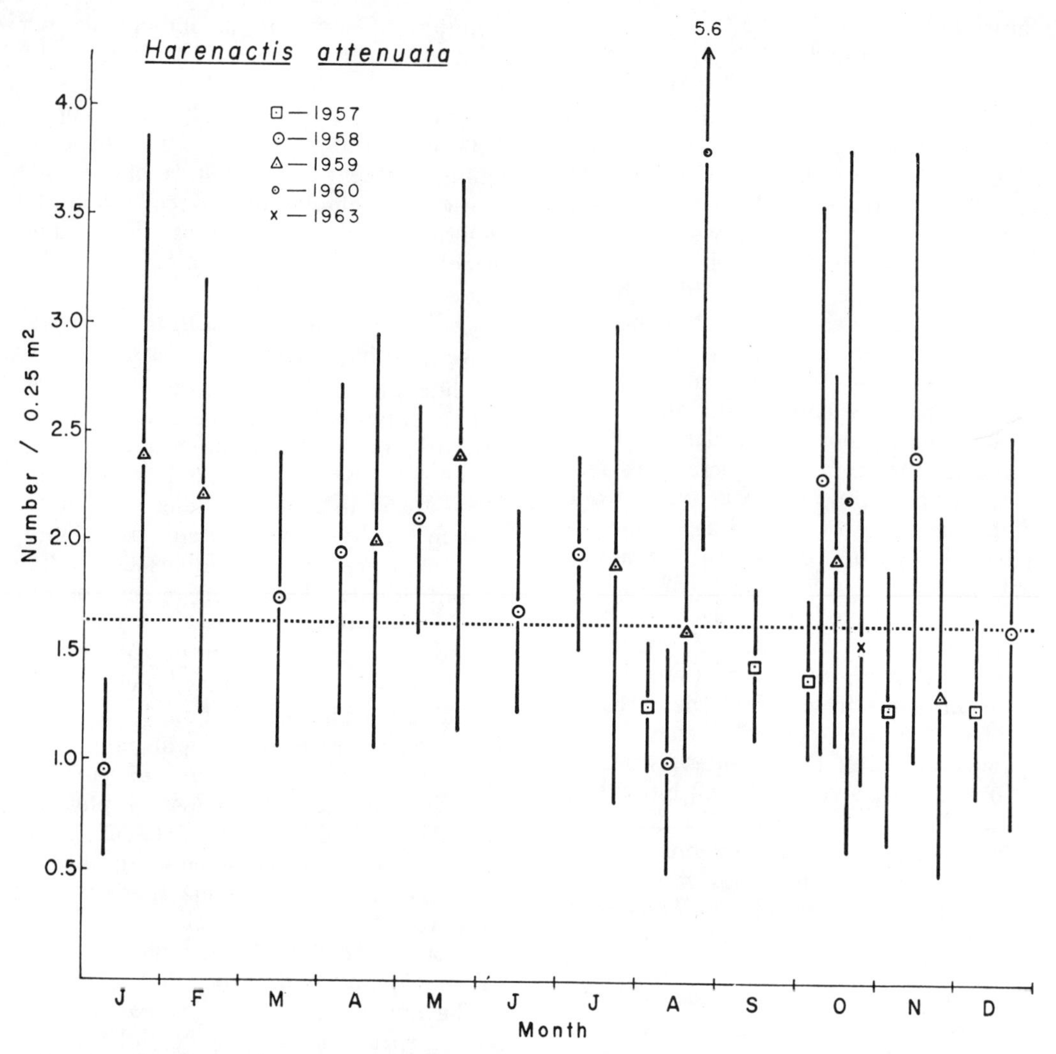

FIG. 2. Abundance of *Harenactis attenuata.* Values given are means and 95% confidence intervals for the mean, $\bar{x} \pm t_{0.05}s_{\bar{x}}$, per m²/4 sample. Dotted line is overall mean.

this measure of longevity is not invalidated by movement, this species is much shorter lived than *Harenactis.* A single caprellid and a young *Lepidopa myops* were the only food items recorded for this species. The position of the tentacle crown well above the sand surface and a more rapid response to stimuli suggest that *Zaolutus* may customarily feed on zooplankton organisms rather than the displaced animals utilized by *Harenactis.* In the laboratory, it easily captured and ingested adult *Artemia,* but did not live long when these were the only food provided.

As was true of the preceding species, the population density of *Zaolutus* remained essentially constant over the six years of observation (Fig. 3); 18 of 24 confidence intervals include the overall mean, 0.44 individual per m²/4. Large numbers of recently settled young were observed in March–July 1958, but young have seldom been seen since at the depths considered. The species was generally more abundant at greater depths, probably because the tube-building polychaete *Owenia fusiformis* to which it attaches was usually more abundant there. Where *Zaolutus* was abundant, a small gastropod *Epitonium tinctum* was frequently seen and appeared to be a specific predator. Individual snails were often found with the proboscis inserted far into the gastrovascular cavity of an anemone. The anemone did not retract under this stimulus.

The distribution of *Zaolutus* individuals was definitely patchy [index of dispersion

TABLE 3. *Characteristics of the abundant epifaunal invertebrates*

	Habit*	Distribution†	Reproduction‡ Mating	Eggs	Young§	Food‖	Predators‖
Harenactis attenuata	S	R	–	–	8–11	Dead crustaceans, polychaetes, and *Dendraster*	—
Zaolutus actius	S	A	–	–	1–8	Crustaceans	*Epitonium*
Renilla köllikeri	S(M)	A	–	–	8–11	—	*Astropecten* *Pleurophyllidia*
Amphiodia occidentalis	S–M	R	–	10	2–11	Plant detritus	—
Nassarius perpinguis	M	A	–	–	7–12	Dead fish and crustaceans	—
Holopagurus pilosus	M	A	5	–	9	Plant detritus	*Polinices*
Polinices recluzianus	M	A	12–3	1–10	1–10	Hermit crabs	—
Nassarius fossatus	M	A?	4	5	5–6	Dead fish, crustaceans, pelecypods, and holothurians	—
Astropecten armatus	M	A?	–	3	11–1	*Dendraster* and *Renilla*	—

* S = sedentary; M = motile.
† R = random ($s^2/\bar{x} \sim 1.0$); A = aggregated ($s^2/\bar{x} > 1.0$, at the 95% level).
‡ Based on field observations; numbers are months in which mating, eggs or gravid females, and young were seen.
§ Individuals were recorded as young if they were less than ¼ the size of the average large individual.
‖ Only relationships actually observed in the field are recorded.

significantly ($p < 0.05$) greater than 1.0]. This may, of course, be a secondary effect, resulting from the patchiness of the solid objects, mostly worm tubes, required for attachment.

Further information on the pattern of distribution was obtained from 100 samples of $m^2/4$ in which the numbers of individuals were recorded separately for each half of the sample. If the density within the patches was appreciably higher than that between and the patches were large relative to sample size ($m^2/4$) or were themselves aggregated, one would expect the counts on the two sides to be generally similar and, therefore, the differences between them to be small. If, on the other hand, the patches were smaller than $m^2/4$ and were randomly distributed, all possible differences would be equally likely. For example: with a set of 4 half-samples with 0 individuals, 5 with 1, and 3 with 2, the theoretical distribution of differences for the latter case would consist of 12 (4×3) differences of 2, 35 ($4 \times 5 + 5 \times 3$) differences of 1 and 19 ($4 \times 3/2 + 5 \times 4/2 + 3 \times 2/2$) differences of 0. In the case of *Zaolutus*, there was a slightly larger proportion of small differences in the observed than in the theoretical, but this was not significant even at the 20% level (Kolmogorov-Smirnov test; Tate and Clelland 1957). It appears, therefore, that the patches in this species are smaller than $m^2/4$ in area and are randomly distributed in relation to each other.

The colonial pennatulid *Renilla köllikeri* was the third most abundant and second most frequent epifaunal organism (Table 4). It lives with the colony flat on the sand surface and the peduncle inserted downwards into the sand. The peduncle can be somewhat inflated and serves as a remarkably good anchor; colonies have been seen capping small mounds of sand 5 cm or so above the general level and holding the mound against violent wave surge. The larger colonies were 7.5 to 8.0 cm across. The data from the permanent stations indicated that this animal, in contrast to *Harenactis*, moves about a good deal. Colonies and groups of colonies seen in one location in a particular $m^2/4$ area on one day would, by the next day, be elsewhere in the circle or gone. In the laboratory, colonies moved on the sand surface at an average rate of 2 mm/min by passing successive waves of contraction back along the edges from the apex. In doing this, they trailed the peduncle on the sand surface. Under the usual conditions in the field, any colony that moved in

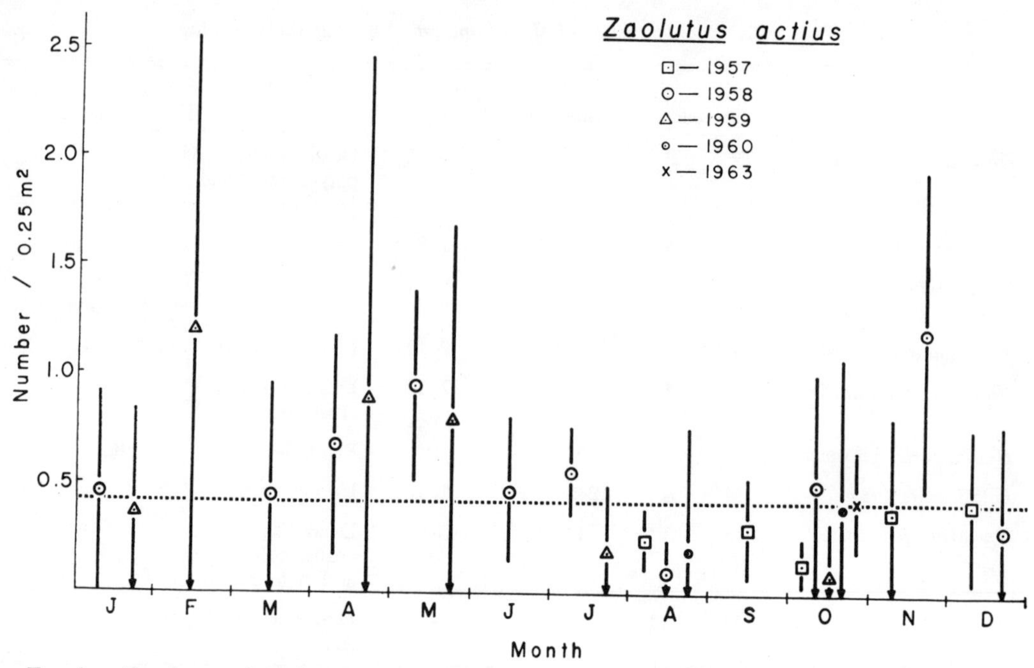

FIG. 3. Abundance of *Zaolutus actius.* Values given are means and 95% confidence intervals for the mean, $\bar{x} \pm t_{0.05}s_{\bar{x}}$, per m²/4. Dotted line is overall mean.

this manner would be quickly tumbled about by the wave surge and finally deposited on the beach; colonies inadvertently kicked out of the sand by a diver suffered this fate. *Renilla* is, however, rarely found cast up on the beach. It must, therefore, have some mechanism of movement in the field that enables a colony to move while holding itself firmly in the sand. The movement has so far precluded the study of longevity of the colonies.

The movement is apparently not random, for the colonies were definitely aggregated [index of dispersion significantly ($p < 0.05$) greater than 1.0]. This aggregation might arise from the response of a colony to changed characteristics of water or sand movement on the bottom caused by the presence of one or more other colonies; the group, perhaps, mutually increasing the individual colony's ability to withstand wave surge.

The observed distribution of differences between numbers of individuals in halves of samples did not depart from the theoretical distribution at the 20% level (cf. discussion under *Zaolutus* for details of method). Most patches of this species are, therefore, apparently smaller than m²/4 in area and randomly distributed in relation to each other.

Colonies have been seen at depths of

TABLE 4. *Community structure*

	Frequency*	Density†	Biomass‡	Area§	Motility‖
Harenactis attenuata	511	1,089 (6.48)	360–410	30–50	—
Zaolutus actius	178	296 (1.76)	10–20	1–3	—
Renilla köllikeri	248	282 (1.68)	80–105	35–50	—
Amphiodia occidentalis	102	109 (0.65)	10–20	20–35	—
Nassarius perpinguis	44	62 (0.37)	1–5	1	75–110
Holopagurus pilosus	28	44 (0.26)	30–40	1–3	30–60
Polinices recluzianus	10	17 (0.10)	5–10	2	75–105
Nassarius fossatus	8	12 (0.07)	1–5	1	200–275
Astropecten armatus	6	7 (0.04)	20–30	5–15	50–70

* Number of m²/4 samples in which species occurred, out of a total of 672.
† Number of individuals observed in 672 m²/4 samples. Values in ( ) are mean numbers of individuals/m².
‡ Estimated milligrams organic nitrogen per m².
§ Estimated area occupied by population, cm²/m².
‖ Estimated times (min) required for the populations of motile species to cover an area equal to that occupied by the *Renilla* population, assuming movement 4% of the time (based on estimates of activity in the population of *O. baetica*, cf. section on methods).

TABLE 5. *Constancy of position of individuals of* Harenactis attenuata *at permanent stations*

Animal	1958 Jul	1958 Oct	1959 Jan	1959 Apr	1959 Aug	1960 Aug	1963 Oct	Location
A	29/14*	30/15	28/17	28/15	30/15	†	27/16	South 1
B	20/8	21/7	19/8	20/6	20/6	†	20/4	South 2
C	40/22	40/23	38/22	‡	40/22	38/23	38/22	North 2
D	12/9	15/9	12/11	12/9	12/10	11/9	10/9	North 3
E	22/9	22/9	‡	23/7	23/6	22/9	21/8	North 4

* Numbers are coordinates of positions at permanent stations, measured in cm from two reference lines. Of the 45 large individuals seen at the permanent stations, 11 were in the same position (never more than ±2 cm from a mean position) for over five years and an additional 13 held their position for over two but less than five years.

† Stakes on the south range were not found on the August 1960 dive.

‡ Individual not observed on this dive, probably retracted beneath sand surface.

30 m or more, but those that occurred out beyond 10-m depth appeared to be unhealthy as judged by the flaccid texture of the colony and the limited numbers of extended polyps. The species was absent at depths less than 5 m.

A heavy annual set of young *Renilla* (2–10 polyps per colony) was usually observed, commencing in August and continuing through part of November. Colonies in the laboratory released planulae in late July (1958). Because of difficulties in making accurate counts of these small colonies, they were not included in the regular population censuses. A set of special counts indicated that their presence often doubled the mean number of colonies per m²/4 but the increases persisted only briefly (Fig. 4, dashed line). As these small colonies could easily be torn out of the bottom by wave surge, most of the loss of young can probably be attributed to physical processes. The loss may be increased by predation because all of the sightings of *Pleurophyllidia californica*, an opisthobranch reputed to be a specific predator on *Renilla* (Ricketts and Calvin 1952), were during the months when the young *Renilla* were present. *Pleurophyllidia* was once seen feeding on a large colony. *Astropecten* is a much more common predator on larger *Renilla*; about 1 in every 10 starfish examined had the peduncle of a colony protruding from the mouth opening. It is not known whether this starfish would eat very small colonies.

If the young are excluded, the population density remained essentially constant over the period of six years (Fig. 4); 22 of 26 confidence intervals include the overall mean, 0.42 individual per m²/4.

In the field, *Renilla* colonies have not been observed feeding. It is suspected that they feed on microzooplankton. In the laboratory, individual polyps caught, but seemed to have some trouble eating, newly hatched *Artemia* nauplii.

The distributions of the preceding three

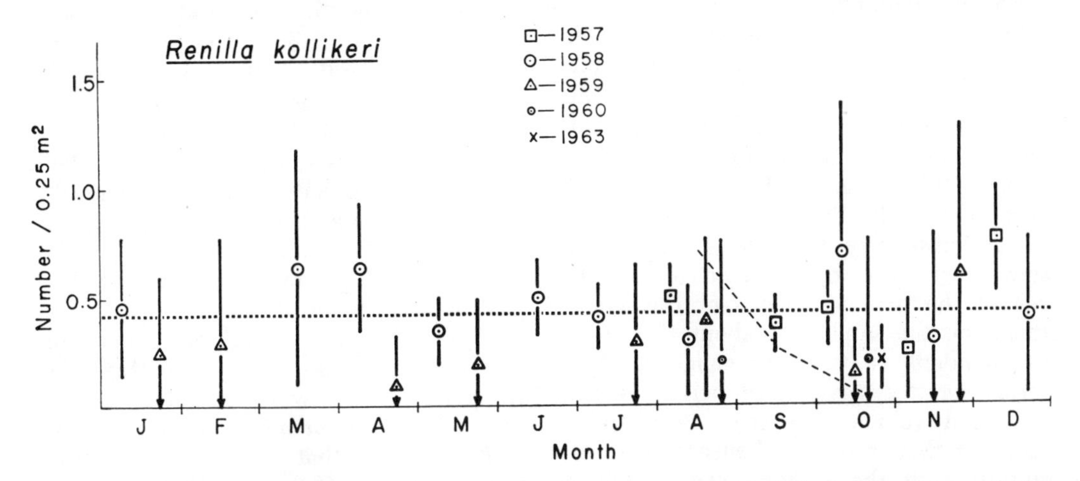

FIG. 4. Abundance of *Renilla köllikeri.* Values given are means and 95% confidence intervals for the mean, $\bar{x} \pm t_{0.05}s_{\bar{x}}$, per m²/4. Dotted line is overall mean. Dashed line is for a special count of newly-settled young that were not included in the regular censuses.

species were examined to see whether the occurrence or numbers of individuals of one species were related to those of another species. None of the three 2 × 2 contingency tables (presence–absence in $m^2/4$ quadrats) differed appreciably from expectation and the abundances of the species were not significantly correlated. There is thus no statistical evidence that the presence, absence, or abundance of one of the numerically dominant species affected the presence, absence, or abundance of the others.

The brittle star, *Amphiodia occidentalis*, the fourth most abundant and frequent species, was always found with the disc buried, oral surface down, a few centimeters beneath the sand surface and, usually, with only two or three arms extending through the surface. Distance between arms ranged from 4 to 12 cm. A water current carrying sand grains and detritus particles moved down one of the arms and another moved up another arm. In deeper water, where wave surge has less effect and, therefore, reducing conditions develop in the sand relatively near the surface, the positions of individual brittle stars were marked on calm days by small conical piles of blackened sand brought up from the reduced sand layer in which the disc was situated.

Individuals of *Amphiodia* were resighted at the same location in the permanent station circles for maximum periods of one month. Animals that were dug out had no difficulty in moving across the sand and reburying. There was no evidence of clumping.

What were recorded as young of this species were observed almost throughout the year. The records may, however, be unreliable because a second, less common, smaller species (*Amphiodia urtica*) does occur in the area and sightings were not all checked by collection of the individuals.

Nassarius perpinguis, the smaller (avg spire height, 1.5 cm; estimated volume, 0.4 cc) of the two gastropods of this genus found on the sand, was about five times as abundant as its larger congener, *Nassarius fossatus* (avg spire height, 4 cm; estimated volume, 5.4 cc). Both spend much of their time just beneath the sand surface with the siphon exerted. Large numbers of both species were actively attacking every fish carcass seen. When a dead fish was staked down on the bottom as bait, the snails emerged from the sand and crawled actively toward it if there was a current and they were on the downstream side. They fed together without apparent interference; *N. perpinguis* tended to crawl into or under the fish more than *N. fossatus* did. Both species reacted to the flesh of the dead fish but not to its carefully separated skin. *N. perpinguis* was also seen feeding on dead *Blepharipoda occidentalis* and on *Balanus* sp., the latter having been recently removed from one of the hose buoys. *N. fossatus* was seen feeding on both of the preceding, plus dead *Portunus xanthusii* and *Inachoides tuberculatus* and moribund *Molpadia arenicola* and *Donax gouldii*. The two species of *Nassarius* have never been seen feeding on active living material. There is no field evidence suggesting a difference in food habits or behavior or any interspecific interference.

For both species, the index of dispersion was greater than 1.0 ($p < 0.05$, in the case of *N. perpinguis*). The cause of aggregations was clear when the gastropods were feeding on larger dead animals. The continued aggregation when not feeding may reflect earlier attraction to a food source followed by limited dispersal after the food was consumed.

Both of the species reacted strongly to being touched by a tube foot of the common sand star *Astropecten armatus*. The reaction involved extrusion of the snail nearly out of its shell and violent threshing about of the foot. This resulted in the animal being thrown considerable distances, in jumps of 5–15 cm. The reaction continued for some time after contact with the tube foot had been lost. *Astropecten* attempted to hold the snails but medium sized snails easily escaped, even when held on the oral surface of the starfish until a good grip should have been obtained. It, therefore, seems unlikely that this starfish is a successful predator on adults of these gastropods (Feder 1963).

N. fossatus has several times been observed laying eggs, always in May. The stalked egg cases were affixed to a worm tube (*Diopatra splendidissima*), to strands of *Phyllospadix* that were embedded in the sand, and once to a polyethylene transect line that had just been put down. *N. perpinguis* was never seen laying eggs. On the other hand, while young of *N. fossatus* have never been common and those seen have always been about a quarter the size of the larger individuals of this species, very small (3–4-mm spire height) individuals of *N. perpinguis* were quite common in sand cores in July 1958 and have been taken frequently in cores in summer months of other years.

The hermit crab *Holopagurus pilosus* seems well-adapted to life in the sand, easily burying itself so that only the eyes, antennae, and a small area in the mouth region are visible, the latter kept clear by motion of the maxillipeds. The crab constantly picks up particles from the sand surface with its chelipeds and passes them to the mouth. When the wave surge was strong, individuals tended to stay buried and pick up food as it came to them; when it was weaker they moved about over the sand. Most of the material they picked up seemed to be of plant origin; they were not attracted to dead fish.

One instance of predation has been observed. A *Polinices recluzianus* had partly drilled through another *Polinices* shell that contained a *Holopagurus.*

The distribution of numbers of individuals per sample of the hermit crab indicated aggregation. Field observations of behavior suggested that individuals were aware of each other's presence at short distances so the aggregation may have arisen through social interaction. On two occasions (Sep 1957 and Jan 1960) quite different, very dense aggregations were observed. In both cases, 100–200 crabs were found packed tightly in several layers, the uppermost one at the sand surface, in an area 10–20 cm on a side. Except for the crabs, there seemed to be no difference from the surrounding sand. Recent observations indicate that such patches may remain in one place for as long as three weeks. There was no evidence of mortality, molting, or breeding; the individuals were active, not soft, small to medium size for the species and none carried eggs. In an unselected sample of 138 individuals, 67 were in *Olivella biplicata* shells, 12 in *Mitrella* sp., three in *Iselica* sp., two in *Amphissa* sp., 41 in *Olivella baetica*, eight in *Nassarius perpinguis*, two in *Acteon punctocaelatus*, and one each in *Epitonium* sp., *Balcis* sp., and *Turbonilla tridentata.* As the first four gastropods in the list do not occur in the immediate area, their shells must have come from some distance yet they comprised 61% of the sample. The larger *Holopagurus* were usually in *Polinices recluzianus* shells but one was found in the rather fragile shell of a land snail, *Helix* sp.

The moon snail *Polinices recluzianus* was frequently seen partly buried in the sand or moving through the surface layer. Its predation on *Holopagurus* has already been mentioned. The snail was attracted to dead fish but it did not stay with the bait as the *Nassarius* did. Perhaps it was unable to feed. The animals on which snails of this genus are known to prey, medium to large pelecypods, were rare in this sand habitat.

Mating pairs of *Polinices* were seen only during winter and early spring months but the characteristic egg collars were present nearly all the year. Very small stages were never recorded, even in the sand cores, but individuals about one quarter the size of adults were present throughout the year.

Astropecten armatus was the only starfish living on sand at these depths. A smaller, more slender, species of the same genus, *A. californicus,* occurred seaward of the distribution of *A. armatus*, being abundant at depths of 15 m and more while *A. armatus* was most abundant at 5–10-m depths and was seldom seen as deep as 15 m.

As noted above, *Astropecten* frequently ate *Renilla* but its commonest prey was the sand dollar *Dendraster excentricus.* These were not taken from the dense beds but were usually individuals displaced shoreward from them. *Astropecten* was attracted to dead fish but did not persist on the bait. Its reaction to the two species of *Nassarius* (*see above*) suggests that even though it could not hold the adults it might be a successful predator on the young and other smaller gastropods.

DISCUSSION

The homogeneity of benthic communities over considerable areas has often been noted (Sanders 1960; Thorson 1957). The community discussed in this paper is a widespread, frequently recurring group of species living along the coasts of southern California and northern Baja California, subtidally on fine sand at depths of 5–10 m and intertidally on sand flats in bays (Ricketts and Calvin 1952). At first glance the two habitats appear dissimilar, but they have certain characteristics in common—considerable water movement but no surf, rapid changes in temperature, sources of appreciable quantities of plant and animal detritus, and a uniform shifting substrate composed of fine, well-sorted sand. The uniformity in the area of this study is reflected in the random distributions of individuals of two of the abundant species, *Harenactis* and *Amphiodia*. Such distributions may be interpreted as indicating that settlement and survival are equally likely at all locations. The other species, being aggregated, impose a biological pattern on the physical uniformity. The pattern is the result of the activities of the species and may be connected with factors influencing their survival. It appears to have little effect on the habitat as it is impermanent, shifting with the movements of the animals, and at any one time occupies less than 2% of the surface area.

The separation of the invertebrate epifauna from the other major sections of the sand-bottom community, the vertebrate epifauna and the invertebrate infauna, is to some extent arbitrary. There are, however, three sorts of justification for the separation. The least important is the size of the animals, although this does indicate something of their requirements and capabilities. Most individuals of the invertebrate epifauna are in the size range 1 to 10 cm and, while young of the most abundant vertebrate, *Citharichthys stigmaeus*, are less than 10 cm long and a few uncommon polychaetes and nemerteans that live in the sand are over 1 cm long, there are surprisingly few overlaps with species in the other two groups. The second, more fundamental, separation is in terms of food. The most abundant epifaunal invertebrate, *Harenactis*, and four of the less abundant ones eat larger pieces of detrital plant and animal material. Two other species, *Astropecten* and *Polinices*, are predators, probably to a large extent within the community. The other two species, *Renilla* and *Zaolutus*, appear to capture small living zooplankton. The infaunal invertebrates, on the other hand, feed on either comminuted detrital material or the bacterial and algal films on the sand grains or on other members of the infauna, and the vertebrates feed mostly on larger living crustaceans and polychaetes or on the smaller vertebrates. Finally, there seems to be little direct interaction between the three sections. There was no evidence that the infauna species were used as prey by the epifaunal invertebrates unless they had been injured and had become part of the general surface detritus. Predation by the vertebrates on the epifaunal invertebrates considered here was not observed in the field. One species, *Z. actius*, was found in stomachs of an uncommon flatfish collected at these depths (Ford 1965).

Table 4 is an attempt to assess the importance of the species in the community from a variety of viewpoints. The reason for the attempt is that it would be convenient if one could simplify the study of communities by only having to consider a relatively few "important" species. In the simple community considered here, *Harenactis* would be the most important to someone "counting heads." In terms of standing crop it could continue to be the most important to a hypothetical omnivore though the low productivity suggested by the observations might move it below species such as *Renilla* that produce relatively large numbers of young each year. From the point of view of an organism looking for a place to settle and not be eaten, it would be no more important than *Renilla* or *Amphiodia*, and all three species might be less important than some of the much rarer, but actively motile, species. As possible competitors for a potentially limited food resource, scattered haphazardly over the sand bottom and moved about by currents and wave surge, the sedentary species may also be less important than the actively motile ones.

Each of the measures used in Table 4 has some limitations as a way of describing the structure of a community. The most commonly used statistics, individuals per species and the indices of diversity that can be derived from them, are probably the least informative ecologically for they indicate little of the demands of a species in terms of food and space or of its possible effects in modifying the environment. Biomass could be more satisfactory both as a measure of the potential contribution of the standing crop as a food resource and, where it can be used as an index of respiration, as an indicator of the demand of the species population on the food resources available in the habitat. By ignoring the activity of the species, however, it too may give an incomplete representation of the species' place in the community. In cases, such as this one, where the activity of the motile species appears to consist mostly in searching for food and does not lead to much alteration of the environment, information on area covered per unit time may provide a satisfactory basis for assessing this aspect of a species' importance. If the species extensively altered the habitat by its activities, such a simple measure would be inadequate. In addition to these limitations, there is evidence that in some communities, species that would not be considered important by any of these or related measures may be essential to the persistence and well-being of the species that dominate both numerically and in terms of biomass and activities affecting the environment (Limbaugh 1961).

The species composition of the community and the abundances of the individual species did not change over the six-year period of study, suggesting that this is a steady-state biological system. Five of the species are dependent on detrital material, much of it coming from rather distant rocky areas. After storms, the sand bottom is littered with a mixture of pieces of formerly attached algae, surf grass, and the animals clinging to these. More than half of the time, however, detrital material is scarce on the bottom and only one or two small pieces of vegetable matter may be seen along a 50-m transect. Pelagic zooplankton killed in the surf might, if the rip currents then transported them seaward, do something to smooth out the fluctuations in food supply but the amount available from this is probably small.

The evidence for population control of the most abundant species, *Harenactis attenuata*, is strong but there is no indication of the mechanism. No predators have been seen feeding on it. The random distribution of individuals indicates that they do not react to each other's presence and that all of the habitat is equally suitable. The evidence from the permanent stations shows that individuals stay in one position and are quite long-lived (estimated maximum exponential death rate of adults = 0.00077/day). As a corollary to the latter, few newly settled young have been seen. If food were the usual control, one might expect a general decrease in population density from the seaward to the shoreward edge of the species distribution because of the transport of food along the bottom in this direction by the relatively persistent inshore current. There was no evidence for such a decrease. Competition with the other abundant anemone, *Zaolutus*, is not excluded but field observations on spacing, behavior, and food suggest that they play well-separated roles in the community. Finally, there is the possibility, although it appears unlikely, that the apparent control is illusory because a period of only six years may be too short to obtain conclusive evidence for instability in a population of such long-lived animals.

The second most abundant of the epifaunal invertebrates, *Zaolutus actius*, requires a small pebble or a rigid polychaete tube for attachment and, as these are uncommon, their abundance may limit the population density. Evidence for this is provided by the 300-fold to 600-fold increase in *Zaolutus* density associated with the transitory presence in the area of this study of a dense bed of the tube-building polychaete *Owenia fusiformis* (Fager 1964). The greater density of adults of the predatory gastropod, *Epitonium* sp., in this bed and the frequency with which strings of its egg cases were found there suggest that this biological control could take over in populations of *Zaolutus* freed from restrictions imposed by the scarcity of suitable substrate. *Zaolutus* is also preyed on by an uncommon flatfish *Pleuronichthys ritteri*. It constituted about 30% of the stomach contents of 12 individuals (Ford 1965).

The third most abundant species, *Renilla köllikeri*, occurs in aggregations, suggesting interaction between individual colonies. There is an annual set of young that for short periods can more than double the population density but these young quickly disappear, probably mostly by destruction by sand movement but also by predation by the opisthobranch *Pleurophyllidia californica*, which has been seen in numbers only during the season when the young *Renilla* are present. The adults are frequently eaten by the relatively abundant starfish *Astropecten armatus*. Perhaps the most likely control mechanism is interaction between this species and its predators, particularly *Astropecten* which has several alternate prey to which it might pay more or less attention depending on their relative abundances.

None of the rest of the epifaunal invertebrates was present in large enough numbers to enable one to state with certainty that their population densities remained constant. However, the available figures suggest constancy and the animals certainly neither became extinct nor very abundant during the six-year period. There is some evidence that the hermit crab population might be limited by the scarcity of suitable empty shells, but no evidence has come to hand even to suggest a plausible mechanism of population control for any of the other five species.

Six years of field observations have not revealed extensive or frequent interspecific interactions. This is reinforced by the fact that the pattern of cooccurrence of the three most abundant species in $m^2/4$ samples indicated no effects on each other's presence or abundance. Interspecific interactions seem to be potentially important in only one (*Renilla*) of the three cases where some information is available to suggest a possible mechanism of population control. In the other two, the most likely mechanism involves a limitation of physically suitable places to live in the environment. The community is comprised of relatively few species (nine with abundances $\geqslant 0.04$ per m^2). The abundances of the nine species range over two orders of magnitude, but the distribution of individuals among species does not seem to be appreciably more skewed than those found for communities living in other habitats (Hairston 1959; Englemann 1961). There is no current evidence that any of the 30 other, rarer, species seen on the sand surface or the invertebrates living in the sand interact substantially with these nine species. The community inhabits an environment that is variable and rigorous, in terms of movements of water and substrate, extent and rapidity of temperature change, and uncertainty of food supply. In view of the generally accepted concept that stability is to be found in communities consisting of large numbers of interacting species living under relatively constant environmental conditions, the constancy of this community and its component species populations is surprising.

REFERENCES

Clark, P. J., and F. C. Evans. 1954. Distance to nearest neighbor as a measure of spatial relationships in populations. Ecology, **35**: 445–453.

———, and ———. 1955. On some aspects of spatial pattern in biological populations. Science, **121**: 397–398.

Englemann, M. D. 1961. The role of soil arthropods in the energetics of an old field community. Ecol. Monographs, **31**: 221–238.

Fager, E. W. 1964. Marine sediments: Effects of a tube-building polychaete. Science, **143**: 356–359.

———, A. O. Flechsig, R. F. Ford, R. I. Clutter, and R. J. Ghelardi. 1966. Equipment for use in ecological studies using SCUBA. Limnol. Oceanog., **11**: 503–509.

Feder, H. M. 1963. Gastropod defensive responses and their effectiveness in reducing predation by starfishes. Ecology, **44**: 505–512.

Ford, R. F. 1965. Distribution, population dynamics and behavior of a bothid flatfish, *Citharichthys stigmaeus*. Ph.D. thesis, Univ. California, San Diego. 243 p.

Hairston, N. G. 1959. Species abundance and community organization. Ecology, **40**: 404–416.

Inman, D. L. 1953. Areal and seasonal variations in beach and nearshore sediments at La Jolla, California. Beach Erosion Board, Tech. Mem. No. 39, U.S. Corps of Engineers, Washington, D.C. 82 p.

Limbaugh, C. 1961. Cleaning symbiosis. Sci. Am., **205**: 42–49.

Paine, R. T. 1964. Ash and calorie determinations of sponge and opisthobranch tissues. Ecology, **45**: 384–387.

Ricketts, E. F., and J. Calvin. 1952. Between Pacific tides, 3rd ed. (Revised by J. W. Hedgpeth). Stanford Univ. Press, Stanford, Calif. 515 p.

Sanders, H. L. 1960. Benthic studies in Buzzards Bay. III. The structure of the soft-bottom community. Limnol. Oceanog., **5**: 138–153.

Shepard, F. P. 1950. Longshore current observations in southern California. Beach Erosion Board, Tech. Mem. No. 13, U.S. Corps of Engineers, Washington, D.C. 54 p.

Tate, M. W., and R. C. Clelland. 1957. Nonparametric and shortcut statistics. Interstate Printers and Publishers, Danville, Ill. 171 p.

Thorson, G. 1957. Bottom communities (sublittoral and shallow shelf), p. 461–534. *In* J. W. Hedgpeth [ed.], Treatise on marine ecology and paleoecology, v. 1. Geol. Soc. Am. Mem. 67.

Ecology of the Deep-Sea Benthos

16

More detailed recent sampling has altered our concepts about the animals living on the deep-ocean floor.

Howard L. Sanders and Robert R. Hessler

Marine benthic communities cover most of the earth's surface. At least 94 percent of the ocean bottom lies below the permanent thermocline, and, in its physical and chemical parameters, this region is remarkably stable and homogeneous. It is constantly dark, with bottom water of constant salinity, oxygen content, and low temperatures. Even the bottom sediments, largely derived from planktonic organisms, are unvarying for hundreds of square kilometers. Great pressures characterize the environment. Probably as a consequence of its separation both spatially and temporally from the primary organic production at the surface of the sea, there is a low rate of food supply to the deep-ocean floor.

There is little information about the kinds of animals living in this vast environment and about the relation of the high environmental stability to the ecology and physiology of the deep-sea fauna.

Because of the sparseness of animal life and the technical difficulties in sampling the deep-sea benthos, relatively few specimens were collected in the past century. Even nonquantitative trawls and dredges traversing appreciable distances on the deep-ocean floor have captured only a few animals. However, understanding of the deep-sea fauna requires that samples contain enough individuals to give statistical support to conclusions. We developed the large deep-sea *Anchor Dredge* (*1*) for quantitatively sampling the infauna (animals living in the bottom) and the *Epibenthic Sled* (*2*) to collect both epifauna (animals living on the bottom) and infauna in large quantities. From 1960 to 1966, we made a study of a transect of the ocean floor, between southern New England and Bermuda (the Gayhead-Bermuda transect, Fig. 1). Subsequently, we extended our study to the tropical Atlantic.

Faunal Composition

Initially, our sampling of the deep-sea benthos was done by the *Anchor Dredge*. These predominantly infaunal samples were dominated by Polychaeta, Crustacea, and Bivalvia. Polychaetes, comprising 40 to 80 percent of these samples by abundance and represented by numerous species, were the most abundant. Hartman (*3*) found from 65 to 77 polychaete species in each of the five quantitative samples which covered 0.5 to 1.0 square meter of bottom at upper slope depths. Crustaceans, the second most common group, formed 3 to almost 50 percent of the fauna and were represented by many isopod, amphipod, cumacean, and tanaid species. Other common faunal elements were glass sponges, sea anemones, pogonophores, sipunculids, echiurids, tunicates, priapulids, brittle stars, and starfish.

In the past most deep-sea collections were made with coarse-meshed trawls which collected predominately larger epifaunal organisms. To provide basic information on the neglected smaller epifaunal animals, as well as to obtain better samples of the total benthic macrofauna, we constructed the *Epibenthic Sled* (*2*). A door which closes the mouth of the net reduces the effect of winnowing of the sample as it is brought up through the long water column. A fine-meshed net in the sled and fine-meshed screens for processing the samples on board ship retain the abundant smaller animals which would otherwise be lost.

The effectiveness of the method is demonstrated by the following collection records.

Malletia abyssorum Verrill and Bush (bivalve); previous record, single specimen; sled samples, 3257 specimens from 19 samples.

Malletia polita Verrill and Bush (= *Malletia bermudiensis* Hass) (bivalve) both known from single empty valve; samples.

Tindaria callistiformis Verrill and Bush (bivalve); previous records, two specimens; sled samples, 1708 specimens from 11 samples.

Serolis vemae Menzies (isopod); previous record, two specimens from South Atlantic; sled samples, 255 specimens from seven samples on transect and in tropical Atlantic.

Desmosoma insigne Hansen (isopod); previous records, six individuals from Davis Strait; sled samples, 294 specimens from four samples.

In contrast to the pronounced abundance of polychaetes in the infaunal *Anchor Dredge* samples, dominance in the sled samples is shared about equally among the crustaceans, polychaetes, bivalves, and brittle stars. When these four groups are compared on the basis of diversity, crustaceans invariably have the largest number of species, followed by polychaetes and bivalves; ophiuroids are always represented by few species. If 1000 individuals of each group are counted for every species of brittle star there are 4.1 bivalve, 9.7 polychaete, and 16.5 crustacean species [see (*4*) for method of rarefaction].

Diversity

Our most unexpected finding was a high faunal diversity in individual samples. Such diversity is far in excess of anything obtained in the past, in contrast to the belief that the deep sea harbors a qualitatively restricted fauna (*5–7*). The results of the first five sled samples taken during the summer of 1964 are given in Table 1; subsequent samplings demonstrated similar diversity.

Only two previous samples from depths greater than 1000 meters have yielded more than 100 species of benthic invertebrates. They are *Challenger* station 320, at 1096 meters, with 124 species and 496 individuals (*8*), and *Galathea* station 716, at 3570 meters, with about 2100 specimens divided among 132 species (*9*).

The greater diversity shown by our samples may be due to a more complete sampling of the total benthic fauna. Earlier samples contained few specimens, and the total number of species was therefore small. The restricted number of specimens in these samples made

Dr. Sanders is senior scientist in biology at the Woods Hole Oceanographic Institution, and Dr. Hessler is an associate professor at the University of California, San Diego.

Reprinted from *Science*, Vol. 163, No. 3874, pp. 1419–1424, March 28, 1969. Copyright 1969 by the American Association for the Advancement of Science.

it impossible to determine the significance of apparently small taxonomic differences. When a specimen resembled a known species, it was usually included in that species. The resultant lumping helped to create the impression of relatively few, broadly distributed species. Our large samples allow us to conclude that most of the major taxa are characterized by numerous closely related species both at a specific locality and among spatially separated localities.

The data obtained from both the *Anchor Dredge* and *Epibenthic Sled* samples indicate that the deep-sea benthos is not impoverished but, instead, is represented by a remarkably diverse fauna. How, then, does the diversity in the deep sea compare with the diversities occurring in other regions of the world?

We answered this question by collecting benthic samples from boreal estuary, boreal shallow marine, tropical estuary, and tropical shallow marine environments. In all cases, the sediments were soft oozes and were therefore comparable in particle size. All sam-

Table 1. The number of species and individuals collected in five *Epibenthic Sled* samples.

Station	Depth (m)	Individuals (No.)	Species (No.)
73	1400	25,242+	365
62	2496	13,425+	257
72	2864	5,897+	208
64	2891	12,083+	310
70	4680	3,737+	196

ples were processed in a similar manner. The analysis is based on total fauna (*2*) and on the polychaete-bivalve fraction of the fauna (*4*). The polychaetes and bivalves comprise about 80 percent of the animals in most of the infaunal samples, and thus we can generalize from the results. Diversity in the deep sea, measured by a rarefaction method (*4*) which allows direct comparison of samples with differing numbers of specimens, is about the same as that in the physically stable, shallow, tropical marine environment and significantly greater than that of the other three environments (Fig. 2).

We believe that the constancy of physical conditions and the long past history of physical stability in the deep sea have permitted extensive biological interactions and accommodations among the benthic animals to yield the diverse fauna of this region. Such communities evolve wherever physical conditions remain constant and uniform for long periods. Other than the deep-sea, tropical shallow-water environments and tropical rain forests best approximate these conditions. Communities found in these environments are characterized by many species and can be termed biologically accommodated communities (*4*).

At the other end of the diversity spectrum are the physically unstable communities where physical conditions fluctuate widely and unpredictably, and thus the organisms are exposed to severe physiological stresses. Here the adaptations are primarily to the physical environment (*4*). Hypersaline bays and temporary ponds exemplify this state, and certain shallow boreal marine and estuarine environments approach such conditions. Physically unstable communities are characterized by a small number of species. A similar paucity of species is found in environments of recent past history, such as most freshwater lakes.

The term "diversity" as used above means "within-habitat" diversity, that is, the number of species in a specific habitat, and not "between-habitat" diversity or the total number of species for all habitats (*10*). (The habitat under study is that of soft, fine-grained sediments.) The pronounced homogeneity of the deep sea permits fewer habitats than do the shallow depths. Thus, although the animal diversity of the soft sediments in the deep sea is well above that of the equivalent inshore boreal habitat, the total between-habitat diversity may be lower.

Zonation and Zoogeography

How are the environmental stability and homogeneity on the deep-ocean floor reflected in faunal vertical zonation? Known abyssal and hadal records of vertical distribution for 1144 species of deep-sea benthic invertebrates, reviewed by Vinogradova (*11*), showed rapid decrease of species from 2000 to 6000 meters, with a much slower reduction at greater depths. (The bathyal region encompasses depths from 200 meters to 2000 or 3000 meters and includes the continental slope; the abyssal zone covers from 2000 or 3000 meters

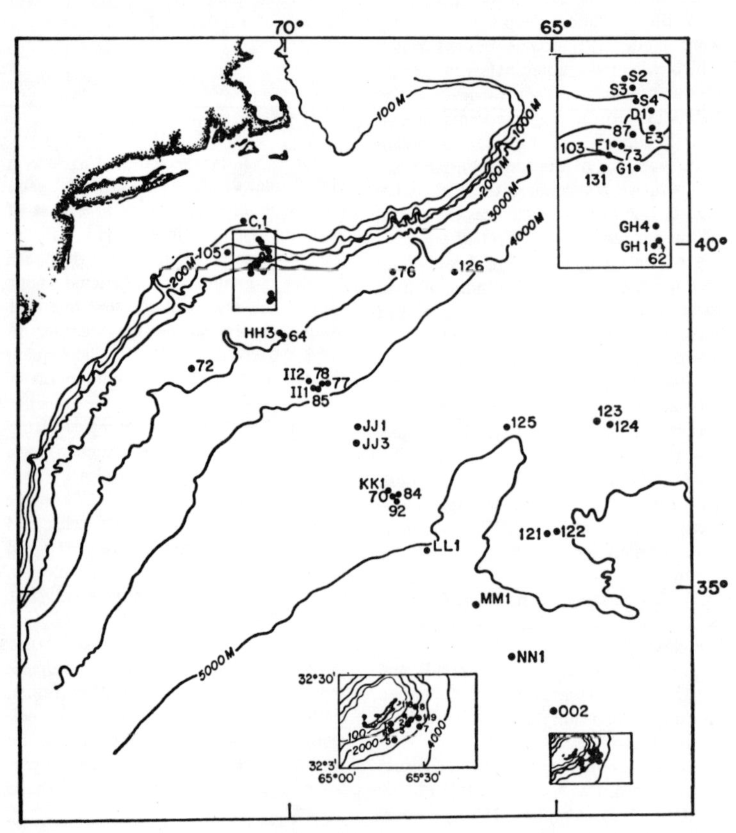

Fig. 1. Depth contours and locations of sampling stations of the Gayhead-Bermuda transect.

to 6000 meters; the hadal zone is at depths greater than 6000 meters and includes the deep-sea trenches.) At about 3000 and 4500 meters, important changes occur in the taxonomic composition of the benthic fauna. Numerous species and higher taxa, broadly distributed on the slope and even shallower, disappear and are replaced by new species, genera, and families found only at greater depths. For these reasons, Vinogradova concluded that 3000 meters represents the true upper limit of the abyssal zone.

We did not find such abrupt boundaries at bathyal and abyssal depths. Our analysis is not dependent on criteria of presence or absence; we measured the percentage of fauna shared by each possible pair of stations on the transect. The data for polychaetes, extracted from Hartman (*3*) and based on 264 species and almost 14,000 specimens, are given in Fig. 3.

The stations are arranged sequentially in a north to south direction traversing, in order, the outer continental shelf, the continental slope, the abyssal rise, the Sargasso abyss, and the Bermuda slope. The diagonal pattern running from the upper left to the lower right corner means that highest faunal indices are always with neighboring stations, that is, there is a gradual and continuous faunal change with depth and distance along the transect.

The very small faunal index values shared by stations C (97 meters) or SL-2 (200 meters) with other stations on the transect marks this portion of the transect as a region of pronounced faunal change. In fact, it is the sharpest zoogeographical boundary encountered. This discontinuity is also true for groups other than the polychaetes. Among the bivalves, the transition of faunal change at the shelf-slope break is even more pronounced. Within the isopods, the typically deep-water subtribe Paraselloidea is absent from station C and extensive samplings in shallower waters. The group makes its appearance on the upper slope at station SL-2 and is a major constituent in all deeper samples.

We believe that this faunal break is related to temperature. In 98 meters at station C, the seasonal temperature change is 10.5°C; in 300 meters at slope station 3, it is 5.1°C; and in 487 meters at station D, only 1.4°C. Therefore, the boreal continental shelf, with highly variable seasonal temperatures, supports a qualitatively impoverished eurytopic (broad physical tolerances) fauna, whereas the neighboring physically stable continental slope harbors a different stenotopic (narrow physical tolerances) benthic fauna of high diversity. Entirely analogous conditions are found in ancient Lake Baikal in Siberia, similarly dominated at shallow depths by a continental boreal climate (*12*). We conclude, on the basis of taxonomy, diversity, and environmental

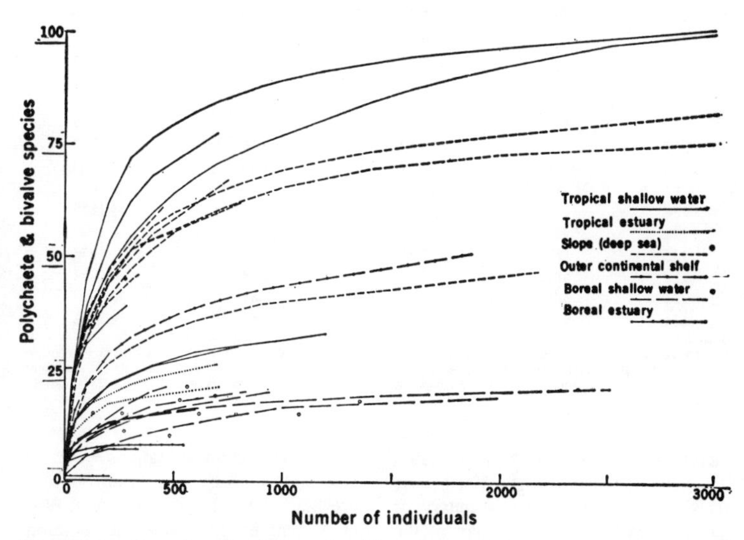

Fig. 2. Diversity values for different benthic environments by the rarefaction method. The lines represent the interpolated curves. The larger circles represent actual samples that have not been rarefied.

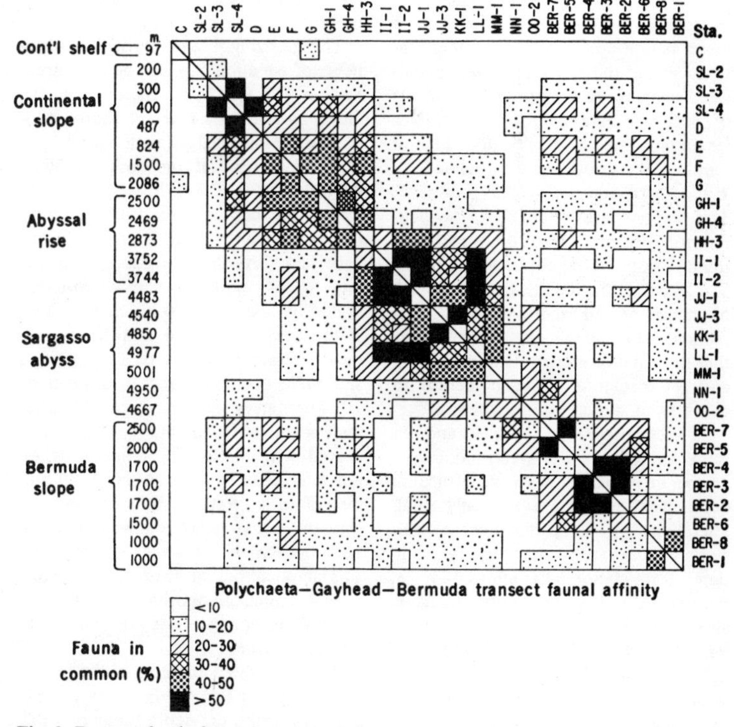

Fig. 3. Degree of polychaete faunal similarity among the stations (*Sta.*) of the Gayhead-Bermuda transect.

Table 2. Degree of bivalve faunal similarity and percentage of bivalve species shared per 500 specimens with station 64 on the Gayhead-Bermuda transect.

Station	Depth (m)	Species shared* (%)	Fauna shared† (%)	Depth deviations‡ (m)
105b	530	7.77	2.61	−2361
87	1102	25.79	4.80	−1789
73	1400	24.47	5.90	−1491
103	2022	22.87	5.80	− 869
131	2178	30.68	6.81	− 713
62	2496	46.41	28.71	− 395
72	2864	60.19	77.74	− 27
64	2891	100.00	100.00	0
126	3806	53.05	9.02	+ 915
85	3834	35.42	6.34	+ 943
70	4680	10.64	1.45	+1789
92	4694	11.65	1.36	+1803
84	4747	16.70	2.68	+1856
121	4800	7.57	1.03	+1909
125	4825	5.75	0.93	+1934
122	4833	13.98	1.38	+1942
123	4853	3.88	0.83	+1962
124	4862	3.88	0.83	+1971

* There are 25.75 species per 500 individuals at station 64. † With station 64. ‡ Minus (—) values mean that the station is the given number of meters shallower than station 64; plus (+) values mean that the station is that number of meters deeper than station 64. Station 64 is at a depth of 2891 meters.

factors, that the self-slope break within the depth range of 100 to 300 meters represents the true upper boundary of the deep-sea benthos in this part of the ocean.

Changes in species composition with depth also can be demonstrated by the use of a modification of the rarefaction method (Table 2). The criterion is the number of species shared between a given station and each of the other stations along the transect when equal-sized samples are compared. The results of such an analysis for the bivalves (excluding the taxonomically difficult family Thyasiridae) are presented in Table 2 for station 64 (2891 meters). The stations are arranged in sequential order by depth from the upper continental slope to the abyssal plain in the Sargasso Sea; the depths, the percentage of species per 500 individuals (by rarefaction), the percentage of the fauna that each station shares with station 64, and the deviation of each station in depth from station 64 are given.

In percentages of species and of fauna shared, there is a continuous decrease in similarity with departure in depth from station 64. The only small alterations in this pattern occur with stations that are widely separated in depth from station 64 (stations 103, 73, 87, and 122). The gradient of change is steeper, with percentage of fauna than with percentage of species shared. This difference in the steepness of the two gradients means that species that are numerically dominant at station 64 are relatively less important or even rare faunal constituents at stations separated from station 64 by more than 800 meters. Station 126 is a case in point. With 53 percent of the species in common, it shares many species with station 64. Yet with only 9 percent of the fauna common to both stations, species that are numerically dominant at one station are not numerically dominant at the other.

The horizontal and vertical components of zonation can also be measured. Station 73 at 1400 meters on the Gayhead-Bermuda transect shares 47.9 percent of its bivalves with station 142 at 1700 meters off the coast of West Africa (10°30.0′N; 17°51.5′W). If we follow the contours of 1000 to 2000 meters in depth along northern North America, Greenland, Iceland, Europe, and Africa, these two stations are separated by about 16,500 kilometers. Yet, if the depth is altered as little as 800 meters either shallower or deeper from station 73 on the transect (a distance of 10 kilometers landward or 12 kilometers seaward), the number of species shared with station 73 falls below 48 percent. Thus a change of only 800 meters in a vertical direction on the transect is at least equivalent to a 16,500-kilometer change horizontally. The implication, at least for bivalves, is that faunal composition is far more sensitive to change in depth than to the effects of distance.

To assess the patchiness of the fauna, we took three samples (stations 72, 64, and 76) within the depth range of 2862 to 2891 meters over a horizontal distance of 366 kilometers. Another series of five samples (121, 122, 123, 124, and 125) was taken within the depth range of 4800 to 4862 meters, at three widely separated localities (Fig. 1). In both series, the faunal index of each pair of stations was very high and, with two minor exceptions, higher than with any station outside the series. These high faunal values imply that many components of the deep-sea benthic fauna at any given depth are uniformly and homogeneously distributed over extensive distances of the ocean floor.

Breeding Patterns

On land and in the shallow boreal and arctic marine regions, reproduction is usually coupled with some cyclical and often seasonal phenomenon such as day length, temperature, or rainfall. This synchrony allows the adult portion of the population to be ready for reproduction at the same time, and the resulting young or larvae are usually present when feeding conditions are best.

The monotonously constant conditions of the deep sea provide no such obvious cyclical environmental phenomena. Even seasonal pulses of primary production in the surface waters are probably dampened and dissipated long before the slow rain of these minute particles reaches abyssal depths.

George and Menzies (*13*) suggest that cyclic reproduction may be present among deep-sea isopods. Their data, derived from studies of the genus *Storthyngura,* are based on only six ovigerous specimens of three species taken at widely differing localities and depths. The statistical inadequacy of these data make their conclusion conjectural.

Schoener (*14*) studied the state of reproduction in two species of brittle stars, *Ophiura ljungmani* and *Ophiomusium lymani,* collected from the continental slope, and abyssal rise on the Gayhead-Bermuda transect. Her analysis of a large number of specimens obtained at different times of the year provides convincing evidence for reproductive periodicity for these deep-sea species.

We are now gradually accumulating information on other deep-sea groups. An August sample from the lower continental slope (station 73, 1400 meters) contained 71 females of an unidentified species of isopod belonging to the genus *Ilyarchna.* Twenty-four percent of the females had well-developed marsupial plates and were therefore in the brooding condition. (As with all peracarid crustaceans, the embryos of isopods are carried in a marsupium.) A December sample (station 131, 2178 meters) yielded 59 females of the same species, with 27 percent bearing well-developed marsupial plates.

Sections were made on 25 adult specimens of the bivalve *Nucula cancellata*, 2.4 to 2.8 millimeters in size, from each of three samples, station 103 (May, 2022 meters), station 73 (August, 1400 meters), and station 131 (December, 2178 meters) (*15*). The findings for each of the stations were essentially identical. At least part of the gonad contained ripe sexual products in almost every individual examined.

These results for *Ilyarchna* sp. and *Nucula cancellata* indicate that reproduction is constant and continuous in the two forms. Schoener's evidence of reproductive periodicity for the two brittle star species suggests that no pattern need be universal. Analysis of numerous additional species will indicate whether there is a characteristic pattern of reproduction in the deep sea.

Animal Abundance and Food

Deep-sea benthos show a decrease in animal life with increasing depth of water and distance from land (*5, 6, 8, 16–20*). Although certain departures from this pattern occur in the shallower water of the sublittoral and upper bathyal zones (*1, 21*), the lower bathyal and abyssal zones invariably support a smaller biomass or density than shallower depths. Island arc trenches are ideal traps for organic material, and, therefore, the high concentration of life occasionally reported from these hadal depths (*22*) are special cases.

In our series of *Anchor Dredge* samples taken along the transect, numerical density of animals increased from near shore (approximately 6000 per square meter) to the edge of the shelf (13,000 to 23,000 per square meter) (*1*). From this depth to the bottom of the continental slope, density dropped precipitously to approximately 500 animals per square meter, with a more gradual decrease on the continental rise, and then a leveling off on the abyssal plain under the Sargasso Sea (25 to 100 animals per square meter). Thus, from the edge of the continental shelf to the abyssal plain, there was a reduction in density by a factor of several hundreds.

The biomass of our samples has never been analyzed; the material was biologically too valuable for mechanical abuse. However, density is an adequate index of biomass in our samples, because the size of animals and the relative proportions of major animal groups do not change between localities.

Table 3. A comparison of the range in amount of benthic faunal density to the range of the amount of organic carbon in the sediment and the annual primary productivity in surface waters.

Area of study	Bathyal-abyssal bottoms		Annual productivity in surface water (cg/m^2)
	Density (No./m^2)	Organic carbon* (%)	
Gayhead-Bermuda transect	21,263–33(*1*)	1.0–0.1(*1*) 2.5–0.8(*20*)	180–72(*25*)
Oregon	2,200–14(*20*)	3.5–0.1(*39*)	152–60(*40*)

* Excluding sand bottoms

The abundance of food is generally assumed to control density and biomass in the deep sea (*5, 17, 19, 23, 24*). Oxygen concentration, sediment type, and temperature either do not show correlated changes or never reach levels regarded as limiting. Hydrostatic pressure is depth-dependent, but although pressure could limit the kinds of species present, there is no apparent way in which it can regulate biomass.

Sources of food fall into two broad categories: either it is produced in the euphotic zone (the relatively narrow, lighted, upper water layer) and conveyed into the deep sea, or it is generated *in situ* in the deep-sea environment. The theory that food is of euphotic origin has until recently dominated our thinking (*6, 23*) because the distribution of biomass in the deep sea correlates with primary production. The highest productivity is in shallow coastal waters (*25*). This concentration of food is enhanced by the outwash from terrestrial sources: the farther from inshore waters of highest production, the lower the benthic biomass. Biomass also decreases with increasing depth, because during its transport the amount of food is progressively diminished by autolysis, bacterial decay, and scavengers.

The rapid decline in animal density on the continental slope, the region of greatest rate of depth change on the transect, labels depth as having more importance than distance from land in determining the amount of available food (*26*). Density decreases 25 times from 200 to 2500 meters of depth. Yet, these localities are so close (61 kilometers) that primary production at the surface barely changes. Stations at 4500 to 5000 meters have densities 50 to 390 times less than at 200 meters, and productivity in surface waters is only half as great (*25*) (Table 3).

The mode of transport of food in the deep sea has long been regarded as a rain of dead plants and animals from the euphotic zone (*6, 23, 27*). This mechanism has often been criticized because settling velocities are so slow and intermediate attrition is so efficient that little of nutritive value would reach the bottom. Inefficiency alone is not a valid reason for rejecting this mechanism. The deep-sea biomass is so much smaller than that of surface waters that an inefficient food supply can account for it. Downward movement by means of diurnal migration would increase the efficiency of food transport (*28*). However, the chain of migrating animals postulated by Vinogradov might be less efficient than passive settling since it entails large energy losses in every change of trophic level.

Turbidity currents provide another possible means of rapid downward movements of organics (*29*). Yet, the infrequency and destructiveness of such phenomena suggest that they are of no more than minor importance in food transport. The presence of layers of carbon associated with turbidity current deposits (*30*) indicates that while organics have been moved, they were never made available to organisms.

Organic aggregates, with considerable surface activity, adsorb dissolved organics and convert them into particulate compounds. Thus, they may serve as an *in situ* source of nutrition for intermediate bacteria (*31*) or directly for metazoans in deep water. However, the concentration of dissolved organics and the amount of carbon bound up in organic aggregates are relatively constant in deep water (*32*) and therefore correlate poorly with the marked decrease in benthic biomass with depth. There is no quantitative data on deep-sea heterotrophic algae (*33, 34*); therefore this food source cannot be evaluated.

Extensive degradation of organics during the slow transport from the surface waters to the deep-sea benthos permits only the most refractory substances to reach the bottom. Here benthic microorganisms (bacteria and, possibly, fungi) are probably a critical intermediate in the food cycle (*6, 7*). The abundance of bacteria has been reported (*35*), and some metazoans show adaptations to utilize an intestinal flora (*36*). The true importance of bacteria and fungi must still be determined.

The amount of organic carbon in deep-sea sediments follows a variety of patterns with relation to depth and distance from land (*1, 20, 28, 29, 37, 38, 39*). Yet, the absolute range of variation is usually less than an order of magnitude, far less than the corresponding variation in animal density (Table 3). Therefore, organic carbon in the sediment is not a good index of available food. If bacteria are nutritional intermediates of major importance, then the ratio of bacterial biomass to metazoan biomass or density should remain constant from place to place, and changes in bacterial biomass would not be correlated with changes in organic carbon content in the sediment. Such a pattern, if true, would suggest that in deep water much of the deposited organic material is not readily available to bacteria.

Feeding Types

According to the principle of competitive exclusion, ecological niches of species living together must have areas of nonoverlap. The uniformity of the deep sea and the limited amount of food indicate that niche separation is especially critical with respect to feeding.

Our knowledge of feeding types in this environment is fragmentary, but interpretation of morphology, analogy with shallow-water relatives, and gut-content analyses show that deposit feeders dominate in our samples. At station 64 (2891 meters), for example, the most diverse groups contain the following percentage of detritus feeders: Polychaeta, 60 percent; Tanaidacea, greater than 90 percent; Isopoda, 90 percent (*23*); Amphipoda, greater than 50 percent (*41*); and Pelecypoda, 45 percent; totaling 47 percent of the species in the whole sample. Many species of other, less diverse groups (Holothuroidea, Sipunculida, Oligochaeta, and others) also eat detritus, and thus this feeding type includes well over half of the fauna.

The high diversity in the deep sea implies a degree of niche fractionation far greater than in equivalent shallow boreal communities. Yet, studies of zonation indicate that appreciable flexibility must exist in intraspecific interactions, for each species has a zone depth preference whose limits vary with the species. Therefore, each species must be associated with a continuously altering assemblage of animals within its depth range.

Intraspecific flexibility, in conjunction with the low density of the benthic fauna, may explain high faunal diversity coupled with environmental homogeneity. Since density is low, only a fraction of the total number of species lives in any one area, for example in a single square meter. An individual organism, because of its small size and presumably restricted mobility, will interact with a relatively small suite of competing species. This hypothesis permits high faunal diversity with an increased, but not necessarily extraordinary, degree of specialization.

Summary

The benthos of the deep sea in a region between southern New England and Bermuda can be characterized by low density but high within-habitat diversity. The low density is probably determined by the amount of food present and is correlated with depth and distance from land. Depth is the more critical variable. The high diversity, about the same order as that found in shallow tropical seas, can be related to the seasonal and geological stability of the deep-sea environment. An unexpectedly large number of deposit-feeding species are found in individual samples. The composition of the benthic fauna gradually and continuously changes with depth throughout the bathyal and abyssal regions, but an abrupt faunal discontinuity is found at the shelf-slope break in from 100 to 300 meters of water. In this region of our study, the fauna shallower than the zone of discontinuity is eurytopic, of low diversity, and taxonomically distinct from the stenotopic, highly diverse populations living below the discontinuity. In comparing faunal composition, a vertical change of a few hundred meters is equivalent to a change of thousands of kilometers horizontally. The few data on reproduction reveal that some elements of the deep-sea fauna breed continuously, whereas others are restricted to a limited period of the year.

References and Notes

1. H. L. Sanders, R. R. Hessler, G. R. Hampson, *Deep-Sea Res.* **12**, 845 (1965).
2. R. R. Hessler and H. L. Sanders, *ibid.* **14**, 65 (1967).
3. O. Hartman, "Deep-Water benthic polychaetous annelids off New England to Bermuda and other Atlantic areas," *Allen Hancock Found. Publ. Occas. Pap. No. 28* (1965), p. 1.
4. H. L. Sanders, *Amer. Natur.* **102**, 243 (1968).
5. S. Ekman, *Zoogeography of the Sea* (Sidgwick and Jackson, London, 1953).
6. N. B. Marshall, *Aspects of Deep Sea Biology* (Hutchinson, London, 1954).
7. A. F. Bruun, *Geol. Soc. Amer. Mem. No. 67* (1957), vol. 1, p. 641.
8. J. Murray, *Challenger Rep.* **1895**, 1608 (1895).
9. T. Wolff, *Galathea Rep.* **5**, 129 (1961).
10. R. H. Whittaker, *Science* **147**, 250 (1965).
11. N. G. Vinogradova, *Deep-Sea Res.* **8**, 245 (1962).
12. M. Kohzov, *Monogr. Biol.* **11**, 352 (1963).
13. R. Y. George and R. J. Menzies, *Nature* **215**, 878 (1967).
14. A. Schoener, *Ecology* **49**, 81 (1968).
15. R. Scheltema sectioned, stained, and interpreted the bivalve material.
16. R. J. Menzies, *Oceanogr. Mar. Biol. Annu. Rev.* **3**, 195 (1965).
17. L. A. Zenkevitch and J. A. Birstein, *Deep-Sea Res.* **4**, 54 (1956).
18. R. L. Wigley and A. D. McIntyre, *Limnol. Oceanogr.* **9**, 485 (1964).
19. N. G. Vinogradova, *J. Oceanogr. Soc. Japan* **20**, 724 (1962).
20. A. G. Carey Jr., *Trans. Joint Conf. Exhibit Ocean Sci. Ocean Eng.* **1**, 100 (1965).
21. A. P. Kuznetsov, *Benthic Invertebrate Fauna of the Kamchatka Waters of the Pacific Ocean and the Northern Kurile Islands* (Moscow, 1963).
22. G. M. Belyaev, *Bottom fauna of the Ultra-abyssal depths of the World Ocean* (Academy Nauk USSR Inst. Okeanol., Moscow, 1966).
23. R. J. Menzies, *Int. Rev. Ges. Hydrobiol.* **47**, 339 (1962).
24. L. B. Slobodkin, F. E. Smith, N. G. Hairston, *Amer. Natur.* **101**, 109 (1967).
25. J. H. Ryther, in *Geographic Variations in Productivity, the Sea*, M. N. Hill, Ed. [Interscience (Wiley), New York, 1963], vol. 2, chap. 17, p. 347.
26. This is in contrast to the findings of G. M. Belyaev [*Trudy Inst. Okeanologii* **34**, 85 (1960)].
27. A. Agassiz, *Bull. Mus. Comp. Zool.* **14**, 1 (1888).
28. M. Vinogradov, *Rapp. Process-Verbaux Reunions Cons. Perma. Int. Explor. Mer.* **153**, 114 (1962).
29. R. C. Heezen, M. W. Ewing, R. J. Menzies, *Oikos* **6**, 170 (1955).
30. D. B. Ericson, M. Ewing, G. Wollin, R. C. Heezen, *Bull. Geol. Soc. Amer.* **72**, 193 (1961).
31. G. A. Riley, D. Van Hemert, P. J. Wangersky, *Limnol. Oceanogr.* **10**, 354 (1965).
32. D. Menzel, *Deep-Sea Res.* **14**, 229 (1967).
33. R. V. Fournier, *Science* **153**, 1250 (1966).
34. J. F. Kimball, E. F. Corcoran, E. J. F. Wood, *Bull. Mar. Sci.* **13**, 574 (1963).
35. C. E. ZoBell and R. Y. Morita, *Galathea Rep.* **1**, 139 (1959).
36. J. A. Allen and H. L. Sanders, *Deep-Sea Res.* **13**, 1175 (1966).
37. D. Ye. Garshanovich, *Okeanology* **5**, 85 (in English translation) (1965).
38. F. A. Richards and A. C. Redfield, *Deep-Sea Res.* **1**, 279 (1954).
39. M. G. Gross, *Int. J. Oceanogr. Limnol.* **1**, 46 (1967).
40. G. C. Anderson, *Limnol. Oceanogr.* **9**, 284 (1964).
41. E. L. Mills, personal communication.
42. We thank L. B. Slobodkin and E. L. Mills for reading the manuscript. Contribution No. 2171 from the Woods Hole Oceanographic Institution. Supported by NSF grants 6027 and 810.

FACTORS CONTROLLING PHYTOPLANKTON POPULATIONS ON GEORGES BANK[1]

By

GORDON A. RILEY

Woods Hole Oceanographic Institution

and

Bingham Oceanographic Laboratory

A complex field such as oceanography tends to be subject to two opposite approaches. The first is the descriptive, in which several quantities are measured simultaneously and their inter-relationships derived by some sort of statistical method. The other approach is the synthetic one, in which a few reasonable although perhaps oversimplified assumptions are laid down, these serving as a basis for mathematical derivation of relationships.

Each approach has obvious virtues and faults. Neither is very profitable by itself; each requires the assistance of the other. Statistical analyses check the accuracy of the assumptioms of the theorists, and the latter lend meaning to the empirical method. Unfortunately, however, in many cases there is no chance for mutual profit because the two approaches have no common ground. Until such contact has been established no branch of oceanography can quite be said to have come of age. In this respect physical oceanography, one of the youngest branches in actual years, is more mature than the much older study of marine biology. This is perhaps partly due to the complexities of the material. More important, however, is the fact that physical oceanography has aroused the interest of a number of men of considerable mathematical ability, while on the other hand marine biologists have been largely unaware of the growing field of bio-mathematics, or at least they have felt that the synthetic approach will be unprofitable until it is more firmly backed by experimental data.

However valid the latter objection may be, the present paper will attempt, in the limited field of plankton biology, to establish continuity between some purely descriptive studies that have been made and mathematical concepts based on what seem to be logical assumptions about plankton physiology. The need for such an attempt has become apparent during the course of several plankton surveys in which the data were analyzed statistically with the idea of correlating plankton populations and their rate of growth with various environmental factors such as solar radiation, temperature, dissolved nutrient salts, *etc.* In each survey a reasonably high degree of correlation was found, but the empirical nature of the relationships was often confusing. For example, temperature affects plankton in several different ways, and the relative importance of these effects varies from time to time and from place to place. The statistical relationship of temperature and plankton represents an average of these different effects. Therefore, it may happen in examining sets of data for particular areas and times that the temperature constant varies widely from set

[1] Contribution No. 353 from the Woods Hole Oceanographic Institution.

Reprinted from *Journal of Marine Research,* 6:54–73, 1946.

to set, and study of the values of their constants does not lead easily to a universally applicable theory.

Furthermore, there is no good reason for assuming that the variations of plankton with environment are always linear. To treat them as such may introduce an error. To evolve nonlinear relationships on a purely empirical basis is possible, but this generally requires a larger set of observations than is readily available.

These limitations of the statistical method will become apparent in the pages that follow. The only way to avoid them is by the opposite approach—that of developing the mathematical relationships on theoretical grounds and then testing them statistically by applying them to observed cases of growth in the natural environment. At present this can be done only tentatively, with over-simplification of theory and without the preciseness of mathematical treatment that might be desired. It is not expected that any marine biologist, including the writer, would fully believe all the arbitrary assumptions that will be introduced. However, the purpose of the paper is not to arrive at exact results but rather to describe promising techniques that warrant further study and development.

ACKNOWLEDGMENTS

My best thanks are due Dean F. Bumpus, who kindly provided me with the necessary zooplankton data, and Henry Stommel, who gave me valuable advice on some of the mathematical treatment.

STATISTICAL SECTION

Phytoplankton studies in the Georges Bank area of the western North Atlantic during the period from 1939 to 1941 have been described in a series of publications (Riley, 1941, 1942, 1943; Riley and Bumpus, 1946). In the first of these papers it was noted that part of the variations that occurred in the distribution of phytoplankton from one part of the bank to another and from one month to the next could be correlated with such factors as the depth of water, temperature and dissolved phosphate and nitrate. Since that time the study of the zooplankton collections has been completed and examination of the data has shown that grazing by zooplankton is important in controlling the size of the phytoplankton population. With the inclusion of the zooplankton material, it is now possible to develop a relatively complete statistical treatment of the ecological relationships of the Georges Bank plankton.

Observations. The original observations made during the 1939–1940 cruises were listed in the papers cited above. This material is briefly summarized in Table I in the form of means and standard deviations for each cruise. Correlations of phytoplankton with its various environmental factors have also been published. These have been used to develop multiple correlation equations by which the variations in horizontal distribution of plant pigments during each cruise are calculated according to the variations in environmental factors. Comparison of calculated values with the actual determinations for plant pigments shows an average error on the different cruises of 20–40%. In other words 60–80% of the variations in phytoplankton on Georges Bank can be accounted for on the basis of variations in depth, temperature, phosphate, nitrate and zooplankton. The multiple correlation equations for each cruise are as follows:

TABLE I. MEANS AND STANDARD DEVIATIONS OF GEORGES BANK PLANKTON AND ENVIRONMENTAL FACTORS

		Sept.	*Jan.*	*Mar.*	*Apr.*	*May*	*June*
Mean	*Depth of water in meters*	247	209	135	209	82	206
	Mean temperature, upper 30 m.	15.24	4.61	2.60	3.81	5.14	9.66
	Mg-atoms phosphate P per m², upper 30 m.	14.4	33.7	34.7	21.9	16.6	19.2
	Mg-atoms nitrate N per m², upper 30 m.	153	209	172	129	285	155
	Number of animals, thousands per m²	135	14	24	32	106	103
	Plant pigments, thousands of Harvey units per m²	560	118	828	2303	871	478
σ	*Depth*	540	371	163	530	357	458
	Temperature	2.14	0.71	0.33	0.37	0.39	2.58
	P	7.2	11.1	8.1	8.9	3.5	8.7
	N	108	209	68	129	79	155
	Zooplankton	126	13	18	38	77	85
	Plant pigments	195	67	781	827	522	233

$$\text{Sept.: } PP = -.011D - 23.8t + 5.20P + .371N - .26Z + 829$$

$$\text{Jan.: } PP = .191D - 61.2t + 7.96P + .956N + .47Z - 115$$

$$\text{Mar.: } PP = -D - 770t - 23.18P - 6.98N - .62Z + 4989$$

$$\text{Apr.: } PP = .469D - 331t - 61.4P + .197N - 5.31Z + 4954$$

$$\text{May: } PP = .007D + 236t + 15.1P - 2.316N - 3.50Z + 437$$

$$\text{June } PP = -.066D - 48.1t - 15.5P + .070N - .55Z + 1300$$

PP is thousands of Harvey units of plant pigments per m², D is depth of water, t is temperature, P is mg-atoms of phosphate P per m² in the upper 30 meters, N is mg-atoms of nitrate N, and Z is thousands of animals per m².

Discussion of the effect of environmental factors on the horizontal distribution of phytoplankton. The equations show, within the limits of error stated above, the amount of variation in the phytoplankton crop that is obtained by varying any one or all of the environmental factors. For example, at a particular station of the September cruise, if all the factors were found to have exactly the mean values as stated in Table I, then the plant pigments would be expected to have the mean value for September. If phosphate were increased one milligram-atom, it would increase the calculated value for plant pigments by the amount of the phosphate constant, or 5.2 thousands of units. If the phosphate varied a "normal" amount, as indicated by the limits of its standard deviation, the plant pigments would be changed ± 6.5%. Use of the standard deviation in this way is a convenient method of rating the importance of a given factor, and in Table II it is applied to all the variables in the equations. Although the standard deviation is a positive or negative variation around the mean, it serves a useful purpose to give the values in the table the same sign as the constant in the equation. The figures then represent the change in plant pigments produced by raising each factor from its mean to the upper limit of its standard deviation.

Table II shows that although one particular factor may be of outstanding significance, such as nitrate during the March cruise, phos-

TABLE II. PERCENTAGE CHANGE IN THE PHYTOPLANKTON CROP PRODUCED BY INCREASING THE VALUE OF EACH ENVIRONMENTAL FACTOR FROM ITS MEAN TO THE LIMIT OF ITS STANDARD DEVIATION

	Sept.	*Jan.*	*Mar.*	*Apr.*	*May*	*June*
Depth	−1	60	−20	11	0	− 1
Temperature	−9	−37	−31	− 5	10	−26
P	7	74	−23	−24	6	−28
N	7	41	−57	1	−21	1
Zooplankton	−6	5	− 1	− 9	−31	−10

phate in April, and zooplankton in May, there is no indication of complete control of the phytoplankton crop by a particular factor. It appears to be a highly complex relationship in which one factor after another gains momentary dominance. This table also shows that each variable has a vastly different significance at different times of the year, which is in accord with our present knowledge of phytoplankton ecology.

It has been shown (Riley, 1942) that the depth of water plays a significant role in the inception of the spring diatom flowering, and it is reasonable to find a strong negative relationship in March. This effect disappears later in the season when radiation becomes strong enough so that vertical turbulence is no longer able to prevent growth by dissipating the surface crop.

Temperature is generally supposed to have a negative effect on phytoplankton because increased respiration uses up part of the store of energy that would otherwise be used in the production of new plant material. The predominantly negative relationship shown in Table II is therefore expected. The positive relation in May is anomalous and not readily explainable.

The nutrient-phytoplankton relationship is one in which cause and effect are not clearly separable; although a large quantity of available nutrients is likely to stimulate growth, the growth-utilization process will reduce the quantity of nutrients so that the relationship becomes negative. The diversified results in Table II come from the complexities of this inter-relation. Probably the relationships are of three main types. First, in January, when the quantity of available nutrients was large and growth was slight because of low light intensity, the observed positive relationship is indicative of a slight stimulating effect by nutrients, which to a slight degree counteracted the inhibiting effect of insufficient light. Second, in spring, when radiation increased and growth became more abundant, negative relationships of the growth-utilization type were established. However, they did not become progressively stronger with the advance of the season, leading to complete exhaustion of nutrients; whether or not the observed partial exhaustion had an inhibiting effect on growth cannot be determined from these data. Third, a situation was established which was particularly apparent in September but which probably began early in the summer; in this situation the total quantity of plant pigments was fairly uniform all over the bank, but with certain localized areas of slightly higher crops accompanied by larger quantities of nutrients. These were sufficiently important to provide a direct relationship of a moderately low order between plankton and nutrients. They included some shallow water stations as well as four in deeper water (50 to 100 m.) on the northern and western edges of Georges Bank. In all cases the vertical distribution of temperature, as well as nutrients, was more nearly uniform than at near-by stations that had smaller crops. It is concluded, therefore, that the observed relationship was due to localized turbulence and upwelling of the nutrient-rich lower waters. Probably such conditions were of transient nature, for it seems likely that a degree of regeneration strong enough to maintain a positive nutrient-phytoplankton relationship for any length of time would lead to a much larger growth than was observed.

The phytoplankton-zooplankton relationship has been discussed in some detail in a previous paper (Riley, 1946) and need be only briefly summarized here. It was concluded that the predominantly negative relationship was due to grazing. The quantities of animals and plants were such as to indicate that the observed relationship could have been established in a very short time, possibly in a day or in a few days. A theory was postulated that tidal currents and turbulent motion of the Georges Bank waters tended continually to destroy the horizontal gradients in phytoplankton, but that the zooplankton, because of their habit of vertical migration, would be absent part of the time from the surface waters where mixing processes are strongest, and hence they would not be so readily dispersed. Therefore, they would tend constantly to reestablish the phytoplankton gradients by their grazing activity.

Effect of Environmental Factors on the Seasonal Cycle of Phytoplankton. It is apparent from the preceding discussion that the relationship of a particular environmental factor with the horizontal distribution of phytoplankton may differ from one month to another both in quantity and in kind. Nevertheless, there are seasonal trends in these factors which are related with plankton variations, as can be observed by inspection of the data in Table I. Thus the seasonal cycle of phytoplankton can be correlated with its environment with a fair degree of accuracy, even though such treatment makes no allowance for special effects that are operative only at particular times during the year.

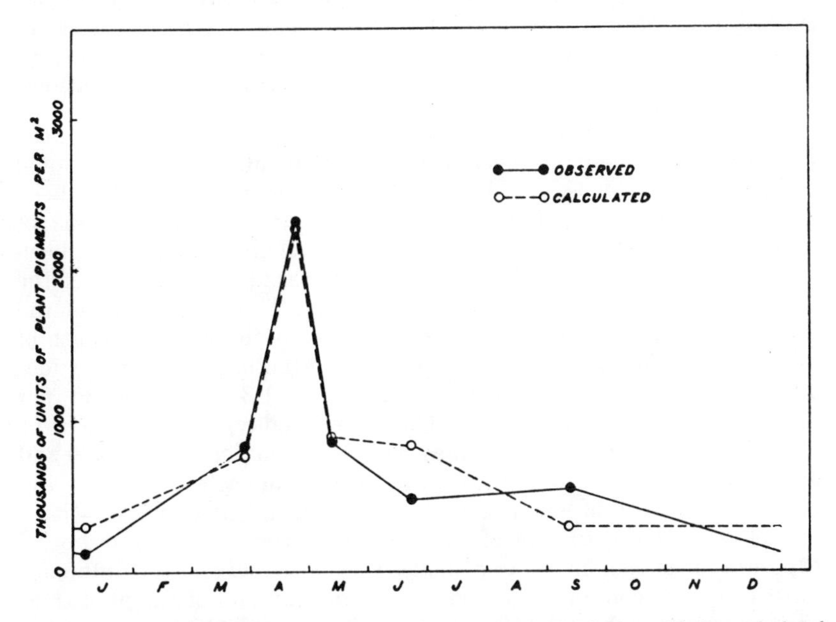

Figure 14. Comparison of the observed phytoplankton population with the calculated population as determined by a multiple correlation analysis of the relationship between phytoplankton and environmental factors.

A multiple correlation equation was developed from the cruise averages in Table I. It is

$$PP = -153t - 120P - 7.3N - 9.1Z + 6713.$$

The symbols are the same as in previous equations. The relation between the observed cruise averages and the values determined from the equation is shown in Fig. 14. The average error between observed and calculated values is 20%, which is slightly less than that obtained in the treatment of individual cruises. Probably the errors caused by ignoring certain special seasonal effects are more than counterbalanced by the reduction in analytical errors when averages are used for the calculation.

THEORETICAL SECTION

The rate of change of the phytoplankton population is determined by the difference in reaction rates between the process of accumulation of energy by the population and the processes of energy dissipation. It will be assumed that all the important reaction rates are included in the equation

$$\frac{dP}{dt} = P(P_h - R - G), \qquad (1)$$

in which the rate of change of the population P in respect to time is determined by the photosynthetic rate per unit of population (P_h), the rate of phytoplankton respiration (R), and the rate of grazing by zooplankton (G). Each of these rates is subject to environmental influences and therefore is continually changing with the seasons. In order to arrive at a practical solution of the equation it is necessary to examine each of its component parts in the light of present day plankton physiology.

A second major assumption will be that seasonal variations in environmental factors used in the analysis can be expressed by smooth curves drawn through the observed cruise averages. Since the latter are relatively incomplete, as far as the whole yearly cycle is concerned, there must be a certain amount of unavoidable error in the calculations.

Photosynthesis. Numerous investigators have reported that the photosynthetic rate in actively growing diatom cultures is proportional to light intensity within wide limits. The lower limit has not been determined accurately due to the insensitivity of the methods of measurement, but values of the right order of magnitude have been detected at depths where the light intensity was about 0.1–1.0% of the surface intensity in summer (Clarke, 1936). The upper limit of the proportionality is variable, depending on the species and the length of exposure; the optimum intensity for photosynthetic activity in particular situations has been reported to range from 1.8 g. cal. per cm^2 per hour (Jenkin, 1937) to 60 g. cal. (Curtis and Juday, 1937).

During the six cruises to Georges Bank between September 1939 and June 1940 two bottles of surface water were taken at each station and suspended in a tub of water on deck, one of them being covered with a bag of several thicknesses of dark cloth. After twenty-four hours' exposure the oxygen in the two bottles was measured and the difference in their oxygen content was used as a rough estimate of the photosynthetic activity of the surface plankton. The inset in Fig. 15 shows the

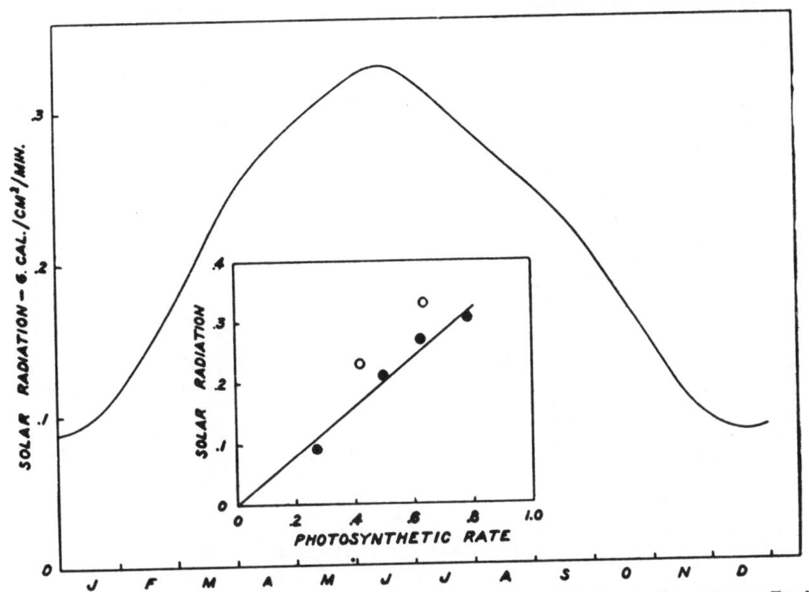

Figure 15. Seasonal variations in average incident solar radiation in the Georges Bank area. Inset shows average observed surface photosynthetic rate plotted against incident radiation. Dots are averages for January, March, April and May; circles are June and September.

average photosynthetic rate[2] obtained on each cruise plotted against the average incident solar radiation in the area at that time of year. Radiation values are obtained from data published by Kimball (1928) and reproduced in Sverdrup, Johnson and Fleming (1942: 103). The plotted points for the January, March, April and May cruises approximate a linear relationship in which the photosynthetic rate equals 2.5 times the incident radiation in g. cal. per cm² per minute. The June and September rates are lower, but it will be shown later that this can be explained as a result of nutrient depletion.

With these facts at hand the mean photosynthetic rate of the population will be estimated according to the following assumptions:

1. When nutrient depletion does not limit photosynthesis,

$$P_h = pI \tag{2}$$

in which P_h is the photosynthetic rate, I is radiation in g. cal./cm²/minute at the depth of the photosynthesizing plankton, and p, the photosynthetic constant, is 2.5.

2. The intensity at the surface, I_o, may be determined for any time in the year from the curve in Fig. 15, which is based on Kimball's data, cited above.

[2] The photosynthetic rate is expressed as grams of carbon produced per day per gram of carbon in the surface phytoplankton crop, using the formula photosynthetic rate $= .375 \times$ oxygen production in g/m³/day $\div$ 17×10^{-6} Harvey units of plant pigments/m³. The conversion factor is based on analyses described in a previous paper on the plankton of Georges Bank (Riley, 1941).

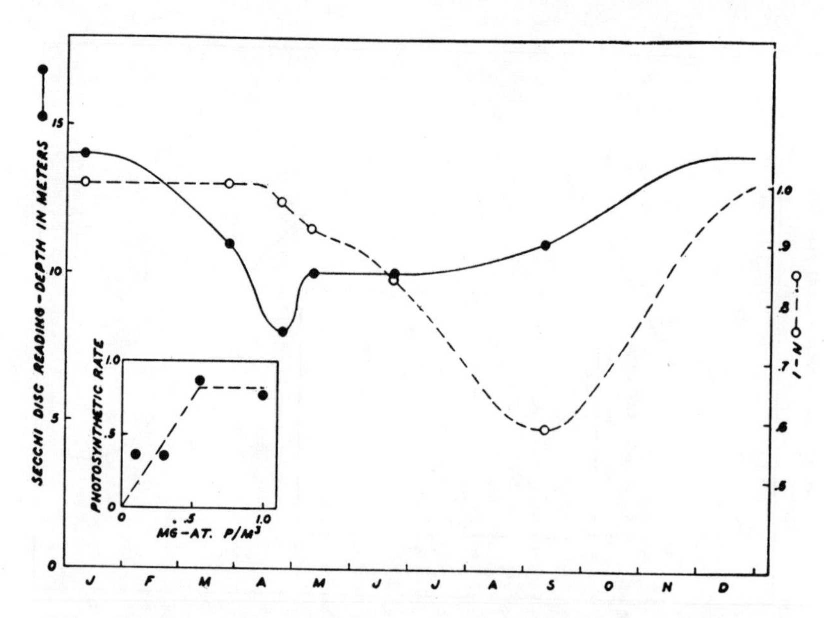

Figure 16. Solid curve is transparency as determined by Secchi disc readings. Dotted line is the estimated reduction in photosynthetic rate caused by nutrient depletion. Inset shows averages of photosynthetic rate plotted against phosphate concentration during June and September.

3. The intensity at any other depth z is determined from the formula

$$I_z = I_0 e^{-kz}$$

and therefore

$$P_{hz} = pI_0 e^{-kz}. \qquad (3)$$

In these formulas the extinction coefficient k is defined as 1.7 divided by the depth of the Secchi disc reading, a rough conversion factor suggested by Poole and Atkins (1929). Secchi disc determinations for the Georges Bank area are shown in Fig. 16.

4. According to equation (3) the photosynthetic rate approaches zero as the depth approaches infinity. But the depth at which a measurable and significant amount of photosynthesis occurs is limited, and so is the depth at which viable phytoplankton can be found. Therefore, it is convenient to set an arbitrary limit to the depth of the euphotic zone. This depth will be called z_1 and will be defined as the depth at which the light intensity has a value of 0.0015 g. cal./cm²/minute. This approximates the intensity at the maximum depth of photosynthesis as reported by Clarke (1936). Calculated values for z_1 are shown in Fig. 17.

5. To find the mean photosynthetic rate in the euphotic zone, equation (3) is integrated from the surface to z_1, and divided by z_1:

$$\overline{P_h} = \frac{pI_0 \int_0^{z_1} e^{-kz}\, dz}{z_1} = \frac{\left. \frac{pI_0}{-k} e^{-kz} \right|_0^{z}}{z_1}$$

$$\underset{0 \to z_1}{\overline{P_h}} = \frac{pI_0}{kz_1}(1 - e^{-kz_1}).$$

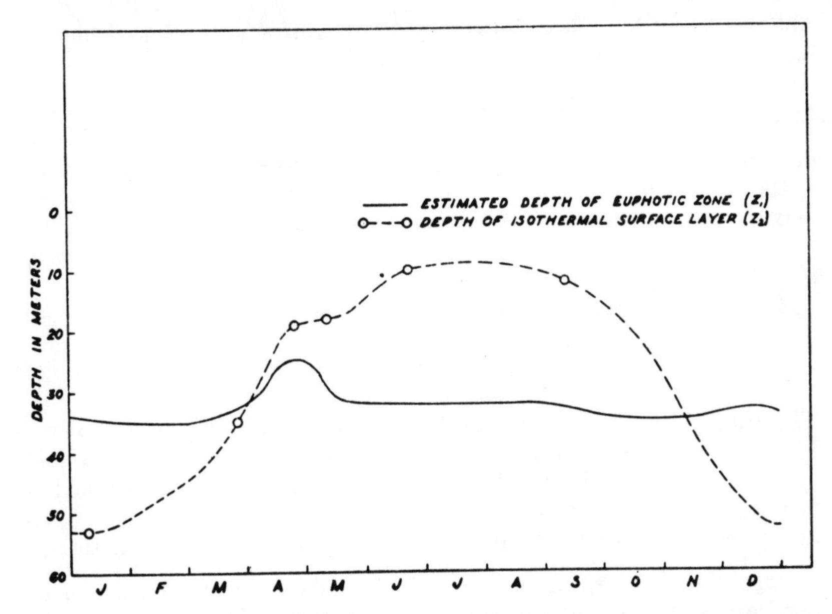

Figure 17. Estimated depth of the layer in which a measurable amount of photosynthesis occurs (z_1), and the depth of the virtually isothermal surface layer (z_2).

6. It was postulated that the proportionality between photosynthesis and radiation holds only when nutrients are abundant. The fact noted previously that the ratio between the photosynthetic rate and light intensity was reduced in June and September led to an investigation of nutrients as a possible cause. Large individual variations were found, and the correlation between photosynthetic rate and phosphate concentration was poor; nevertheless, when the rates were averaged for different ranges of phosphate concentration, there was a pronounced reduction in the average rate when the phosphate fell below about 0.5 to 0.6 mg-atom of P per m^3 (Fig. 16, inset). Ketchum (1939) reported a decrease in the growth rate of experimental cultures of *Nitschia closterium* when the phosphate concentration was less than 50 gamma of PO_4 per liter (0.55 mg-atom per m^3). Therefore it seems reasonable to assume that the mean photosynthetic rate as determined in equation (4) should be multiplied by a factor $(1 - N)$, in which N is the reduction in rate due to nutrient depletion. According to the facts above,

$$N = \frac{0.55 - mg\text{-}at.\ P/m^3}{0.55} \qquad \text{when } P \leqq 0.55.$$

Mean values for $(1 - N)$ are shown in Fig. 16.

7. Several investigators have pointed out the importance of vertical turbulence in reducing phytoplankton crops by carrying the breeding stock below the euphotic zone. That this is an important phenomenon on Georges Bank has been demonstrated (Riley, 1942). If turbulence is such that each phytoplankton cell spends only a certain proportion of its time in the euphotic zone, then the mean photosynthetic rate of the population as a whole will be reduced. Therefore equation (4) should be multiplied by still another factor $(1 - V)$, in which V is the reduction in rate produced by vertical water movements. It is impos-

sible to define V in any simple way that will be entirely satisfactory, but as an approximation,

$$(1 - V) = \frac{z_1}{z_2} \text{ when } z_1 \leqq z_2,$$

in which z_1 is the depth of the euphotic zone as previously defined and z_2 the depth of the mixed layer, which is arbitrarily defined as the maximum depth at which the density is no more than 0.02 of a σ_t unit greater than the surface value. Fig. 17 shows the estimated values for z_1 and z_2 for Georges Bank.

The final equation for the mean photosynthetic rate is now

$$P_h = \frac{pI_0}{kz_1}(1 - e^{-kz_1})(1 - N)(1 - V). \tag{5}$$

The application of the equation to the Georges Bank data is shown in Fig. 18. The upper curve shows the primary calculation of the photosynthetic rate based on light intensity. The reduction obtained by introducing $(1 - N)$ and $(1 - V)$ is indicated by hatched areas. The heavy lower curve is the final estimate of the mean photosynthetic rate.

Phytoplankton respiration. The few available measurements of the respiration of pure diatom cultures have not yielded precise results. Observed rates have varied from one species to another as well as during different stages of growth of the same culture. The recorded values differ by a factor of 10 to 20, and there are not enough of them to draw a good average.

No direct measurements have been made of the respiration of a natural phytoplankton population, since the measured oxygen consumption also includes zooplankton and bacterial respiration. Statistical estimates have been made by the writer on the basis of the

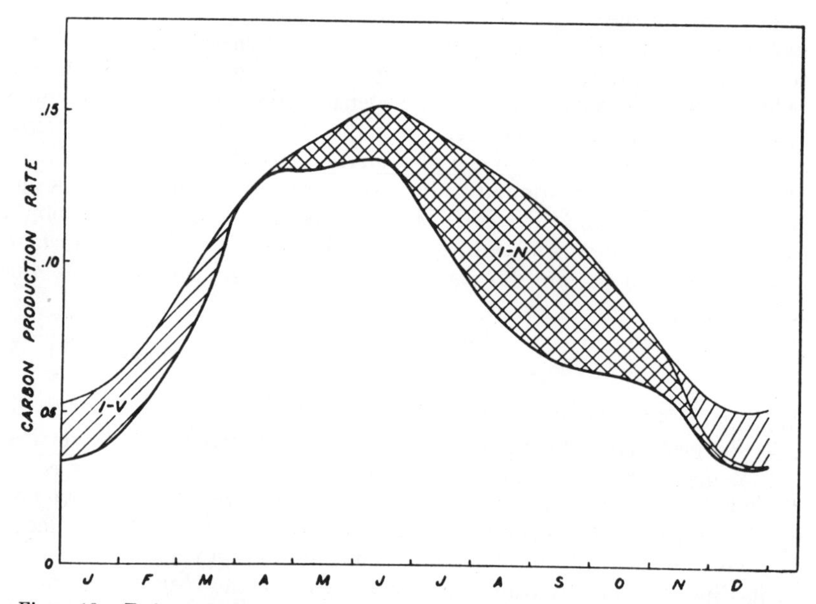

Figure 18. Estimated mean photosynthetic rate. Upper curve is the maximum possible rate as determined by incident radiation and transparency. Lower curve is the estimate obtained by introducing correction factors for the effects of vertical turbulence and nutrient depletion.

observed correlation between phytoplankton and total oxygen consumption. The best of these, judging by quantity and homogeneity of data, was obtained in Long Island Sound (Riley, 1941a). It was estimated that in winter (average temperature 2.05° C.) the respiration rate was 0.024 ± 0.012 mg. of carbon consumed per day per mg. of phytoplankton carbon. Calculations based on summer observations suggested that the respiration increased with higher temperatures (0.110 ± 0.007 at 17.87° C.), the rate being approximately doubled by a 10° increase in temperature.

On the basis of these rather scanty data it will be assumed that:

1. The temperature effect can be stated as

$$R_T = R_o e^{rT} \tag{6}$$

in which R_T is the respiratory rate at any temperature T, R_o is the rate at 0° C. and r is a constant expressing the rate of change of the respiratory rate with temperature. The value of r is 0.069 when the rate is doubled by a 10° increase in temperature. The seasonal cycle of Georges Bank surface temperatures used in computing respiratory rates is shown in Fig. 19.

2. The value chosen for R_o will be 0.0175. This is the mean of the two estimates derived from the Long Island Sound data mentioned above, in which the calculated values of R_o for winter and for summer are respectively 0.020 and 0.015.

Grazing. The greater part of the zooplankton population consists of filter-feeding organisms which tend to strain a relatively constant volume of water in a given time irrespective of the quantity of food material in it. Therefore a fixed proportion of the phytoplankton population will be consumed in successive units of time. This is stated as

$$G = gZ\,, \tag{7}$$

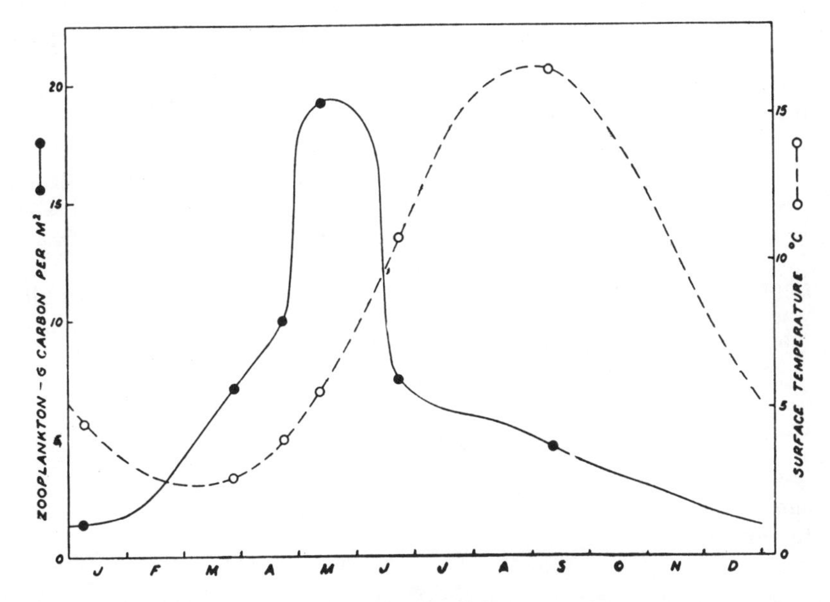

Figure 19. Solid line is the seasonal cycle of zooplankton. Measurements of zooplankton volume by the displacement method are treated by a conversion factor (wt. in g. = 12.5% × vol. in cc.) to derive a rough estimate of the carbon content. Dotted line is the mean surface temperature.

in which G is the rate of grazing, g is the rate of reduction of phytoplankton by a unit quantity of animals and Z is the quantity of zooplankton in grams of carbon per m^2.

There is some question as to whether g is nearly constant over long periods of time or undergoes a marked seasonal change. On the one hand there are the experiments of Marshall, Nicholls and Orr (1935) which showed that the respiratory rate of *Calanus finmarchicus* increases with increasing temperature, implying a greater food requirement at higher temperatures and possibly an increased filtering rate. On the other hand, feeding experiments by Fuller (1937) showed variations of a more complex nature. The grazing rate of *Calanus finmarchicus* was greater at 8° C. than at 3° or 13° C., and animals captured in the late summer, when the natural breeding stock was declining, had lower feeding rates than those studied earlier in the summer. Thus a factor that can be called "depressed physiological state" for want of a more precise term appears to counterbalance the expected effect of high temperature late in the summer. It is clear that the whole process of zoöplankton feeding requires much more thorough study. However, lacking the necessary information to describe the process accurately, it is believed that it can best be approximated by the simple form of equation (7).

The value of g must be of an order of magnitude which will at least satisfy the minimum respiratory requirements of the zooplankton population at times when the latter is stable. According to Marshall, Nicholls and Orr (1935), the daily food requirement of *Calanus* in winter (5° C.) lies between 1.3 and 3.6% of the carbon content of the animals. This estimate applied to the January plankton on Georges Bank yields a grazing constant of 0.0091 to 0.0252. In summer (15° C.), these authors suggest a food requirement of 1.7–7.6%, for which the corresponding values of g are 0.0084 to 0.0374 in September on Georges Bank. The latter estimates are perhaps too high, since the zooplankton was decreasing at a rate of 0.5% per day and therefore probably was not getting enough food to satisfy the minimum respiratory requirements. If it is assumed that the food intake equaled the food requirement minus the rate of population decrease, then the food intake was 1.2–7.1% of the animals' carbon content per day, and the corresponding values for g are 0.0059 and 0.0350.

Within these wide limits it is difficult to choose a correct value for the grazing constant, and again the need for more experimental work is apparent. On a purely empirical basis, a good fit for the data is obtained by using the average of the minimum values of g for the September and January cruises, namely 0.0075. This factor, multiplied by the quantity of zooplankton, shown in Fig. 19, estimates the Georges Bank grazing rate.

Conclusions. The original equation

$$\frac{dP}{dt} = P\,(P_h - R - G)$$

can now be expanded by substituting the right hand terms of equations (5), (6) and (7):

$$\frac{dP}{dt} = P\left[\frac{pI_o}{kz_1}\,(1 - e^{-kz_1})\,(1 - N)\,(1 - V) - R_o e^{rT} - gZ\right]. \quad (8)$$

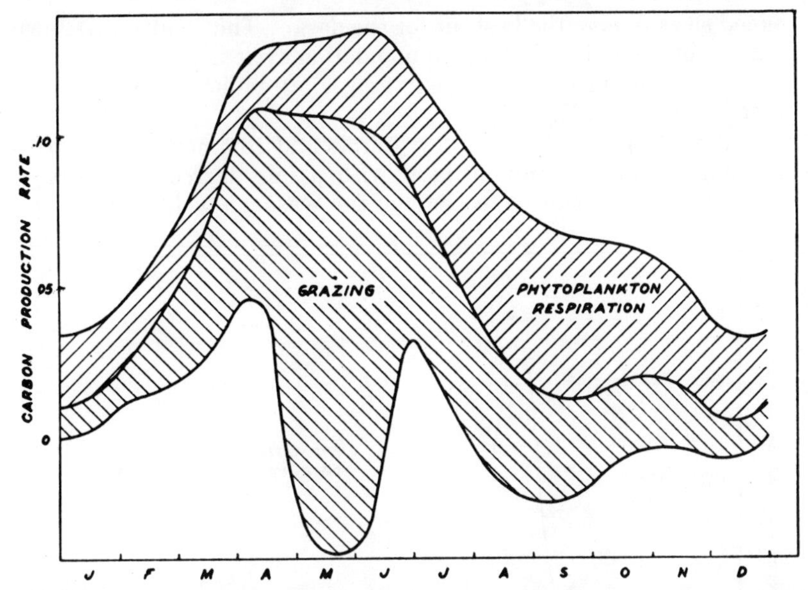

Figure 20. Estimated rates of production and consumption of carbon. Curve at top is the photosynthetic rate. By subtracting the respiratory rate the second curve is obtained, which is the phytoplankton production rate. From this is subtracted the zooplankton grazing rate, yielding the curve at the bottom, which is the estimated rate of change of the phytoplankton.

The rate of change of the population is dependent on six ecological variables: the incident solar radiation, transparency of the water, the quantity of phosphate, the depth of the mixed layer, the surface temperature and the quantity of zooplankton. The results of the application of equation (8) to the Georges Bank data are shown graphically in Fig. 20. The curve at the top is the photosynthetic rate previously illustrated (Fig. 18). The second curve is the photosynthetic rate minus the respiratory rate of the phytoplankton, or in other words the phytoplankton production rate. By subtracting the grazing rate from the production rate, the rate of change of the population is obtained. This is the bottom curve of Fig. 20. Numerical values used in drawing these curves are shown in the appendix.

Equation (8) cannot be integrated readily, but an approximation is obtained by integrating over successive short intervals of time and assuming for each variable a constant, average value during that time. Thus the change in population in the time interval 0 to t is determined by

$$lnP_t - lnP_o = \overline{P_h} - \overline{R} - \overline{G}\,.$$

Therefore, by a series of integrations the quantity $lnP_t - lnP_o$ can be approximated for the whole seasonal cycle. This quantity indicates the relative size of the population from one part of the cycle to another. To convert to absolute terms requires evaluation of the integration constant P_o (the size of the population at the minimum point in the cycle), which is readily obtained if the quantity of plankton (P_t) is measured at one or more times during the year. In the present case, in which six cruise averages are available, P_o was statistically determined so as to give the best fit for the data. The results are shown in Fig. 21, in which the theoretical population cycle is shown by a smooth curve, and the observed cruise averages are indicated by dots. The average error is 27%.

It is now apparent that a few simple and commonly measured environmental factors can be used with a fair degree of accuracy to evaluate the quantitative aspects of the seasonal cycle of phytoplank-

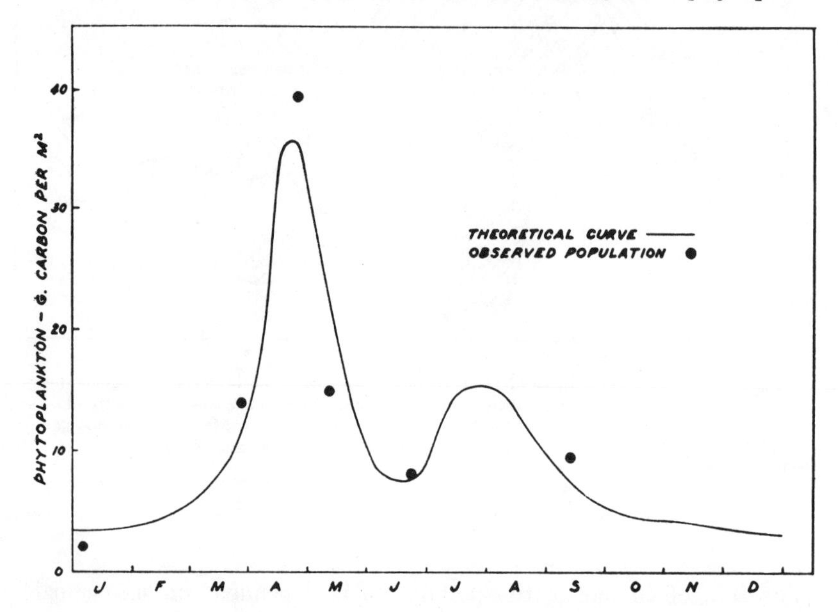

Figure 21. Curve shows the calculated seasonal cycle of phytoplankton, determined by approximate integration of the equation for the rate of change of the population. For comparison the observed quantities of phytoplankton are shown as dots.

ton. Furthermore, this can be done in two different ways: first by statistical comparison of the simultaneous variations of phytoplankton and environmental characteristics in a particular locality; second, by theoretical evaluation of the way in which changes in the environment might affect growth as evidenced by the results of various physiological experiments. Each of these methods has something to contribute to plankton biology. The statistical method is useful in determining whether a particular factor is significant; the theoretical method carries on from there, discriminating between cause and effect and helping to establish certain quantitative relationships that are not likely to be derived empirically.

While these methods are obviously crude at the present time and need to be developed further, both by examination of other areas and by better experimental evaluation of constants, it does not seem too much to hope that they will eventually solve some of the problems of seasonal and regional variations that puzzle marine biologists today.

SUMMARY

1. Variations in the phytoplankton population of the Georges Bank area are correlated with various environmental factors. Equations are developed statistically by which the size of the population can be calculated on the basis of such factors as temperature, depth of water, and the quantities of nitrate, phosphate and zooplankton. Calculated

horizontal variations in the plankton crop at various times in the year differ from observed values by about 20–40%. Calculations of the seasonal variation of the average crop in the area are accurate within about 20%.

2. The seasonal cycle of phytoplankton is also evaluated from a more theoretical standpoint. It is postulated that the rate of change of the phytoplankton population is equal to the photosynthetic rate minus the phytoplankton respiratory rate minus the grazing rate of the zooplankton. Factors affecting these rates are discussed, and the ones that are considered particularly important are solar radiation, temperature, transparency of the water, the depth of the isothermal surface layer, phosphate, and zooplankton. The observed variations of these factors are combined with appropriate constants derived from experimental data to develop an equation that expresses the seasonal rate of change of the phytoplankton population. Approximate integration of the equation yields a population curve of the same order of accuracy as the statistical estimate.

REFERENCES

CLARKE, G. L.
1936. Light penetration in the western North Atlantic and its application to biological problems. Rapp. Cons. Explor. Mer, *101*, Pt. 2: 1–14.

CURTIS, J. T. AND C. JUDAY
1937. Photosynthesis of algae in Wisconsin lakes. III. Observations of 1935. Int. Rev. Hydrobiol., *35:* 122–133.

FULLER, J. L.
1937. Feeding rate of *Calanus finmarchicus* in relation to environmental conditions. Biol. Bull. Wood's Hole, *72:* 233–246.

JENKIN, P. M.
1937. Oxygen production by the diatom *Coscinodiscus excentricus* Ehr. in relation to submarine illumination in the English Channel. J. Mar. biol. Ass. U. K., (N. S.) *22:* 301–342.

KETCHUM, B. H.
1939. The absorption of phosphate and nitrate by illuminated cultures of *Nitzschia closterium*. Amer. J. Bot., *26:* 399–407.

KIMBALL, H. H.
1928. Amount of solar radiation that reaches the surface of the earth on the land and on the sea, and methods by which it is measured. Mon. Weath. Rev. Wash., *56:* 393–398.

MARSHALL, S. M., A. G. NICHOLLS AND A. P. ORR
1935. On the biology of *Calanus finmarchicus*. Part VI. Oxygen consumption in relation to environmental conditions. J. Mar. biol. Ass. U. K., (N. S.) *20:* 1–27.

POOLE, H. H. AND W. R. G. ATKINS
1929. Photo-electric measurements of submarine illumination throughout the year. J. Mar. biol. Ass. U. K., (N. S.) *16:* 297–324.

RILEY, G. A.
1941. Plankton studies. IV. Georges Bank. Bull. Bingham oceanogr. Coll., *7* (4): 1–73.
1941a. Plankton studies. III. Long Island Sound. Bull. Bingham oceanogr. Coll., *7* (3): 1–93.
1942. The relationship of vertical turbulence and spring diatom flowerings. J. Mar. Res., *5* (1): 67–87.
1943. Physiological aspects of spring diatom flowerings. Bull. Bingham oceanogr. Coll., *8* (4): 1–53.

RILEY, G. A. AND D. F. BUMPUS
1946. Phytoplankton-zooplankton relationships on Georges Bank. J. Mar. Res., *6* (1): 33–47.

SVERDRUP, H. U., M. W. JOHNSON AND R. H. FLEMING
1942. The oceans. Prentice-Hall, Inc., New York. 1060 pp.

APPENDIX

NUMERICAL VALUE OF QUANTITIES USED IN THE CALCULATION OF THE SEASONAL RATE OF CHANGE OF THE GEORGES BANK PHYTOPLANKTON POPULATION AS DEVELOPED IN THE THEORETICAL SECTION OF THIS PAPER

Date	I_o	k	z_1	$1-N$	z_2	$1-V$	Ph	T	R_T	Z	G	dP/dt	lnP_t-lnP_o	lnP_t	P_t
1/1	.088	.121	34	1.00	53	.64	.034	5.2	.024	1.3	.010	.000	.000	1.217	3.4
1/15	.094	.121	34	1.00	53	.64	.036	4.1	.023	1.4	.011	.002	.030	1.247	3.5
2/1	.112	.124	35	1.00	51	.69	.044	3.2	.021	1.7	.013	.010	.180	1.397	4.0
2/15	.138	.128	35	1.00	48	.73	.055	2.7	.021	2.6	.020	.014	.390	1.607	5.0
3/1	.174	.136	35	1.00	45	.78	.071	2.4	.021	4.2	.031	.019	.675	1.897	6.7
3/15	.212	.145	34	1.00	40	.85	.091	2.5	.021	5.8	.043	.027	1.080	2.297	10.0
4/1	.247	.159	32	1.00	32	1.00	.120	2.7	.021	7.5	.056	.043	1.725	2.942	19.1
4/15	.272	.200	26	1.00	23	1.00	.130	3.4	.021	8.9	.067	.042	2.355	3.572	35.5
5/1	.290	.205	25	.95	19	1.00	.131	4.5	.024	17.2	.129	−.022	2.025	3.242	25.7
5/15	.306	.170	31	.92	18	1.00	.132	5.9	.026	19.3	.145	−.039	1.440	2.657	14.2
6/1	.321	.170	32	.90	15	1.00	.134	7.6	.030	18.8	.141	−.037	.885	2.102	8.2
6/15	.329	.170	32	.88	11	1.00	.134	9.7	.035	14.0	.105	−.006	.795	2.012	7.5
7/1	.319	.170	32	.82	10	1.00	.122	11.8	.038	6.9	.052	.032	1.275	2.492	12.1
7/15	.302	.170	31	.76	9	1.00	.108	13.9	.045	6.2	.047	.016	1.515	2.732	15.4
8/1	.284	.165	32	.69	9	1.00	.093	15.5	.051	6.0	.045	−.003	1.470	2.687	14.8
8/15	.267	.162	32	.63	10	1.00	.081	16.3	.054	5.7	.043	−.016	1.230	2.447	11.5
9/1	.250	.159	32	.60	11	1.00	.073	16.6	.056	5.1	.038	−.021	.915	2.132	8.4
9/15	.230	.154	33	.59	13	1.00	.067	16.4	.054	4.5	.034	−.021	.600	1.817	6.1
10/1	204	.145	34	.63	16	1.00	.065	15.5	.051	3.9	.029	−.015	.375	1.592	4.9
10/15	.174	.138	34	.69	20	1.00	.063	14.2	.045	3.3	.025	−.007	.270	1.487	4.4
11/1	.144	.131	35	.77	27	1.00	.060	12.4	.040	3.2	.024	−.004	.210	1.427	4.2
11/15	.115	.126	35	.85	37	.95	.053	10.5	.037	2.6	.020	−.004	.150	1.367	3.9
12/1	.094	.121	34	.92	45	.76	.039	8.5	.031	2.0	.015	−.007	.045	1.262	3.5
12/15	.086	.121	33	.97	50	.66	.033	6.7	.028	1.6	.012	−.007	−.060	1.157	3.2

I_o = incident solar radiation in g. cal./cm² 1 /cm²/minute
k = extinction coefficient = 1.7/Secchi disc reading in meters
z_1 = depth of euphotic zone in meters, defined as depth where the light intensity is .0015 g. cal.
$1 - N$ = correction factor for nutrient depletion = mg-at. phosphate P/0.55 when $P \leqq 0.55$.
z_2 = depth of mixed layer = maximum depth at which $\sigma_{tz} - \sigma_{t_0} \leqq 0.02$.
$1 - V$ = correction factor for vertical turbulence = z_1/z_2 when $z_1 \leqq z_2$.
Ph = estimated mean photosynthetic rate according to equation (5) in text.
T = mean surface temperature.
R_T = estimated phytoplankton respiration according to equation (6).
Z = g. of zooplankton carbon/m², estimated on the assumption that the weight of carbon in grams = 12.5% of the volume (by displacement) in cubic centimeters.
G = grazing rate = $0.0075Z$
dP/dt = rate of change of the phytoplankton population = $Ph - R - G$.
$lnP_t - lnP_o$ = the summation of 15 × dP/dt (since the rate determinations are at approximately 15-day intervals)
$lnP_t = lnP_t - lnP_o + 1.217$. The latter is a value for P_o determined statistically as the best fit for the observed population data.
P_t = estimated population in g. of phytoplankton carbon/m² (considered equivalent to 17 × Harvey units of plant pigments × 10^{-6}/m²).

MEASUREMENTS OF PRIMARY PRODUCTION IN COASTAL SEA WATER USING A LARGE-VOLUME PLASTIC SPHERE

18

C. D. McAllister, T. R. Parsons, K. Stephens and J. D. H. Strickland

Fisheries Research Board of Canada,
Pacific Oceanographic Group, Nanaimo, British Columbia, Canada

ABSTRACT

A free-floating 20-ft diameter thin transparent sphere, with its center 5.5 m below the sea surface, was filled with nutrient rich water filtered free from plants and animals and inoculated with a natural population of coastal phytoplankters. The resulting phytoplankton "bloom" was then studied in this well-mixed water mass under near-natural conditions of temperature and illumination with the minimum of complications arising from grazing, sinking or lateral transport of the plant cells.

Daily measurements were made of temperature, light and photosynthesis and the water was analyzed for oxygen, carbon dioxide and all common micronutrients. The production of plant material was followed by cell counts and pigment values and by analyses for particulate carbon, nitrogen, phosphorus, carbohydrate, protein and fat.

The results have a bearing on nearly all aspects of marine primary productivity and are discussed in detail. In particular, the growth kinetics and chemical composition of a natural mixed crop have been related to the environment with the minimum of ambiguity. Growth was finally limited by nitrogen deficiency, although the plant biomass nearly doubled after nitrate became depleted, with a shift of metabolism to carbohydrate and fat synthesis. The composition of the plankton at various stages in a bloom is summarized by ratios involving carbon and chlorophyll *a*.

Differences between the results for photosynthesis obtained by the oxygen and C^{14} method have emphasized the importance of knowing photosynthetic quotient values and possible unlabeled sources of plant carbon. During an initial period of low illumination there was strong evidence for heterotrophic growth.

The principal species found in the "bloom" were *Skeletonema costatum, Thalassiosira nordenskiöldii, T. rotula, Gyrodinium fulvum, Glenodinium danicum, Nitzschia delicatessima* and *Asterionella japonica.* The last two grew very slowly and the two dinoflagellates only appeared after the nitrate was depleted.

INTRODUCTION

It is unnecessary to stress the difficulty of studying photosynthetic productivity in sea water that arises because of the impossibility of observing any one discrete body of water for a sufficient length of time. Valuable information can be obtained by observing the behavior of pure cultures of marine phytoplankton but such work is no adequate substitute for the study of the growth and decay processes that occur with mixed populations under natural conditions.

An obvious solution to the problem would be to impound sea water in a tank and valuable exploratory experiments of this nature were undertaken by Edmondson and Edmondson (1947) and Edmondson (1956). However, the use of a tank has several disadvantages unless it is very deep. The light-

Reprinted from *Limnology and Oceanography*, 6:237–258, 1961.

ing of the water mass as a whole can be unrealistically high, and undesirable temperature increases usually occur. Furthermore, no precise "balance" of dissolved oxygen or carbon dioxide can be readily computed when there is a large surface area of water in contact with the atmosphere.

The thin plastic tube experiment proposed by Margalef (1956) would avoid some of this trouble, but in order to reduce wall-effects (the undesired growth of epigenous organisms, especially bacteria) one must have as large a volume to surface area ratio as possible. It is now common experience that a wall-effect from bacteria can be noticed within a very few hours in a small bottle and we may therefore assume that a container must be many feet in diameter before wall-effects can be neglected for a period of many days. A large container has the additional advantage that errors from accidental contamination are reduced.

To overcome most of the difficulties outlined above we have constructed a free-floating thin plastic sphere, 20 ft in diameter, suspended with its center a constant distance (5.5 m) beneath the sea surface. This plastic bag was connected to the surface by a comparatively narrow neck through which samples could be taken. The state of the water contained in the sphere approximated closely that of the surrounding sea water, both in temperature and in the conditions of illumination. An account of the construction of this equipment has already been published by Strickland and Terhune (1961). It was used in Departure Bay (49°20 N, 123°50′ W), near Nanaimo, on the east coast of Vancouver Island, British Columbia.

The purpose of the present paper is to describe a phytoplankton bloom which was observed in the plastic sphere under these near-natural conditions. The production of such a bloom has enabled growth conditions to be studied without any of the ambiguities normally introduced by water movement. Further, the detailed analysis of the resulting crop has given the first extensive set of results applicable to detritus-free marine phytoplankton grown under natural conditions, that has so far been reported. Phenomena such as the succession of species and the effect on phytoplankton of nutrient environment were studied with the minimum of complications.

In one experiment the plastic sphere was torn after only 5 days but in the other (June–July 1960) 22 days of growth and decay were recorded before the equipment commenced to leak and this experiment forms the basis of most of the results reported in the present communication. We propose to repeat the experiment at least once more, and expect to be able to prolong it for many more weeks so that the decay processes of an initial bloom can be studied in greater detail. There is really no end to the number of such experiments that could be undertaken as each time different species of plant cells would doubtless develop. The present work, however, measures the production of a diatom crop which is quite typical of the spring bloom in temperate coastal seas.

In this experiment some difficulty was encountered with the stirring mechanism and it was not possible to keep all dead and dying plant cells in uniform suspension within the bag. In these circumstances one cannot equate the formation of soluble metabolites to the decrease of particulate matter sampled at the center of the bag. However, this limitation applied only to the last few days of the decay processes which followed peak phytoplankton production.

The authors would like to record their thanks to L. D. B. Terhune and others at the Pacific Biological Station, Nanaimo, for their work in designing, assembling and launching the equipment. We also wish to acknowledge the assistance given by the undergraduate student assistants, Wing Wai and John Yu, in the analytical program.

PROCEDURE AND METHODS

General procedure

The plastic bag was filled with water taken near the sea floor, about 350 ft from the low-tide shore line. The depth of the water at the intake varied with the tide between some 15 and 20 m. The water was pumped to land, filtered through a cloth-plate filter coated with diatomaceous earth, and returned to the location of the bag (250 ft out in 11–16 m of water) by an all-plastic pipe system. Volumes were recorded with a water meter and a total of 125 m^3 of sea water was required to fill the bag.

After preliminary analyses the contents of the bag were "inoculated" with 4 m^3 of the surrounding sea water drawn from a depth of about 4 m. This water was pumped through an all-plastic assembly, and was strained through 320-μ mesh nylon netting to remove all but the smaller zooplankters.

The initial water in the bag was entirely free from animals and contained few if any plant cells. The chlorophyll *a* content of about 0.15 mg/m^3 was probably detrital, although there could have been a few microflagellates present. The inoculum was high in pigment (an average chlorophyll *a* content of 11 mg/m^3) and undoubtedly introduced some animal population into the bag. However, no significant zooplankton population developed during the course of the experiment and no chitin was detected analytically at any time.

The initial "background" level of particulate organic matter was undesirably high. The material consisted mainly of protein and we cannot explain its presence. Most probably there was a leak of diatomaceous earth at one stage in the filling, or detrital material of near-colloidal dimensions may have escaped filtration. The amount was very much less than in unfiltered sea water.

The inorganic nutrient content of the bottom water used to fill the plastic sphere was large and representative of the winter levels of concentration found in this area, which is one of the most fertile coastal regions of the world. Therefore, no artificial fertilization with silicate, nitrate, or phosphate was attempted.

As a daily routine, the depth of the plastic sphere was adjusted to a constant level by the suitable addition or removal of air from the buoyancy bell. The temperature and salinity of the water at the center of the bag and at the same depth in the sea alongside the bag was measured first thing each morning. A plastic Nansen reversing bottle with protected thermometer was employed to measure temperature and to take samples for salinity determination.

At about 0900 hr each day samples were taken in a plastic Van Dorn bottle and used for water analyses. The water was filtered at once through a Millipore HA membrane filter and suitable aliquots of the filtrate stored at –20°C in the dark until ready for analysis. Separate samples were drawn for oxygen, pH and alkalinity determinations, which were performed immediately.

Measurements of the attenuation of light in the sea and in the bag were carried out in the mid-morning and mid-afternoon. At about 1030 hr an 8-L sample was taken in Van Dorn bottles and well mixed in a polyethylene aspirator. Nitrogen was passed through the water, if necessary, to reduce the concentration of dissolved oxygen to below the saturation level and then the water was passed into 300-ml B.O.D. bottles for the measurement of photosynthesis. Some bottles were placed in a stainless steel holder suspended at a point 2 ft above the center of the bag on the stirring line and were left in the bag for 24 hr (1200 to 1200). Other bottles were illuminated in a laboratory incubator under constant (artificial) light conditions for a period of less than 6 hr.

At 1330 the water in the bag was sampled for the analysis of particulate matter. Samples were taken at the mid-point of the plastic sphere, using a Van Dorn bottle, and were mixed and stored in a 20-L polyethylene bottle. Filtration of this sample was commenced without delay.

Occasionally samples were taken from several depths and analyzed for pigment and dissolved oxygen, *etc.* in order to check on the uniformity of distribution of metabolites within the sphere.

Determination of dissolved substances

Full details of all analytical methods are given by Strickland and Parsons (1961). The following brief summary indicates the general techniques that were used.

Salinity was measured by an electrical conductivity salinometer, oxygen by the Winkler method, pH using the Beckman model GS meter and a glass electrode standardized by standard phosphate buffer at pH 6.87, alkalinity by the Anderson-Robinson method (Anderson and Robinson 1946), silicate by the method of Mullin and Riley (1955a), nitrite by the method of Bendschneider and Robinson (1952), nitrate by a modification of the procedure described by Mullin and Riley (1955b), and phosphate-phosphorus by a modification of the method of Robinson and Thompson (1948). Total soluble phosphorus was determined as phosphate after wet oxidation wtih perchloric acid (*cf.* Hansen and Robinson 1953). Soluble organic phosphorus was measured as the difference between the total phosphate in a sample passed through a Millipore HA filter and the inorganic phosphate in the same sample. Particulate and soluble reactive iron and ammonia were estimated as outlined by Strickland and Austin (1959). The method used for manganese was based on the catalyzed oxidation of the leucobase of malachite green by periodate, as developed by one of the present authors (see also Yuen 1958). Copper was determined as the colored diethyldithiocarbamate complex which was extracted directly into carbon tetrachloride. Soluble organic nitrogen was estimated by evaporating 25 ml of filtered water and subjecting the residue to Kjeldahl oxidation. The ammonia in the residue was then determined absorptiometrically, as above, and the result corrected for any ammonia found in the initial sample.

The change of total carbonate-carbon in the bag was obtained from precise measurements of the change in pH, knowing the total alkalinity of the system, and using a modification of the Buch tables given by Strickland and Parsons (1961) with graphical interpolations. As only changes in carbonate concentration were being measured in an isolated system, a very precise alkalinity determination was not necessary (*cf.* Strickland 1960).

Determination of particulate matter

Details of all methods (with the exception of the enumeration of phytoplankton cells) are described by Strickland and Parsons (1961). A brief summary of the procedures will be given here for the convenience of the reader.

All sea water samples were first filtered through a nylon net with a mesh size of 300 μ to remove any large particles of extraneous detritus introduced during sampling. The concentration of the particulate matter was accomplished by filtering the water through a Millipore type AA membrane filter, previously treated with magnesium carbonate (Parsons and Strickland 1959). When sufficient sea water had been filtered the Millipore filter was removed and the concentrated particulate material was washed from the surface with a small quantity of 3.0% saline. The suspended particulate matter and magnesium carbonate were then further concentrated by centrifugation and the supernatent liquid was discarded. The resulting residue was used for the determination of carbon, carbohydrate, crude fibre carbohydrate, protein, fat, nitrogen, and phosphorus.

Plant pigments

Chlorophyll *a*, *b*, *c* and carotenoids were measured as described by Richards with Thompson (1952) using the Millipore filter technique to concentrate the phytoplankton cells described by Creitz and Richards (1955). Extinctions were measured in 10-cm cuvettes. A correction for turbidity was made using the extinction at a wave length of 7,500 A, multiplied by a predetermined factor to allow for increasing light scatter at decreasing wave lengths.

Carbon

Particulate carbon was determined by wet oxidation with potassium dichromate in concentrated sulphuric acid by a modification of the procedure described by Johnson (1949). The decrease in the extinction of dichromate solutions at a wave length of 4,400 A served as a measure of the amount of oxidizable carbon.

Protein and nitrogen

Particulate protein was determined by two methods. The first involved Kjeldahl digestion of the sample and an estimation of the ammonia using Nessler's reagent, Johnson (1941). Nitrogen values determined by this method were converted to protein values by multiplying by a factor of 6.25. The second method involved the hydrolysis of the particulate protein with 6 N HCl at 100°C for 6 hr, followed by a colorimetric determination of the hydrolyzed protein using 2,5-hexanedione and Ehrlich's reagent, as described by Keeler (1959). The method was standardized using pure dry casein. Nitrogen values were obtained by dividing by a factor of 6.35. Unlike the tyrosine methods for protein, the Keeler technique involves the reaction of 8 amino acids and ammonia in a hydrolysate. However, results still depend on the amino acid spectrum of the algal proteins which differs from species to species and which cannot be expected to resemble that of casein exactly.

Carbohydrate

Particulate carbohydrate was determined by the anthrone reaction using the reagents employed by Hewitt (1958) and measuring the light absorption of the colored solution developed at a wave length of 6,200 A. Additional absorptions were measured at wave lengths of 5,500 A and 6,500 A in order to determine, qualitatively, the presence of hexuronic acids and pentoses. The method was standardized with glucose. "Crude Fibre Carbohydrate" was also determined by the anthrone reagent on a concentrated residue after consecutive treatments with 1.25% sulphuric acid and 1.25% sodium hydroxide in a boiling water bath.

Fat

The amount of fat present in the particulate matter was determined by a modification of the method described by Mukerjee (1956). The method specifically measures fatty acids. Results obtained by the use of this method may be less than values obtained by an ether extraction technique. The method was standardized with stearic acid.

Particulate phosphorus

Particulate phosphorus was determined as phosphate after a wet oxidation with perchloric acid (*cf.* Hansen and Robinson 1953). The particulate phosphorus is defined as that retained on a Millipore HA filter.

Enumeration of phytoplankton cells

Phytoplankton cells were counted using a Unitron inverted biological microscope. Samples of sea water were preserved in 2% neutral formalin and counts were made on 5- or 10-ml aliquots of sea water without prior concentration of the cells. Samples were counted at magnifications of 50×, 100×, and 400×. Duplicate counts of 20 fields made at equal intervals across the diameter of the cylinders (2.6 cm in diameter) normally agreed within 10%. An approximate estimate of the average cell volume of each of the principal species present was made by direct microscopical measurement of the dimensions and shapes of the cells. All cells were counted, living or dead, provided that they were sufficiently intact for ready identification.

Phytoplankton cell numbers, as such, were of limited interest because of the enormous size differences between one species and another. For better comparison total cell numbers were converted to total cell volume (μ^3) by the following experimentally determined factors: *Nitzschia delicatessima*, 500; *Skeletonema costatum*, 1,900; *Glenodinium danicum*, 3,600; *Thalassiosira nordenskiöldii*, 11,000; *Thalassiosira rotula*, 35,000, and *Gyrodinium fulvum*(?), 150,000. (In the other experiment, of only 5 days duration, the following additional factors were used: *Coscinodiscus radiatus*, 12,500; *Coscinodiscus centralis*, 7,350,000.) To convert the resulting algal volumes to organic carbon a factor of 0.1 was employed (*cf.* Strickland 1960). The evaluation of total carbon from cell counts by the above procedure is, at best, very approximate. However, results of cell counts alone can be most misleading unless some such attempt is made to translate cell numbers into a measure of biomass.

Measurement of radiant energy

Incident radiation was measured with an Eppley pyrheliometer situated on a hill about 400 yd from the plastic sphere. Photosynthetically active radiation (3,800–7,200 A) was assumed to be 50% of that measured by the pyrheliometer, and is reported in cal/cm^2/min (ly/min, *cf.* Strickland 1958).

Mean daily light intensities are the averages of the values recorded between sunrise and sunset.

Optical extinction coefficients of the water in the bag and the surrounding sea were determined from measurements with an underwater light meter and deck cell. Both cells were fitted with 2 mm of Schott BG. 12 blue glass, giving a maximum response of photocell-filter combination near to a wave length of 4,300 A. The attenuation of photosynthetically active light was estimated from these extinction coefficients using the data collected by Jerlov (1951).

Mean daily light intensities at the center of the plastic bag were calculated using the extinction coefficient of the surrounding water through the first 3 m, and then the extinction coefficient of the water in the sphere through a further 3 m. When the morning and afternoon extinction coefficients differed markedly their mean value was used. It must be emphasized that the radiation intensities obtained by this method are only very approximate but relative values from day to day are strictly comparable.

Measurement of photosynthesis

Measurements of photosynthesis were carried out by both the light and dark bottle oxygen method and a modification of Nielsen's C^{14} technique (see Strickland and Parsons 1961). All rates were measured using duplicate "light" bottles and a single "dark" bottle. The initial oxygen concentration was determined so that the net oxygen changes and respiration values in dark bottles could be evaluated.

The constant light intensity incubator used in this work held the B.O.D. bottles lying horizontally in black painted trays and cooled by running tap water. Samples were illuminated from above by blue and green fluorescent lamps emitting 0.1 ly/min of radiation as measured by a pyrheliometer.

No account was taken of possible periodicity effects in photosynthesis as these are thought to be quite small at latitude 49°N (Doty 1959).

One would suppose from the work of Fogg (1958) and others that $C^{14}O_2$ uptake experiments of more than about 6-hr duration would give errors due to the secretion of labelled material. It was impracticable, however, to conduct a series of short exposures in the sphere throughout each day and extrapolation from a single 6-hr period to a full 24 hr of light and dark was too inaccurate to be of use. Therefore we resorted to 24-hr radiocarbon experiments, on the assumption that at low light intensities the results would be tolerably accurate, as already reported by Rodhe (1958). We confirmed this experimentally by comparing the sum of the uptakes in two half-day exposures with that of a full 24-hr exposure in the bag. At the higher light intensity in the incubator there was very poor agreement between the rates found with 6- and 24-hr exposures, again supporting the findings of Rodhe (1958). At this higher light intensity plants were metabolizing many times more rapidly in the incubator than in the bag, and dividing once or twice in the 24-hr period of continuous illumination. Under such conditions the secretion of metabolites and "light fatigue" of the plant cells would be expected to be pronounced.

Experiments to measure the relative rate of photosynthesis as a function of illumination were carried out with the C^{14} method and the wire-screen attenuators described by McAllister and Strickland (1961). Natural daylight was used as a source of illumination during the period between about 1000 and 1400 on cloudless days. The light intensity over this period was nearly constant.

RESULTS

The salinity of the water in the bag was 27.00‰ and its temperature initially 11°C. The transmission of light through the polyvinyl chloride plastic and its supporting coarse-mesh nylon netting was greater than 75%, excluding any effects from direct shading by the supporting buoyancy bell and cradle. Temperature changes in the water inside the sphere closely followed those of the surrounding sea water. This is illustrated by Figure 1 which shows the total daily amount of solar radiation at the sea surface, the temperature at 5 m depth in the sea, and

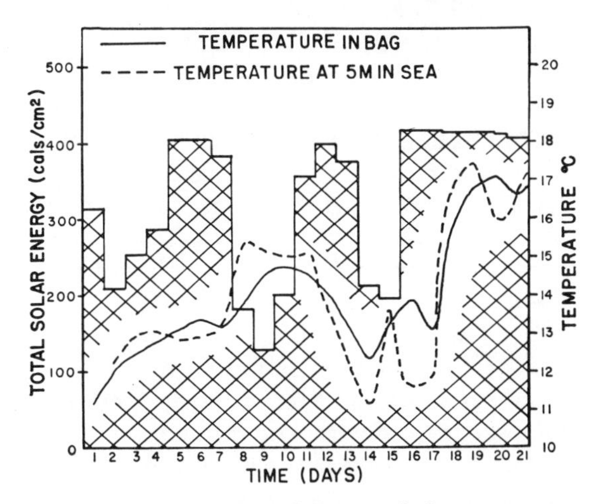

FIG 1. Total solar radiation and the water temperature inside and outside the plastic sphere.

the temperature at 5.5 m in the bag. It is interesting to note the delay of some 4 days before the sea temperature showed the influence of large changes of incoming radiation. The temperature of the water in the bag followed that of the sea outside with a lag of only about one day.

Most of the data obtained during the present experiment are summarized graphically, and generally the limits of variation are indicated by vertical lines or by the size of the plotted points. The confidence limits embrace the 95% probability region or are thought to approximate this value. Certain important ratios of metabolites and of metabolites to photosynthetic rates, *etc.* are collected in the tables and will be discussed later. The precision of these ratios varies greatly but their significance may generally be judged by the number of significant figures quoted. Some other results appear to have little obvious significance and will be summarized briefly in this section.

No chlorophyll *b* was detected in any analyses and the level of animal carotenoid (astacin) pigments was very low and erratic. The apparent presence of the latter may have been an analytical artifact. Copper was present in a concentration of 0.048 μg–at Cu/L which did not change throughout the three weeks of the experiment. The concentration of reactive manganese varied between 0.01 and 0.04 μg–at Mn/L with little obvious pattern, although there was a maximum just before the main phytoplankton growth period started (day 8). Soluble reactive iron changed little in concentration being in the range 0.07 to 0.12 μg–at Fe/L. The amount of particulate reactive iron was quite large. Initially present at 0.7 μg–at Fe/L it decreased in quantity to one-third of this value by day 12. The concentration thereafter increased to the initial level. The significance of this behavior, if any, is not clear. The decrease roughly paralleled that of soluble nitrogen but more experimental work is necessary before definite conclusions are possible.

The initial concentration of soluble organic phosphorus in the sphere was appreciable (0.3 μg–at/L) and seems to have increased slowly during the growth of the phytoplankton crop and then more rapidly as the crop decayed, reaching a value as high as 0.55 μg–at/L. Three anomalous results were obtained in this series of determinations and more work is required before any definite pattern can be stated. The concentration of soluble organic nitrogen showed a similar increase throughout the experiment (about 2 up to 7 μg–at N/L) but there was no marked correlation with the algal growth behavior. The precision of this method is relatively low (±0.8 μg–at N/L or worse) which may have masked any subtle relationships.

In a well-stirred conserved system the decrease of total dissolved nitrogen or phosphorus must be equal to the increase found in the particulate matter. The analytical results of losses and gains in the present work gave tolerable agreement until after day 17. Then serious discrepancies resulted which showed that stirring was insufficient to keep dead and dying cells in suspension. The decrease of particulate matter after day 17, therefore, represents both the true dissolution of solids and their removal from the bag center by settling. Although analytical results on the particulate material, *per se*, are still valid, no conclusions can be drawn from *changes* of concentration (as sampled at the center of sphere) for the last 5 days of the experiment.

The useful life of the plastic bag was ter-

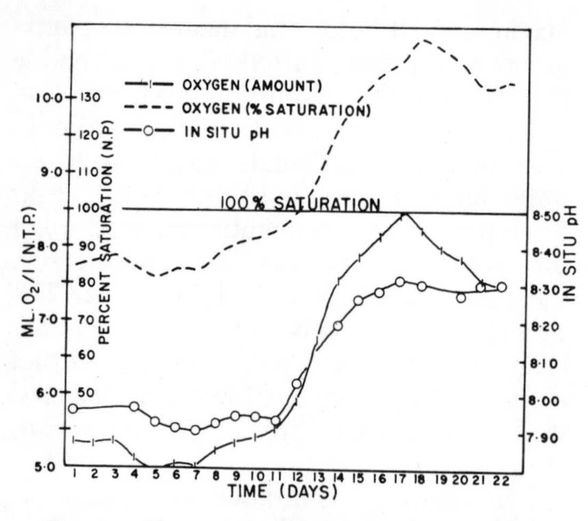

FIG. 2. Changes in pH and dissolved oxygen.

minated by tears near the neck (at day 5 in one trial and day 23 in the main experiment). As the salinity of the surrounding sea water was generally 2 or 3‰ less than the water in the bag the entry of outside water was easily detected by precise (conductivity) salinity measurements.

DISCUSSION

The net production of organic matter during the course of the main experiment is illustrated by Figure 2, which records the change of pH and oxygen concentration in the plastic sphere. The percentage saturation of oxygen (at atmospheric pressure) shows the striking degree of supersaturation often reported in the literature and attributed to the effect of photosynthesis. A value of nearly 150% was recorded in the present work. The subsequent decrease of oxygen concentration (after day 17) was mainly due to physical desaturation processes as no corresponding decrease of pH occurred.

The main phytoplankton species present at various stages in the experiment are shown in Figure 3. The use of algal weight (derived from algal volume assuming a mean density of unity) enables one to comprehend the relative contribution of each species to the overall primary production. The total cell count exceeded that of the named species by up to 25% but the unnamed species represented a much smaller addition to the biomass than numbers alone would indicate. *Asterionella japonica* was the most conspicuous of the minor constituents.

The growth rate of each species was not uniform but over the period of most constant growth mean division times could be evaluated. Values for the diatoms *Skeletonema costatum*, *Thalassiosira rotula*, *Thalassiosira nordenskiöldii*, and the dinoflagellate *Glenodinium danicum* were very similar (30–38 hr). The dinoflagellate *Gyrodinium fulvum* (?) was noticeably more rapid in growth (22 hr) and for an initial period *T. rotula* increased with a doubling time of only 19 hr. The cell numbers of all species eventually showed a decrease which we attribute to the rapid sinking of dead or dying cells (ref., *e.g.*, Gross and Zeuthen 1948; Steele and Yentsch 1960). *T. rotula* passed its peak of metabolic efficiency somewhere between days 8 and 11. This could reflect either a marked sensitivity to the concentration of nitrate or be the result of an increase of water temperature from 12 to 15°C. One cannot rule out the possibility of an inhibition by the increasing concentration of *Skeletonema* cells. *Nitzschia delicatessima* was the only organism to grow to any degree under markedly unfavorable conditions (a doubling time of over 80 hr) and this species commenced to die off soon after the depletion of nitrate at the end of day 13 (Fig. 4). *Asterionella japonica* persisted in small numbers throughout most of the experiment but was obviously living under conditions so unfavorable as to bring cell division almost to a standstill.

The growth rate of *S. costatum* decreased as soon as nitrate disappeared from solution but, statistically, the cells were capable of one further division. *T. nordenskiöldii* grew more vigorously and divided twice before the cells commenced to die. By contrast the two dinoflagellates only began to increase rapidly after the nitrate was exhausted and *Gyrodinium fulvum* survived in fair condition until the end of the experiment. The above observations give a striking illustration of the kind of conditions that bring about a "succession of species" in nature (*cf.* Conover 1956; Margalef 1956). We are unable to explain the kinks in the curves for *T.*

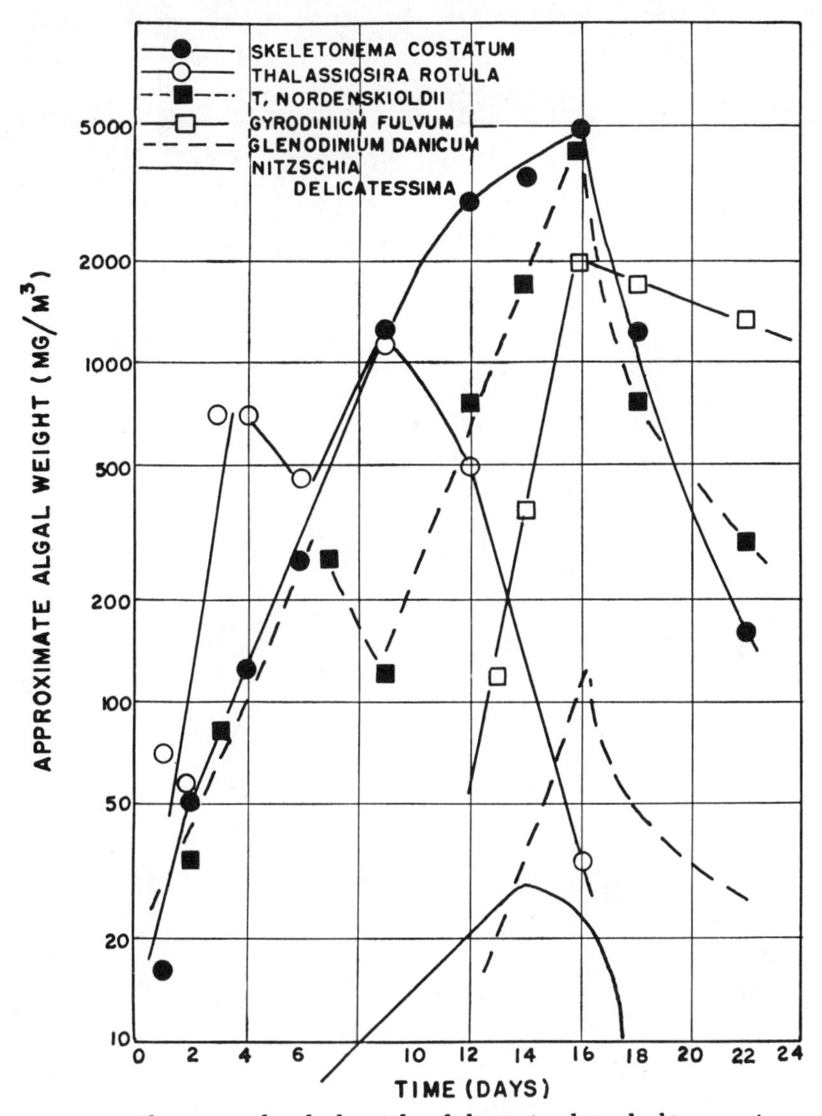

FIG. 3. Changes in the algal weight of the main phytoplankton species.

rotula and *T. nordenskiöldii* (there was a similar very high point for *Skeletonema* at day 3, not plotted). There must have been an error either in sampling or in counting brought about by a change in mean chain length, and hence a contagion factor, for the species concerned.

Iron, copper, and manganese were clearly not limiting nutrients for plant growth and reactive silicate was initially present at such a concentration that it never became depleted, despite the crop being predominantly diatoms. One would infer from our results that the initial Si:N ratio in sea water would have to be less than 2:1 (on an atom basis) for silicon to become a limiting nutrient in a diatom bloom. It is interesting to note the small but significant rise in silicate concentration after day 16 which indicates that a fraction of the silicon in a diatom is very labile (*cf.* Jørgensen 1955a). Subsequently there was only very slow remineralization. In considering this fact and the remineralization of nitrogen (see later) it must be remembered that the present experiment is unrepresentative of nature in that no regeneration processes could occur as a result of grazing (*cf.* Cushing 1959).

The initial ratio of nitrogen to phosphorus in solution was 9 (on an atom basis) or 10 if

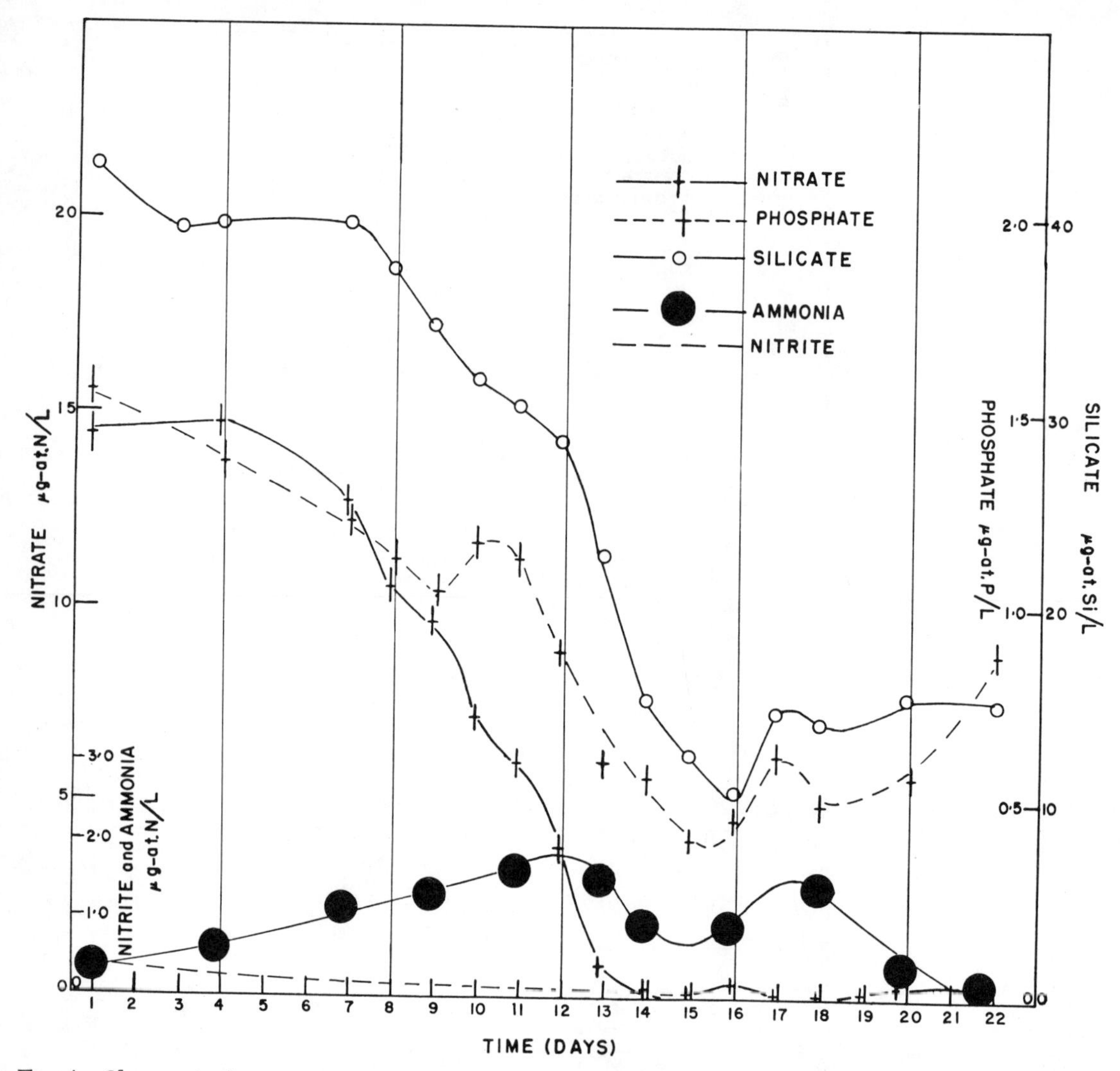

Fig. 4. Changes in the concentration of nitrate, nitrite, ammonia, reactive phosphate and reactive silicate.

we consider only nitrate and inorganic phosphate. On the assumption that growing plants withdraw these elements from sea water in the mean atomic ratio of 15:1 (Cooper 1937, 1938) one would expect to find nitrate removed before phosphate, as was indeed the case (Fig. 4). Furthermore, the tendency for phosphate to continue to decrease in concentration after all the nitrate has gone, as found in cultures of open Pacific Ocean water by McAllister, Parsons, and Strickland (1960), was also shown in this experiment with coastal water. It will be seen from Figure 4 and from the analytical data given in Table 2, that the ratio of nitrogen to phosphorus in the growing cells was variable, being quite low during the first part of the experiment and then increasing to approach the "classical" 15:1 value.

Phosphorus was not a "limiting nutrient." The comparatively labile nature of its incorporation in phytoplankton (Gest and Kamen 1948; Goldberg, Walker and Whisenand 1951; Rice 1953; Rigler 1956) and the rapidity with which it can leave dead cells (Hoffmann 1956; Golterman 1960) is reflected by irregularities in the curve in Figure 4. In particular there was a small but definite *increase* in the amount of the element in solution between days 9 and 10. Cells appear to have stored an abnormally large amount of phosphorus (see Table 2), some of which

TABLE 1. *Ratios involving photosynthetic rates*

Day	Ratio: Rate. (O_2 gross) / Rate. (C^{14}) ($PQ=1$)		Ratio: Rate. mg C/hr (O_2 gross) / mg chlorophyll *a* ($PQ=1.3$)	Productivity index[2] from gross O_2 ($PQ=1.3$)
	Bag	Incubator	Incubator	Incubator
1	–	–	1.3[1]	–
2	–	–	1.2[1]	–
5	–	3.3	1.65	–
6	2.4	–	–	–
7	1.8	–	–	–
8	2.1	3.3	1.6	–
9	2.7	–	–	–
10	1.9	2.9	1.75	0.9
11	1.9	3.3	1.6	0.75
12	1.9	3.3 *Mean 3.3*	1.45	0.75
14	1.7	–	1.6	0.75
15	2.2	2.0	1.65	0.75
16	1.5	1.7	1.05	0.55
17	2.0	1.4	1.05	0.4
18	1.5	1.9 *Mean 1.75*	0.95	0.2
19	1.7 *Mean 1.95*	–	–	–

[1] Values using carbon-14 rate measurements.
[2] Carbon measured by direct chemical method.

was then liberated during a period of low illumination and increasing temperature. Phosphorus was obviously lost more rapidly than nitrogen from dead cells and this is also shown by the ratios given in Table 2.

Nitrite nitrogen, initially present at 0.35 μg–at N/L decreased to one-fifth of this value in the first 14 days. It is not possible to say whether or not nitrite was incorporated into phytoplankton (*cf.* Harvey 1953; Bongers 1956) or oxidized by bacteria.

Whereas nitrate nitrogen was consumed rapidly and in a more or less regular fashion it is interesting to see that there was an increase in ammonia concentration (Fig. 4) until day 13 and a significant increase in the amount of soluble organic nitrogen. Ammonia was clearly not used in preference to nitrate when the latter was in great excess (*cf.* Harvey 1940) although when nitrate became depleted there was some indication that ammonia could be scavenged from solution and might, in this instance, have been providing *Gyrodinium fulvum* with its most readily accessible supply of nitrogen. There was little or no indication of any nitrogen remineralization, but this is not surprising considering the time scale suggested by the work of Von Brand, Rakestraw and Renn (1939), Spencer (1956) and others and the fact that some plant cells were still living and removing available forms of the element until the end of the experiment.

As noted by several workers (*e.g.*, Harvey 1953; Rodhe 1957) the growth rate of phytoplankton can be quite rapid even when all nitrate has been removed from a medium and cells exist for one or more divisions on their reserves of protein. The present work gives striking illustration of this fact as nearly half of the final organic matter in the phytoplankton bloom was produced after nitrate depletion. Any attempt to predict the kinetics of natural population growth using the concept of a "limiting concentration" of nitrogen (or phosphorus) therefore has little meaning unless the past history of the phytoplankton crop is also known. There may, of course, be conditions where the supply of nutrients is just balancing consumption but for how long such an unstable equilibrium could exist in nature, especially in coastal regions, remains to be shown.

There is a shortage in the literature of direct comparisons between the oxygen and C^{14} methods of measuring photosynthesis and many of the results quoted are for pure cultures. In the present work ratios were measured by the two methods whenever practical, both at the center of the plastic bag for 24 hr and in the constant light incubator for shorter periods. In Table 1 the ratios between *gross* oxygen liberation and radioactive carbon uptake are presented with no assumptions as to what is being measured.

The ratios are disconcertingly large, but they can be reduced if a large photosynthetic quotient (*PQ*) is allowed and if the $C^{14}O_2$ method is assumed to measure net photosynthesis in cells which have a high respiration rate. The highest *PQ* and respiration values normally used are about 1.35 and 15% respectively which would give a ratio of 1.6 at moderate light intensities. (This subject is fully reviewed by Strickland 1960.)

The ratios measured by the bottles at the center of the bag averaged 1.95. They fell into two distinct groups in the incubator, where there was from 3 to 10 times more illumination. Values before day 14 (and nitrate depletion) averaged 3.3 and the four values after day 14 had a mean of 1.75. It is generally supposed that nitrogen deficient cells have a high respiration rate and hence low rate of net photosynthesis (Ketchum, *et al.* 1958) but these last four results do not indicate that the C^{14} results are usually low, unless *PQ* values are also very small. The ratio of 3.3 in a nitrate rich medium is even less easy to explain. One is compelled to assume very high values for the *PQ* of diatom cells taken from a nitrate rich medium at low illumination and exposed for a few hours to comparatively high light intensities. The whole question of a correct *PQ* value is of much greater importance in marine studies than is generally admitted, especially if one wishes to make an accurate estimate of carbon uptake from oxygen measurements.

It is interesting to note that the depletion of nitrate had no significant effect on the 24-hr ratios determined in the plastic bag. These values may be slightly high due to the prolonged period of $C^{14}O_2$ uptake.

With all these results we suspect that ratios are greater than they should be, even having regards to likely experimental errors and high *PQ* and respiration values. It is

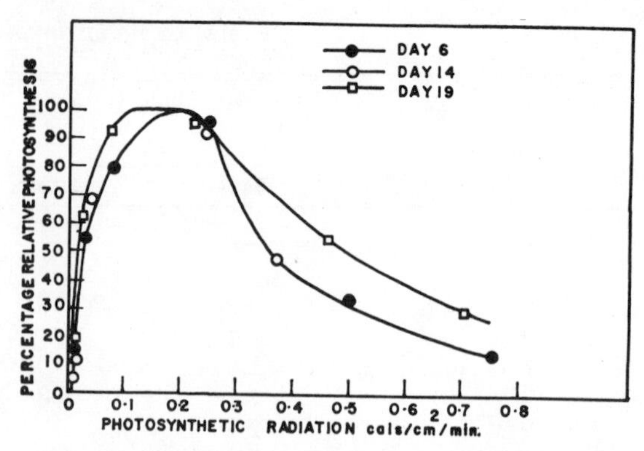

FIG. 5. Relative rate of C^{14} assimilation as a function of radiation intensity.

possible that in coastal water the C^{14} method is giving low results due to the photosynthetic assimilation of non-labeled forms of carbon (Smith, Tatsumoto and Hood 1960; and see also Strickland 1960 for a discussion).

The relative photosynthetic rates (f) *vs.* illumination values (I) for the populations in the sphere on days 6 and 14 have been plotted on the same curve in Figure 5. This is very similar to that given by Ryther (1956). At day 19, when *Gyrodinium fulvum* comprised about half the population, the values lay on a slightly different curve. There was then the greater tolerance to high light intensities to be expected from a dinoflagellate (*e.g.*, Ryther 1956) but the curve at lower intensities still resembled that for the diatom population. It will be seen that the light intensity of 0.1 ly/min used in the incubator was at approximately the optimum value. Mean light intensities at the center of the bag did not exceed 0.055 ly/min and photosynthesis was therefore always roughly proportional to light intensity.

Back extrapolation of f *vs.* I curves to zero light intensity should indicate the amount of plant respiration, if the assumptions made by Nielsen and Hansen (1959) are justified. The values so obtained, about 10%, are reasonable.

Figure 6 shows the increase in carbon in the plastic bag from day 1 to day 22, as measured by both oxygen and C^{14} methods, again with no assumption of *PQ* values or the significance of the C^{14} measurements. The increase on day 13, and hence the total for all subsequent days, may be low as there

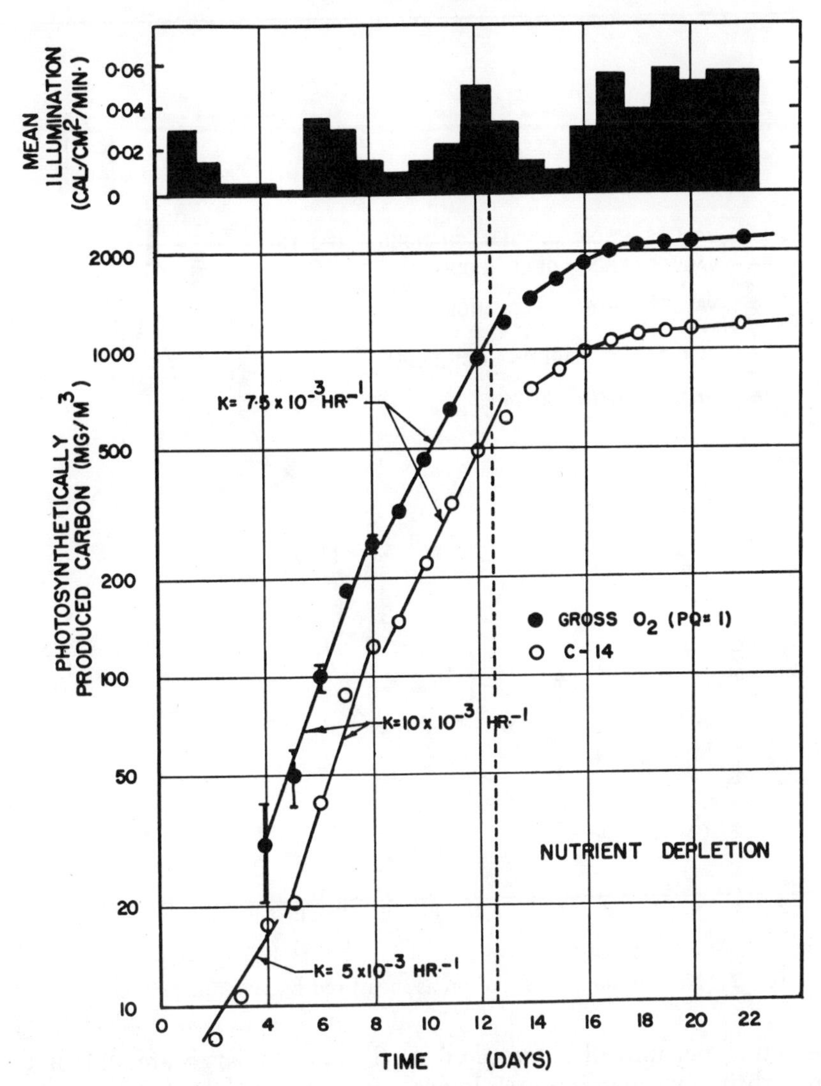

FIG. 6. Photosynthetic carbon fixation in the plastic sphere as a function of time under natural lighting.

was poor stirring that day with a temporary loss of plant cells. The curves are parallel and their slope can be used to evaluate logarithmic growth constants. The values given in Figure 6 (base 10) probably reflect changes in temperature and light intensity (see the histogram for mean daily illumination at the bag center). After the end of day 15, severe nitrogen shortage brought about a reduced growth rate and results no longer have much meaning, because cells were sinking below the sampling point in the bag.

The most fruitful way to examine the effect of nutritional changes on the photosynthetic metabolism of algal cells is to expose them to constant light intensity in an incubator and find either the rate of carbon increase per unit of chlorophyll *a*, or the productivity index, defined by Strickland (1960) as:

$$\frac{\text{hourly rate of carbon increase}}{(\text{amount of plant carbon}) \times (\text{illumination in ly/min})}$$

Both these values are given in Table 1, using the gross photosynthetic rate measured by oxygen evolution and an assumed *PQ* of 1.3. There was no significant change in the photosynthetic potential of the cells from day 1 to day 15, having regards to the general magnitude of experimental errors.

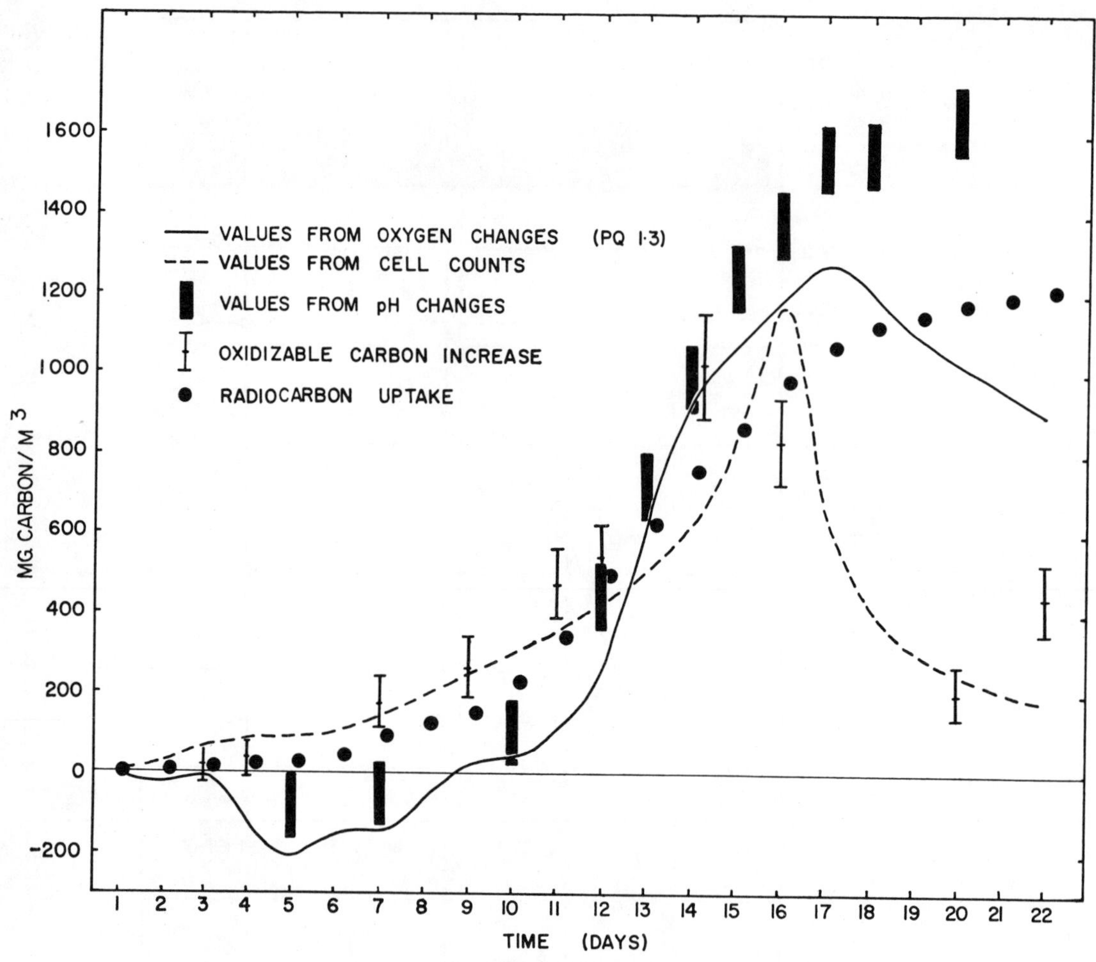

FIG. 7 Net production of carbon as measured by five independent methods.

The gross rates, per unit of chlorophyll *a* at optimum lighting, were very much lower than the rates reported by Ryther and Yentsch (1957), although they were within the range found in the literature (Strickland 1960) and comparable to the values found by McAllister, *et al.* (1960) in the open Pacific Ocean.

These productivity index values are the first of their kind to be reported, and are near to many of the values inferred from the literature (Strickland 1960). The high value at day 10 may be significant but was not reflected in the corresponding figure for the rate per unit of chlorophyll *a*. Similarly the decrease of productivity index at the end of the experiment was much more pronounced than the corresponding decrease in the chlorophyll ratios, but this would be expected because a large amount of detrital carbon was accumulating. In particular it should be noted that the loss of nitrate at the end of day 13 had no effect on either ratio. Only at day 16, when the mean nitrogen content per cell was probably less than a half of its peak value, did the photosynthetic processes, as reflected by these ratios, become seriously impaired. At the same time cells commenced to settle out rapidly (Fig. 3).

There was no evidence in our experiment for the situation reported by Ryther, *et al.* (1958) where the growth of a diatom bloom in nitrogen deficient water was static as the result of a sizable amount of photosynthesis during the day being compensated for by respiration during the hours of darkness.

The net production of organic particulate carbon in the plastic bag was estimated by five methods and the results are shown together for comparison in Figure 7. The full line shows the values calculated from changes of dissolved oxygen in the bag, assuming a *PQ* of 1.3. Results are probably low after day 15 or 16 due to physical losses of oxygen from the supersaturated system. The dotted line is a measure of plant carbon derived from cell counts and, as mentioned earlier, is only very approximate. The bars show the range of individual results estimated from changes of total carbon dioxide content. This is the most direct method of all but unfortunately not as precise as could be wished. The solid circles mark uncorrected data calculated from the uptake of radiocarbon. These results were used in preference to the gross oxygen photosynthesis data because they formed a more complete set, and did not involve assumptions as to respiration rates and photosynthetic quotients. However, it is by no means certain that C^{14} data represent true net photosynthesis; they may well be too low. Finally, there are the experimentally measured oxidizable carbon results corrected for initial detritus. These results are entered as horizontal dashes with the 95% confidence limits indicated by vertical lines. Any excretion of soluble organic compounds during plant growth would make these results lower than they should be and this would also influence the ratios in Table 2.

Ignoring the measurements derived from dissolved oxygen and pH, the agreement between the other three estimates was as good as could be expected for much of the time. The two direct methods, carbon analysis and algal volume, finally showed decreases due to the settling of cells, although there was appreciable net synthesis going on inside the bag until at least day 20.

The various measurements of carbon during the first few days of this experiment gave the most anomalous results and they merit closer investigation. The dissolved oxygen content of the bag first *decreased*, representing the consumption of over 200 mg C/m³, but there was no corresponding decrease of particulate carbon. A wall effect would not be expected so soon in equipment of this size (in the other experiment the oxygen decreased steadily from the first day) so presumably, much of the oxidation was by nonproliferating bacteria in suspension, using a substrate of dissolved organic carbon. This was borne out by the high but fairly constant respiration values found in the dark bottles used in photosynthesis experiments.

Not only was there no decrease of particulate matter, corresponding to the net consumption of oxygen in the bag, but the increase in photoplankton carbon, as estimated by cell volumes and changes of chlorophyll (see later) was appreciable and very rapid during the first 4 to 5 days. This increase, mainly in *T. rotula* and *Skeletonema*, could not be accounted for by photosynthesis, as the radio carbon uptake data predicted very much smaller values for carbon (one quarter or less). The most likely explanation, therefore, is that during the period of very low illumination at the start of this experiment (see Fig. 6) heterotrophic growth from dissolved organic carbon was a large fraction of the total growth. The effect was not detected in the 5 days of the second experiment, when the illumination was quite high. But at the beginning of the main experiment, a very heavy bloom of flagellates in the surrounding water prevented light entering the bag for several days. It is interesting to note that what little photosynthesis did occur during this period took place in a normal manner. The photosynthetic rate, per unit of chlorophyll *a*, in the incubator had about the same value as found during the rest of the experiment (Table 1).

Heterotrophic growth of algae is well established (Saunders 1957) and has been studied in marine diatoms by Lewin and Lewin (1960). Conversion of dissolved substrate to algal carbon can be high if the behavior of *Chlorella* is representative (Myers and Johnston 1949; Samejima and Myers 1958). Several workers have stressed the probable importance of heterotrophy in phytoplankters (Rodhe 1955; Wood 1956; Bernard 1958) but we believe the present work to be the most direct demonstration of its occurrence in marine phytoplankton under

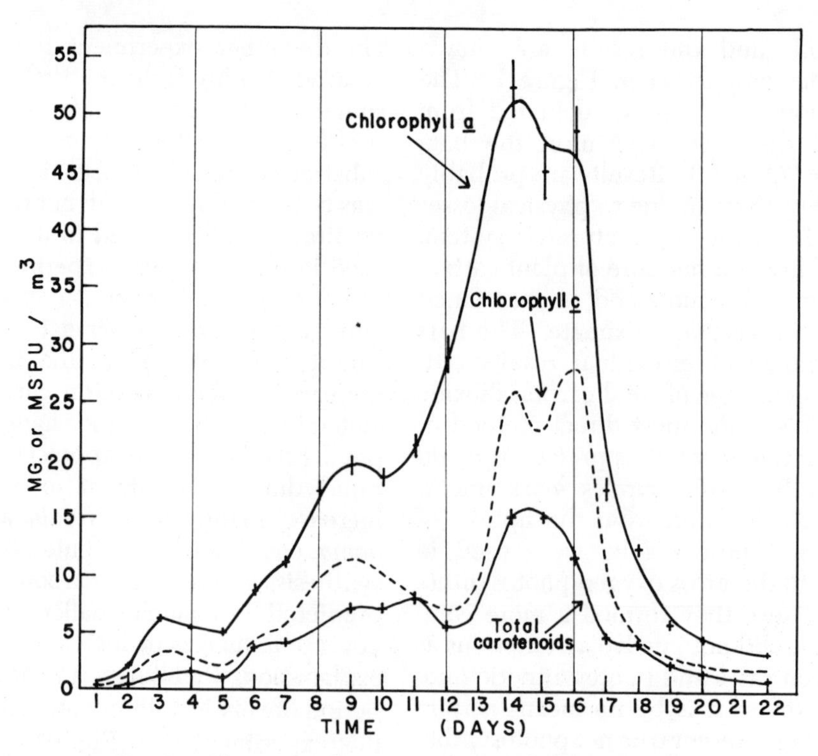

FIG. 8. Changes in the concentration of plant pigments.

conditions resembling those found at the base of the euphotic zone.

Variations of chlorophyll *a*, chlorophyll *c*, and total carotenoid concentration at the center of the bag are shown in Figure 8. The chlorophyll *c* curve roughly parallels that for chlorophyll *a* for most of the experiment, giving values about 50–60% of the chlorophyll *a*, in terms of MSP units and milligrams respectively. Between days 11 and 13, however, the ratio of chlorophyll *c* to chlorophyll *a* was much smaller, decreasing to 25%. This period coincided with the predominance of *Skeletonema costatum* in the bloom and we have found (unpublished work using pure cultures) that this species has what is probably an abnormally high chlorophyll *a* content and a low chlorophyll *c* to chlorophyll *a* ratio for a diatom. The large chlorophyll *a* content of *Skeletonema* would explain the very high maximum value found in the present experiment (50 mg/m^3). In the open ocean, a natural culture in water of essentially the same initial nitrate concentration bloomed to only 26 mg/m^3 chlorophyll *a* at its peak (McAllister, Parsons and Strickland 1960).

The first few days of the experiment are particularly interesting. In 3 days the chlorophyll *a* concentration increased from 0.4 to 6 mg/m^3. If we estimate carbon from cell volumes, the ratio of carbon to chlorophyll *a* is found to change from 25 to 11 in the first 3 days, which is reasonable for a surface population recovering its pigment content in a nutrient-rich, relatively dark environment. The remainder of the chlorophyll increase must be associated with a rapid development of cells during this initial period. Ratios calculated using photosynthetically produced carbon data were impossibly low which, as mentioned earlier, supports the suggestion of some heterotrophic growth.

The sudden decrease of chlorophyll after day 16 was striking and much more than would be expected from a decrease of cell numbers alone. The possible presence of "detrital chlorophyll" in nature is a matter of some importance in field studies. Odum, McConnel and Abbott (1959) could find no

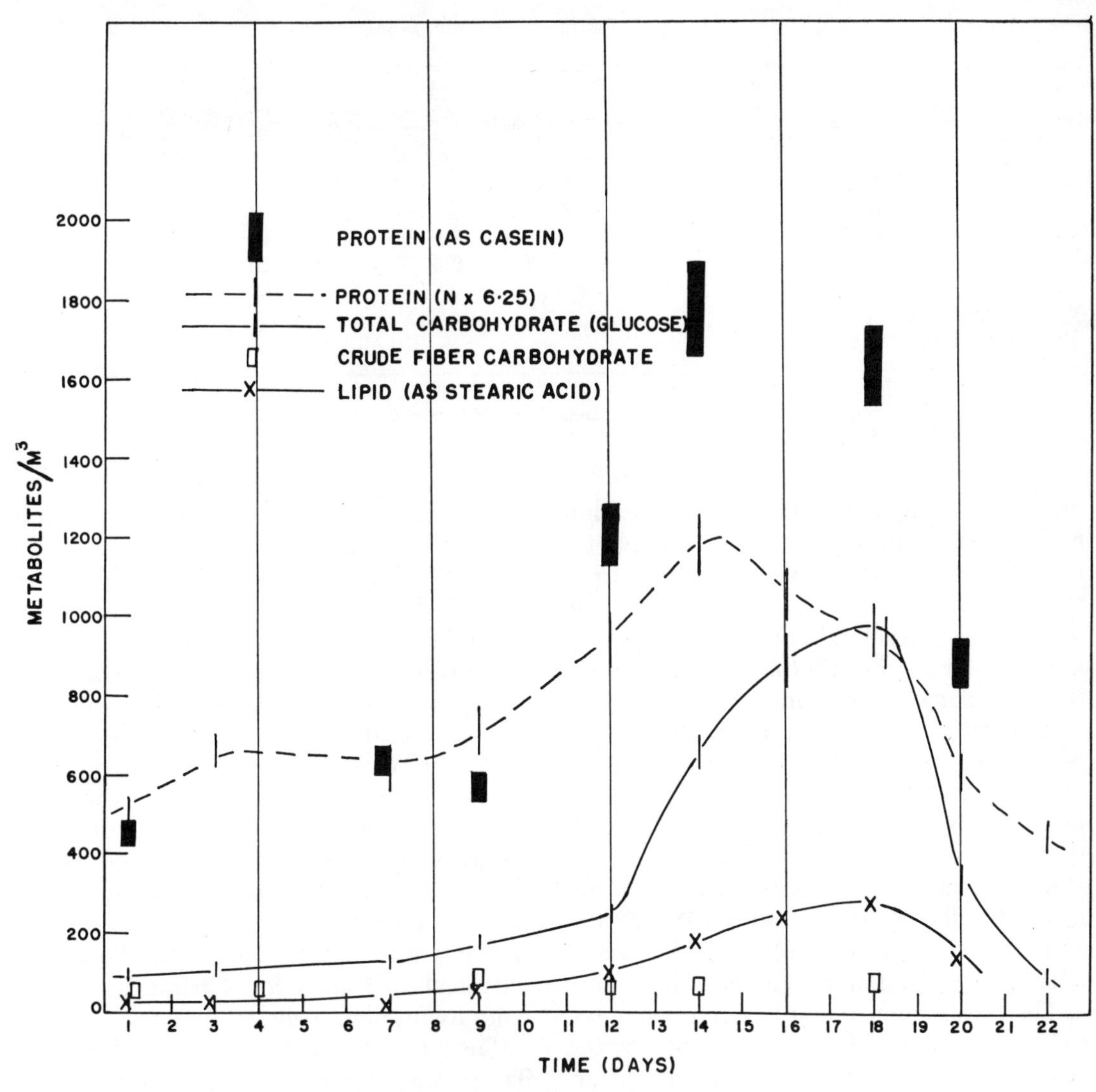

FIG. 9. Changes in the concentration of major metabolites.

spectrophotometric indication of such material. The present experiment seemed to be an excellent opportunity to investigate the matter. We measured the complete pigment spectrum, in 90% acetone, of the populations on day 12 when cells were growing with full vigor and day 18 when many cells were dead or dying but found no evidence for phaeophytin or phaeophorbide pigments. There was a noticeable broadening of the long-wave-length chlorophyll *a* peak in the extract made on day 18, and a slight shift to shorter wave lengths (6,650 A to 6,630 A). We have seen this effect in old laboratory cultures and it is reminiscent of the *in vivo* behavior reported by French and Elliott (1958). There is little likelihood, however, that such small changes would have any practical significance in marine pigment analysis. However, chlorophyllides, which could conceivably occur in decomposing plant tissue, have similar absorption spectra to chlorophylls and it also remains to be proven whether the detrital chlorophyll from faecal pellets is in the form of phaeopigments and hence may give misleading results in nature.

The composition of the particulate matter in the sphere, in terms of major metabolites, is summarized in Figure 9. In order to esti-

TABLE 2. *Ratios involving carbon*

Day	Si[1]/C	Protein/C	Protein[2]/C	Carbohydrate/C	Lipid/C	C/P	C/N	C/N[2]	N[2] (atoms)/P
7	–	0.66	0.89	0.2	–	19	9.5	7	6
9	1.0	0.72	0.76	0.3	–	17	9	8.5	5
12	0.6	0.78	1.4	0.3	0.15	23	8	4.5	11[3]
15	0.65	0.65	1.3	0.7	0.2	28	10	5	12.5[3]
18	–	0.65	1.8	1.35	0.4	40	10	4	25
20	–	0.43	1.9	0.7	0.55	35	15	3.5	23
Value suggested by Strickland (1960)	0.8	–	–	–	–	40±15	–	6±2	14.5

[1] Calculated from the decrease of reactive silicate.
[2] Values from the absorptiometric protein method, standardized with casein.
[3] The ratio had a maximum of about 15 on day 13.

mate the amounts of carbon, protein, carbohydrate, phosphorus, *etc.* in living phytoplankton it was necessary to subtract from the amounts found on a given day, the amounts of each metabolite present initially in the water used to fill the sphere and, further, to assume that this initial "background" concentration of detrital material did not change throughout the experiment. Almost certainly such an assumption is not fully justified, although there is reason to suppose that settling of detritus was comparatively slow. Errors from this source were most significant in the earlier part of the experiment and ratios such as those given in Tables 2 and 3 had little precision until at least day 7 or 9.

There was a considerable discrepancy between the amount of protein obtained by the two chemical methods. Some of this difference undoubtedly arises because the amino acid spectrum of algal protein is not the same as that of casein, but we would not expect from our work on pure phytoplankton cultures (unpublished) that the differences would have been so great. Evidence from the "balance" between dissolved and particulate nitrogen in the bag pointed to the Kjeldahl nitrogen values being too low. A similar poor balance between nitrate lost from solution and nitrogen found in growing algae has been reported by Yentsch and Vaccaro (1958) but we are at a loss to explain the matter and more work is desirable. In the meantime, we have reported both sets of data in Figure 9, Table 2 and Table 3 and suggest that the colorimetric protein values are probably nearer to the "correct" results until at least day 12.

Crude fibre carbohydrate in the bag never exceeded a few per cent of the total phytoplankton carbohydrate. This confirms our (unpublished) findings for pure phytoplankton cultures and for natural coastal populations at the time of heavy diatom blooms. A quite different analysis is found for water containing much detritus, when as much as 80% of the total carbohydrate may be present as crude fibre. (There was about 60% in the detrital carbohydrate initially in the water of the present experiment.) An analysis for crude fibre is one of the best rapid chemical indicators for detritus, as distinct from living plant cells, in the sea.

The most striking feature shown by Figure 9 is the way in which the lipid and the carbohydrate (mainly glucose) in the plant cells increased when nitrate became depleted from solution on day 13 (*cf.* Spoehr and Milner 1949; Bongers 1956; Fogg 1957). Between about days 13 and 17 fats and sugars were almost the sole products of photosynthesis.

The classical conception of diatoms containing large reserves of fat is not borne out by the present work. When cells were in a dying condition in nutrient depleted water their lipid content was higher than earlier in the experiment, but was still less than 20% of the dry weight (see Table 2, assuming about one-third of the dry weight of a diatom is carbon). In the sea, near the surface, nitrate depletion doubtless occurs frequently and

TABLE 3. *Ratios involving chlorophyll* a

Day	C / chl. *a*	C / Carotenoids[2]	N / chl. *a*	N[1] / chl. *a*	P / chl. *a*	Carotenoids[2] / chl. *a*	Protein / chl. *a*	Protein[1] / chl. *a*	Carbohydrate / chl. *a*	Lipid / chl. *a*
7	15	43	1.6	2.2	0.81	0.36	10	14	3	–
9	13	37	1.5	1.6	0.80	0.36	11	10	4	–
12	19	100	2.4	4.2	0.83	0.19	15	27	5.5	3
15	22	70	2.2	4.5	0.78	0.31	14	28	15	4
18	56	160	5.9	16	1.4	0.36	37	100	76	22
20	49	150	3.3	15	1.4	0.33	21	93	33	27
Values suggested by Strickland (1960)	30	40	–	7±3	0.75±0.2	–	–	–	–	–

[1] Values from the absorptiometric protein method, standardized with casein.
[2] Total carotenoids in MSPU.

hence diatoms collected by net tows may often be expected to have their maximum lipid content (*cf.* Barker 1935). However, it is difficult to believe from the present work that the presence of fat assists diatoms in maintaining their buoyancy to any great extent.

The ratios of elements and metabolites to carbon and chlorophyll *a* at various stages in the phytoplankton bloom are given in Tables 2 and 3 respectively. The ratios involving carbon (the direct oxidizable carbon data were used) illustrate the relative changes in composition of the phytoplankton with time. The values involving chlorophyll *a* are important mainly because this pigment is the only convenient measure of plant material in sea water, as distinct from animals and detritus. Data such as those in Table 3 show the limits within which the plant carbon, nitrogen, metabolites, *etc.* can be estimated, knowing the pigment content of a water sample. Clearly each possible mixed algal population in nature, at every stage of its growth, will give rise to different ratios so that the value of any particular one is limited. Nevertheless, as pointed out by Strickland (1960) there are still very few relevant data of this kind available for marine phytoplankton. The results given here are a useful beginning, especially as they apply to a mixed diatom population growing under near-natural conditions. For comparison, the approximate mean values for marine crops suggested by Strickland (1960) are given at the bottom of each table.

It is unfortunate that the precision of the ratios measured during the early part of the experiment is not as high as could be desired, but the data for days 7 and 9 are approximately correct for a diatom population growing under low illumination in nutrient rich waters. The data for day 12 apply to a vigorously growing and rapidly dividing bloom, about to strip the nutrients from the water. On day 15 the results apply to algae at the end of their growth period in nitrogen-deficient water. By day 18 there were many dead and dying cells, and on day 20 little living material was present at the bag center, except dinoflagellates.

The Si:C ratios given in Table 2 are some of the few to be reported for a live natural population and are of the order to be expected from results cited in the literature. The decrease from day 9 to day 12 may be real, supporting the observations made by Jørgensen (1955b).

Most of the other ratios in Tables 2 and 3 are self-explanatory and illustrate the trends that have already been discussed. The increase in the ratios involving chlorophyll *a* after day 15 reflects the rapid loss of this pigment from dead or dying plant material, a loss more nearly paralleled by that of phosphorus and total carotenoid, than by any other of the measured variables.

The co-variation of chlorophyll *a* and the carotenoids is indeed remarkable and supports the detailed observations made by Haskin (1941) on *Chlorella* and our own (unpublished) work on pure marine cultures.

Except for day 12, when there was a preponderance of *Skeletonema costatum*, which we have found to have a relatively low carotenoid content, the ratios were very constant and similar to the values found for mixed coastal populations in Departure Bay at all times of the year. There was no evidence of either the particulate nitrogen values or the ratios in Table 1 being functions of the carotenoid-to-chlorophyll ratio, as has been reported by Ketchum, *et al.* (1958) and Yentsch and Vaccaro (1958). Why there should be this discrepancy between their well-established laboratory results and the present large-scale experiment is not clear. Of course, conditions were not identical and this emphasizes the dangers of extrapolating from laboratory-scale results or even, to a lesser extent, from the present experiment to phenomena taking place in the sea under completely natural conditions.

REFERENCES

ANDERSON, D. H., AND R. J. ROBINSON. 1946. Rapid electrometric determination of the alkalinity of sea water using a glass electrode. Industr. Eng. Chem. (Anal.), **18:** 767–769.

BARKER, H. A. 1935. Photosynthesis in diatoms. Arch. Microbiol., **6:** 141–156.

BENDSCHNEIDER, K., AND R. J. ROBINSON. 1952. A new spectrophotometric method for the determination of nitrite in sea water. J. Mar. Res., **11:** 87–96.

BERNARD, M. F. 1958. Comparison de la fertilité élémentaire entre l'Atlantique tropical african, l'Ocean Indien et la Médeterranée. Compt. rend. Acad. Sci. Paris, **247:** 2045–2048.

BONGERS, L. H. J. 1956. Aspects of nitrogen assimilation by cultures of green algae. Mededel. Landbouwhogeschool Wageningen, **56:** 1–52. C. A., **54:** 18662. 1960.

CONOVER, S. A. 1956. Oceanography of Long Island Sound 1952–1954. IV Phytoplankton. Bull. Bingham Oceanogr. Coll., **15:** 63–112.

COOPER, L. H. N. 1937. On the ratio of nitrogen to phosphorus in the sea. J. Mar. Biol. Ass. U.K., **22:** 177–182.

———. 1938. Redefinition of the anomaly of the nitrate-phosphate ratio. J. Mar. Biol. Ass. U.K., **23:** 179.

CREITZ, G. I., AND F. A. RICHARDS. 1955. The estimation and characterization of plankton populations by pigment analysis III. A note on the use of Millipore membrane filters in the estimation of plankton pigments. J. Mar. Res., **14:** 211–216.

CUSHING, D .H. 1959. On the nature of production in the sea. Fish. Invest. Lond., Ser. 2, **22:** 1–40.

DOTY, M. S. 1959. Phytoplankton photosynthetic periodicity as a function of latitude. J. Mar. Biol. Ass. India, **1:** 66–68.

EDMONDSON, W. T. 1955. Factors affecting productivity of fertilized salt water. Deep Sea Res., **3:** (Suppl.) 451–464.

———, AND Y. H. EDMONDSON. 1947. Measurements of production in fertilized salt water. J. Mar. Res., **6:** 228–246.

FOGG, G. E. 1957. Relationships between metabolism and growth in planktonic algae. J. Gen. Microbiol., **16:** 294–297.

———. 1958. Extracellular production of phytoplankton and the estimation of primary production. Rapp. Cons. Explor. Mer, **144:** 56–60.

FRENCH, C. S., AND R. F. ELLIOTT. 1958. The absorption spectra of chlorophyll in various algae. Ann. Rept. Dept. Plant Biology, Carnegie Institution of Washington Year Book 1957–1958, pp. 278–286.

GEST, H., AND M. D. KAMEN. 1948. Studies on the phosphorus metabolism of green algae and purple bacteria in relation to photosynthesis. J. Biol. Chem., **176:** 299–318.

GOLDBERG, E. D., T. J. WALKER, AND A. WHISENAND. 1951. Phosphate utilization by diatoms. Biol. Bull., **101:** 274–284.

GOLTERMAN, H. L. 1960. Studies on the cycle of elements in fresh water. Acta. Bot. Neerl., **9:** 1–58.

GROSS, F., AND E. ZEUTHEN. 1948. The buoyancy of plankton diatoms: a problem in cell physiology. Proc. Roy. Soc. Lond. (Ser. B), **135:** 382–389.

HANSEN, A. L., AND R. J. ROBINSON. 1953. The determination of organic phosphorus in sea water with perchloric acid oxidation. J. Mar. Res., **12:** 31–42.

HARVEY, H. W. 1940. Nitrogen and phosphorus required for the growth of phytoplankton. J. Mar. Biol. Ass. U.K., **24:** 115–123.

———. 1953. Synthesis of organic nitrogen and chlorophyll by *Nitzschia closterium*. J. Mar. Biol. Ass. U.K., **31:** 477–487.

HASKIN, H. H. 1941. The chloroplast pigments in *Chlorella pyrenoidosa*. Doctorate Thesis, Dept. of Biology, Harvard University.

HEWITT, B. R. 1958. Spectrophotometric determination of total carbohydrate. Nature, **182:** 246–247.

HOFFMANN, C. 1956. Untersuchungen über die Remineralisation des phosphors im plankton. Kiel. Meeresforsch., **12:** 25–36.

JERLOV, N. G. 1951. Optical studies of ocean waters. Rep. Swedish Deep-Sea Exped. 1947–1948, **3:** 1–59.

JOHNSON, M. J. 1941. Isolation and properties of a pure yeast polypeptidase. J. Biol. Chem., **137:** 575–586.

———. 1949. A rapid method for estimation of non-volatile organic matter. J. Biol. Chem., **181**: 707–711.

JØRGENSEN, E. G. 1955a. Solubility of the silica in diatoms. Physiol. Plantarum, **8**: 846–851.

———. 1955b. Variations in the silica content of diatoms. Physiol. Plantarum, **8**: 840–845.

KEELER, R. F. 1959. Colour reactions for certain amino acids, amines and proteins. Science, **129**: 1617–1618.

KETCHUM, B. H., J. H. RYTHER, C. S. YENTSCH, AND N. CORWIN. 1958. Productivity in relation to nutrients. Rapp. Cons. Int. Explor. Mer, **144**: 132–140.

LEWIN, J. C., AND R. A. LEWIN. 1960. Auxotrophy and heterotrophy in marine littoral diatoms. Canadian J. Microbiol., **6**: 127–133.

MCALLISTER, C. D., T. R. PARSONS, AND J. D. H. STRICKLAND. 1960. Primary productivity at Station "P" in the northeast Pacific Ocean. J. Cons. Int. Explor. Mer, **25**: 240–259.

———, AND J. D. H. STRICKLAND. 1961. Light attenuators for use in phytoplankton photosynthesis studies. Limnol. Oceanogr., **6**:226–228.

MARGALEF, R. 1958. Spatial heterogeneity and temporal succession of phytoplankton. Perspectives in Marine Biology. U. of California Press, pp. 323–349.

MUKERJEE, P. 1956. Use of ionic dyes in the analysis of ionic surfactants and other ionic organic compounds. Analyt. Chem., **28**: 870–873.

MULLIN, J. B., AND R. P. RILEY. 1955a. The colorimetric determination of silicate with special reference to sea and natural waters. Analyt. Chim. Acta., **12**: 162–176.

———, AND J. P. RILEY. 1955b. The spectrophotometric determination of nitrate in natural waters with particular reference to sea water. Analyt. Chim. Acta., **12**: 464–480.

MYERS, J., AND J. A. JOHNSTON. 1949. Carbon and nitrogen balance of *Chlorella* during growth. Plant. Physiol., **24**: 111–119.

NIELSEN, E. STEEMAN, AND V. KR. HANSEN. 1959. Measurements with the carbon-14 technique of the respiration rates in natural populations of phytoplankton. Deep Sea Res., **5**: 222–233.

ODUM, H. T., W. MCCONNEL, AND W. ABBOTT. 1959. The chlorophyll *a* of communities. Pub. Inst. Mar. Sci. (Univ. Texas), **5**: 65–96.

PARSONS, T. R., AND J. D. H. STRICKLAND. 1959. The proximate analysis of marine standing crops. Nature, **184**: 2038.

RICE, T. R. 1953. Phosphorus exchange in marine phytoplankton. Fish. Bull. U.S. Fish and Wildl. Serv., **54**(80): 77–89.

RICHARDS, F. A., WITH T. G. THOMPSON. 1952. The estimation and characterization of plankton populations by pigment analysis. II. A spectrophotometric method for the estimation of phytoplankton pigments. J. Mar. Res., **11**: 156–172.

RIGLER, F. H. 1956. A tracer study of the phosphorus cycle in lake water. Ecology, **37**: 550–562.

ROBINSON, R. J., AND T. G. THOMPSON. 1948. The determination of phosphates in sea water. J. Mar. Res., **7**: 33–41.

RODHE, W. 1955. Can plankton production proceed during winter darkness in subarctic lakes? Verh. Int. Ver. Limnol., **12**: 117–119.

———. 1958. The primary production in lakes, some results and restrictions of the ^{14}C method. Rapp. Cons. Int. Explor. Mer, **144**: 122–128.

RYTHER, J. H. 1956. Photosynthesis in the ocean as a function of light intensity. Limnol. Oceanogr., **1**: 61–70.

———, AND C. S. YENTSCH. 1957. The estimation of phytoplankton production in the ocean from chlorophyll and light data. Limnol. Oceanogr., **2**: 281–286.

———, C. S. YENTSCH, E. M. HULBURT, AND R. F. VACCARO. 1958. The dynamics of a diatom bloom. Biol. Bull., **115**: 257–268.

SAMEJIMA, H., AND J. MYERS. 1958. On the heterotrophic growth of *Chlorella pyrenoidosa*. J. Gen. Microbiol., **18**: 107–117.

SAUNDERS, G. W. 1957. Interrelations of dissolved organic matter and phytoplankton. Bot. Rev., **23**: 389–410.

SMITH, J. B., M. TATSUMOTO, AND D. W. HOOD. 1960. Carbamino carboxylic acids in photosynthesis. Limnol. Oceanogr., **5**: 425–431.

SPENCER, C. P. 1956. The bacterial oxidation of ammonia in the sea. J. Mar. Biol. Ass. U.K., **35**: 621–630.

SPOEHR, H. A., AND H. W. MILNER. 1949. The chemical composition of *Chlorella*; effect of environmental conditions. Plant. Physiol., **24**: 120–149.

STEELE, J. H., AND C. S. YENTSCH. 1960. The vertical distribution of chlorophyll. J. Mar. Biol. Ass. U.K., **39**: 217–226.

STRICKLAND, J. D. H. 1958. Solar radiation penetrating the ocean. A review of requirements, data and methods of measurement, with particular reference to photosynthetic productivity. J. Fish. Res. Bd. Canada, **15**: 453–493.

———. 1960. Measuring the production of marine phytoplankton. Bull. Fish. Res. Bd. Canada, **122**: 1–172.

———, AND K. H. AUSTIN. 1959. The direct estimation of ammonia in sea water with notes on reactive iron, nitrate and inorganic phosphorus. J. Cons. Int. Explor. Mer, **24**: 446–451.

———, AND T. R. PARSONS. 1961. A manual of seawater analysis (with special reference to the more common micro-nutrients and to particulate organic material). Bull. Fish. Res. Bd. Canada, **125**: 1–185.

———, AND L. D. B. TERHUNE. 1961. The study of in-situ marine photosynthesis using a large plastic bag. Limnol. Oceanogr., **6**: 93–96 von Brand, T., N. W. Rakestraw, and C. E. Renn. 1939. Further experiments on the decomposition and regeneration of nitrogenous organic matter in sea water. Biol. Bull., **77**: 285–296.

Wood, E. J. F. 1956. Considerations on productivity. J. Cons. Int. Explor. Mer, **21**: 280–283.

Yentsch, C. S., and R. F. Vaccaro. 1958. Phytoplankton nitrogen in the oceans. Limnol. Oceanogr., **3**: 443–448.

Yuen, S. H. 1958. Determination of traces of manganese with leucomalachite green. Analyst, **83**: 350–356.

Limnology of Canyon Ferry Reservoir

II. Phytoplankton Standing Crop and Primary Production

JOHN C. WRIGHT

Department of Botany and Bacteriology, Montana State College, Bozeman, Montana

ABSTRACT

In general, one μg chlorophyll was equivalent to 0.5 mm^3 cell volume and 0.12 mg ash-free dry weight. The average seston content consisted of 34.5% phytoplankton, 9.8% zooplankton, and 55.7% detritus. Optimal photosynthesis per unit chlorophyll and cell volume averaged 0.39 μ mole $O_2/\mu g$ chl/hr and 0.86 μ mole $O_2/mm^3/hr$, respectively. The average euphotic zone photosynthetic rate was 52% of the optimal rate. Chlorophyll content per unit volume of cells and photosynthesis per unit chlorophyll or cell volume decreased with increase in population size. Evidence was found of an interacting effect of temperature, light intensity, and phosphates on photosynthesis. An empirical method was described for estimating photosynthesis on the basis of chlorophyll content, optimal rate of photosynthesis per unit chlorophyll, extinction coefficient, and the ratio of euphotic zone photosynthesis to optimal photosynthesis. Determinations of phytoplankton respiration were made, and net euphotic zone photosynthesis was computed for various sizes of chlorophyll standing crop. It was concluded that the most frequently occurring values of chlorophyll standing crop were most likely to produce a maximum net production and to be at a steady state level.

INTRODUCTION

The attempt to measure primary production in relation to phytoplankton standing crop has involved three measures: (1) cell volume, (2) seston, and (3) chlorophyll (Verduin 1956b). Data on rate of photosynthesis based on cell volume and ash-free dry weight have been presented by Verduin (1952, 1954, 1956a, and 1956b) and McQuate (1956). Information on chlorophyll: photosynthesis ratios has been provided by Edmondson (1955), Manning and Juday (1941), Riley (1941), and Ryther and Yentsch (1957). Photosynthesis as related to seston has been discussed by Jackson and McFadden (1954). To my knowledge, only Riley (1941) has presented data in which all three measures of standing crop are related to primary production.

One of the objectives of the present study was to obtain information on the relationship of these three quantities to primary production. The information presented was collected during the periods April to October, 1957, and April to September, 1958.

These studies were supported by a National Science Foundation grant, G. 3894. The writer is particularly indebted to his wife, Sally, and Blaine LeSeur, graduate assistant, for their invaluable aid in collecting and processing the data and gratefully acknowledges the advice and suggestions of John Ryther and Jacob Verduin.

METHODS

Samples were collected and handled in a manner similar to that described in a previous paper (Wright 1958), except that clear bottles were suspended at two and one-half

Reprinted from *Limnology and Oceanography*, 4:235–245, 1959.

foot intervals from the surface to 20 feet, and one bottle was placed at the 25-foot depth.

Volume of phytoplankton was determined by counting and measuring the algal cells with a Zeiss-Winkel inverted microscope. The plankton which was fixed with Lugol's solution was allowed to sediment overnight in a five ml cylindrical chamber. Counts were made at 20, 80, 320, and 800× depending upon the size of the alga. Dimensions of the cells or colonies were estimated with a Whipple micrometer disk, and volumes were computed using appropriate formula. When large cells or colonies were enumerated at 20× magnification, the entire surface of the counting chamber was scanned. At higher magnifications, either 50 or 100 fields were counted, depending upon the density of the species present.

Samples for chlorophyll determination were obtained by filtering 500 ml of the sample through a type AA 047 mm "millipore" filter. The filter was treated as described by Creitz and Richards (1955). Upon return to the laboratory, the filter was dissolved in five ml of 90% acetone and held in the refrigerator overnight. The following morning, the acetone extract was separated from the residue by centrifugation in a clinical centrifuge. Chlorophyll concentrations were determined with a Klett-Summerson colorimeter using a red filter. In 1958, the method of chlorophyll determination using a Beckman DU spectrophotometer as described by Richards with Thompson (1952) was employed. It was found that 0.28 μg chlorophyll *a*/L was equivalent to one Klett unit. A comparison was made between chlorophyll content of "millipore" filter samples and centrifuge samples on several occasions.

Samples of seston were obtained by twice centrifuging one liter of water through a Foerst continuous-flow centrifuge at a rate of one liter per six minutes. The samples obtained during 1958 were dried for 24 hours or longer at 100°C in "Vycor" crucibles, weighed, and then ashed at 600°C for 30 minutes and weighed again.

A series of net plankton samples was obtained by towing a #20 plankton net. The concentrate was mixed with lake water from which a 250-ml glass-stoppered black bottle was filled and suspended in the lake for six hours. The oxygen decline in the black bottle was used as a measure of respiration. A 125-ml aliquot of the sample was filtered through a "millipore" filter for chlorophyll determination, and another 125-ml aliquot was preserved with formaldehyde and later filtered through a filter crucible with a maximum pore size of three microns. Dry weight and ash-free dry weight were determined from this sample. A third aliquot was fixed with Lugol's solution for zooplankton counts.

Zooplankton samples were collected by oblique tows with a Clarke-Bumpus plankton sampler from 25 feet to surface using a #10 net. The formalin fixed samples were made up to 500 ml, and all animals in one ml aliquots were counted.

Phosphate was determined on "millipore" filtered samples, using the ammonium molybdate-stannous chloride method. The intensity of color formation was determined colorimetrically, using a 4-cm cell.

Temperatures were measured with a Tri-R electronic thermometer in 1957 and with a bathythermograph in 1958. Light penetration data were obtained with a G. M. Manufacturing Company, model 15M-02 submarine photometer which employs a Weston photronic cell. During each photosynthesis run, surface light intensity was estimated by averaging readings taken at one-half hour intervals with the photometer. Average light intensity at each clear bottle depth was computed on the basis of the extinction coefficient and average surface light intensity.

RELATIONSHIP BETWEEN CHLOROPHYLL, CELL VOLUME, AND ASH-FREE DRY WEIGHT OF PHYTOPLANKTON

The histogram (Fig. 1) shows the frequency at which various chlorophyll/cell volume ratios occurred. Approximately 60% of the samples was included in the range between 1-3 μg chlorophyll/mm^3 cells. The median ratio was 2 μg chl/mm^3 cells.

The regression equation between chlorophyll and cell volume, $C = 1.68V + 2.05$, indicates that there is a non-linear relation-

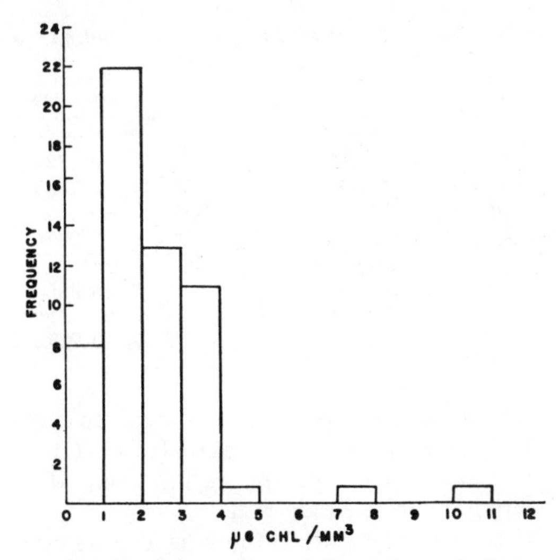

FIG. 1. Frequencies of the values of chlorophyll/cell volume ratios.

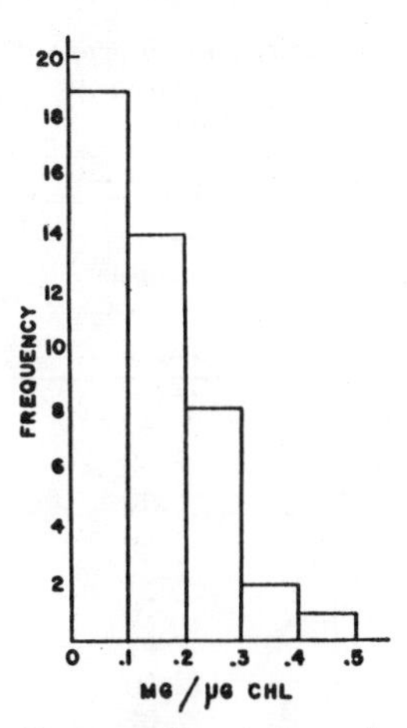

FIG. 2. Frequencies of the values of ash-free dry weight/chlorophyll ratios.

TABLE 1. *Size of standing crop compared with chlorophyll content and rate of optimal photosynthesis per unit cell volume*

Standing crop mm³/l		Average Chlorophyll/cell volume ratio µg chl/mm³	Average rate of optimal photosynthesis µmole O_2/mm³/hr
Range	Average		
0–2.5	1.75	3.53	1.48
2.6–5.0	3.72	2.09	0.81
5.1–7.5	6.18	1.80	0.55
7.6–10.0	8.38	1.76	0.46
>10.1	16.10	1.42	0.48

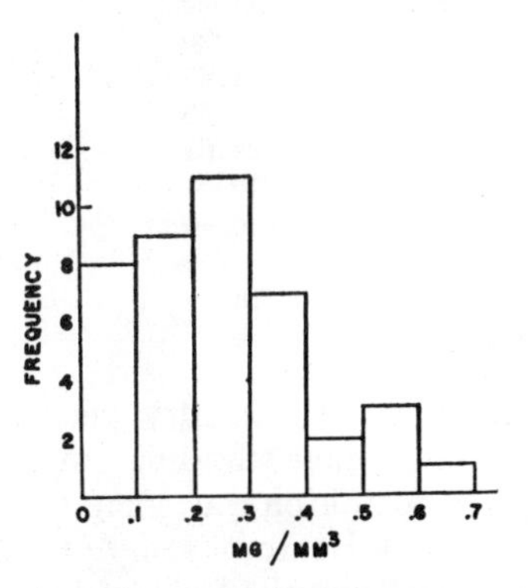

FIG. 3. Frequencies of the values of ash-free dry weight/cell volume ratios.

ship between chlorophyll and cell volume, since chlorophyll would be 2.05 µg/L when cell volume was zero. Examination of Table 1 shows that, as the size of the standing crop in terms of cell volume increased, the ratio of chlorophyll to cell volume decreased. This relationship suggests that chlorophyll synthesis becomes limited by some factor (perhaps nitrogen) as the population increases.

The frequency distribution of the ratios of ash-free dry weight to chlorophyll is shown in Figure 2. Of the 44 samples, 73% ranged between 0.02 and 0.20 mg ash-free dry weight/µg chl. The median ratio of 0.12 is very close to the average ratio of 0.11 for marine plankton computed from the data of Harris and Riley (1956).

Figure 3 shows the frequency at which various ratios of ash-free dry weight to cell volume occurred. The most frequently occurring ratios were in the range between 0.2 and 0.3 mg ash-free dry weight/mm³.

On the basis of the data presented, the best estimate of the average relationship between the three indices of phytoplankton standing crop would be 1.0 µg chl = 0.5 mm³ cells = 0.12 mg ash-free dry weight.

Millipore filters were considerably more efficient in retaining chlorophyll than the centrifuge method (Table 2). The average of 20 millipore samples was 13.5 compared to 9.7 µg/L for the centrifuge samples.

COMPOSITION OF THE SESTON

The average ash-free dry weight of seston samples collected during 1958 was 2.35 mg/L. The average chlorophyll content of these samples was 6.8 µg/L, which would be

TABLE 2. *Comparison of chlorophyll retention by Foerst plankton centrifuge and AA type millipore filters*

	μg Chlorophyll/L		Per cent difference
	Centrifuge	Millipore	
	12.88	12.88	−00
	7.00	12.88	−46
	6.16	8.40	−27
	6.72	8.40	−20
	7.56	8.40	−11
	2.24	5.60	−60
	5.74	7.84	−27
	5.60	7.70	−27
	6.16	7.67	−20
	6.25	12.50	−50
	7.56	7.67	− 3
	6.16	10.64	−42
	16.80	21.84	−23
	8.68	11.20	−23
	8.12	11.20	−28
	2.80	1.68	+40
	54.32	70.56	−23
	12.88	24.08	−47
	6.72	10.08	−33
	3.36	8.40	−60
Mean	9.69	13.48	−27

equivalent to an ash-free dry weight of 0.81 mg/L. The average standing crop of larger species of zooplankton was computed on the basis of Clarke-Bumpus samples from the euphotic zone. On the basis of the calculations presented in Table 3, the average ash-free dry weight of zooplankton was 0.23 mg/L. This figure does not include the smaller zooplankton forms, but their weights would be negligible and would not alter the order of magnitude. On the basis of these figures, 34.5% of the seston was phytoplankton, 9.8% was zooplankton and the remainder (55.7%) could be considered as detritus.

Davis (1958) presented data on the volume of phytoplankton, zooplankton, and detritus for western Lake Erie. On the basis of his figures, phytoplankton comprised 50.9%, zooplankton 28.0%, and detritus 21.1% by volume of the suspended organic material of the Lake Erie samples. Riley (1940) concluded that phytoplankton in Linsley Pond averaged 18.5% of the seston, zooplankton 2.0, and detritus 79.5. Pennak (1955) estimated that detritus com-

TABLE 3. *Average 1958 zooplankton standing crop*

Organism	Average number/liter	Average dry weight/ individual (mg)	Average dry weight (mg)/ liter
Daphnia	20.0	0.0086	0.172
Cyclops	46.5	0.0040	0.186
Diaptomus	2.1	0.0086	0.018
Asplanchna	6.7	0.0008	0.005
Total dry weight			0.381
Total ash free dry weight: 0.60 × 0.381 =			0.229

prised 65% of the seston content in eight Colorado lakes. Birge and Juday (1934) considered 50% of the seston content of Trout Lake to be detritus.

It is evident that there is considerable variation among the estimates of detritus content of the few lakes in which an attempt has been made to estimate its magnitude. On the basis of these scanty data, there seems to be a fair degree of correlation between the size of the lake and the detritus/seston ratio. In Table 4 the areas of the lakes mentioned are compared with the detritus/seston ratio. It will be noted that the greater the area of the lake, the smaller the ratio of detritus to seston. This is of importance in view of the divergence of opinion expressed by Pennak (1955) and Davis (1958) concerning the importance of detritus in secondary production.

As Edmondson (1957) pointed out, detritus may have three principal sources, (1) terrestrial, (2) littoral, (3) planktonic. If the detritus is of planktonic origin, then discussion as to the relative importance of detritus vs. plankton is absurd. If the source is littoral, the problem of food consumption by the zooplankton is still a problem of primary production within the lake. However, if the major source of detritus is terrestrial, the energy economy of the zooplankton could proceed at a faster rate than could be supported by phytoplankton production. It would be difficult to separate littoral- and terrestrial-derived detritus from each other; however, the importance of both would diminish with increase in volume of the lake, as the volume would tend to increase as the square of the circumference. Therefore one would expect the ratio of

TABLE 4. *Detritus/seston ratios for lakes of various size*

Lake	Area in hectares	Detritus/ seston ratio
Western Lake Erie	3.3×10^5	0.21
Canyon Ferry Reservoir	1.4×10^4	0.56
Trout Lake	6.5×10^2	0.50
Colorado Lakes	1.5–205.2	0.65
Linsley Pond	9.4	0.80

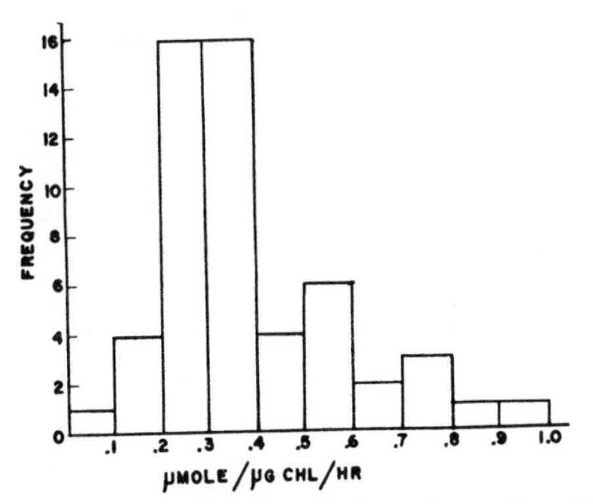

FIG. 4. Frequencies of the values of optimal photosynthesis/chlorophyll ratios.

detritus to seston to vary with lake size as indicated in Table 4. It would appear that the disagreement in conclusions reached by Pennak and Davis is more apparent than real, since they were working with lakes of different sizes.

OPTIMAL RATE OF PHOTOSYNTHESIS

Optimal photosynthesis averaged 0.39 µ mole O_2/µg chl/hr and ranged from 0.10 to 0.91. The average value is equivalent to the value of 3.7 g C/g chl/hr (Ryther and Yentsch 1957), assuming a photosynthetic quotient of 1.25. The frequency distribution of chlorophyll-based ratios is shown in Figure 4. Over 59% was in the range between 0.20 and 0.40 µmole O_2/µg chl/hr. The correlation coefficient (r = 0.910) between chlorophyll and optimal photosynthesis was highly significant. On the basis of the chlorophyll/ash-free dry weight ratio, optimal photosynthesis would be 3.25 µmoles O_2/mg ash-free dry weight/hr.

Volume-based rates averaged 0.86 µmoles O_2/mm^3/hr and ranged between 0.20 to 3.86. The frequency distribution of ratios based on cell volume is shown in Figure 5.

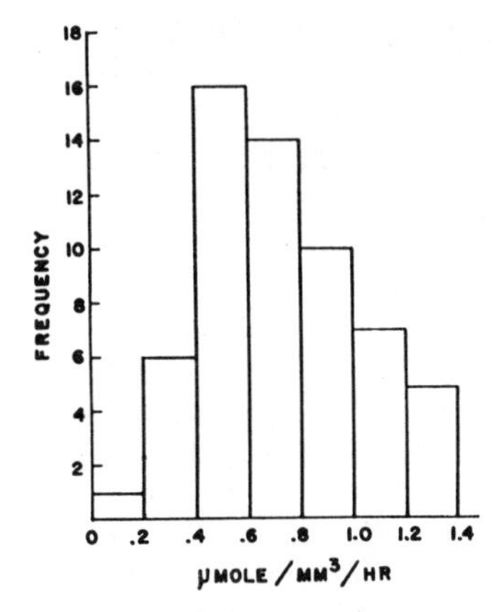

FIG. 5. Frequencies of the values of optimal photosynthesis/cell volume ratios.

TABLE 5. *Chlorophyll standing crop compared with average rates of optimal and euphotic zone photosynthesis and average daily rate of change of chlorophyll*

Number of samples	Chlorophyll µg/l		Photosynthesis per unit chlorophyll µmole O_2/µg chl/hr			Rate of change of chlorophyll standing crop µg chl/µg chl/day
	Range	Average	Optimal	Euphotic zone	Ratio	
4	0–2.5	1.68	0.70	0.45	0.64	+0.230
10	2.6–5.0	4.37	0.48	0.23	0.48	+0.024
20	5.1–10.0	7.51	0.38	0.18	0.47	−0.006
13	10.1–15.0	11.65	0.34	0.15	0.44	−0.035
6	15.1–25.0	18.03	0.25	0.11	0.44	−0.067

The correlation coefficient (r = 0.913) between cell volume and optimal photosynthesis is highly significant.

If a comparison is made between cell volume per liter and optimal photosynthesis per unit cell volume, it is apparent that, paralleling the decrease in chlorophyll per unit cell volume with increasing cell volume, there is also a decrease in the photosynthetic rate per unit volume (Table 1).

In Table 5, standing crops of chlorophyll are compared with the corresponding chlorophyll-based rates of optimal and euphotic zone photosynthesis and average daily rate of change of chlorophyll. Although there was considerable scatter in the data on which these averages were based, nevertheless there was a consistent trend in their relationship. Low standing crops of chlorophyll generally had high rates of photosynthesis per unit chlorophyll, and as noted previously (Table 1) a high concentration of chlorophyll per unit volume. On the average such crops had a tendency to increase very rapidly, although it must be remembered that for any particular date various limiting factors may prevent this, such as lack of nutrients, grazing, low light intensity, etc. Larger standing crops of chlorophyll had progressively smaller photosynthetic rates and less chlorophyll per unit volume. The average rate of growth also became less with larger standing crops. The decrease in photosynthetic activity with increasing size of standing crop has also been observed in Sandusky Bay (McQuate 1956) and in Lake Erken (Rohde *et al.* 1958).

The data in Table 5 seem to indicate that when one randomly samples phytoplankton populations at intervals over an extended period of time, some of the populations sampled are young and actively growing, others are in a steady-state condition, and others senescent. Populations ranging between 5 and 10 μg chl/L seem to be at an approximate steady-state level corresponding to an average set of conditions occurring during the time these populations were selected. It is interesting to note (Table 5) that the most frequently occurring populations had chlorophyll concentrations ranging between 5.1 and 10.0 μg/L. Sixty-four per cent of the populations had less than 10 μg chl/L, which is within the range of standing crops having photosynthetic rates that generally resulted in an increase of standing crop or the maintenance of a steady state.

EUPHOTIC ZONE PHOTOSYNTHESIS

The average euphotic zone photosynthetic rate was 52% of the optimal rate, or 0.20 μmoles O_2/μg chl/hr. The correlation coefficient between optimal photosynthesis and the corresponding euphotic zone rate was high ($r = 0.850$). Size of standing crop affected the average rate of euphotic zone photosynthesis as well as the optimal rate (Table 5). The greater the standing crop the less was the average rate of photosynthesis per unit chlorophyll in the euphotic zone. With increase in chlorophyll standing crop, the ratio of euphotic zone photosynthesis to optimal photosynthesis declined also.

TABLE 6. *Effect of light intensity on the average rate of euphotic zone photosynthesis per unit chlorophyll with respect to temperature*

Temperature °C	Average surface light intensity (microamperes)	Average euphotic zone photosynthesis μmole O_2/μg chl/hr
Low temperature series 0–14°	8606	0.092
	5932	0.129
	4556	0.231
	3234	0.324
High temperature series 16°–23°	9209	0.263
	8003	0.199
	6603	0.138
	4651	0.139

TABLE 7. *Average light intensity at which optimal photosynthesis occurred compared to average water temperature of the euphotic zone*

Temperature °C	Light intensity microamperes
4	1180
9	1424
14.5	2093
18.5	3510
21.3	3634

An interesting interaction between surface light intensity, temperature, and euphotic zone photosynthesis was found. The populations were separated into two groups on the basis of temperature—a low temperature group (4–14°C) and a high temperature group (16–23°C). The effect of increasing light intensity on euphotic zone photosynthesis in the low temperature series was just the reverse of that in the high temperature series (Table 6). High light intensities markedly inhibited the rate of photosynthesis of cold temperature populations and increased the rate of photosynthesis in warm temperature populations.

The interaction between light and temperature is illustrated further when a comparison is made between temperature and the average light intensity at which optimal photosynthesis occurred (Table 7). As the temperature increased, optimal photosynthesis occurred at higher light intensities.

Rohde *et al.* (1958) observed a similar relationship in Lake Erken, and concluded that "the phytoplankton is able to compensate to a very large extent, seasonal changes in light conditions by means of its successive adaptation. Owing to this ecological as well as other physiological reasons, the concept of a constant light optimum has to be replaced by a concept of the movable optimum."

RELATIONSHIP BETWEEN AVERAGE EUPHOTIC ZONE PHOTOSYNTHESIS AND OPTIMAL PHOTOSYNTHESIS

It is apparent from the data discussed above that there is no simple constant relationship between standing crop and photosynthesis. Instead there is a complex relationship between variable populations of phytoplankton which undergoes a seasonal succession correlated with changing environmental factors such as temperature and light intensity. Any attempt to derive an expression which will enable one to predict primary production on the basis of standing crop must involve the collection of a large number of samples covering a wide range of standing crops and environmental conditions. From an empirical consideration of such data it should be possible to derive expressions of relationship that will at least enable one to predict an average rate of production on a monthly or perhaps weekly basis. On the basis of the limited data available, an attempt has been made to predict average monthly euphotic zone photosynthesis on the basis of the following equation:

$$\sum P = \frac{4.6}{k} \times F \times P_{opt} \quad (1)$$

in which $\sum P$ is euphotic zone photosynthesis, k is the extinction coefficient, P_{opt} is optimal photosynthesis, and F is the ratio of euphotic zone photosynthesis to optimal photosynthesis on a unit volume basis. The expression $\frac{4.6}{k}$ defines the depth of the euphotic zone as the depth of a layer of water lying above the plane at which light intensity is 1% of surface light intensity. If P_{opt} is expressed as mmoles O_2/m^3, then P would be in terms of mmoles/m^2.

Numerical values of F were computed from observed values of optimal and euphotic zone photosynthesis and the extinction coefficient for each sampling date. All the values for each month were then averaged (Table 8).

TABLE 8. *Average monthly values for euphotic zone photosynthesis* ($\sum P$), *optimal photosynthesis* ($P_{opt.}$), *extinction coefficient* (k), *and the ratio of euphotic zone to optimal photosynthesis* (F)

These values are compared with average light intensity, temperature, and phosphate phosphorus concentration.

Month	$\sum P$ mmole $O_2/m^2/hr$	k	$P_{opt.}$ mmole $O_2/m^3/hr$	F	Light microamperes	Temperature °C	Phosphorus mmole P/m^3
April	10.18	0.64	4.04	0.34	4873	4.8	1.06
May	7.82	0.64	2.27	0.44	5985	10.7	0.54
June	11.27	0.66	2.75	0.49	7804	16.8	0.48
July	10.27	0.74	2.92	0.58	6704	18.9	0.59
Aug.	10.81	0.67	2.69	0.59	7902	20.9	0.12
Sept.	12.80	0.79	3.68	0.60	6315	18.6	0.70
Oct.	5.91	0.71	1.44	0.59	3839	13.8	0.61
Mean	9.87	0.69	2.83	0.52			

It will be noted that there is no clearly defined relationship between light intensity and F when these two variables are considered alone. The increase in F seems to be better correlated with temperature from April to July. However, the F value for August should be higher if the temperature trend was extrapolated. The extremely low phosphate concentrations during August may have been responsible for this discrepancy. Furthermore, the higher phosphate concentrations in September and October may possibly have counteracted the effect of declining radiation and temperature, resulting in higher values of F than could be expected on the basis of the trend from April to July. Admittedly the data are inadequate to substantiate this hypothesis, but certainly some relationship is indicated between temperature, light, nutrients, and photosynthesis that warrants further study.

On the basis of 24 hr C^{14} measurements, Rohde *et al.* (1958) found the following simple relationship between daily euphotic zone photosynthesis and optimal photosynthesis:

$$\sum a = (2.4 \text{ to } 2.7) \frac{a_{opt}}{\epsilon} \quad (2)$$

TABLE 9. *A comparison between monthly values of total relative photosynthesis as computed by the method described, using a 12-hour photoperiod and Ryther's (1956) method*

Month	Total relative photosynthesis	
	4.6 *FT*	R_s (Ryther)
April	18.8	21.0
May	24.3	25.0
June	27.0	26.0
July	32.0	25.0
Aug.	32.5	24.0
Sept.	33.1	22.0
Oct.	32.6	20.0
Mean	28.6	23.3

The constant (2.4–2.7) is equivalent to 4.6 F in Equation (1). The average value of 4.6 F would be 2.4, which agrees very nicely with the values reported by Rohde and his co-workers. However the range of values between 1.6 for April and 2.8 for August is greater than the range of 2.4–2.7 which they found.

If one assumed that the factor F applied to the entire photosynthetic period as well as to the six-hour photosynthetic periods used in this study, daily euphotic zone photosynthesis could be calculated from the following equation, where T is the number of hours of daily photosynthesis:

$$\sum P = \frac{4.6}{k} \times F \times T \times P_{\text{opt}} \qquad (3)$$

The expression 4.6 FT is equivalent to total relative euphotic zone photosynthesis (R_s) as used by Ryther (1956).

Since Doty and Oguri (1957), Verduin (1957), and Yentsch and Ryther (1957) have found variations in the hourly rate of photosynthesis during the day, one cannot simply substitute the number of hours of daylight for T with any degree of confidence. Furthermore, as pointed out previously in the discussion concerning the effect of light and temperature on photosynthesis, there is no fixed light optimum for photosynthesis. In view of this uncertainty it seems reasonable to assume a 12-hour photoperiod for the months of April to October as Verduin (1956b) has suggested. Values of 4.6 FT computed on a 12-hour basis are shown in Table 9 and are compared with values of R_s according to Ryther's (1956) method. These latter values were computed on the basis of radiation measurements from the Twin Falls, Idaho, weather station (Duggar 1936). The two estimates agreed closely for April, May, and June, during the period when diatoms were dominant in the phytoplankton; but Ryther's method gave lower values for July, August, September, and October. One reason for the discrepancy may be that Ryther's treatment overestimates surface inhibition by high light intensities during this period when blue-green algae dominate the plankton. If the values for total relative photosynthesis (R_s) were adjusted to correspond to the relationship between photosynthesis in dinoflagellates and light intensity (Ryther 1956), they would more nearly correspond to the values found in this study.

CALCULATION OF EUPHOTIC ZONE PHOTOSYNTHESIS ON THE BASIS OF CHLOROPHYLL CONTENT

Euphotic zone photosynthesis has been calculated from the relationship

$$\sum P = \frac{4.6}{k} \times F \times T \times C \times R_{\text{opt}} \qquad (4)$$

where C is chlorophyll concentration (mg/m^3), and R_{opt} is the optimal rate of photosynthesis per unit chlorophyll (mmole O_2/mg chl/hr).

Euphotic zone photosynthesis was calculated two ways. The average rate of optimal photosynthesis (0.39 mmole/mg chl/hr) was used in one set of calculations. Variable rates of optimal photosynthesis dependent upon size of standing crop (Table 5) were used in another set. The results expressed as average values per month are compared with average measured values in Table 10.

The average monthly deviation from actual measurements as calculated by using a constant rate of optimal photosynthesis was 23%. When variable rates were used, the average deviation of calculated values from measured values was 13%. It appears that better estimates of euphotic zone photosynthesis may be secured by correcting for the effect of size of standing crop on the rate of photosynthesis.

In Table 11 are given daily photosynthesis values from *in situ* light and dark bottle oxygen experiments (Ryther and Yentsch 1957) and from *in situ* changes in dissolved oxygen in the water column or a combination of the two methods (Riley 1956, Conover 1956, and Riley 1957). The data of Ryther and Yentsch, which were given in terms of g C/m^2/day, were converted to mmoles O_2/day by assuming a photosynthetic quotient of 1.25, which was used by them to convert oxygen production to carbon production. These values are compared with values of daily photosynthesis calculated from Equation (4), assuming a 12-hour photoperiod. In the case of Riley's (1957) data, the average rate of optimal photosynthesis per unit volume (0.86 mmole/ml/hr) was used instead of chlorophyll-based rates. On the whole, there is close agreement between photosynthesis values calculated by the method outlined above and the measured values. Agreement is best where a large number of values is averaged as in the case of the Long Island Sound and Sargasso Sea data.

TABLE 10. *A comparison of measured values of euphotic zone photosynthesis with values estimated on the basis of chlorophyll concentration, optimal rate of photosynthesis per unit chlorphyll, and the extinction coefficient*

Month	Number of samples	Average chlorophyll mg/m³	Averages by month of hourly euphotic zone photosynthesis (mmole O_2/m^2/hr)		
			Average rate of optimal photosynthesis used	Variable rate of optimal photosynthesis used	Actual measured rates
April	5	10.0	8.8	8.1	10.2
May	7	8.7	10.8	9.8	7.8
June	9	9.4	12.5	11.3	11.3
July	12	16.7	15.5	12.4	10.9
Aug.	10	6.1	9.6	10.4	10.8
Sept.	7	9.3	12.7	11.4	12.8
Oct.	4	2.1	4.0	6.2	7.6
Mean			10.6	9.9	10.2

PHYTOPLANKTON RESPIRATION AND NET PRODUCTION

The determination of phytoplankton respiration under natural conditions is most difficult since any measurement of respiration of the phytoplankton community will include respiration of the zooplankton and bacteria as well. An approximate estimate of phytoplankton respiration was obtained by measuring the oxygen uptake of net concentrates placed in black bottles and suspended in the lake for six hours.

The average rate of respiration of the samples of low zooplankton content was 0.033 μmoles O_2/μg chl/hr and ranged from 0.004 to 0.118. Most of the samples (62%) were between 0.02 and 0.04 μmoles/μg chl/hr (Figure 6).

In Table 12, gross photosynthesis, respiration, and net photosynthesis are compared for several ranges of euphotic zone standing

TABLE 11. *A comparison of measured rates of photosynthesis from various sources and values computed from equation 4*

Place	Reference	K	Chlorophyll mg/m³	Cell volume ml/m³	Photosynthesis mmole O_2/m^2/day	
					computed	measured
Long Island Sound						
1952–53	Riley (1956)	0.63	6.20		110	113
1953–54	Conover (1956)	0.68	6.74		110	106
Sargasso Sea	Riley (1957)	0.11		0.132	30	23
Friday Harbor, Wash.	Ryther & Yentsch (1957)	0.13	1.0		86	35
East Sound, Wash.	Ryther & Yentsch (1957)	0.30	15.0		589	521
Gulf of Alaska	Ryther & Yentsch (1957)	0.10	2.5		280	156
Woods Hole Harbor	Ryther & Yentsch (1957)	0.27	3.35		139	169
Woods Hole Harbor	Ryther & Yentsch (1957)	0.27	1.69		70	82
Woods Hole Harbor	Ryther & Yentsch (1957)	0.24	2.80		131	114
Average					168	147

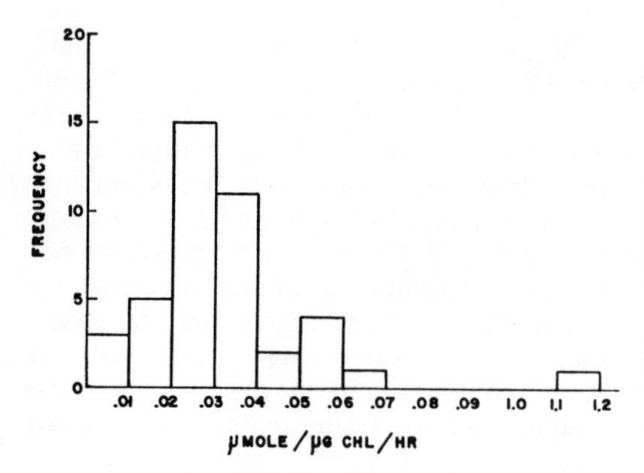

FIG. 6. Frequencies of the values of respiration/chlorophyll ratios.

TABLE 12. *A comparison of euphotic zone chlorophyll standing crop, gross photosynthesis, respiration, and net photosynthesis*

Chlorophyll mg/m³		Gross photosynthesis mmole O_2/m^2/hr	Respiration mmole O_2/m^2/2 hr	Net photosynthesis mmole O_2/m^2/hr
Range	Average			
0–2.5	1.7	4.6	1.0	3.6
2.6–5.0	4.2	7.9	2.3	5.6
5.1–10.0	7.9	10.7	4.3	6.4
10.1–15.0	12.2	12.4	6.7	5.7
15.1–20.0	15.7	14.4	8.6	5.8
20.1–25.0	23.0	16.9	12.6	4.3
70.6	70.6	11.7	38.7	−27.0

crop. The values for gross photosynthesis are averages of clear bottle measurements of photosynthesis for each range of standing crop. The respiration figures were computed by averaging the products of standing crop and average respiration rate over a two-hour period. This time interval was used in order to correct for night time respiration.

Crops ranging from 5 to 10 mg chl/m³ had the highest average rate of net photosynthesis. This is significant in view of the previous discussion in which it was pointed out that the most frequently occurring populations had chlorophyll contents within this range (Table 5) and would be most likely to be in a steady state condition. This may indicate that there is a tendency for standing crops of phytoplankton to become adjusted to a level which will deliver maximum net production, as Odum and Pinkerton (1955) have postulated. The smaller crops which were the most efficient per unit chlorophyll delivered a smaller net production than did the larger crops which were only about 50% as efficient.

It will also be noted that a wide range of crops smaller and larger than the optimal size crop had nearly the same net photosynthetic rate as the optimal size crop. After crops became larger than 20 mg chl/m³, the rate of net photosynthesis rapidly decreased. The largest crop encountered in the study (70.6 mg chl/m³) had a negative rate of net photosynthesis and rapidly declined. The plankton at this time was predominantly composed of *Aphanizomenon flos-aquae*, which had rapidly built up to bloom proportions from a level of 1.7 mg chl/m³ five days previously. The rapid increase followed a mass mortality of *Daphnia* and a great increase in soluble phosphate and was partially the result of concentration by advection in the vicinity of the sampling station.

Objection might be raised against the foregoing argument on the grounds that the limiting effects of nutrient deficiency were ignored. On the basis of culture experiments, Ryther (1956) found that the photosynthesis:respiration ratio approached unity after nutrients were depleted and growth had stopped. However, growth of algae in culture experiments of this type is not equivalent to the growth of algae in lakes. In a natural situation the observed rate of growth is a balance between actual rate of growth and rate of removal by grazing, currents, sinking, etc. The natural situation is more comparable to the continuous flow culture experiments carried out on *Chlorella* by Cook (1950). In his experiments, population density was kept constant by adjusting the rate of overflow from the culture. Hence the growth rate and the yield (growth rate × density) could be determined from the rate of overflow. It was found that growth rate decreased with increase in density as would be expected. However, yield reached a maximum value of 0.48 g/L/day at a population density 0.36 g/L. Population densities above or below 0.36 g/L gave lower yields. Yield, as used by Cook, is equivalent to net photosynthesis in Table 12, if the values in Table 12 were converted to dry matter equivalents of net photosynthetic oxygen production. On this basis net production in relation to standing crop (density) behaved similarly in the laboratory experiment and in the lake.

The major difference is that environmental factors affecting growth and removal rates vary in a natural situation in contrast to a laboratory experiment where they can be controlled. In discussing the results in Table 12 it was implied that during the period of time the samples were collected, an average set of conditions existed, and that within the usual range of these conditions an optimal size population could be defined as one giving maximum net production.

REFERENCES

Birge, E. A., and C. Juday. 1934. Particulate and dissolved organic matter in inland lakes. Ecol. Monogr., **4**: 440–474.

Conover, S. A. M. 1956. Oceanography of Long Island Sound, 1952–1954. IV. Phytoplankton. Bull. Bingham Oceanogr. Coll., **15**: 62–112.

Cook, P. M. 1950. Large-scale culture of *Chlorella*. In: The culturing of algae. A symposium, Brunel, J., G. W. Prescott, L. H. Tiffany (Ed.) Antioch Press, Yellow Springs, Ohio. 114 pp.

Creitz, G. I., and F. A. Richards. 1955. The estimation and characterization of plankton populations by pigment analyses. III. A note on the use of "millipore" membrane filters in the estimation of plankton pigments. J. Mar. Res., **14**: 211–216.

Davis, C. C. 1958. An approach to some problems of secondary production in the western Lake Erie region. Limnol. Oceanogr., **2**: 140–142.

Doty, M. S., and M. Oguri. 1957. Evidence for a photosynthetic daily periodicity. Limnol. Oceanogr., **2**: 37–40.

Duggar, B. M. 1936. Biological effects of radiation. Vol. 1. McGraw-Hill, New York. 676 pp.

Edmondson, W. T. 1955. Factors affecting productivity in fertilized salt water. Pap. Marine Biol. Deep-Sea Res. Oceanogr., Supp. Vol. 3: 451–564.

———. 1957. Trophic relations of the zooplankton. Trans. Amer. Micros. Soc., **76**: 225–245.

Harris, E., and G. A. Riley. 1956. Oceanography of Long Island Sound 1952–1954. VIII. Chemical Composition of the Plankton. Bull. Bingham Oceanogr. Coll., **15**: 315–323.

Jackson, D. F., and J. McFadden. 1954. Phytoplankton photosynthesis in Sanctuary Lake, Pymatuning Reservoir. Ecology, **35**: 1–4.

Manning, W. M., and R. E. Juday. 1941. The chlorophyll content and productivity of some lakes in northeastern Wisconsin. Trans. Wis. Acad. Sci. Arts Lett., **42**: 127–133.

McQuate, A. G. 1956. Photosynthesis and respiration of the phytoplankton in Sandusky Bay. Ecology, **37**: 834–839.

Odum, H. T., and R. C. Pinkerton. 1955. Times speed regulator, the optimum efficiency for maximum output in physical and biological systems. Amer. Scient., **43**: 331–343.

Pennak, R. W. 1955. Comparative limnology of eight Colorado mountain lakes. Univ. Colorado Stud., Ser. Biol., **2**: 1–75.

Richards, F. A., with T. G. Thompson. 1952. The estimation and characterization of plankton populations by pigment analyses. II. A spectro-photometric method for the estimation of plankton pigments. J. Mar. Res., **11**: 156–172.

Riley, G. A. 1940. Limnological studies in Connecticut. Part III. The plankton of Linsley Pond. Ecol. Monogr., **10**: 279–306.

———. 1941. Plankton studies. IV. Georges Bank. Bull. Bingham Oceanogr. Coll., **7** (4): 1–73.

———. 1956. Oceanography of Long Island Sound, 1952–1954. IX. Production and utilization of organic matter. Bull. Bingham Oceanogr. Coll., **15**: 324–334.

———. 1957. Phytoplankton of the North Central Sargasso Sea. Limnol Oceanogr., **2**: 281–286.

Rohde, W., R. A. Vollenweider, and A. Nauwerk. 1958. The primary production and standing crop of phytoplankton. In: Perspectives in marine biology, A. A. Buzzati-Traverso (Ed.). Univ. Calif. Press, Berkeley. 621 pp.

Ryther, J. H. 1956. Photosynthesis in the ocean as a function of light intensity. Limnol. Oceanogr., **1**: 61–70.

———. 1956. Interrelation between photosynthesis and respiration in the marine flagellate, *Dunaliella euchlora*. Nature, **173**: 861–862.

———, and C. S. Yentsch. 1957. The estimation of phytoplankton production in the ocean from chlorophyll and light data. Limnol. Oceanogr., **2**: 281–286.

Verduin, J. 1952. Photosynthesis and growth rates of two diatom communities in western Lake Erie. Ecology, **33**: 163–169.

———. 1954. Phytoplankton and turbidity in western Lake Erie. Ecology, **35**: 550–561.

———. 1956a. Energy fixation and utilization by natural communities in western Lake Erie. Ecology, **37**: 40–50.

———. 1956b. Primary production in lakes. Limnol. Oceanogr., **1**: 85–91.

———. 1957. Daytime variations in phytoplankton photosynthesis. Limnol. Oceanogr., **2**: 333–336.

Wright, J. C. 1958. The limnology of Canyon Ferry Reservoir. I. Phytoplankton-zooplankton relationships in the euphotic zone during September and October, 1956. Limnol. Oceanogr., **3**: 150–159.

Yentsch, C. S., and J. H. Ryther. 1957. Short-term variations in phytoplankton chlorophyll and their significance. Limnol. Oceanogr., **2**: 140–142.

20

THE *PISASTER–TEGULA* INTERACTION: PREY PATCHES, PREDATOR FOOD PREFERENCE, AND INTERTIDAL COMMUNITY STRUCTURE

ROBERT T. PAINE
Department of Zoology, University of Washington, Seattle, Washington 98105

(MS received May 26, 1969; accepted September 19, 1969)

Abstract. The herbivorous gastropod *Tegula funebralis* is not highly ranked in a food preference hierarchy of its major predator, the starfish *Pisaster ochraceus,* and exhibits a persistent broad overlap with it in the rocky intertidal zone at Mukkaw Bay, Washington. Observations on *Tegula* over a 5-yr period indicate that it settles high intertidally, lives there for 5-6 yr, and then tends to migrate lower into contact with *Pisaster. Tegula* lays down an annual growth line permitting it to be aged and a growth curve constructed. Analysis of relative growth and reproduction indicates that beyond a certain size (16 mm) large individuals perform less well in the upper than those in the lower intertidal zone. *Pisaster* consumes 25–28% of the adult *Tegula* per year in the area of spatial overlap, based on analysis of the age structure of 6–17 yr old *Tegula,* and by direct estimates of the percentage of the standing crop consumed annually. The relationship between *Pisaster* and sex ratio, relative energy limitation and reproductive output (fitness) of *Tegula* is discussed for three subpopulations.

It is suggested that the implied results of the interaction is typical of that between a major predator and one of its less preferred prey. The prominent zonation exhibited by preferred prey, the observed intimacy of association of predator and less preferred prey, and the zoogeographic homogeneity of the Pacific rocky coastline community are discussed in relation to three intermeshing ecological processes.

An increasing, though still small, body of data leads to the following hypothesis about the structure of some natural communities or associations of organisms: the patterns of species occurrence, distribution and density are disproportionately affected by the activities of a single species of high trophic status. I have labelled these "keystone species," and have given some arguments to support the validity of the definition for certain marine situations (Paine 1969). The same reasoning is easily extended to those freshwater communities described by Brooks and Dodson (1965) and Maguire, Belk and Wells (1968); the activities of, respectively, a fish in a lake and mosquito larvae in banana bracts, determine the structure of the community, a conclusion substantiated by both natural comparisons and artificial experimentation.

The generality of keystone species can only be determined by further observational and experimental work under a variety of natural conditions. However, there are some logical extensions of the notion that deserve elaboration. The species important in determining community structure have been identified from communities in which some primary consumer (1) is capable of monopolizing a basic resource and outcompeting or at least excluding other species and (2) is itself preferentially consumed by the keystone species. This suggests that the rank in a food preference hierarchy occupied by any particular prey species will be important in the biological ordering of the association, for in the presence of the keystone species advantage, in the sense of successful occupation, maturation and reproduction, is switched to those species that are less preferred, or those that are completely unattractive as prey. I have used the term less- or secondarily-preferred, although partially subjective, to refer to prey that would not usually be selected by a predator if alternative, more highly preferred prey were available.

This paper examines in some detail the biological consequences of being a secondarily-preferred prey in a keystone-dominated system and attempts to provide some information not only about the general structure of the whole system, but also the ecological processes that serve to shape it. The system is that widespread association of primarily intertidal organisms characteristic of exposed rocky shorelines existing along the Pacific coast of North America from Alaska to Baja California (Ricketts and Calvin 1952). The keystone species is the carnivorous starfish, *Pisaster ochraceus;* the secondarily preferred prey the trochid gastropod *Tegula funebralis.*

THE STUDY AREA

The majority of these observations were made at Mukkaw Bay, just south of Neah Bay, at the western tip of Washington State during the period June 1963–July 1968. Because the area was visited once or twice per month, the period can be subdivided into five single year intervals (July through the following June) that are approx-

Reprinted from *Ecology,* 50:950–961, 1969.

imately equivalent in sampling effort, providing annual replicates through time. This area, although appearing unprotected, does not receive the brunt of wave action and shock characterizing adjacent localities, and as such is classified as protected outer coast but one characterized by many attributes of more exposed areas (Ricketts and Calvin 1952). All detailed samples and observations on the interaction between *Tegula* and *Pisaster* were made on a flat bench extending about 100 m from high water to the extreme low water mark, lying between two stacks. The upper reaches are sandy, grading into small cobbles and boulders covered with *Fucus distichus* and *Gigartina papillata.* Lower down a bed of the tracheophyte *Phyllospadix scouleri* eventually gives way to more algal covered rocks. Bounding this area is one usually accepted as more typical of exposed coastline conditions. The substratum is a solid rock basement overgrown extensively by a variety of macroscopic, often sessile, plants and animals, characteristic of which are the algae *Endocladia muricata, Gigartina papillata* and *Hedophyllum sessile,* and among animals, *Balanus glandula, B. cariosus, Mytilus californianus, Strongylocentrotus purpuratus* and *Pisaster ochraceus.* A species list from both areas would be relatively consistent, but changes in the substratum induce biologically meaningful variations in the proportionate representation of the various species. These variations, although contributing to the mosaic nature of the environment, are not of sufficient magnitude to permit either a realistic subdivision of the system or to alter the relative importance of the ecological processes that shape it. *Tegula* lives throughout much of the entire area and is confined to the intertidal zone. *Pisaster* ranges from mid-tidal levels to about —30 m, although peak abundances are reached in the lower intertidal zone.

A series of measurements were also made in 1967 and 1968 on *Tegula* inhabiting the exposed western shore of Waadah Island, situated in the entrance to Neah Bay. At this locality *Pisaster* is extremely rare, and the intertidal zone very broad, covered with dense algal growth and superficially comparable to the Mukkaw Bay environment. *Tegula* extends throughout the area but is much less dense (20/m^2 as opposed to average values of 400–800/m^2) and much larger. I will suggest later that this may be a population declining toward local extinction.

Tegula funebralis

This species is widely but rather patchily distributed along the Pacific coast, and is seemingly confined to moderately exposed, rocky shorelines. At Mukkaw Bay *Tegula* is particularly abundant in those areas characterized by the presence of small cobbles rather than solid bench rock. It is an herbivorous species grazing on benthic diatoms and algae (Best 1964); it also consumes large kelp drifting through the intertidal zone if a sufficient number of snails aggregate to capture and anchor this potential food by their combined weight. *Tegula* settles high in the intertidal zone above the level exploited by *Pisaster,* and remains there for a significant length of time. *Tegula* density here is difficult to estimate as the animals are clumped, highly motile and probably migratory to some extent (mean of 77 samples taken between June 1964 and March 1966 was 792/m^2; SD 95; range 0–3,588/m^2). I have used a density estimate of 800/m^2 for the spatially variable upper intertidal population.

Sampling data taken throughout the 5-yr period are given in Table 1 for that portion of the *Tegula* population living lower down and showing overlap with *Pisaster.* They suggest an average density

TABLE 1. Census data for *Tegula funebralis* having a spatial overlap with *Pisaster ochraceus*

Date	Number of samples	Avg	SD	Number/m^2
Apr. 14, 64	6	1.5	1.4	59
June 10, 64[a]	7	36.1	24.0	353
June 12, 64	12	13.7	19.5	544
Oct. 4, 64	11	14.1	data lost	551
Nov. 2, 64	5	13.8	9.7	539
Apr. 17, 65[a]	8	28.1	21.0	275
July 29, 65	7	14.0	15.6	547
Oct. 23, 65	4	8.8	5.3	342
Mar. 20, 66	21	5.9	5.5	231
Mar. 18, 68	12	13.0	9.6	508

[a]Quadrat size 32 by 32 cm. All others 16 by 16 cm.

of 412/m^2. The high standard deviations for all samples are attributable to the aggregative behavior of *Tegula* when exposed. The range of densities encountered varies greatly throughout the geographic range but other populations seem capable of attaining comparably high densities. For instance, at Pacific Grove, California, Wara and Wright (1964) give maximum densities of 1,400/m^2 and 1,200/m^2 for high intertidal (+1–1.7 m) samples, and values ranging to about 800/m^2 for what I would term lower areas (ca. 0 m level).

A density estimate of 19.6/m^2 for the Waadah population was obtained from 93 haphazard tosses of a 15 by 15 cm frame. Subjective surveys of the entire area indicate that this value is representative and certainly that no patches of high density exist.

Another attribute of *Tegula,* characteristic of

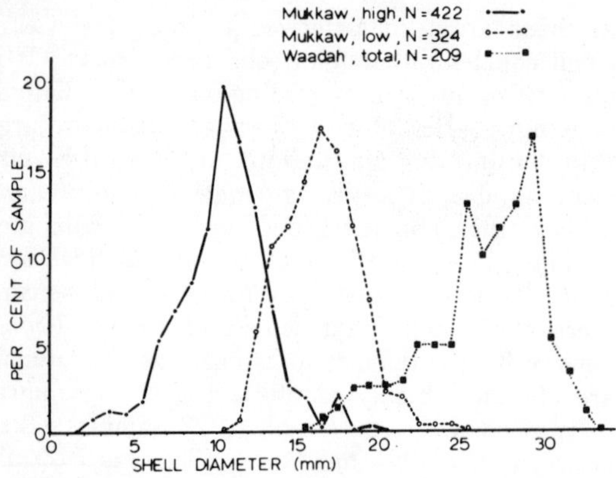

FIG. 1. Representative size-frequency distributions of *Tegula*.

many organisms of indeterminate growth form, is an absence of uniformity in their size (age) distribution. Rather, different subpopulations have persistently different modal values and are often characterized by a significant underrepresentation of certain age classes (see below) extending over periods of years. Three frequency distributions representative of their respective sampling sites are shown in Figure 1 for samples taken high (+2–1 m) and low (+0.7 to −0.15 m) in the intertidal zone at Mukkaw Bay in April and June 1965, and throughout the zone at Waadah in January and July 1968. Shell size was measured with vernier calipers to the nearest 0.1 mm, and was taken as the maximum diameter across the umbilical region between the upper margin of the shell lip and the other side (Frank 1965a).

Perhaps the most ecologically convenient attribute of *Tegula* is that it lays down an annual ring, as the precise studies of Darby (1964) and Frank (1965a), based on the recovery of individually marked specimens, have shown. I have demonstrated that annulus formation holds for the Mukkaw Bay population as well (Fig. 2). For the period October 1964–December 1966, samples of from 32 to 180 individuals were collected from the lower intertidal zone, and the amount of current shell deposition estimated as a percentage of the total growth attained during the previous year. Figure 2 shows that *Tegula* stops adding shell about November and initiates the process again in February or March. In the interval of no growth a conspicuous annulus is produced, owing either to erosion of the old shell margin, or to structural differences between old and new shell.

I have used annulus formation to estimate age for all the resident snails only in the low area at Mukkaw. In the high area a combination of slower growth and resulting congestion of annuli in the larger individuals, and shell erosion, make

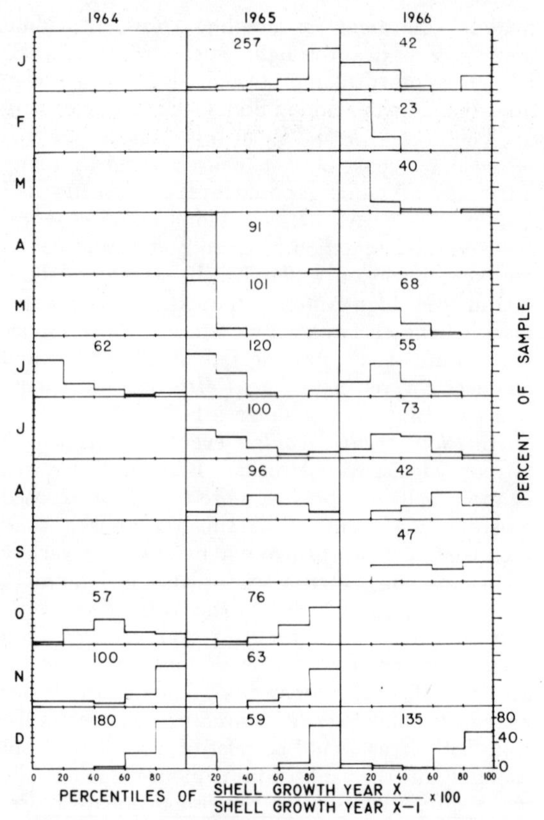

FIG. 2. The relative proportions of new shell characterizing *Tegula* at different times, suggesting that shell accretion ceases in October–November and is reinitiated in February–March. The numbers indicate sample sizes.

proper evaluation difficult. At Waadah most of the shells had *Ralfsia* or *Lithothamnion* encrustations that obscured the lines, in addition to being so large (old) that no direct count of all the annuli was possible even on the smallest individual. A growth curve (Fig. 3) characteristic of all small (< 13 mm) Mukkaw Bay *Tegula* but only for those larger individuals inhabiting the lower area was constructed as follows, assuming that the annuli are separated by approximately 1 year's growth. Since small snails live primarily in the upper intertidal zone, and begin to occur in number lower down above a size greater than 12 mm (Fig. 1), the shell diameter–age relationship between 5.9 and 13.0 mm was established primarily from high intertidal snails. For larger individuals as many annuli as possible were counted on the externally visible portion of the shell, this count being added to an estimate of the age of the remaining, internal, shell. It appears that the low portion of the Mukkaw *Tegula* population, despite growing at a reduced tempo when compared to populations at Sunset Bay, Oregon (Frank

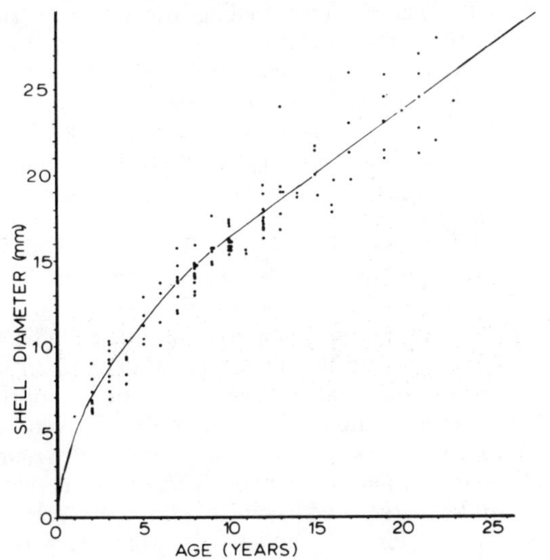

FIG. 3. A growth curve for low intertidal Mukkaw Bay *Tegula*. The line was eye-fitted. See text for method.

1965a), is capable of attaining comparable longevities.

This curve (Fig. 3) also facilitates an interpretation of the heavy skew towards larger individuals characterizing the Waadah populations (Fig. 1). Assuming that Waadah and low intertidal Mukkaw *Tegula* grow at comparable rates, the largest individual, 32.6 mm shell diameter, would be about 32 yr old and the smallest, 16.8 mm, about 11 yr old. The majority of the population with a shell diameter between 25.5 and 29.5 mm would have been between 22 and 28 yr old when sampled in 1968. Waadah Island was connected to the mainland by a breakwater in 1942–43 (Rigg and Miller, 1949), and it seems likely that this event 25 yr ago altered the ecological situation to the extent that subsequent *Tegula* recruitment was minimized and the density of *Pisaster* greatly reduced. If there is a cause-effect relationship between the breakwater construction and the population skew, these *Tegula* are a diminishing population subject primarily to the attrition of physiological death, and can be expected, in another 10–15 yr, to become rare components of the local fauna.

I have made an arbitrary division of *Tegula* at Mukkaw Bay into two groups with reference to *Pisaster's* activities: high intertidal (unexploited) and low intertidal (exploited). Size-frequency distributions (Fig. 1) indicate that Mukkaw Bay *Tegula* spend the initial 6 yr of their lives removed from *Pisaster,* and first appear in the lower intertidal zone in quantity as individuals 12–13 mm in shell diameter (6 yr old). Thus *Tegula* has a refuge that cannot be or at least is not entered by *Pisaster,* and, reciprocally, the effect of *Pisaster* predation is borne entirely by the larger size classes. Mortality in the upper intertidal zone, of unknown intensity, is more of the unpredictable catastrophic sort such as storm waves washing individuals onto shore, or sand deposition over the animals. Gulls also take their toll. Evidence will be given that at these relatively great densities ($800/m^2$) the larger individuals ($>$ 16 mm) that fail, for whatever reason, to emigrate are subject to a food shortage manifested in reduced growth and reproductive output. This could provide a proximate reason for the migratory movements, although Frank (pers. comm.) has pointed out that as *Tegula* age, they tend to move downward regardless of density, thus implicating a more ultimate cause for the movements.

Pisaster ochraceus

This species is the commonest large intertidal starfish of rocky shores along the Pacific coast, and various aspects of its biology have been covered by Feder (1959), Paine (1966), Mauzey (1966), Mauzey, Birkeland and Dayton (1968) and Landenberger (1968). At Mukkaw Bay in those flat, cobbly areas where it is associated with *Tegula, Pisaster*'s movements and activities are seasonal, and its density is significantly lower than in other portions of the environmental mosaic. To interpret the interaction with *Tegula, Pisaster*'s food preference, rate of feeding and density must be known.

Food preference

Extensive information is available on this topic, and despite variations in the techniques and their relative precision used to assess food preference trends in *Pisaster ochraceus,* the conclusions are remarkably homogeneous. In the most recent and complete laboratory study, Landenberger (1968) has shown that *Mytilus edulis* and *M. californianus* are ranked ahead of *Thais, Tegula* and a chiton in a preference hierarchy. Much the same conclusion was reached by Mauzey et al. (1968) and Mauzey (1966), the latter paper even suggesting that *Pisaster* has probably evolved many of its ecological traits in response to the abundant coastal populations of *M. californianus.* My own casual laboratory observations indicate that few other prey are consumed as long as mussels are readily available.

Statements about preference based on field studies are much less conclusive. All three investigations of the natural diet of *Pisaster* (Feder 1959, Paine 1966, Mauzey 1966) indicate barnacles as the numerically predominate prey. Feder has suggested that this may be due to an avail-

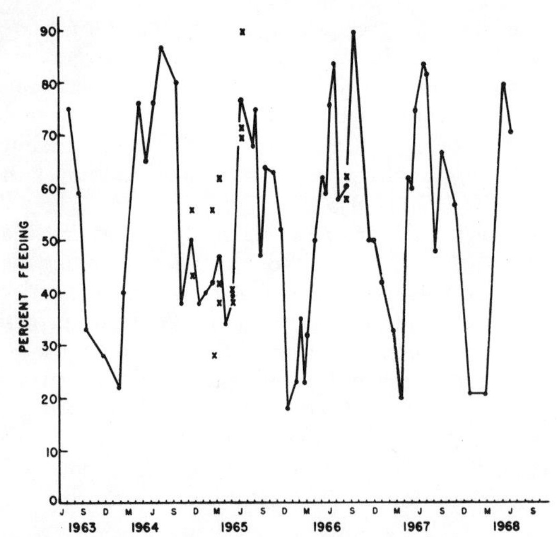

FIG. 4. The relationship between season and the strongly cyclical incidence of feeding in *Pisaster*. These data only apply to *Pisaster* feeding primarily on *Tegula*. Solid circles (●) are either individual points or the means of multiple observations (x).

ability factor, and I have shown that while barnacles are numerically the most important, they are only 30% as significant as *Mytilus* as an energy source. Extensive field observations (Paine, unpublished) further suggest when *Pisaster*'s feeding activities are accelerated in the spring (Fig. 4), that *Mytilus* is selectively taken before all other prey, and therefore that *Pisaster* must move over and bypass most of the secondarily-preferred items, including *Tegula*.

It seems conclusive that *P. ochraceus* has a strong preference for *Mytilus* spp., and that *Tegula* is lowly ranked (i.e., Landenberger 1968). At Mukkaw Bay *Tegula* forms a significant fraction of *Pisaster's* diet only when *Mytilus*, and to a lesser extent *Balanus* spp., are uncommon, a situation occurring in the area of overlap being discussed.

Rate of feeding

Pisaster is known to have a seasonal cycle in many of its activities. For instance the pattern of feeding (Fig. 4) based solely on observations in the cobbly area characterized by high *Tegula* densities, indicates that a maximum of 75–88% of all individuals observed feeding was reached in June–September of each year, and a minimum of close to 20% in December–March of all years except 1964–65. These observations were made by examining what the animals had in the folds of the cardiac stomach or adjacent to the mouth at low tide. *Tegula* consistently formed 90–100% of the observed diet and often more than one (up to five) was being digested. The ratio *Tegula* being consumed: total *Pisaster* feeding per census varies seasonally (5-yr monthly averages: July, 1.20; August, 1.34; September, 1.24; October, 1.16; November, 1.06; December, 1.02; January, 1.00; February, 1.00; March, 1.18; April, 1.12; May, 1.16; June, 1.20) and has been used as a factor in calculating the rate at which *Pisaster* consumes *Tegula*.

Field observations suggest that *Pisaster* feeds twice a day corresponding to the two periods of immersion. That is, observations made in the early morning following a night of feeding indicate no difference in the incidence of feeding from comparable observations made the same day in the late afternoon after a day of feeding. Another approach utilized was to mark all *Pisaster* sampled with a small scratch on the madreporite plate, record the percentage feeding of that sample, return at the next low tide, and record what percentage of the previously marked individuals were again feeding and the state of digestion (usually completed) of the prey. Representative pairs of percentage feedings are: July 1963 (84–84; 27–32); July 1964 (80–93); June 1965 (91–70; 70–68); August 1965 (36–40); September 1966 (80–77). These data, plus general observations on the state of digestion, indicate that there are two cycles of feeding per day. A final approach, not applied to *Tegula*, has been to calculate the rate at which populations of sessile organisms (*Mytilus* and *Balanus*) inhabiting a surface are consumed by known numbers of *Pisaster*. Very close agreement is attained between the expected and observed rates of decimation when *Pisaster* is assumed to feed twice a day. Laboratory measurements indicate that *Pisaster* can locate, catch and completely digest a *Tegula* within 6–9 hr, depending on the size relationships of prey and predator. Although *Pisaster* cannot pursue prey when exposed at low tide, the digestion process appears to continue, with the predator's cardiac stomach everted into the prey's shell. The report (Landenberger 1968) of continual feeding and voraciousness in laboratory *Pisaster* suggets that this species will feed whenever the opportunity arises under natural conditions. The implications are, then, that a similar portion of the population feeds at each opportunity and that the incidence of feeding and the number of *Tegula* per feeding starfish varies seasonally. To calculate a predation rate, I have fitted a smoothed curve to the percent feeding data (Fig. 4), determined a monthly percentage, multiplied this by the two daily feedings, the number of days in that month, the mean number of *Tegula* being eaten by those starfish actually feeding, and summed over the year. The computation for each year is

TABLE 2. Summary of data used to estimate per cent of *Tegula* standing crop consumed per year by *Pisaster* in the lower intertidal zone

Year	*Tegula* eaten yr^{-1} *Pisaster*$^{-1}$	*Pisaster* density (m^2)	*Tegula* eaten m^{-2} yr^{-1}	Avg annual *Tegula* density (m^2)	% *Tegula* eaten/yr
1963-64	413	0.23	95	369	26
1964-65	496	0.15	74	457	16
1965-66	477	0.20	95	314	30
1966-67	475	0.27	128	411[a]	31
1967-68	525	0.24	126	508	25
Avg			104	412	25

[a]Estimated by averaging 1965-66 and 1967-68 mean density figures.

given in Table 2, and has the dimension *Tegula* consumed per year by those starfish observed to be feeding. It includes an adjustment for the fact that all starfish do not appear to feed at every opportunity; it bears no relationship to the actual density of predators.

Density

Pisaster can be highly mobile, and shows a seasonal movement into the middle intertidal zone in spring and summer and a retreat to lower levels during the winter months. Throughout this period its effective density varies considerably in the area of spatial overlap with *Tegula.* Two methods have been used to evaluate density. Beginning in July 1963, a 78-m^2 plot was outlined, and the *Pisaster* within its boundaries simply counted without physical disturbance at each subsequent date when the area was accessible. Probably all individuals were noted at each census, especially during winter when algal cover was minimal, because no physical hiding places (crevices, extensive under rock surfaces) are present. Because this site was near the upper limit of *Pisaster*'s annual excursions, considerable annual variations in density are apparent: 1963–64 max. $0.65/m^2$, mean 0.18; 1964–65 max. 0.45, mean 0.10; 1965–66 max. 0.40, mean 0.13; 1966–67 max. 0.50, mean 0.10; 1967–68 max. 0.37, mean 0.08. The minimum density in all years was 0. The second method was to lay a rope marked at random intervals normal to the low water mark, and to count the *Pisaster* in 1-m^2 quadrats, care being taken to overturn rocks and search each quadrat thoroughly. Between 40 and 150 of these were taken each year, in the spring or early summer, and tended to traverse the zone co-occupied by *Pisaster* and *Tegula* at a time when the predators were active. These values, by year, of $0.23/m^2$, 0.15, 0.20, 0.27, and 0.24 though greater than the mean annual figures given above for the permanent area, have been accepted as more representative of the total area of overlap and are used in Table 2 to translate the number of *Tegula* consumed per year into a more meaningful rate, the number of *Tegula* consumed *Pisaster*$^{-1}$ m^{-2} yr^{-1}.

IMPLIED CONSEQUENCES OF THE INTERACTION FOR *Tegula*

Mortality

The data on number of *Tegula* consumed *Pisaster*$^{-1}$ m^{-2} yr^{-1} are available in Table 2, as are the estimates of *Tegula* density low in the intertidal where they are exposed to this mortality (Tables 1 and 2). When those consumed are expressed as a percent of those available, *Pisaster* apparently eats from 16–31% (5-yr average 25%) of the available *Tegula* per year. There is no way of evaluating whether the annual variation in mortality is biologically meaningful, or whether it simply reflects the fact that all the estimates are highly derived. There is a technique, however, to assess the magnitude of mortality that is independent of field observation on *Pisaster.* Because *Tegula* can be aged, I have constructed the tabular equivalent of "catch curves" (Ricker 1958). In five instances the *Tegula* density data obtained in randomly placed quadrats (Table 1) were supplemented by all the individuals found in more haphazard sampling from the same part of the intertidal zone. The specimens were aged from the growth curve (Fig. 3) appropriate for low intertidal *Tegula.* These age distributions and their derived implications are given in Table 3. The instantaneous death rate, *d,* was calculated from the relationship for an exponential decrease, $N_t = N_o e^{-dt}$ over the indicated interval of years. The percent annual mortalities equivalent to the instantaneous mortalities were taken from the appropriate table in Ricker; they indicate an average annual mortality of about 28%/yr between the age of 6–17 yr. The close agreement between these two independently derived estimates greatly strengthens the impression that *Pisaster* annually

TABLE 3. The age structure of 5 samples of *Tegula* inhabiting the lower intertidal zone. Estimates of the instantaneous death rate, d, for each sample have been calculated for the indicated interval, and then converted to an estimate of per cent annual mortality

Sampling time	Age structure of sample (yr)																							N	*d*	% annual mortality	Age interval for *d* estimate	Number of years
	0	1	2	3	4	5	6	7	8	9	10	11	12	13	14	15	16	17	18	19	20	21	22					
June 1964		0	1	1	2	11	27	57	37	26	20	22	15	9	12	5	5	2	1	0				253	0.27	24	7–16	9
Oct. and Nov. 1964				0	12	28	55	85	61	43	23	25	6	5	3	3	3	0	0	1	0			353	0.37	31	7–16	9
April and June 1965					0	3	10	30	44	31	47	38	38	34	22	12	5	7	1	0	1	1	0	324	0.27	24	10–17	7
Sept. 1965						0	18	34	47	36	23	10	14	4	3	3	0	1	0					193	0.39	32	8–15	7
March 1966		0	2	3	2	15	67	48	43	27	24	12	10	7	4	1	4	2	1	1	0			273	0.31	27	6–17	11

consumes a significant portion of the *Tegula* standing crop.

Sex ratio

One aspect of the predation is particularly puzzling. Both Frank (pers. comm.) and I have found sex ratios of unity in some *Tegula* populations. For instance, in one sample from Waadah examined in July 1968, with a mean individual diameter of about 29 mm, 46 females were found in a sample of 85. At Mukkaw Bay, however, a definite bias exists with a significant tendency for larger individuals to be female (Fig. 5). Thus up to a diameter of at least 18 mm (age about 12 yr) and probably 22 mm (18 years old) there is no discernible difference in sex ratio. Beyond this point, however, the proportion of females increases rapidly. Because the large individuals are relatively rare at Mukkaw (Fig. 1, Table 3) this age-specific skew will have little effect on the sex ratio of the entire population. Comparisons of the exploited Mukkaw population and unexploited Waadah one suggest, though far from prove, that some attribute of *Pisaster* predation may be causal to the phenomenon. *Tegula* exhibits a well developed, though seemingly relatively ineffectual in nature, escape response to *Pisaster* (Feder 1963, Yarnall 1964), but why or how this might contribute to the observed skew is enigmatic, especially so because of the suggestion of male altruism. The finding of similar rates of growth in males and females (Frank 1965a) eliminates differential growth as a possibility.

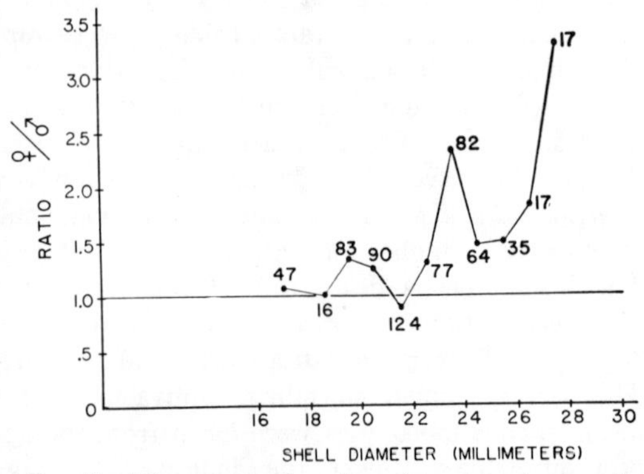

FIG. 5. The progressive skew in sex ratio favoring female *Tegula* collected in the low area at Mukkaw Bay. This size (age) specific skew is unlikely to cause the population ratio to depart significantly from unity (see text).

Growth

Through its migration *Tegula* enters a portion of the environment in which food is apparently much more abundant, as indicated by the greater algal standing crop. In Figure 6 the width of the last complete growth ring as measured along the body whorl is plotted against size for samples from both high and low populations. This should permit a comparison of the relative amount of growth in the last complete year for individuals of a given size. An analysis of covariance (Snedecor 1959) indicated that the lines are not parallel ($F_{(1, 125)} = 18.8$) and no further analyses were attempted of these highly variable data. Two tendencies are noticeable. Small *Tegula* up to a diameter of 16–17 mm (about 10 yr old) grow better higher up than they will lower down. Beyond the point of intersection of the lines, however, it is increasingly advantageous in terms of growth potential for an individual to be situated lower in the intertidal zone, albeit at times directly exposed to predation. This difference in growth probably can be attributed to a greater algal productivity lower down, since demand, as indicated by *Tegula* standing crop biomass, is 2.4 times greater than in the higher area.

Reproduction

I have chosen to present the data on reproduction elsewhere because they form a primary component in an analysis of *Tegula* energetics. These data, for the present purposes given only for females, suggest one spawning annually between the months of May and September, and indicate that different sexual categories are characterized by different caloric values: all immature individuals (5.2 Kcal/ash-free g), intermediately mature females (5.8) and fully mature females (6.1).

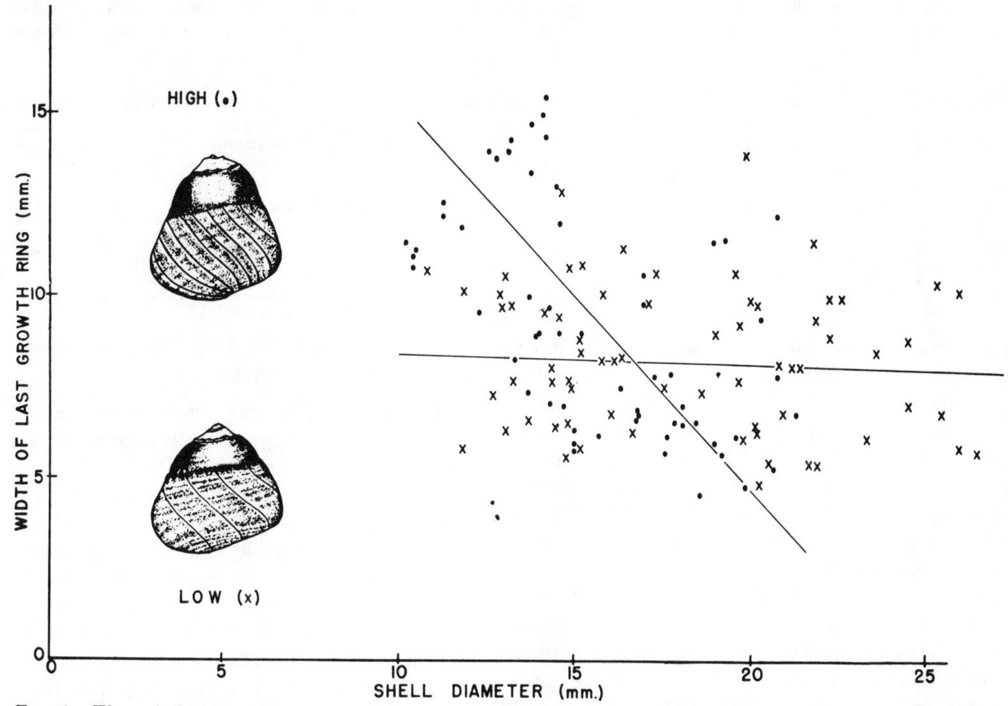

FIG. 6. The relationship between shell diameter and the width of the last complete year's shell growth, measured along the body whorl. Lines of best fit determined by the median procedure. The *Tegula,* suggesting the magnitude of difference in the position of the annuli, were drawn from preserved specimens by Miss Sue Heller.

Further, in large fully mature female *Tegula* (> 18 mm diameter) the caloric content decreases 8.2% on spawning owing to the release of energy rich gametes. Smaller individuals (14–18 mm) were estimated to lose 3.4% of their pre-spawning value. For any population then, the average caloric decrease on spawning is a function of size structure; in the high area at Mukkaw 1.9% is lost while at Waadah, where almost all individuals encountered are fully mature, a near maximal value of 8.1% is reached.

These and other values are shown in Table 4. Column A is based in part on Tables 1 and 2. Column B is derived from the data in Figure 1 and other representative samples. The gonad in *Tegula* becomes detectable at 12 mm and has been assumed to become functional at 14 mm. On this basis 83% of the individuals from the low, 7% of those from the high, and 100% of the Waadah populations shown in Figure 1 would be reproductively capable. Column C shows the percentage of those in the reproductive class judged to be fertile. I never found a female *Tegula* in the low zone before spawning that did not have a full gonad and the same is assumed to apply at Waadah. In the high area, on the basis of the caloric studies, 13/15 (87%) were so classified. Column D gives the mean shell diameter of mature individuals from the areas, E their mean pre-spawning values in Kcal and F, the average percent caloric loss on spawning, adjusted for the reproductive structure of the population. In Column G is an estimate of the Kcal spawned from an average sexually mature female. Column H represents the product of Columns A (assuming a sex ratio of unity), B, C and G, and suggests the relative reproductive or genetic contribution of populations in different parts of the environment. One aspect of reproduction correlated with the contact with *Pisaster* at Mukkaw Bay is very clear. Those snails may not live as long, but on an individual basis (Table 4, Column G) they produce about four times the caloric value of ova per spawning. This figure would have been greater if the "representative" sample being considered (Fig. 1) had included more of the larger individuals, and if the skew in sex ratio favoring females (Fig. 5) had been taken into account. On a population basis as well (Table 4, Column H) the reproductive advantage lies with the low population with about 28 times the caloric value of ova/m² being generated. To interpret this latter figure fairly,

TABLE 4. Comparison of mean individual and population reproductive performances of female *Tegula*

Intertidal location	*A* Female *Tegula* density (m^2)	*B* % sexually mature	*C* % fertile of mature size	*D* Avg size mature individual (mm)	*E* Kcal of avg reproductive individual	*F* Avg % caloric loss on spawning	*G* Kcal ova per avg individual	*H* Kcal ova per m^2
Mukkaw low	206	83	100	17.2	1.40	5.6	0.078	13.34
Mukkaw high	400	7	87	16.0	1.03	1.9	0.020	0.49
Waadah	10	100	100	26.4	6.10	8.1	0.494	4.94

though (and ignoring the group selection attributes), the relative proportions of the whole area occupied by high and low *Tegula* would have to be known, to give an indication of the total reproductive output from different portions of the environmental mosaic. These data are not yet available, although I have estimated that the high areas are three times as prevalent as the lower ones at Mukkaw Bay.

The situation at Waadah is less easily interpreted. If the assumptions of percent sexually mature, percent caloric loss on spawning, and individual caloric value can be satisfactorily established from Mukkaw material, then the average female reproductive performance, six times greater than for low Mukkaw populations, would indicate no detrimental influences affecting unexploited populations. However, the Waadah population may be atypical because of the highly skewed age structure and low density. Further, these *Tegula* seemed to be exerting some grazing pressures on the environment, patches of which appeared overexploited as indicated by the conspicuous presence of coralline algae, suggesting that some diminution of reproductive output could be expected. Final interpretation of the reproductive contribution (the relative fitness) of female *Tegula* will depend on direct measures from unexploited populations of high density inhabiting low areas.

The caloric analysis also suggested that large (> 20 mm) female *Tegula* in the upper intertidal zone were characterized by a reduced caloric content in comparisons with individuals of similar body size and reproductive condition collected lower down. Thus the 13 high-area, large, prereproductive *Tegula* combusted averaged 5.8 Kcal/ash-free g whereas the mean for comparable low zone individuals was 6.1 Kcal. The implication is that fewer calories are translatable into reproductive products in the high area at Mukkaw, providing both another proximate and perhaps also ultimate reason for the tendency for large individuals to migrate lower into the intertidal area.

RELEVANCE OF THE INTERACTION TO THE SYSTEM IN GENERAL

The above data suggest a seemingly paradoxical situation. *Tegula* settles in areas relatively free of known predators, develops there to the age of sexual maturity, and then tends to migrate, entering a portion of the environment co-occupied by a predator capable of consuming a significant fraction of its population per year. Such behavior would seem maladaptive unless mitigating circumstances were present. I believe one explanation is that, historically, the relatively reduced growth and reproduction demonstrated for dense, high intertidal populations has proven conducive to the evolution of migratory behavior. That such movements tend to be lower rather than higher, as in the case of *Acmaea digitalis* (Frank 1965b), has probably been reinforced by the predatory activities of *Pisaster ochraceus*. Thus, the argument would go, any unexploited, dense *Tegula* population would place such a demand on the local resources that the demonstrated relative advantages would not occur. The fitness of individual adults, then, would be independent of their intertidal position, and no selective advantage would accrue through migration. Moderate exploitation localized to one segment of the population would alter this condition. The interaction between *Pisaster* and *Tegula* is capable of enhancing this directional tendency for two reasons. (1) *Pisaster* is probably unable to attain densities in this cobbly area that would normally lead to over-exploitation (elimination) of the prey, because of the grinding action of the substratum during storms, and perhaps also due to the heavy algal cover impeding their movements. Thus if the density of 1–3 *Pisaster*/m^2 that characterizes the *Mytilus-Balanus* area existed in the *Tegula* patch, and the calculated rate of predation (Table 2) held, *Tegula* at the observed densities would be eliminated within a year. In addition to *Pisaster* such other macroscopic invertebrates as chitons, sea anemones and sea urchins that are abundant on immediately adjacent solid substrata but which are relatively vulnerable to crushing are also uncommon. (2) *Tegula* is not a highly preferred prey (Feder 1959, Landenberger 1968), therefore deriving a modicum of immunity because in nature *Pisaster* seems to search for other prey first (Paine, unpublished). I cannot calculate the extent to which this contributes to their spatial coexistence, but it must have some effect. This type of behavior is particularly noticeable in those areas where *Pisaster* in its preferential search for *Mytilus* leaves behind clumps or individuals of *Balanus cariosus, Pollicipes polymerus* and *Katherina tunicata,* all of which eventually serve as important energy sources (Paine 1966).

The general significance of a prey species' relative position in a food preference hierarchy may be seen in a compilation of the geographic ranges of *Pisaster* and its main prey (Table 5), judged to be those open coast species that form the bulk of the published feeding observations (Feder 1959, Paine 1966, Mauzey 1966). These show a considerable zoogeographic homogeneity, characteristic of the entire assemblage (Ricketts and Calvin 1952, Glynn 1965), although there are species replacements and diversity increases toward the south. Many of the species that characterize the association throughout its entire range can be shown to be affected either directly or indirectly by *Pisaster.* As a vast oversimplification I believe that the major characteristics of this widespread community, especially in the middle and lower intertidal zone, can be explained by the intermeshings of three ecologically different processes.

The first has been described in detail for *Tegula,* in which the reproductive output (fitness) of mature individuals in an exploited subpopulation was higher than in an adjacent unexploited one. Further, an extensive literature on exploited populations, for instance Nicholson (1954) or Watt (1955), often shows a maximization of certain population processes under intermediate levels of exploitation. In these cases the predation effect is not measured by numerical size or weight of the standing crop, but rather by its turnover and/or productivity, including reproductive contribution. It is thus reasonable to suppose, as long as the prey does not reach a more preferred status with the main predator, that coexistence is not only possible but also selected for. That is, there is an advantage to *Tegula* to live lower in the intertidal as long as this advantageous behavior is not negated by either an increased effect of the predator or increased "crowding." Two species in Table 5, *Tegula* and *Katherina,* both of which are important energy sources for *Pisaster* (Paine 1966) probably are of this sort, and *Acmaea t. scutum* may also be; all three are characterized by a perplexing intimacy of association with their major predator.

TABLE 5. The geographic distribution of the major prey species and associates of *Pisaster ochraceus.* An asterisk (*) indicates those species not known to be preyed upon by *Pisaster.* The sources for the ranges are (1) Ricketts and Calvin (1952); (2) Scagel (1957); (3) Abbott (1954); (4) Lance (1961); (5) Pilsbry (1916)

Species	Source	Northern limit	Southern limit
Pisaster ochraceus	1	Sitka	Baja California
Mytilus californicus	1	Alaska	Mexico
Pollicipes polymerus	1	Bering Straight	Middle Baja California
Balanus glandula	1	Aleutian Islands	Ensenada, Baja California
B. cariosus	5	Bering Sea	Oregon
Tetraclita squamosa	5	Farallones, Calif.	Cape St. Lucas, Baja California
Acmaea digitalis	3	Aleutian Islands	San Diego, California
A. t. scutum	3	Alaska	Baja California
A. scabra	3	Vancouver, B. C.	Baja California
Tegula funebralis	3	Sitka	Baja California
*Searlesia dira**	3	Alaska	Monterey, California
Thais emarginata	3	Alaska	Baja California
*Archidoris montereyensis**	4	Alaska	San Diego, California
*Halichondria panicea**	1	Cosmopolitan	
Haliclona sp.*			
Katherina tunicata	3	Alaska	Baja California
Tonicella lineata	3	Aleutian Islands	San Diego, California
Mopalia muscosa	3	Alaska	Baja California
*Anthopleura xanthogrammica**	1	Unalaska	Panama
*Endocladia muricata**	2	Alaska	Mexico

The second process is a direct interaction that is completely negative for the prey species. Paris (1960) and Paine (1966) have described the local destruction of mussels by *Pisaster,* an interaction resulting in a contiguous yet generally nonoverlapping arrangement of the prey and predator in space, i.e., a vertical zonation with a well-defined lower limit, since the probability of an individual member of the prey species surviving to sexual maturity in the overlap zone is limited. This suggests that the major portion of the reproductive output of preferred prey species is derived from those individuals inhabiting a refuge; their continual settlement lower down and hence momentary exposure to a devastating predator is surely related to the incomplete nature of natural predation in the areas of spatial overlap. Observations and caloric calculations on *Mytilus californianus* and *Balanus cariosus* demonstrate that because the probability of being eaten increases rapidly with age very few, unlike *Tegula,* attain sexual maturity in the zone of co-occurrence and that their mean contribution is lower than those living in the refuge. Thus there is no demonstrable reproductive advantage for the prey correlated with the association. At Mukkaw Bay and elsewhere *Mytilus* and *Balanus* are the preferred prey and an apparently stable, large scale, coexistence is only attained because they are capable of recruitment, growth and reproduction in a portion of the environment inaccessible to *Pisaster*. Without the refuge the situation would probably change into one in which all major participants were both rarer and more patchily distributed.

The final important process is one in which living space is kept available for a complex of other species through the activity of the keystone species. Here the keystone species preferentially removes a prey on which or with which the others cannot coexist. Organisms in this category may or may not be consumed by the system's primary carnivore. They often support carnivores of their own, but these, though of equivalent trophic status to *Pisaster,* lack its effect on community structure. These species are closely and predictably associated with *Pisaster* within the zone of maximum foraging, but seem unable to maintain their presence in its absence as indicated by the experimental removal of the keystone species (Paine 1966). For instance, the sponges *Haliclona* and *Halichondria* and their predator *Archidoris* (incorrectly identified as *Anisodoris* in Paine 1966) all require solid surfaces for occupancy. The sponges are outcompeted for space by mussels, and in the absence of its prey, the nudibranch leaves. The anemone *Anthopleura xanthogrammica* feeds extensively on mussels that have been dislodged from the clumps and washed into surge channels. The feeding activities of *Pisaster* increase the probability that mussels will be detached and transported by wave action. Further, the anemones lose in space competition with *Mytilus,* a condition ameliorated by *Pisaster:* both P. K. Dayton (pers. comm.) and I have observed anemones cut in half by mussels pushed through them by forces generated within a mussel clump. The filamentous red alga *Endocladia muricata* persists in part because *Pisaster* consumes small *Mytilus* that have settled on this plant in immense numbers, and which would, through growth, lift it from the rock surface. *Endocladia* foraged over by *Pisaster* are usually unharmed, and have many fewer small mussels than those in other areas. More qualitative examples could be given, demonstrating the broad influence of *Pisaster*. Thus, although it is possible that the highly congruent geographical ranges occupied by those species known to be affected by *Pisaster,* either directly or indirectly, are simply a historical accident, it is also possible that the ecological interactions and mechanisms discussed above have played a meaningful role in the development and deployment of this set of organisms.

Acknowledgments

Many friends have generously provided field help and companionship during the course of this research, and they deserve special recognition. R. M. Cassie gave advice on certain statistical matters, and the manuscript has benefited from critical examinations by J. M. Emlen, P. W. Frank, A. J. Kohn and G. H. Orians, to all of whom I am grateful. Prof. J. E. Morton kindly placed facilities at the University of Auckland, where this paper was written, at my disposal. The research has been supported by the National Science Foundation under grants GB-341 and GB-2950.

Literature Cited

Abbott, R. T. 1954. American seashells. D. van Nostrand Co., Inc., Princeton, N. J. 541 p.

Best, B. A. 1964. Feeding activities of *Tegula funebralis* (Mollusca: Gastropoda). The Veliger **6**(suppl): 42–45.

Brooks, J. L. and **S. I. Dodson.** 1965. Predation, body size, and composition of plankton. Science **150:** 28–35.

Darby, R. 1964. On the growth and longevity in *Tegula funebralis* (Mollusca: Gastropoda). The Veliger **6**(suppl.): 6–7.

Feder, H. M. 1959. The food of the starfish, *Pisaster ochraceus,* along the California coast. Ecology **40:** 721–724.

———. 1963. Gastropod defensive responses and their effectiveness in reducing predation by starfishes. Ecology **44:** 505–512.

Frank, P. W. 1965a. Shell growth in a natural population of the turban snail, *Tegula funebralis.* Growth **29:** 395–403.

———. 1965b. The biodemography of an intertidal snail population. Ecology **46:** 831–844.

Glynn, P. W. 1965. Community composition, structure and interrelationships in the marine intertidal *Endocladia muricata-Balanus glandula* association in Monterey Bay, California. Beaufortia **12:** 1–198.

Landenberger, D. E. 1968. Studies on selective feeding in the Pacific starfish *Pisaster* in southern California. Ecology **49:** 1062–1075.

Lance, J. R. 1961. A distributional list of southern California opisthobranchs. The Veliger **4:** 64–69.

Maguire, B., D. Belk and **G. Wells.** 1968. Control of community structure by mosquito larvae. Ecology **49:** 207–210.

Mauzey, K. P. 1966. Feeding behavior and reproductive cycles in *Pisaster ochraceus.* Biol. Bull. **131:** 127–144.

Mauzey, K. P., C. Birkeland, and **P. K. Dayton.** 1968. Feeding behavior of asteroids and escape responses of their prey in the Puget Sound region. Ecology **49:** 603–619.

Nicholson, A. J. 1954. Compensatory reactions of populations to stresses, and their evolutionary significance. Aust. J. Zool. **2:** 1–8.

Paine, R. T. 1966. Food web complexity and species diversity. Amer. Naturalist **100:** 65–75.

———. 1969. A note on trophic complexity and community stability. Amer. Naturalist **103:** 91–93.

Paris, O. H. 1960. Some quantitative aspects of predation by muricid snails on mussels in Washington Sound. The Veliger **2:** 41–47.

Pilsbry, H. A. 1916. The sessile barnacles (Cirripedia) contained in the collections of the U. S. National Museum; including a monograph of the American species. Bull. U. S. Nat. Mus. **93:** 1–357.

Ricker, W. E. 1958. Handbook of computations for biological statistics of fish populations. Bull. Fish. Res. Bd. Canada **119:** 1–300.

Ricketts, E. F. and **J. Calvin.** 1952. Between Pacific tides. 3rd edition, revised by J. W. Hedgpeth. Stanford Univ. Press. 502 p.

Rigg, G. and **R. Miller.** 1949. Intertidal plant and animal zonation in the vicinity of Neah Bay, Washington. Proc. Calif. Acad. Sci. **26**: 323–351.

Scagel, R. F. 1957. An annotated list of the marine algae of British Columbia and Northern Washington. Nat. Mus. Canada Bull. **150**: 1–289.

Snedecor, G. W. 1959. Statistical methods. 5th edition. The Iowa State College Press, Ames, Iowa. 534 p.

Wara, W. M. and **B. B. Wright.** 1964. The distribution and movement of *Tegula funebralis* in the intertidal region of Monterey Bay, California. The Veliger **6**(Suppl.): 30–37.

Watt, K. E. F. 1955. Studies on population productivity. I. Three approaches to the optimum yield problem in populations of *Tribolium confusum.* Ecol. Monogr. **25**: 269–290.

Yarnall, J. L. 1964. The responses of *Tegula funebralis* to starfishes and predatory snails. The Veliger **6**: (suppl.): 56–58.

21 COMMUNITY METABOLISM IN A TEMPERATE COLD SPRING

JOHN M. TEAL[1]
Biological Laboratories, Harvard University, Cambridge, Massachusetts

INTRODUCTION

The study of community metabolism is one means of making a functional analysis of an ecosystem. Essentially it consists of the study of energy transformation by the organisms of an ecosystem. It provides a measure of the total activity of a community just as a study of individual metabolism does for an individual organism.

The present study of the relatively simple ecosystem of a cold spring was undertaken to provide a more exact measurement of community metabolism than had been available. It should be emphasized, however, that in the present state of our knowledge of community metabolism considerably more assumptions have to be made in order to present a complete picture than would be the case in many other fields.

Studies of community metabolism have been generally made either in terms of energy or of biomass (either as biomass itself or in terms of a portion of the biomass such as protein or fat). The author follows the lead of Macfadyen (1948) in believing energy units to be preferred in studies of community metabolism.

Biomass units are less suitable because there is recirculation of matter in the ecosystem and because the rates of turnover are so different for different sizes and species of organisms. Macfadyen (1948) has shown that confusion often results from the fact that many authors fail to see the distinction between the cycle of matter in a community and the flow of energy through a community. For example, Gerking (1954) states that the variability in quantity of fat in organisms makes calories an unsuitable unit for production studies.

Energy enters the organic world in the form of sunlight which is absorbed by the green plants and this energy is then used by those plants and by the organisms which feed upon the plants to do their internal and external work. The energy which enters the ecosystem in the form of heat is not usually important. Warm-blooded animals, if their body temperature is maintained by their environment expend less energy in keeping themselves warm. However, this effect is not important for several reasons: (1) the temperature range of all animals is small; (2) most animals are not warm-blooded; (3) related animals which live at and are adapted to different temperatures tend to have a similar rate of metabolism (Bullock 1955); (4) experiments with large domestic animals have shown that the amount of energy saved by environmental heating of the animal's body is negligible.

There are two essential points about the transfer and use of energy that need emphasis. The first is that according to the law of conservation of energy, whatever energy is used by the organisms in doing work will appear as a definite amount of heat which is lost as far as the organisms are concerned. The second is that whatever path the chemical reactions follow, an identical amount of energy is released in the oxidation of a unit amount of an organic compound.

In practice it is not possible actually to measure the increase in heat that results from the organisms' transformation of energy. It is necessary to calculate the calories transformed from data obtained through respiratory rate measurements. This can be done with the aid of the average oxycalorific coefficient determined by Ivlev (1934) since all of the energy that an aerobic organism uses is derived from the oxidation of organic compounds. Allowance must be made for the respiration of that biomass which was produced within one sampling interval and which also died within that interval and so did not appear in any sample (Birch & Clark 1953).

The trophic level concept of Lindeman (1942) has been used in this paper although it has been modified to meet the objections of Ivlev (1945) by considering each important species separately and by dividing the energy flow, for a population that functions to an important extent on more than one trophic level, among the levels concerned. Lindeman's methods of quantifying his trophic level analysis, used also by Dineen (1953), are open to many criticisms (Birch & Clark 1953, Macfadyen 1948) and in this study more accurate methods have been used.

The calculation of various ratios is of value in comparing the energy flow in different species and different communities. The terms in these ratios are defined here as follows: "Assimilation" is the rate of energy assimilation by a population; "Energy Transformed" or "Respiration" is the rate of energy use; "Net Production" is the difference between the previous two. "Gross Production" is used only in reference to primary producers and refers to the energy fixation (Odum 1956).

The ratio, $\frac{\text{net production}}{\text{assimilation}}$, is commonly used by those interested in the amount of potential food that a population can produce. This is the efficiency with

[1] Present address: Marine Biological Laboratory, University of Georgia, Sapelo Island, Georgia.

Reprinted from *Ecological Monographs*, 27:283–302, 1957.

which energy is fixed in the organic matter of a population or trophic level and made potentially available as food to other populations or trophic levels.

The ratio of respiration to assimilation is also used. To the energy assimilated in a time unit must be added the energy equivalent of any decrease in standing crop within the period since such a decrease represents a mobilization of energy previously assimilated and stored in the bodies of organisms.

The ratio of respiration to assimilation is also used. R. S. Miller for their interest and helpful suggestions, also to Selwyn Roback, H. K. Townes, C. J. Goodnight and Arthur Clarke for help in identifying the fauna. The cooperation of the Root family who own the spring is gratefully acknowleded.

The Cold Spring

Small, constant temperature springs are as nearly perfect systems for the study of community metabolism as can be found in nature. They have the advantage of a comparatively unchanging chemical and physical environment, which reduces the difficulties of measurement and makes laboratory experiments simpler, for it is easier to duplicate constant conditions than varying ones. Also, the biota in cold springs has fewer species than do most communities.

In spite of these advantages, the ecology of springs, especially cold springs, has not received much attention in the United States. The faunas of the hot springs of this country were studied by Brues (1928). Dudley (1953) has investigated the faunas of some springs of varying temperature and Odum has worked with the rich cold springs of Florida (Odum 1957).

The spring chosen for this study of community metabolism is a limnocrene, a spring in which the water emerges into a basin, located on Intervale Farm belonging to the Root family in Concord, Massachusetts. It lies at the foot of a bank of glacial till which extends laterally for about one mile and from which emerge a number of rheocrenes, springs which form brooks immediately. The basin of Root Spring is about 2 m in diameter and the water, which comes out of the ground around the uphill edge and flows out in a springbrook on the opposite side, is 10-20 cm deep. Most of the bottom is covered with mud and it is in the mud that the organisms are found. As is the case with most springs there is no true plankton.

The spring was sampled from June 1953 until November 1954, although general observations extended from February 1953 to March 1955.

November 1953 through October 1954 was the period chosen for the analysis because the emergence of the insects was over for the year by November and hence egg laying by the insects was also completed.

Environmental Conditions.—Figure 1 gives a summary of the conditions within the spring from August 1953 to July 1954. Since there was no very unusual weather during this year, it seems likely that the observed conditions were typical. The fauna and flora were subject to little in the way of changes in their physical and chemical environment. The temperature varied at most 2° C from the mean annual air temperature for the Concord region, 9.5° C. The high concentration of CO_2, 20-30 ppm, had no adverse effect on the spring fauna as far as could be observed although it may have had an important effect in excluding intolerant species. The same may be said for the oxygen concentration which ranged from 26-65% of saturation.

The Flora.—The flora of Root Spring from June 1953 to November 1954 consisted entirely of benthic algae and the duckweed, Lemna. During November and December there were only a very few diatoms present. In January the flora began to increase and the first species to appear in abundance was one of the diatoms, Eunotia, which grew on all of the available solid surfaces. As the amount of light per day increased, filamentous green algae appeared in masses all over the bottom of the spring. *Stigeoclonium stagnatile* was the principal species, along with a smaller amount of Spirogyra sp. By May these species had decreased in numbers and a colonial green alga, *Tetraspora lubrica,* and a tiny diatom, *Nitzschia denticula* (?), made up the main biomass of the flora. As summer progressed the green algae all diminished in abundance and the benthic diatoms were the only plants of importance. In the autumn Spirogyra, and Oedogonium and *Coleochaete soluta* formed a considerable part of the green plant flora along with the diatoms.

Lemna covered the edges of the water out of the current in the early part of the year and was mostly gone by April. It did not seem to contribute much to the spring community because, instead of sinking to the bottom of the springpool, it was washed out of the outlet and down the springbrook when the rains of March and April agitated the water.

While the algae contributed a considerable amount of energy to the animal populations, the main source of food, as will be shown below, came in the form of plant debris, mostly leaves from apple trees, which collected on the bottom of the spring.

The Fauna.—Although over 40 species of animals were identified from Root Spring, there were relatively few species that occurred in numbers and sizes large enough to be important in the energy balance. The most abundant animals were those which fed on debris and algae, taking mud into their gut and assimilating the digestible material. These included the oligochaete, *Limnodrilus hoffmeisteri;* and the chironomid larva, *Calopsectra dives.* Also feeding upon detritus and debris were the isopod, *Asellus militaris,* the amphipod, *Crangonyx gracilis,* and the fingernail clam, *Pisidium virginicum.* The snail, Physa, feeds on detritus and on algae which it scrapes off of the surface of the mud. The caddis larvae, *Frenesia difficilis, F. missa,* and Limnophilis sp. eat larger bits of vegetation.

Another chironomid, *Anatopynia dyari,* eats other animals but when this food is in short supply it can get along on plant material. Other predators were the planarians, *Phagocata gracilis* and *P. morgani.*

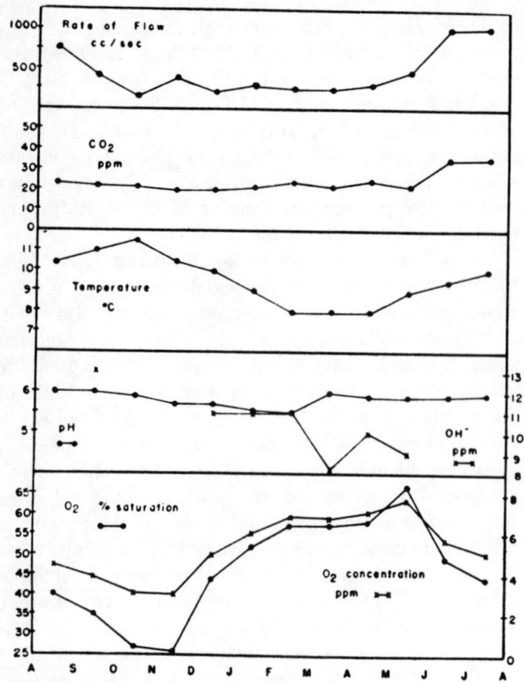

FIG. 1. Physical and chemical characteristics of the water in Root Spring, Concord, Mass., 1953-54. All data are plotted as monthly averages.

Root Spring differed from the characteristic cold spring described by Pennak (1954) in lacking leeches and black fly larvae and having planarians in abundance.

A list of all of the organisms found in the spring which were identified follows:

Algae
- Chlorophyta
  - *Stigeoclonium stagnatile* (Hazen)
  - *Coleochaete soluta* (Brebisson)
  - *Tetraspora lubrica* (Roth)
  - *Spirogyra* sp.
  - *Oedogonium* sp.
  - *Closterium* sp.
- Chrysophyta
  - *Cymbella aspera* (Ehrenberg)
  - *Nitzschia denticula* Grunow
  - *Stauroneis phoenicentron* (Nitzsch)
  - *Eunotia* sp.
- Cyanophyta

Higher plants
  - *Lemna minor* L. (?)

Protozoa
  - *Euplotes* sp.
  - *Paramecium* sp.
  - *Rhabdostyla* sp.
  - *Vorticella* sp.

Platyhelminthes
  - *Phagocata gracilis woodworthi* Hyman
  - *Phagocata morgani* (Stevens and Boring)

Gastrotricha
  - *Chaetonotus* sp.

Rotatoria
  - *Trichocerca* sp.
  - *Lepadella* sp.

Nematoda
  - *Monhystera* sp.
  - *Plectus* sp. (?)

Nematomorpha
  - *Gordius* sp.

Annelida
- Naididae
  - *Chaetogaster langi* Bretscher
- Tubificidae
  - *Limnodrilus hoffmeisteri* Claraparede
  - *Peloscolex* sp.

Mollusea
  - *Pisidium virginicum* (Gmelin)
  - *Physa* sp.

Crustacea
- Copepoda
  - *Cyclops vernalis* Fischer
  - *Canthocamptus staphylinoides* Pearse
  - *Eucyclops agilis* (Kock)
  - *Paracyclops fimbriatus* (Fischer)
- Ostracoda
  - *Eucypris* sp.
- Isopoda
  - *Asellus militaris* Hay
- Amphipoda
  - *Crangonyx gracilis* Smith

Arachnoidea
- Hydracarina
  - (Parasite of *C. dives*)

Insecta
- Collembola
  - *Tomocerus* sp.
- Ephemeroptera
- Megaloptera
  - *Sialis* sp.
- Trichoptera
  - *Frenesia difficilis* (Milne)
  - *Frenesia missa* (Walker)
  - *Limnophilis* sp.
  - *Lepidostoma* sp.
- Coleoptera
  - *Agabus* sp.
  - *Bidessus* sp.
- Diptera
  - *Calopsectra dives* (Johannsen)
  - *Calopsectra xantha* Roback
  - *Anatopynia dyari* (Coquillett)
  - *Anatopynia brunnea* Roback
  - *Pentaneura carnea* (Fabr.)
  - *Prodiamesia olivacea* (Meigen)
  - *Tendipes tuxis* Curran
  - *Brillia parva* Johannsen

Metriocnemus hamatus (Johannsen)
Hydrobaenus obumbratus (Johannsen)
Tanytarsus fuscicornis (Malloch)
Erioptera septemtrionis O.S.
Psychoda sp.
Eucorethra sp. (?)
Pericoma sp.
Culex apicalis Adams
Mansonia perturbans (Walker)
Hymenoptera
(A parasite of *C. dives*)

METHODS

Population Estimates.—A Dendy sampler (Welch 1948) was used to obtain samples of the bottom animals. A map of the spring was covered with a numbered grid and successive sampling sites were chosen from a table of random numbers. The samples were placed in a 16 mesh per centimeter sieve and the mud and fine debris removed by washing in the spring.

The animals were separated by hand under a magnifying lens from the mass of tubes and debris that remained. The individuals of each species were counted and the live weight determined after removal of excess water with filter paper.

The Phagocata presented a special problem because they secreted mucus when handled and because their epidermis permitted rapid water loss. They were placed on a fine screen and excess water was removed with filter paper. The screen with the animals was then weighed, the animals quickly removed with forceps, and the screen reweighed.

After counting and weighing the animals, with the exception of less than 1% used in experiments, were returned to the spring in order to minimize disturbance of the community.

Weight was converted to calories by oxidation with potassium dichromate in sulfuric acid (Ivlev 1934). The nitrogen content of the animals, needed for this method, was taken from the literature or found by Nesslerization.

The numbers of adult insects that emerged from the water from April to October were measured with tent traps of ⅓ m^2 or ⅒ m^2 area.

Some attempt was made to determine the relative number of bacteria in the spring. This was done following the method of Henrici (1936) by putting clean slides about one-fourth of their length into the mud and determining the time required for a coating of bacteria to grow.

Respiratory Measurements.—Rates of respiration of the principal animals were measured by the method of Ewer (1941), with the spring serving as a constant temperature bath. Animals were taken from the spring with as little disturbance as possible and quickly transferred to 20 cc syringes filled with spring water. The oxygen content of the water was then measured with the Micro-Winkler technique (Fox & Wingfield 1939). The syringes were placed in the spring and after sufficient time had elapsed for a measurable change to occur but not so long that oxygen tension was appreciably lowered (1-3 hrs) the oxygen content of the water was again determined. During the interval the water was usually kept sufficiently stirred by the activities of the animals themselves but if this was not the case, the syringes were turned over at regular intervals. This procedure measured the respiration under the same conditions of temperature, oxygen, pH, alkalinity, etc. as those which the animals normally experienced. At the end of the measurement the animals were taken to the laboratory and weighed. Oxygen consumption was converted to calories with the average oxycaloric coefficient of Ivlev (1934), 3.38 calories per milligram of oxygen. This average coefficient was used as the respiratory quotient was not known.

Since two of the species studied, Calopsectra and Limnodrilus, normally live in tubes, measurement of their oxygen consumption with the animals out of their tubes and in the syringes could be subject to error. Walshe-Maetz (1953) found, however, that while the oxygen consumption of *Chironomus plumosus* was different at oxygen concentrations below 25% saturation if the animals were removed from their tubes, at higher concentrations there was no significant difference. Since the oxygen concentration in the spring never fell below 25% saturation, the above possibility of error may be safely neglected.

Molting Losses.—When an arthropod molts it leaves a certain amount of energy behind in the material of the cast off exoskeleton. Midge larvae were kept in the laboratory and weighed daily. The loss of weight after a larva-to-larva molt represents the amount of material lost with the cast skeleton. The larva-to-pupa-to-adult transition was treated as one process. The loss of weight between the prepupal larva and the adult represents the material lost in the larva-to-pupa molt, the material lost in the pupa-to-adult molt and loss due to pupal respiration.

Oxygen Changes Due to Respiration and Photosynthesis.—Changes in oxygen due to the combined activities of all organisms living in the spring were measured by determining the changes in oxygen in the water enclosed by glass cylinders 17 cm in diameter which were pushed into the mud until they projected only 2-4 cm above the water surface. They were then filled to the top with water and covered with a glass plate. The top edge of the cylinder was ground so that the fit would be air tight. The water was sampled from all levels within the cylinder and the oxygen content measured by the standard Winkler technique at the beginning and again after a period of about 24 hours. Oxygen gradients that may have formed within the cylinder caused no appreciable sampling error. Their possible effect on the organisms was ignored.

Respiration was then measured during a period of several hours with a black box covering the cylinder to prevent any photosynthesis. By subtracting the respiration due to the macrofauna, the respiration of the decomposers, microfauna, and algae was determined.

Portion of Food Assimilated.—To find out how much potential food was not assimilated in the feeding process, Anatopynia larvae, kept at 9° C in spring water, were weighed and given weighed amounts of live Calopsectra or Limnodrilus every second or third day. The respiration of Anatopynia during the experiment was measured.

The portion of food assimilated was calculated from these measurements on the assumption that the difference between the energy contained in the food and the energy contained in the tissue added by growth plus the animal's respiration represented loss on the form of feces and in the form of material from the prey which was killed but which never entered the predator's gut.

The feeding efficiency of Phagocata was determined in a similar manner. There was a difference, however, in that most of the loss which occurred during the feeding activities of Phagocata was due to the large quantities of mucus that were secreted by the animals in their movements in search of food and in their actions in subduing their prey.

Anatopynia larvae are stated to be carnivorous (Johannsen 1937, Pennak 1954) but the present observations showed that they also digested plant food. Larvae were anesthetized each month, the contents of their guts examined under a compound microscope and the portion of animal and plant food estimated. As much could be seen in the gut of the live animal as could be observed by examining the guts of intact, cleared animals or examining food removed from larval guts. Only a relative measure of the amounts of the various types of food that were taken could be obtained with any of these methods. Difficulties were encountered in identifying the contents of the gut and in estimating the amount of food that was represented by the portions that remained.

Measurement of Mortality.—In order to determine completely the energy flow through a population it is necessary to know what quantity of energy is lost by the death of organisms from whatever cause between periods of sampling.

Each month's sample mean was assumed to represent the actual population on the midday in the month. A sampling interval extended, therefore, from the 16th of one month to the 15th of the next.

An estimate of the mortality between samples was obtained with the use of the equation given by Ricker (1946): $P_t = P_o e^{(k-i)t}$ in which k is rate of growth or net production; i is rate of mortality; P_o is weight of population at time zero; P_t is weight of population at time t; t is taken to equal one month.

Values for k were determined by raising animals in a 9° C cold room and by estimation from the greatest rate of increase that occurred in the natural population. In those populations which did not reproduce during the summer, k was determined from the average increase in weight of surviving individuals.

Once k was determined it was used as a constant (except in the cases of Calopsectra and the Trichoptera where it was determined each month by the last method mentioned above). Net production was probably about the same at all seasons of the year for most animals in the constant conditions of the spring. It makes no difference in calculating energy flow whether the net production of a population takes the form of growth of individual organisms or of increase in numbers of individuals (i.e., reproduction).

A value for i was found for each sampling interval from the above equation and multiplied by the average population (geometric mean) during the interval to give an estimate of mortality. This mean is not so exact as that given by Ricker (if i is constant for the interval considered) but it is easier to calculate. Actually, i is a continuous varying quantity rather than constant for any month and use of the geometric mean is accurate enough for this study.

Since the data for energy flow were compiled on basis of the calendar months, the total mortality was divided between the two months involved according to the ratio of the geometric means of the populations present in the second half of the first month and the first half of the second month (assuming each month to contain 30 days). The mortality is a minimum estimate since the mean of the populations present during each was used in the calculations as if it were the true value on the fifteenth of the month. This use tends to smooth out the curve of population size.

Input of Energy in the Form of Organic Debris.—The amount of debris that fell into the spring was estimated by placing a box having sides 30 cm high and a bottom area of 1/10 m^2 on the ground next to the spring and collecting the material that accumulated in the box. This material, mostly leaves, twigs, fruit, etc., was dried and weighed and its caloric content determined.

The organic matter in the inflow and outflow of the spring was measured following the method of Pennak (1946).

SPECIES POPULATIONS

The results will be presented and discussed separately for each species that was individually studied. These data will then be combined to give a picture of the community metabolism for the spring as a unit. The actual sampling data may be found in a PhD thesis deposited in the Biology Library, Harvard University.

Calopsectra (= Tanytarsus) dives (Johannsen).—This species is one of the two abundant chironomids in the spring and is a typical member of this genus characteristic of oligotrophic waters. The population of Calopsectra during the study is shown in Fig. 2 and Table 1. There were no larvae of this species during the winter. A maximum standing crop of 87.6 KC/m^2 was reached in August after which pupation and emergence of adults reduced the population again to zero. The population means for June,

TABLE 1. Population of larvae and emerging adults of *Calopsectra dives* in Root Spring, Concord, Mass., in 1954. Energy content is estimated on the basis of 1.58 cal/mg as determined for *Anatopynia dyari.*

	Jan.-April	May	June	July	Aug.	Sept.	Oct.-Nov.
Larvae							
Number in thousands/m^2	..	1.7	89.5	65.0	57.0	0.2	......
Weight in gms/m^2	..	3.0	58.5	82.4	127.0	0.2	......
Energy content in KC/m^2	..	2.1	40.4	56.8	87.6	0.1	......
Adults							
Number/m^2	..	13	170	953	3464	13250	533
Weight in mg/m^2	..	12	156	876	3170	12200	490
Energy content in KC/m^2	..	0.019	0.246	1.38	5.00	19.30	0.775

July and August differ significantly at P = 0.01. The calories contained in the standing crop were found to be 0.69 KC/gm ± 0.020 (12 determinations).

Fig. 2 also shows the number of adults that emerged and the monthly totals are given in Table 1. The sex ratio among the adults was 53% female and 47% male (based on 1500 adults).

Because mating and egg laying take place so soon after emergence it was estimated by observing the insects that about two-thirds of the females laid their eggs in the spring from which they had just emerged and the others dispersed. By counting the eggs contained in 5 virgin females it was found that they laid an average of 250 eggs each.

The number of eggs that gave rise to the 1954 population was calculated from emergence data for 1953. Fifty-eight hundred females emerged at that time and these females at 250 eggs apiece laid roughly 980,000 eggs per square meter.

The mortality of larvae calculated from Table 1 for July to August was 15% and for June to July, 28%. Thirty percent, therefore, will be used for mortality from May to June when the larvae were smallest. This figure agrees with the mortality rates of young chironomid larvae given by Borutzky (1939). Most of these larvae did not appear in the samples in May because they were so small that they passed through the meshes of the sieves. On the basis of 30% larval mortality 128,000 larvae emerged from eggs in May. As 980,000 eggs were laid, 87% of these did not survive to hatch.

As no new larvae hatched during the summer, all of the individuals are of about the same age and mortality is easily calculated with the formula:

$$\text{Mortality} = \text{antilog}\left(\frac{\log P_o - \log P_t}{2}\right)\left(\frac{\log P_p - \log P_t}{\log e}\right).$$

The first expression gives the average population present during the interval and the second gives the death rate. P_o is the size of the population at the beginning of the instar; P_t the size at the end of the instar; and P_p the theoretical size of the population at the end of the instar had there been no mortality. To facilitate the calculation, Table 2 has been set up.

Mortality was divided among the calendar months as described previously except for that calculated for the last instar which was divided among August, September and October in proportion to the adult emergence. Most of the deaths during this interval occurred as the pupae tried to wriggle out of the larval tubes and were caught in the mucus which the increased numbers of planarians had spread over the tube mouths.

The energy lost by respiration is, in a sense, the most important part of the energy flow through the population since this is the energy that the animals actually use in their life processes. From 9 measure-

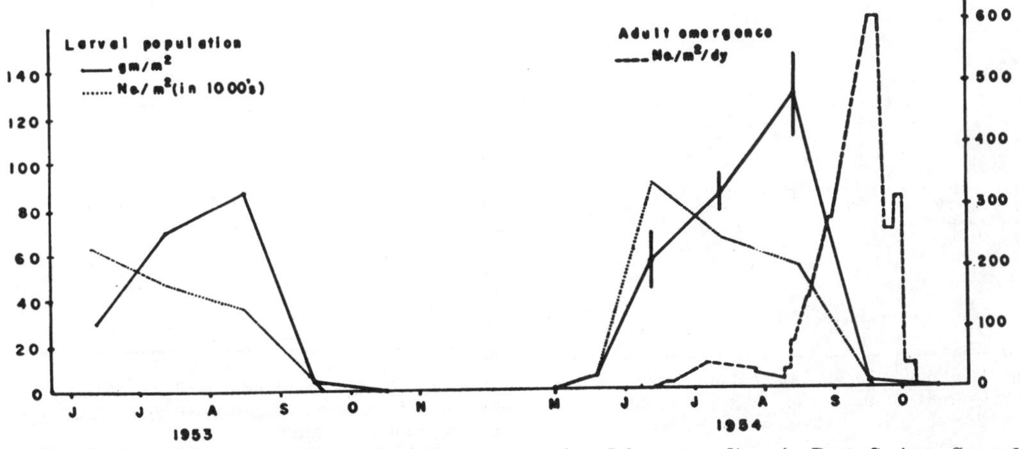

FIG. 2. Size of larval population and adult emergence for *Calopsectra dives* in Root Spring, Concord, Mass. Vertical lines represent 95% confidence limits.

TABLE 2. Data for calculation of mortality of *C. dives* larvae (per square meter). The value of P_o for May is taken from the total estimated larvae hatching multiplied by the calories per egg (.002). The value for P_t in August is the total calories contained in the larvae that successfully emerged as adults.

	Instar	No. of larvae	Po	Gram cal. larvae at end instar	Pt	Pp
May 15 to Jan 15...	1	128,000	0.256KC	.45	40.4KC	57.6KC
Jun 15 to Jul 15....	2	98,500	40.4	.86	56.8	76.5
Jul 15 to Aug 15...	3	65,000	56.8	1.54	87.6	103.1
Aug 15 to Sep 15...	4	57,000	87.6	2.09	28.8	119.1

ments the respiratory rate of Calopsectra was found to be 0.475 ± 0.022 mg 0_2/gm/hr or 1.67 ± 0.08 cal/calorie of larva per month.

With the data presented above it is possible to draw up an energy balance sheet for the population of *Calopsectra dives* during the summer of 1954 (Table 3). Calculation of the respiration of animals that died between samplings was based on an average life of one-half month (i.e. mortality uniform with time). The sum of the energy assimilated by the population of *Calopsectra dives* was 520.3 kilo-calories per square meter. The larvae transformed 389.6 KC, about 75% of the input.

Some of the energy carried out of the system in the bodies of the adults was used in the life processes of the adults and some of the energy contained in the female adults was in the eggs and would have been used by the developing embryos. Therefore, about 80% of the input energy was transformed to heat by the midges.

The efficiency, $\frac{\text{net production}}{\text{assimilation}}$, of the Calopsectra was 100% minus 80% = 20% since this is the amount of energy that the population passes on to other populations.

Anatopynia dyari (Coquillett).—Anatopynia is the second most abundant species of midge occurring in the spring. Figs. 3 and 4 and Table 4 show the size of the population of Anatopynia larvae which varied from a low of 4.4 KC/m^2 in September during pupation and adult emergence to a high of 28.2 KC/m^2 in October. These figures include two species other than *A. dyari; Anatopynia brunnea* Roback and *Pentaneura carnea* (Fabr.). Both of these latter species are similar in habits to *A. dyari* and were less than ⅕ as numerous.

The energy content of the larvae was found from 6 determinations to be 0.88 ± 0.056 KC/gm.

The differences in numbers of individuals between May and June, June and July, July and September 1954 and October and November 1953 were significant at $P = 0.01$ and between September and October 1954 at $P = 0.05$.

It was estimated from field observations that about ¼ of the emerging females lay their eggs in the spring. Since there are no other springs in the vicinity from which Anatopynia adults could reach Root Spring, 264 females which emerged in May and June deposited 23,100 eggs and 980 females which emerged in August and September deposited 86,100 eggs. (Five virgin females were examined and found to contain 325, 340, 350, 370, and 375 eggs, or an average of 350 eggs each.)

By examination of the graph of the numbers of individuals of *A. dyari* (Fig. 3) the mortality of the eggs and the very young larvae can be estimated. From November through May 15 there were about 7500 larvae/m^2. From May 15 through July, 510 adults emerged but between May and June there was a decrease in the population of 4500 larvae/m^2. This indicates that 3990 larvae died in this interval. This decrease is reflected in the curve for weight/m^2. Newly hatched larvae did not appear in the samples until about one month after the eggs were laid.

In July the numbers/m^2 increased to 5000, representing an addition of 2000 new larvae from the

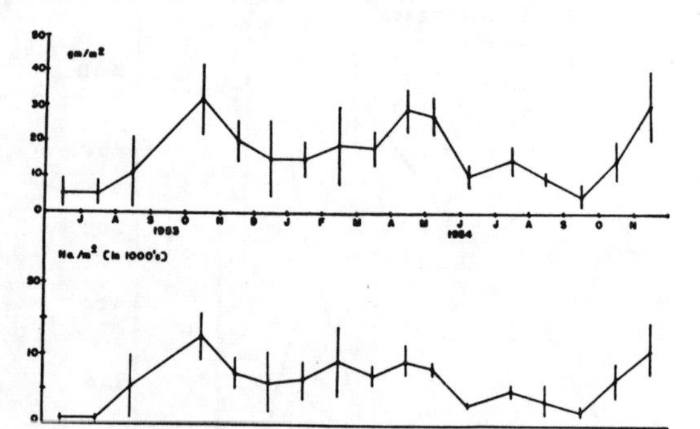

FIG. 3. Population of Anatopynia larvae in Root Spring, Concord, Mass. (The vertical line indicates the 90% confidence limits of the mean.)

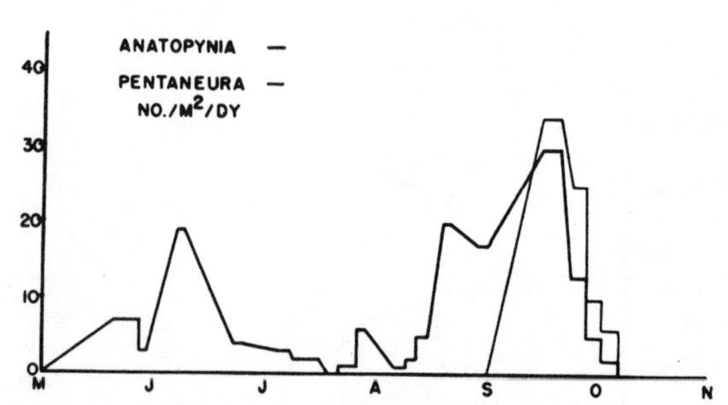

FIG. 4. Emergence of adults of Anatopynia (heavy line) and Pentaneura (light line) from Root Spring, Concord, Mass. in 1953-54.

TABLE 3. Balance sheet for energy flow in *Calopsectra dives* in Root Spring, Concord, Mass., in kilo-calories per square meter. Column 4 shows the calories respired before molting by the biomass represented in the deposited larval skins. Values for deposits were measured for *A. dyari* and assumed proportional for *C. dives.*

Month	Standing crop	Respiration of S. C.	Respiration of animals that died	Respiration of deposit	Emergence	Pupal deposit	Larval deposit	Mort.
October to April.....	....	.....	....	....	....	....	....	....
May.......	2.1	3.5	....	....	0.02	0.01	....	....
June.......	40.4	67.5	0.6	3.3	0.25	0.11	4.0	0.7
July.......	56.8	95.0	1.0	4.8	1.38	0.61	5.7	1.2
August.....	87.6	146.2	12.4	7.3	5.00	2.2	8.8	14.9
September..	0.1	0.2	45.8	....	19.30	8.5	....	54.8
October....	....	.....	1.8	....	0.78	0.34	....	2.1
		312.6	61.6	15.4	26.7	11.8	18.5	73.7

Net change in standing crop....0
Total respiration............389.6
Total assimilation...........520.3

23,100 eggs laid during May and June. A mortality of 91.4% occurred among the eggs and young larvae. In November somewhat more than 10,000 new larvae appeared indicating a mortality of 88% of the autumn eggs and young larvae. On the basis of a three to one ratio for the mortality of eggs to the mortality of young larvae (from the data of Borutzky 1939 for another species of chironomid), it was calculated that these larvae stored a total of about 0.4 KC/m^2 in their bodies before they died, and this amount is included in the tabulation of energy flow. This calculation is very rough since exact data are not available, but it is certain that the effect of these young larvae did not materially affect the total energy flow for this species, 239.2 KC/m^2.

A few determinations were made of the amount of energy lost in the molted exoskeletons of the larvae. *A. dyari* was used for these measurements because of their large size and good viability under repeated handling. Three larva-to-larva molts were successfully measured. The loss was 11.5%, 10% and 9%, or an average of 10% loss of energy/molt in terms of the final larval weight. This agrees well with the figure of Borutsky (1939) for *Chironomus plumosus* who found a loss of 16.8% in all three molts in terms of the final larval weight. A loss of 10%/molt is equal to 16.3% loss in three molts.

There is a similar loss of energy in the transformation from prepupal larva to pupa and then to adult. It was found that an adult midge came from a larva which contained 1.44 times as many calories as the adult. Thus 0.44 cal/cal of adult were deposited with each adult emergence.

It was apparent, after examination of larvae from

TABLE 4. Monthly summary of larval population and adult emergence of *Anatopynia dyari* and *Pentaneura carnea* in Root Spring, Concord, Mass., 1953-54. Larvae of the two species were not distinguished.

	1953			1954										
	Oct.	Nov.	Dec.	Jan.	Feb.	Mar.	Apr.	May	Jun.	Jul.	Aug.	Sep.	Oct.	Nov.
Larvae														
Number in 1000's...	12.4	7.2	6.0	6.3	8.9	6.8	9.1	8.1	2.6	5.1	3.8	2.0	6.6	10.9
Energy content in KC/m^2.........	28.2	17.6	13.2	12.3	15.8	15.8	25.5	16.7	9.7	13.2	8.8	4.4	13.2	27.3
Adult														
Anatopynia														
No/m^2.........	....	....	....	....	....	....	....	128	308	73	308	633	15	....
Mg/m^2..........	....	....	....	....	....	....	....	311	599	129	642	1188	31	....
Pentaneura														
No/m^2..........	....	....	....	....	....	....	....	...	...	...	...	646	40	....
Mg/m^2..........	....	....	....	....	....	....	....	...	...	...	...	667	41	....

the spring and raising larvae successfully on a diet of Spirogyra, that *A. dyari* larvae are both herbivores and carnivores. The larvae may not, however, be able to exist solely upon detritus. During December and January when algae were scarce and the larvae were not obtaining much animal food, their average weight decreased. The approximate portion of the food of *A. dyari* larvae which came from animal sources was as follows: none in January; one-eighth in April, one-quarter in November, December, March, July and October; one half in May, June and August; three-quarters in February and September.

It will be seen when the predator-prey balance for the entire community has been presented that the estimation of the proportions of foods from the two sources must be at least approximately correct.

When detritus-feeding forms eat, they take into their guts a variety of material usually in small bits and their digestive enzymes act on the mass of material to make available for absorption all the potential food. The only organic material that passes through is that which is indigestible, considered by many workers to be the crude fiber portion of the organic matter (Lindeman 1942, Birch & Clark 1953). When the food is animal prey, however, unless the predator can swallow the prey whole, a certain amount of it will be lost without ever entering the alimentary canal. This waste was measured for *A. dyari* feeding on Calopsectra and Limnodrilus.

Thirty duplicate experiments revealed thac when feeding on animals *A. dyari* are able to assimilate, on the average, only 30% of their prey (29% ± 2%). The remaining 70% is lost in the form of (1) blood of the prey which passes into the water, (2) flesh which is not swallowed by the predator, and (3) indigestible material which passes through the predator in the form of feces. This figure seems reasonable when it is realized that Anatopynia grab their prey in their mandibles and draw it into their esophagus bit by bit. The skin of the prey is ruptured in the process and blood escapes. Since the mouth of Anatopynia is smaller than the usual prey, a large amount of the flesh is also lost. The loss may be less when Anatopynia feeds on animals which are small enough to be swallowed whole.

Measurements of respiration of the larvae of this species were made in Root Spring throughout the year. Eighteen measurements gave a mean of 0.45 ± 0.04 mg O_2/gm fresh wt/hr or 1.24 calories respired/calorie of larvae/month. The figures for respiration in the balance sheet (Table 5) are corrected for mortality and deposit as explained for *C. dives.* The net change in the standing crop for the year was found by subtracting from the increases the decreases due to deposit, emigration, and mortality. Mortality was calculated from the differences in the standing crop and deposit and emigration. Mortality and increases in the standing crop were divided between months in proportion to their standing crops. Energy obtained from cannibalism was figured as ⅓ mortality on the basis of the feeding efficiency. Except for August when there were a large number Phagocata present, the mortality of Anatopynia was probably due to cannibalism since this chironomid was the most abundant predator present. In December and January the mortality may only be respiratory loss as there was no significant decrease in numbers during those months.

The total assimilation of energy by Anatopynia larvae for the year of the study, 249.1 KC/m², was found by adding losses due to respiration, death, deposits of larval and pupal exuvia, and emigration of adults. Of this total assimilation 9.9 KC came from cannibalism and 3.2 KC from a decrease in the stand-

TABLE 5. Energy balance sheet for Anatopynia in Root Spring, Concord, Mass. for 1953-54 (in kilo-calories per square meter).

Month	Standing crop	Larval deposit	Pupal deposit	Emigration	Mortality	Energy from Cannib.	Respiration	Increase in S. C.	Energy from animals	Energy from algae
Oct. '53	(28.2)	(1.4)	...	...	....	...	.....	....	....	.....
Nov.	17.6	1.0	...	...	6.1	2.0	21.4	....	4.9	14.5
Dec.	13.2	...	...	...	2.5	2.5	16.4	....	3.5	10.4
Jan. '54	12.3	...	...	...	0.4	0.4	15.3	1.6	....	16.5
Feb.	15.8	...	...	...	....	...	19.6	1.9	16.1	5.4
Mar.	15.8	...	...	...	....	...	19.6	4.3	6.0	17.9
Apr.	25.5	1.6	...	...	4.5	1.5	32.6	8.0	4.8	33.3
May	16.7	1.0	0.2	0.5	6.9	2.3	21.3	....	9.5	9.5
Jun.	9.7	...	0.4	1.0	1.3	0.4	12.0	1.7	6.6	6.7
Jul.	13.2	...	0.1	0.2	2.6	0.8	16.6	2.3	4.5	13.6
Aug.	8.8	...	0.4	1.0	3.1	(1.0)	10.8	....	5.4	5.4
Sep.	4.4	...	1.3	2.9	....	...	5.5	5.5	8.2	2.8
Oct.	13.2	1.3	...	0.1	0.2	...	17.4	13.1	7.6	22.9
Nov.	(27.3)	...	...	...	....	...	.....	....	....	.....
	−3.2	4.9	2.4	5.7	27.6	9.9	208.5	37.4	77.1	158.9

TABLE 6. Population sizes in numbers per square meter and kilogram calories per square meter for animals other than chironomids in Root Spring, Concord, Mass. in 1953-54.

	1953			1954										
	Oct.	Nov.	Dec.	Jan.	Feb.	Mar.	Apr.	May	Jun.	Jul.	Aug.	Sep.	Oct.	Nov.
Limnodrilus														
No. of animals	?	7600	7700	9000	10600	4800	4200	7300	4300	3700	3900	3450	3200	1300
Energy	23.0	32.0	30.4	32.0	53.0	17.6	18.0	30.6	16.7	16.4	23.2	21.3	18.7	3.7
Asellus														
No. of animals	1700	270	500	1100	400	100	400	100	100	200	2400	1700	900	1200
Energy	16.6	4.3	8.6	26.5	8.4	1.8	4.9	1.9	2.3	2.8	11.4	10.0	8.1	19.4
Phagocata														
No. of animals	1000	200	500	3300	1800	1100	300	400	500	900	2400	4600	1500	100
Energy	7.1	2.3	2.7	22.4	5.1	3.9	2.4	0.8	1.3	3.1	16.2	27.9	6.8	1.2
Trichoptera														
No. of animals	....	....	....	980	390	270	470	710	350	80	200	....	....	....
Energy	....	....	....	0.78	0.78	1.14	3.41	9.73	7.45	2.39	10.8	....	....	....
Pisidium														
No. of animals	1000	500	470	1300	2700	1500	2700	1500	1100	1900	1400	2400	3600	800
Energy	4.0	1.7	2.0	6.8	6.5	4.8	9.3	5.5	6.4	6.6	4.7	5.7	6.8	3.9
Physa														
Energy	0.2	0.8	1.9	3.6	1.4	1.1	1.1	0.7	1.3	....	....	....	0.1	....
Crangonyx														
Energy	....	....	0.2	1.4	0.9	0.6	1.2	0.7	0.8	0.3	4.1	0.7	....	....

ing crop, leaving 236.0 KC which came from sources outside the Anatopynia population.

The proportion of energy transformed by Anatopynia is $\frac{\text{208.5 KC respired}}{\text{249.1 KC assimilated}}$ or 84%. Their efficiency at passing energy containing material to other populations is $\frac{\text{30.7 KC passed on to other populations}}{\text{236.0 KC assimilated from outside sources}}$ or 13%.

Since Anatopynia larvae are both herbivores and carnivores, the energy flow through them may be divided into two parts in proportion to the energy obtained from animal and plant sources. This was done in the balances for these two trophic levels presented below.

Limnodrilus hoffmeisteri Claraparede.—Limnodrilus is a typical aquatic, tube dwelling oligochaete, which reproduces throughout the year in Root Spring. It feeds upon detritus.

Table 6 gives the monthly population of Limnodrilus. The caloric content was found to be 0.76 ± 0.026 cal/mg fresh wt. The great decrease in population in November 1954, just after the study period, was probably caused by the action of the several hurricanes of that fall, which disturbed the spring by uprooting nearby trees.

The amount of energy transformed by Limnodrilus, measured by respiration, was calculated to be 1.22 ± 0.35 cal/cal fresh wt/month (0.83 ± 0.024 mg O_2/mg/hr) from 15 determinations.

The mortality was calculated using the formula of Ricker (1946) by making the assumption that the rate of increase was constant throughout the year. Since the temperature of the water is nearly constant the year round and since it makes no difference for this calculation whether the increase takes the form of reproduction or growth, this assumption of constant rate of increase is probably quite reliable. It may be noted that the worms had full guts at all seasons.

To find the rate of increase, k, the two methods mentioned previously were used. Weighed animals confined in strained mud in the laboratory cold room were reweighed after two weeks. When one or more animals in an experiment died, the experiment was discarded. Of the 5 successful experiments the one with the maximum rate of increase, in which k equaled 0.474 for a 30 day month, was taken as the significant one since laboratory conditions were not as conducive to growth as were conditions in the spring to which this population of worms was adapted.

A value for the rate of increase was also obtained from the increase of the natural population in the spring from April to May, k = 0.530. Even though the P value lay between 0.05 and 0.10 for this population increase, the larger value for k was used as it agreed fairly well with the maximum value obtained in the laboratory experiments and as k obtained from the fluctuations of a natural population will have a minimum value. The balance sheet for the energy flow through the population of Limnodrilus was set up in Table 7.

These tube worms assimilated a total of 644.1 KC/m^2 during the year of investigation. Some energy assimilated previously also flowed through the population since the standing crop decreased by 13.3 KC/m^2.

TABLE 7. Energy flow figures for the animals of Root Spring, Concord, Mass., 1953-54, with the exception of the chironomids. All data in kilocalories per square meter per year.

Species	Change in S. C.	Respi-ration	Immi-gration	Morta-lity	Mucus loss	Canni-balism	Outside assimi-lation	Total energy flow	Net Produc-tion
Limnodrilus......	−13.3	483.6	....	173.8	....	....	644.1	657.4	173.8
Asellus..........	3.8	486.1	....	104.5	....	....	604.4	604.4	104.5
Phagocata.......	−0.9	18.7	....	48.2	89.1	23.9	131.2	156.0	113.4
Trichoptera......	....	67.5	18.3	39.2	....	....	88.4	106.7	39.2
Pisidium.........	5.1	90.9	....	76.7	?	....	172.7	172.7	81.8
Physa and Crangonyx.....	....	90	....	30	....	....	120	120	30

The oligochaetes used, in their life processes, $\frac{173.8\ KC}{657.4\ KC}$ or 74% of the energy assimilated. They passed on to other populations 173.8 KC, an efficiency of $\frac{173.8\ KC}{657.4\ KC}$ or 26%.

Asellus militaris Hay.—*A. militaris* is the most common and widespread of the American species of aquatic isopods. It feeds on anything edible that it encounters although it does not ordinarily kill prey. In the Root Spring the isopods reproduced throughout the year. The population size is given in Table 6.

The figure for the conversion of moist weight to calories was calculated from the analysis of the chemical composition of Asellus by Ivlev (1934).

Mortality was figured using a value for k calculated from the increase in the population from July to August. The difference between these two means is significant at $P = 0.05$. An error could arise from migration into the spring since this is one of the species that could crawl up the springbrook from the pond below. However, no Asellus were even found in the springbrook and migration was probably insignificant. Losses due to respiration were found to be 1.22 ± 0.05 mg O_2/mg/hr.

There is an error in the energy balance due to the fact that no allowance was made for energy lost by molting in this crustacean. The error would not come to more than 10% of the energy passed through the population and would probably be less to judge from the molting of the chironomids for which this calculation was made.

The energy balance for this species is presented in Table 7. Some energy, 3.8 KC, was stored in an increase in the standing crop during the year of study. About 80% of the energy intake of Asellus was used for its life processes.

Phagocata gracilis woodworthi Hyman and *P. morgani* (Stevens and Boring).—These two planarians are the only important exclusively predatory animals in the spring and feed on live or recently dead animals. Since they are able to suck in only the softer parts of their prey, the harder parts, such as the exoskeletons of arthropods, are left behind. These species were observed to have no definite breeding season and reproduced throughout the year.

Both species of Phagocata are considered together in the energy flow calculations because they are ecologically similar. They feed on the same sorts of material and live in the same habitat. The fact that they are members of the same genus also indicates probable similarity in ecology. They differ in size, however; *P. morgani* is seldom found to weigh more than 1-2 mg while *P. gracilis* attains a weight of 20 mg. They also differ in that *P. gracilis* has many pharynxes while *P. morgani* has only one, and in that most of the apparent "cannibalism" among flatworms in the spring is really due to *P. morgani* feeding on *P. gracilis*. The size of the combined populations is given in Table 6.

The calories per unit weight were calculated from data gathered for *P. gracilis* and assumed to be the same for *P. morgani* (1.33 ± 0.02 cal/mg). The rate of respiration was also calculated using *P. gracilis* as the experimental animal and was 0.0735 ± 0.005 mg O_2/gm fresh wt/hr. Starvation of planarians causes their respiratory rate to vary (Hyman 1919) but was not a factor in these determinations since the animals were taken at random from the natural population.

Flatworms differ from other sorts of animals in that they secrete a great deal of mucus in their activities. They lay down a film of mucus whenever they move over objects and use it to ensnare their prey. For this reason it was necessary to measure the amount of energy lost by the Phagocata in the form of mucus.

Food in the form of weighed amounts of live oligochaetes was given to the worms at intervals far enough apart so that as far as could be determined by observation all of the food was consumed. It was then assumed that the difference between (1) food supply and (2) respiration, growth, and inedible parts of the food represented mucus secreted by the animals. The assimilation of food given to the worms was figured to be 90% of the prey biomass on the basis of the analyses of Birge & Juday (1922) which showed that the crude fiber content of the sort of animal that the Phagocata were fed was about 10%. The results showed that Phagocata secreted an amount of energy in the form of mucus that was nearly equal to their body caloric content each month (0.94 ± 0.10 cal mucus/cal/month).

It is possible with the foregoing data to construct the energy balance sheet for the Phagocata (Table 7). Energy available from cannibalism was taken as one-half of the decrease in population of Phagocata. There are no other macroscopic animals which will eat flat-worms (Hyman 1951); therefore, all of the dead Phagocata were either eaten by their relatives or decomposed by microorganisms. It was not possible to measure mortality except by comparing the size of the standing crop in successive months. There is not enough information to explain the fluctuations in population size of these animals and no reason to assume that the rate of increase is constant throughout the year. Therefore, it is not possible to calculate mortality with Ricker's (1946) formula.

The fraction of assimilated energy transformed by Phagocata is very low, $\frac{\text{18.7 KC transformed to heat}}{\text{156.0 KC total assimilation}}$ or 12%. This is probably an adaptation made necessary by the large amounts of energy they lose in mucus which they constantly secrete. If, in calculating the fraction, the energy in the mucus is included with the energy transformed as energy "used" by the animals, the result is $\frac{\text{107.8 KC}}{\text{156.0 KC}}$ or 69%, which is much closer to the fraction of energy transformation by other animals.

Including the energy secreted in mucus with the energy transformed gives a valid basis for comparison with other animals of the community as the mucus secreted is a necessary part of a planarian's existence and demands a large proportion of the energy assimilated.

The efficiency of net production compared to assimilation is $\frac{\text{113.4 KC net production}}{\text{131.2 KC assimilated from sources outside the population}}$ or 87%.

Caddis Fly Larvae.—The caddis fly larvae in Root Spring, *Frenesia missa* (Milne), *F. difficilis* (Walker), *Limnophilis* sp., and *Lepidostoma* sp., are considered as a unit in the energy flow picture.

Frenesia difficilis and *Limnophilis* sp. were the species most commonly found. *Frenesia difficilis* was raised from the larva and identified. Limnophilis was not raised to adulthood and could not be determined to species. *Frenesia missa* was collected as the adult flying around the spring and was not associated with a larva in the spring. It may have not been present in Root Spring. The Lepidostoma larva was rare and also not associated with an adult.

The preferred habitat of the caddis larvae was not the spring but the springbrook and there were always more larvae in the latter place. The animals in the spring were individuals that had wandered into the pool due to their orientation to the current. Very few of the pool larvae emerged as adult insects and none was collected in the tent trap set over the spring.

Caddis larvae in general are herbivorous although they will eat each other if crowded conditions prevail. The larvae of Frenesia feed almost entirely on roots and leaves of higher plants (Lloyd 1921) and in the spring they fed on those objects which fell into the water from the surrounding vegetation. Cannibalism among caddis larvae was not important in the present study as they were never present in numbers large enough to constitute crowding.

The population figures for the Trichoptera (Table 6) were checked by direct counting of the caddis larvae on the spring bottom. The errors are less than 10%.

All reproduction of the population occurs in the late fall and young larvae first appear in mid-winter. The increases in the number of individuals during the summer were due to immigration of larvae from the springbrook.

The caloric value of the Trichoptera was determined from one measurement which gave a value of 0.98 cal/mg fresh weight. This agrees with the data of Birge & Juday (1922) on the chemical composition of Trichoptera larvae.

The value for respiratory losses for the Trichoptera larvae was taken from the work of Fox & Baldes (1935), who found that at 10° C, larvae of *Limnophilus vittantus* consumed 0.73 mg 0_2/gm fresh wt/hr. The animals used by Fox & Baldes averaged 4.6 mg. Because the average weight of the Trichoptera larvae in Root Spring was 14 mg, the average rate of respiration would be approximately 0.50 mg 0_2/gm fresh wt/hr (based on the theory that rate of respiration is proportional to body surface, Zeuthen 1953).

For the calculation of energy flow it was assumed that from the middle of January to the middle of March there was no immigration. The larvae at that time were very small and did not move about much. (Error from this assumption could not exceed 1% of the total energy flow for these species.) Again from the middle of May to the middle of July there was no immigration because the outlet of the spring was blocked as far as the caddis larvae were concerned. The mortality rates were calculated for those intervals when there was no immigration and assumed to be the same for the months when larvae did enter the spring and direct calculation was not possible. The constant environmental conditions give a basis for this assumption with which the energy flow balance was constructed (Table 7).

The fraction of energy transformed by the caddis larvae was rather less than for the other populations in the spring, $\frac{67.5 \text{ KC}}{106.7 \text{ KC}}$ or 64%. The value is lower for the trichoptera in the springpool than it would be for those in the brook, since energy transformation which was not measured occurred in immigrant larvae before they entered the pool. The efficiency in terms of the energy passed on to other populations, net production over total energy flow, was 36%.

Pisidium virginicum Bourguignat and *Musculium partumeium* Say.—These fingernail clams live completely buried in the mud and feed on organic matter which they filter out of the water. While it was difficult to separate these two genera, it was believed that most of the population (Table 6) belonged to Pisidium.

The value for calories per unit weight of live tissue was obtained from the determinations of the chemical composition of Sphaerium (a fingernail clam) performed by Ivlev (1943). The respiration of these two species of molluscs was measured in three experiments with 23 animals and found to be 0.36 mg 0_2/gm/hr.

The total mortality was calculated using the increase in the natural population from December to January to find a value for i in the formula for mortality. The difference in the mean populations for these two months was significant at $P = 0.05$.

The energy flow chart for the Pisidium and Musculium is given in Table 7. Respiration over total energy intake was 53%. The efficiency in passing organic matter on to other populations was 100% minus 53% or 47%.

Crangonyx gracilis Smith and *Physa* sp.—The amphipods and gastropods were relatively unimportant in the economy of the spring. *C. gracilis* feeds on all sorts of organic matter, both animal and plant but rarely kills its own prey. Physa is omnivorous. Unlike many of the animals in the spring, it breeds only during those months of the year when the length of daylight is more than 13½ hours (Jenner 1951). Table 6 gives the population of these two species.

The rates of respiration and of calories per unit weight for Physa were taken from the data for Pisidium. The data for Crangonyx were taken from one measurement of each variable: calories per unit weight equaled 0.81 cal/mg fresh wt; respiratory rate equaled 1.15 mg 0_2/gm fresh wt/hr or 3.4 cal/cal fresh wt/month. The latter figure is reasonably close to the rate found for the other crustacean in the spring, Asellus, which respired 3.3 cal/cal/month.

Physa and Crangonyx respired at least 47.3 KC/m^2/yr and had a mortality of 9.9 KC/m^2/yr (the total amount built into their tissues during the year since at the end of the period both populations were practically absent). Assuming the efficiencies of these species to be similar to those of the other molluscs and crustaceans in the spring Physa and Crangonyx together respired roughly 90 KC/m^2/yr and passed 30 KC/m^2/yr to other populations in the system (Table 7).

ENERGY EXCHANGE BETWEEN THE SPRING COMMUNITY AND SURROUNDING AREAS

There were several forms of energy exchange between the spring and its surroundings: (1) dissolved and particulate organic matter contained in the water which entered and left the springpool, (2) the organic matter that entered the spring in the form of leaves, other pieces of vegetation, and animals that fell into the water, (3) the adult insects that left the system when they emerged (there was no other emigration), (4) the immigration of caddis larvae, and (5) the sunlight which was used by algae for photosynthesis. (Since the heat from sunlight was of no use to the spring organisms, it was not considered.) Of these, the energy of the emergence and immigration of insects has already been calculated.

The results of determinations of the organic content of the water entering and leaving the spring showed no significant difference in kind or amount between the organic matter being carried into the system and that being carried out. Rain did not affect these determinations as surface water did not drain into the spring and there were no noticeable short-term changes in ground water flow following rains or drought.

The most important source of energy for the spring community consisted of the leaves and other plant material that fell into the water from the surrounding land. This occurred mostly in the autumn and came to approximately 2350 kilocalories/m^2 during the year under consideration.

MICRO-ORGANISM METABOLISM

The rates of respiration given for the microflora and microfauna were obtained from the total respiration of the benthos minus the calculated rates of respiration of the known average biomass of macrofauna (Table 8). There were two periods of maximum activity of micro-organisms in Root Spring, one early in spring and the other in autumn.

The total respiration of these micro-organisms was 350 KC/m^2/yr. A rough estimation of the portion of this amount which represents the respiration of algae may be obtained by adding three-fourths of the respiration in April when the algae were most active to one-half of the respiration in March, May, June and July; this gives 55 KC/m^2/yr. The respiration

TABLE 8. Photosynthesis and respiration of micro-organisms in Root Spring, Concord, Mass.

Month	Mean rate of photosynthesis	Mean rate of respiration
	KC/m²/month	KC/m²/month
November 1953	...	17
December	...	17
January 1954	8	33
February	56	50
March	138	25
April	250	14
May	102	25
June	68	25
July	45	20
August	23	42
September	11	50
October	10	32
	710 KC/m²/yr.	350 KC/m²/yr.

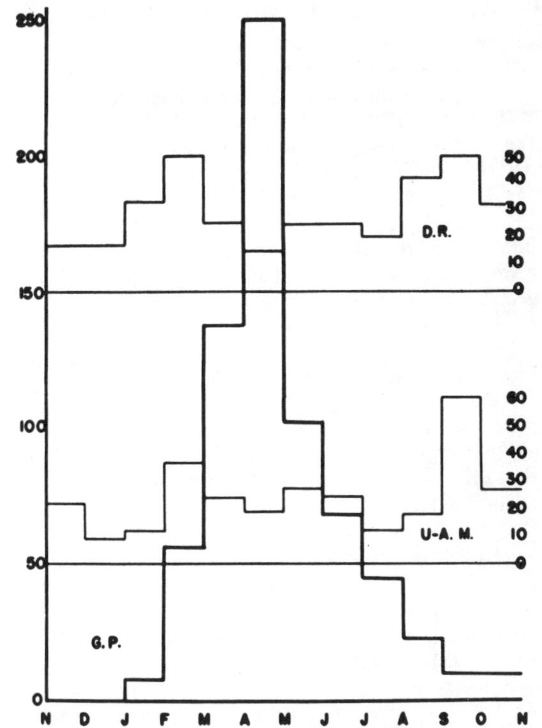

FIG. 5. Comparison of Gross Photosynthesis (G.P.); Mortality of macroscopic animals which is not assimilated by macroscopic carnivores (Unassimilated Mortality, U-A.M.); and Respiration of all micro-organisms (D.R.) for Root Spring, Concord, Mass. in 1953-54.

of micro-organisms was lowest in April and, since the algae were most active then, most of this respiration must have been due to algae. Since the algae were also active in the four other months mentioned, they must have accounted for a considerable portion of the respiration then as well.

Micro-organism metabolism throughout the year was compared with the various sources of energy upon which these organisms could have drawn. Fig. 5 gives a picture of relative rates of photosynthesis in the spring and respiration of micro-organisms and the energy available to micro-organisms from animal sources (non-predatory mortality and prey killed but not assimilated by carnivores; the chitin in cast larval and pupal skins was not included). It is apparent from the figure that the respiration of micro-organisms is much more closely correlated with the variation in usable energy lost by the macrofauna. This would indicate that the micro-organisms were probably mostly bacteria and fungi and not algae-eating animals.

The nematodes and protozoa, although not abundant, were the only other micro-organisms observed with any frequency. They probably fed mostly upon bacteria. The two genera of nematodes identified had only bacteria in their guts and members of these genera are reported to feed only upon bacteria (Nielsen 1949).

PHOTOSYNTHESIS

The amount of photosynthesis occurring in the spring was negligible in November and December and very slight throughout the autumn and in January, due to lack of light at those times. It was not until February that the duration and strength of the illumination became sufficient for a large crop of algae to grow in the spring. In February all of the solid surfaces in the water became covered by the filamentous green alga Stigeoclonium. This alga increased in amount and a scum of diatoms became visible to the naked eye on the surface of the sandy part of the bottom as the illumination increased. The occurrence of maximum photosynthesis of the algae in April (Table 8) is explained by the fact that after April the terrestrial plants around the spring leafed out and shaded the water. By summer this shading greatly reduced photosynthesis.

The gross production of the algae was 710 KC/m²/yr and since the algal respiration was estimated at 55 KC, net algal production was approximately 655 KC/m²/yr.

A possible source of error in these values lay in the fact that the measurements extended over 24 to 32 hours. It is possible that in this time the available phosphate or some other essential nutrient was exhausted and that the measured rate of photosynthesis was lower than the true value. However, one experiment in May was allowed to run for five days while another ran only one day in the same week and there was no significant difference in the results. Perhaps these algae are able to use organic phosphate for their growth as Rice (1949) reported for Chlorella but failed to find for Nitzschia.

BACTERIAL GROWTH ON GLASS PLATES

The numbers of bacteria which grew on the exposed parts of glass slides stuck in the middle of Root Spring allow a rough comparison to be made with other aquatic areas. One slide in the spring for 20 days showed a growth of 625 bacteria/mm^2/day (21 fields counted). Two others in the water for 30 days showed 847 and 1070 bacteria/mm^2/day (45 and 56 fields counted). Slides placed in a pond into which the springbrook flowed but where the water temperature was about 25° C showed more growth in the pond in two weeks than occurred in the spring in a month.

Comparison may be made with Silver Springs, Florida (Odum 1957) where slide counts at the mud surface gave 3280 bacteria mm^2/day, about three times the Root Spring value. Since Silver Springs has a temperature of 23° C, it would be expected that growth there would be two to three times as fast as in Root Spring with a temperature of 9° C.

Probably some of the bacteria in Root Spring were photosynthetic but the slides were not examined for colored forms. Henrici (1939) indicates that photosynthetic bacteria are more common in shallow water than elsewhere and all of the slides in the spring were in full daylight.

Since respiration of the bacteria is included with that of the micro-organisms and their importance to the community metabolism thus evaluated, they were not considered further.

PREDATOR-PREY BALANCE

From the data for Anatopynia and Phagocata the amount of energy assimilated by carnivores each month can be found. However, a larger biomass of prey is killed than is assimilated. It has been seen that Anatopynia assimilated only about one-third of the amount that it kills and that Phagocata assimilated about 90% of its food on the average, as 10% is indigestible. Table 9 lists under "Predation" the biomass of killed prey, given in energy terms, needed to support the carnivores.

The mortalities from all causes of the various prey species are added in the first column of Table 9. Anatopynia and Phagocata are not included as prey, even though they did feed on members of their own species, because cannibalism has already been taken into account in the energy flow balances.

By comparing the total mortality of the herbivores with the predation it can be seen that for the trophic level as a whole the non-predatory mortality is 19.4% of the total mortality $\left(\frac{90.9}{467.9}\right)$

It seems logical to believe that non-predatory mortality would have been smaller in the spring than in other aquatic or terrestrial soils since the constant environmental conditions would result in fewer non-predatory deaths in the spring than would be the case

TABLE 9. Comparison of herbivore mortality and loss due to predation (KC/m^2) in Root Spring, Concord, Mass.

Month	Mortality	Predation	Difference Non-pred. mortality
November 1953	24.0	16.8	7.2
December	21.4	20.6	0.8
January 1954	39.4	37.1	2.3
February	63.0	51.5	11.5
March	32.9	22.1	10.8
April	20.8	16.3	4.5
May	33.2	29.6	3.6
June	32.7	22.7	10.0
July	23.1	22.9	0.2
August	53.8	54.1	−0.3
September	93.8	58.7	35.1
October	29.8	24.6	5.2
	467.9	377.0	90.9

in environments subject to freezing, anaerobiosis, etc.

It may be concluded, then, that non-predatory mortality will be one-fifth or more of the total mortality of a species or trophic level in many communities. It is not, therefore, safe to neglect this part of the energy flow through a trophic level as was done by Lindeman (1942).

RELATIVE IMPORTANCE OF VARIOUS SPECIES

An energy balance sheet for the herbivores and carnivores is given in Table 10. From this table we can compare the various groups of herbivores with respect to the amount of energy which they assimilated and the amount of energy which they transformed to heat. The oligochaetes assimilated more energy than any other group and transformed more than any other group with the exception of the isopods. The Calopsectra, which had by far the largest biomass in the summer, were third in importance in amount of energy assimilated and transformed.

These data emphasize the difficulty in making valid comparisons of the metabolic importance of various kinds of animals from subjective impressions, counts or measurements of biomass, or even energy flow determinations if these are confined to only one season. Statements of the relative importance of various kinds of animals to a community have, however, often been based on just such evidence as individual size and apparent abundance at one time of year. Indeed, the entire scheme of classifying animals as major and minor influents in a community (Clements & Shelford 1939) seems based mostly on the inadequate criterion of individual size. The use of size as a criterion of community importance has been criticized by E. P. Odum (1953) who suggests that total biomass of a species is a better indicator of the

TABLE 10. Balance sheet for herbivores and carnivores in Root Spring, Concord, Mass., 1953-54. Data for Anatopynia are divided between both trophic levels as explained in the text.

	Assimilation	Respiration
Herbivores	KC/m²/yr.	KC/m²/yr.
Limnodrilus	644	484
Asellus	604	486
Calopsectra	520	390
Anatopynia	159	138
Pisidium	173	91
Trichoptera larvae	88	67
Physa & Crangonyx	120	90
Total Herbivores	2300	1746
Carnivores		
Phagocata	131	19
Anatopynia	77	70
Total Carnivores	208	89

importance of various animals in a community. An even better notion of the relative influence of different populations in an ecosystem is obtained from a study of energy flow.

The planarians were the most important carnivores even though they transformed less energy than the Anatopynia because the latter obtained more energy from plant than animal sources.

By combining all of the data so far presented, the community energy balance chart (Fig. 6) was constructed.

On the credit side of the energy balance is:

Organic debris	2350 KC/m²/yr	(76.1%)
Gross photosynthetic production	710	(23.0%)
Immigration of caddis larvae	18	(0.6%)
Decrease in standing crop	8	(0.3%)
	3086	

This total is divided in the following way on the debit side of the balance:

Transformation to heat	2185 KC/m²/yr	(71%)
Deposition	868	(28%)
Emigration of adult insects	33	(1%)
	3086	

PORTION OF ENERGY TRANSFORMED TO HEAT

The fraction of assimilated energy which various groups of animals in Root Spring transformed to heat during the year studied is summarized in Table 11. The fraction is close to 50% for all groups except the clams and planarians. (The value for carnivores is low because of the influence of the planarians.)

It was suggested that the reason for the very low ratio of energy transformation in the planarians is their necessity, in an evolutionary sense, to compensate for their large loss of energy in the form of mucus. However, since the energy in mucus serves

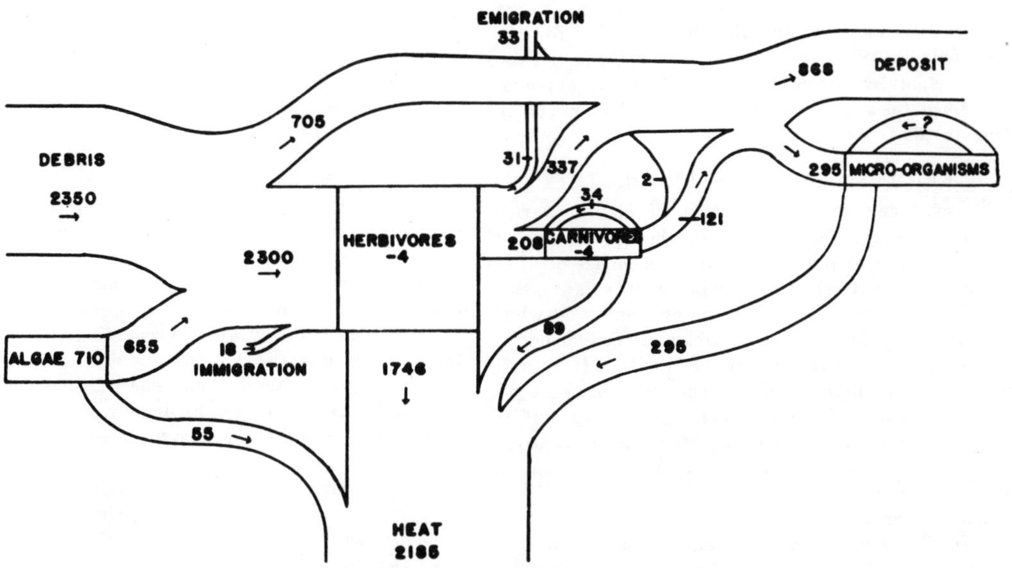

FIG. 6. Energy flow diagram for Root Spring, Concord, Mass. in 1953-54. Figures in KC/m²/yr; numbers inside boxes indicate changes in standing crops; arrows indicate direction of flow.

TABLE 11. Energy ratios for various groups of organisms from Root Spring and from other sources. Efficiencies for Mytilus were calculated for season of maximum growth and would be lower if figures on an annual basis as were Root Spring data.

	Net Production / Assimilation	Transformed to Heat / Assimilation
Root Spring		
Calopsectra	20%	80%
Anatopynia	13	84
Limnodrilus	26	74
Asellus	20	80
Trichoptera larvae	36	64
Pisidium	47	53
Planarians	87	12
All herbivores	25	75
All carnivores	59	37
Entire Community	29	71
Jorgensen (1952)		
Veligers, marine	60 - 70	..
Mytilus, 40.9-49.0mm	54	..
Mytilus, 90mm	11	..
Harvey (1950)		
Calanus	70	..
Brody (1945)		
Embryos	50 - 65	..

the worms in their activities and is subsequently lost to them just as the energy used for doing work and transformed to heat is used and lost, the energy in mucus may logically be included with the energy transformed to heat in calculating the ratio. The portion of energy transformed plus mucus energy was 69% which is comparable to the value for the other animals. The relatively low transforming ratio of the clams may possibly have a similar explanation since these animals lose considerable mucus in their pseudofeces.

EFFICIENCY OF NET PRODUCTION

The efficiency of net production (Table 11) varied around 30% for the animals in the spring except for the clams and planarians with efficiencies of 47% and 87% respectively. These latter two values are close to those for larval animals and embryos while values from 20-30% are more common for efficiencies of post natal growth (Jørgensen 1952). It is possible that the high efficiency in the planarians is associated with their exceptional regenerative ability.

The net production efficiency for the entire community was 29%. In other words, there would be a continual accumulation of usable energy in the ecosystem if organic matter were not removed.

A small amount of organic matter was removed in the bodies of the insects that emerged, 33 KC/m², in comparison to the deposition of 868 KC/m². The pool was prevented from filling completely by shifts in the position of the boils which stir up and wash out some mud.

COMPARISON WITH DIFFERENT COMMUNITIES

With the data available in the literature comparisons can be made, not only of the primary productivity in a series of communities but also of the energy flow through the trophic levels of a much smaller number of systems.

Table 12 gives values for energy available to organims other than primary producers (primary net production plus accumulated energy in organic matter) and the ratio of this to total incident light. Incident light in the Concord region was calculated to be 1.095 x 10^6 KC/m²/yr from weather bureau data (U.S. Dept. of Commerce 1951). In aquatic communities some light is absorbed by the water but in order to make comparisons between land and water systems as producers on the earth, no correction was made.

TABLE 12. Efficiencies and productivities of various communities compared with those of Root Spring, Concord, Mass.

	Energy available to consumers	Column One / Light	Gross Prod. / Light	Source
	KC/m²/yr.			
Root Spring	3005	0.27%	0.2%*	
Eniwetok Reef	21800	1.8	5.8	Odum & Odum 1955
Cedar Bog Lake	879	0.074	0.1	Lindeman 1942
Lake Mendota**	3730	0.31	0.4	Lindeman 1942
Minnesota Pond	394	0.033	0.04	Dineen 1953
Corn field in summer	6170	1.2	1.6	Transeau 1926
Georges Bank	.....	......	0.3	Clarke 1946
Silver Spring, Florida	.....	......	8.0	Odum 1957
Average Terrestrial plant	.....	......	0.09	Riley 1944
"Best forests"	.....	......	0.25	Riley 1944

*For April only.
**Values considered erroneously high by Lindeman.

Root Spring provided more energy per unit area for consumers than average temperate communities because of its action in accumulating energy that had been fixed by plants outside of the spring. The spring community itself expended no energy to bring about this accumulation.

As far as primary gross production is concerned, the algae of Root Spring were as efficient as the better temperate communities. The efficiency was calculated for April before shading by trees was important, since no direct measurement of light intensity was made.

Comparisons of the different trophic levels of a few communities are made in Table 13. The ratio, $\frac{\text{assimilation of one trophic level}}{\text{net production of next lower level}}$, is used for comparison. This represents the efficiency of utilization of available food by a trophic level. Values were calculated from the data given by the various authors.

The most striking difference shown in Table 13 between the various communities is the large assimilation and efficiency of herbivores in Root Spring. (The large carnivore assimilation appears to be a result of the large food supply in the form of herbivores since the efficiency of carnivores in Root Spring was about the same as in the other communities.)

The reason that a large portion of the available plant energy was assimilated probably lies in the fact that the Root Spring conditions were more favorable to life than those in the other environments of Table 13. For example, water currents in the spring provided sufficient oxygen and rapid removal of metabolic wastes at all times. It must be noted, however, that the transformation by a community of as much or more than three-fourths of the incoming usable energy to heat is not unusual.

If in any community the primary net production is not accumulating to any appreciable extent, as is the case in the woods in the south of England (Pearsall 1948), it is because the efficiency of utilization of primary net production is close to 100%. In other words, if the primary consumers consume all the primary net production there will be no accumulation of organic matter in the community. This contradicts Lindeman's (1942) thesis that the utilization efficiency will be progressively greater for higher trophic levels.

In Root Spring the primary consumers are mostly macroscopic animals. These animals are herbivores, detritus feeders, and scavengers. In terrestrial soils, on the other hand, the most important primary consumers are probably microscopic, heterotrophic plants. Data summarized from Stöckli (1946) indicate that the microflora—fungi, bacteria, actinomycetes (and algae)—compose most of the biomass of terrestrial soil organisms, and since these organisms are also very small, they are undoubtedly the most active metabolically. These plants are "decomposers," but are still primary consumers as long as they are burning fuel from green plant sources. It makes no difference in an energy flow analysis, whether organisms which obtain energy from plants are herbivores, saprophytes, detritus feeders or decomposers; they are still all primary consumers.

A complicating factor arises from the fact that decomposers, etc. feed upon consumers as well as producers (i.e. are secondary and tertiary as well as primary consumers). This complication is common to all soil communities, aquatic and terrestrial, and does not invalidate the comparisons made from Table 13.

The data from this investigation raise a number of interesting questions: Is it a general rule that the ratio of energy flow of secondary consumers to that of primary consumers remains about the same for all communities? Does the stability of a community affect the aforementioned ratio? What relation might there be between assimilation by primary consumers and producer net production? Is there a correlation between body size of consumers and deposit or removal of a sizable fraction of primary net production? These and other questions can only be answered when more data from both experimental and natural communities are available.

SUMMARY

1. A study of community metabolism in terms of energy flow was undertaken in 1953-54 in order to provide a more accurate picture of energy flow through the populations of a community than had

TABLE 13. Efficiencies and assimilations for higher trophic levels of various communities compared with those of Root Spring, Concord, Mass.

	Root Spring		Cedar Bog Lake		Lake Mendota		Minnesota Pond		Silver Spring
			Lindeman (1942)		Lindeman (1942)		Dineen (1953)		Odum 1957
	Assim.*	Effic.**	Assim.	Effic.	Assim.	Effic.	Assim.	Effic.	Assim.
Primary Consumers (Herbivores)....	2300	76%	148	16.8%	416	11.2%	92	23%	1280
Secondary Consumers (Carnivores)....	208	36%	31	29.8%	23	8.7%	34	47%	28
Tertiary Consumers (Secondary Carnivores)....	....	....	...	......	3	23.0%	..	....	....

*All assimilations are KC/m^2yr.

**All efficiencies are $\frac{\text{Assimilation}}{\text{Net production of next lower trophic level}}$

hitherto been available. Root Spring, a temperate cold spring in Concord, Massachusetts, was chosen for study because of the relatively constant environmental conditions and simple biota.

2. The energy flow through the larger animal populations was studied in detail. These animals included two chironomids, *Calopsectra dives* and *Anatopynia dyari,* a tubificid oligochaete, an isopod, two planarian species, three species of caddis larvae, two species of fingernail clams, one snail, and one amphipod. To minimize artificial disturbance of the community the animals in the samples were kept alive and returned to the spring after being weighed and counted.

3. Respiration of the larger species of animals was measured in containers filled with spring water and placed in the spring so that conditions were as nearly natural as possible. Photosynthesis and respiration of the micro-organisms were determined by measuring oxygen changes in water over bottom mud enclosed in light and dark, covered glass cylinders.

4. Non-predatory mortality was calculated as the difference between the total mortality determined with the formulae published by Ricker (1946) and the mortality due to predation. Non-predatory mortality accounted for nearly 20% of the mortality occurring in the herbivore trophic level. Reasons were given for believing that this was a lower rate of non-predatory mortality than exists in most communities and that non-predatory mortality may not be safely neglected in studies of productivity.

5. The Anatopynia were found to assimilate only 30% of the prey which they killed. This carnivorous midge larva actually obtained about two-thirds of its food from plant sources which enabled it to exist during periods in which it could not obtain sufficient animal food.

6. Phagocata, the planarian, lost more energy in the mucus which it secreted than it used in doing external and internal work as measured by respiration. Mucus secretion took 79% of this species' net production.

7. 76% of the energy transformed by the organisms entered the spring in leaves, fruit and branches of terrestrial vegetation. Photosynthesis in the spring accounted for 23% and less than 1% entered in the bodies of immigrating caddis larvae.

8. The micro-organisms respired about 350 KC/m^2/yr. Of this only 55 KC were respired by algae, the remainder by bacteria, fungi, protozoa and nematodes. Evidence was given to indicate that most of the micro-organisms were feeding upon animal remains.

9. The most conspicuous group of herbivores, Calopsectra, was only third in importance in energy flow. It was suggested that the relative amount of energy which flows through each population is a better criterion than size or biomass of the importance of various animal species in a community.

10. The energy balance sheet for the community shows that, of the year's total energy input, 71% was transformed to heat, 28% was deposited in the community and only 1% emerged in adult insects.

11. The efficiency of energy transformation, energy used for internal and external work as measured by respiration divided by assimilation, was close to 80% for all animals but the planarians and clams. The clams transformed only 53% of their assimilation, the planarians only 12%. These low rates of respiration may be an adaptation connected with the large amounts of mucus lost by these animals.

12. The ratio $\frac{\text{assimilation}}{\text{net production of algae plus energy of organic debris}}$ was 76% in Root Spring, higher than other aquatic communities investigated but probably not as high as for some soils.

13. The efficiency $\frac{\text{algal photosynthesis}}{\text{available solar energy}}$ in Root Spring in April when shading of the water at a minimum was 0.2% which is comparable to any temperate area. If the efficiency of the spring community is calculated on the basis of total inflow of energy instead of just photosynthesis, the efficiency for the year was 0.27%, higher than for most temperate latitude communities which have been studied.

14. The assimilation by macroscopic primary consumers (over 1 mg) was five times that reported for any other temperate latitude aquatic community. The steady flow of constant temperature water bringing oxygen, removing waste, and enabling the animals to be active all year was probably responsible for this high assimilation.

LITERATURE CITED

Birch L. C. & D. P. Clark. 1953. Forest soil as an ecological community with special reference to the faunas. Quart. Rev. Biol. **28**: 13-36.

Birge, E. A. & C. Juday. 1922. The inland lakes of Wisconsin. The plankton, Part I: Its quantity and chemical composition. Bull. Wis. Geol. Nat. Hist. Surv. **64**: 1-222.

Borutzky, E. V. 1939. (Dynamics of *Chironomus plumosus* in the profundal of Lake Beloie.) Proc. Kossino Limnol. Stat. **22**: 156-195. (Russian with English summary p. 190-195.)

Brody, S. 1945. Bioenergetics and growth. New York: Reinhold Publ. Corp. 1023 pp.

Brues, C. T. 1928. Studies on the fauna of hot springs in the Western United States and the biology of thermophilous animals. Proc. Am. Acad. Arts Sci. **63**: 139-228.

Bullock, T. H. 1955. Compensation for temperature in the metabolism and activity of poikilotherms. Biol. Rev. **30**: 311-342.

Clarke, G. L. 1946. Dynamics of production in a marine area. Ecol. Monog. **16**: 321-335.

Clements, F. E., & V. E. Shelford 1939. Bio-ecology. New York: John Wiley & Sons. 425 pp.

Dineen, C. F. 1953. An ecological study of a Minnesota pond. Am. Midland Nat. **50**: 349-376.

Dudley, Patricia L. 1953. A faunal study of four springbrooks in Boulder County, Colorado. Master's thesis, Univ. of Colorado.

Ewer, R. F. 1941. On the function of haemoglobin in *Chironomus*. J. Exper. Biol. **18**: 13-205.

Fox, H. M., & E. J. Baldes. 1935. Quoted in Prosser, C. L., 1950. Comparative Animal Physiology. Philadelphia: W. B. Saunders Co.

Fox, H. M., & C. A. Wingfield. 1938. A portable apparatus for the determination of oxygen dissolved in a small volume of water. J. Exper. Biol. **15**: 437-445.

Gerking, S. D. 1954. The food turnover of a bluegill population. Ecology **35**: 490-497.

Harvey, H. W. 1950. On the production of living matter in the sea off Plymouth. J. Mar. Biol. Ass. U.K. **29**: 97-137.

Henrici, A. T. 1936. Studies of freshwater bacteria. III. Quantitative aspects of the direct microscopic method. J. Bact. **32**: 265-280.

Hyman, L. H. 1919. Oxygen consumption in relation to feeding and starvation. Amer. Jour. Physiol. **49**: 377-402.

Hyman, L. H. 1951. The Invertebrates. Vol. II. Platyhelminthes and Rhynchocoela, the acoelomate Bilateria. New York: McGraw-Hill Book Co. 550 pp.

Ivlev, V. S. 1934. Eine Mikromethode zur Bestimmung des Kaloriengehalts von Nahrstoffen. Biochem. Ztschr. **275**: 49-55.

Ivlev, V. S. 1945. The biological productivity of waters (Translation by W. E. Ricker) Adv. Mod. Biol. **19**: 98-120.

Jenner, C. E. 1951. Photoperiodism in the fresh-water pulmonate snail, *Lymnaea palustris*. Ph.D. thesis, Harvard University.

Johannsen, O. A. 1937. Aquatic Diptera Pt. III Chironomidae: Subfamilies Tanypodinae, Diamesinae, and Orthocladiinae. Mem. Cornell Univ. Agr. Exp. Sta. **205**: 1-84.

Jørgensen, C. B. 1952. Efficiency of growth in *Mytilus edulis* and two gastropod veligers. Nature **170**: 714-715.

Lindeman R. L. 1942. The trophic-dynamic aspect of ecology. Ecology **23**: 399-418.

Lloyd J. T. 1921. The biology of North American caddis fly larvae. Bull. Lloyd Libr. **21**: Ento. Ser. No. 1.

Macfadyen, A. 1948. The meaning of productivity in biological systems. J. Anim. Ecol. **17**: 75-80.

Nielsen, C. Overgaard, 1949. Studies on the soil microfauna. II. The soil inhabiting nematodes. Natura Jutlandica **2**: 1-131.

Odum, E. P. 1953. Fundamentals of Ecology. Philadelphia: W. B. Saunders Co. 384 pp.

Odum, H. T. 1956. Efficiencies, size of organisms, and community structure. Ecology **37**: 592-597.

Odum, H. T. 1957. Trophic structure and productivity of Silver Springs, Florida. Ecol. Monog. **27**: 55-112.

Odum, H. T. & E. P. Odum. 1955. Trophic structure and productivity of a windward coral reef community on Eniwetok Atoll. Ecol. Monog. **25**: 291-320.

Pearsall, W. H. 1948. An ecologist looks at soil. Symposium on Soil Biology, British Ecological Society, April, 1948. J. Ecol. **36**: 328.

Pennak, R. W. 1946. Annual limnological cycles in some Colorado reservoir lakes. Ecol. Monog. **19**: 233-267.

Pennak, R. W. 1954. Fresh-water invertebrates of the United States. New York: Ronald Press. 769 pp.

Rice, T. 1949. The effects of nutrients and metabolites on populations of planktonic algae. PhD. thesis, Harvard University.

Ricker, W. E. 1946. Production and utilization of fish populations. Ecol. Monog. **16**: 375-389.

Riley, G. A. 1944. The carbon metabolism and photosynthetic efficiency of the earth as a whole. Amer. Sci. **32**: 129-134.

Stöckli, A. 1946. Die biologische Komponente der Vererdung, der Gare und der Nährstoffpufferung. Schweiz. Landw. Monatsh. **24**.

Transeau, E. N. 1926. The accumulation of energy by plants. Ohio J. Sci. **26**: 1-10.

U. S. Department of Commerce. 1951. Weather Bureau, climatological data for the U. S. by sections. Vol. **38**. Washington.

Walshe-Maetz, B. M. 1953. Le metabolisme de *Chironomus plumosus* dans des conditions naturelles. Physiol. Comp. e. Ecologia **111**: 135-154.

Welch, P. S. 1948. Limnological Methods. Philadelphia: Blakiston Press. 381 pp.

Zeuthen, E. 1953. Oxygen uptake as related to body size in organisms. Quart. Rev. Biol. **28**: 1-12.

Part 5

BIOGEOCHEMICAL CYCLES

In the previous section, the flow of energy through ecosystems was discussed and exemplified. Since this flow of energy depends ultimately on photosynthesis, it can proceed no faster than photosynthesis. In turn, photosynthesis depends not only upon the flux of radiant energy, but also upon the availability of matter to be incorporated into the products of photosynthesis. An understanding of energy alone, therefore, is necessary but not sufficient for an understanding of ecosystems. If solar radiation were the only factor limiting productivity, there would be a gradient in productivity, decreasing from the equator toward the poles; this is obviously untrue, and other limiting factors merit study. Thus, in deserts, where solar radiation is high, productivity is kept low by lack of water.

In terrestrial systems, water is commonly a substance that limits productivity, whereas in aquatic systems the principal macronutrients which have been studied as limiting factors are carbon, hydrogen, oxygen, nitrogen, sulphur, phosphorus, silicon, and some of the more common minerals. The first four substances in this list are the main constituents of fats and carbohydrates; these, together with nitrogen and sulphur, make up amino acids, and, therefore, proteins. Phosphorus is a structural component of the nucleic acids and is needed for the transfer of chemical energy within organisms. Silicon is an important constituent of diatoms.

Unlike energy, which is not *cycled* but passed through the biosphere from higher to lower energy states until it is finally re-radiated into space, the important chemical substances do cycle in nature. Each of these substances circulates through air, land, sea, and living systems in a vast biogeochemical cycle. The circulation of water (hydrological cycle), the erosion and uplift of the continents (geological cycle), and the opposing and complementary processes of photosynthesis and respiration (ecological cycle) each play a role in the larger biogeochemical cycle. These cycles have been characterized as more or less perfect, depending upon the extent to which substances re-enter the "pools" constituting the parts of each cycle. Because there is a continuous flow, with substances moving through a system, steady state conditions prevail, rather than those typical of equilibrium. Energy must be constantly supplied for steady states, whereas systems in equilibrium require no input of energy.

A pool is the quantity of a chemical substance in some component of an ecosystem; for example, in a pond the dissolved inorganic phosphate constitutes one pool, the dissolved organic phosphate another pool, and the quantity of phosphorus in the zooplankton yet another. These pools and the others that may be described are interrelated by physical and biological processes which transfer each substance from pool to pool, the processes occurring at different rates which depend on physical, chemical, and biological characteristics of the environment. These rates of transfer are the flux rates, and they have been studied extensively for phosphates in a number of ecosystems. To understand the relationships in a biogeochemical cycle more fully, it is also necessary to know the flux functions, which mathematically describe how flux rates change as the environment changes. Examples of such environmental factors are (1) temperature, (2) concentration of the substance, (3) concentration of some interfering chemical, and (4) the presence of a plant requiring the substance in

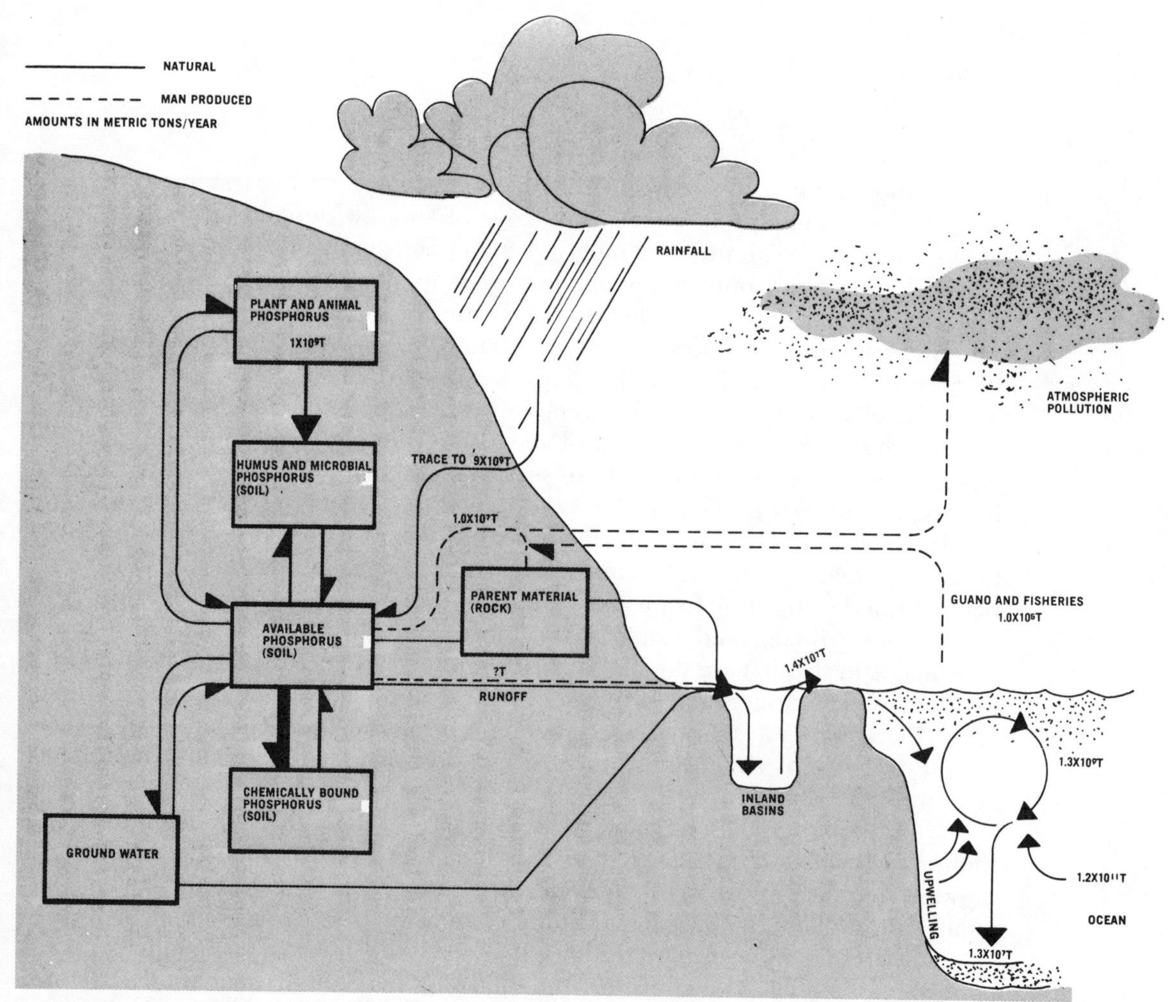

Figure 1. The phosphorus cycle. (From Inger, R. F., A. D. Hasler, F. H. Bormann, and W. F. Blair, Co-Principal Investigators, *Man in the Living Environment* [Madison: The University of Wisconsin Press; © 1972 by the Regents of the University of Wisconsin], p. 49.)

its nutrition. Examples of chemical compounds that interfere in the cycling of phosphorus are the ferric ions present in oxygenated lakes which lead to a precipitation and removal of ferric phosphates from lake ecosystems. Most models of biogeochemical cycles presume the flux functions to be linear, but Dugdale (this collection) suggests that the nonlinear Michaelis-Menton equations may be appropriate. This aspect of biogeochemical systems, however, needs further study with sets of differential equations used to explicate the reactions of the system.

Among the examples examined here, the phosphorus cycle is an imperfect cycle, since even when human beings do not interfere with it, phosphates in the earth's mantle erode to the sea and are recovered on land only in the slowness of geological time. In contrast, nitrogen, with its several oxidative states, its presence in the solid, liquid, and gaseous parts of the earth, and its many feedback mechanisms, exemplifies a nearly perfect cycle.

The general features of the phosphorus cycle are well known (Fig. 1). Phosphorus occurs in the biosphere almost exclusively in its fully oxidized state, phosphate. Because of the abundance of calcium, aluminum, and iron in the earth's mantle, all of which form phosphate salts with very low solubilities in water, most of this element is immobilized in rock, soil, or sediment. The natural movement of phosphorus is therefore slow, but life as we know it in the biosphere depends on this movement. The cycle in the hydrosphere begins as phosphates, which are leached as dissolved salts or eroded as particles, find their way to streams and lakes, where they are precipitated or enter living matter. The remainder enter the ocean, where the same processes occur. As the bodies of planktonic plants and animals sink through the water column, the surface waters become depleted of phosphorus. The deep waters are nearly saturated with calcium phosphate, so that additions to the deep ocean pool are balanced by precipitation to the sediment. Upwelling of deep water returns a portion to the surface, but the return rate is slower than the sinking rate. Return of phosphorus to the land depends therefore upon the geological uplifting of sediments. On a human scale, this rate is virtually zero—phosphorus is a nonrenewable resource. The significant exceptions to this return by geological processes are the different fisheries and deposits of guano by fish-eating birds. On a local and short-term scale, ecological systems on land and in water accumulate and cycle phosphorus, where it is, however, often a substance limiting productivity. A budget of phosphorus input to a lake is shown in Table 1.

Table 1. Nutrient Source and Percentage of Total Nutrients Entering Lake Mendota*

Nutrient Source	*Total Nutrients Entering Lake (%)*	
	Nitrogen	Phosphorus
Precipitation	17	2
Groundwater	45	2
N fixation	14	–
Runoff		
Rural lands (not manured)	1	12
Rural lands (manured)	8	30
Urban area	5	17
Waste waters (municipal and industrial)	8	36

*From Biggar, J. W., and Corey, R. B. 1969. Agricultural drainage and eutrophication. *In*: Eutrophication: Causes, Consequences, Correctives—Proceedings of a Symposium. National Academy of Sciences (Wash.). 267 pp.

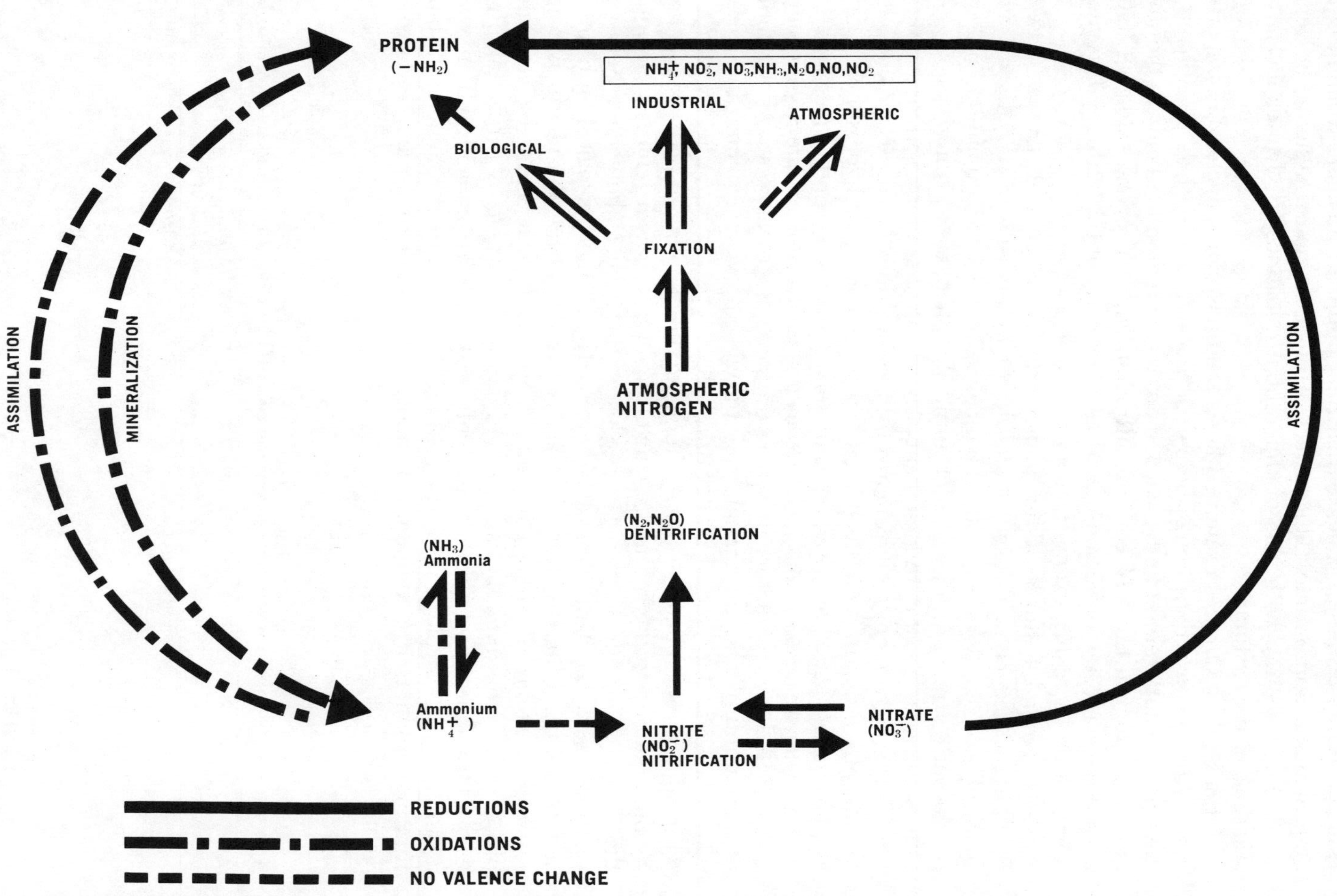

Figure 2. The nitrogen cycle. (From *Man in the Living Environment.* Institute of Ecology, Workshop in Global Ecological Problems. 1971, p. 63.

In contrast, the global cycling of nitrogen is the most complex of the biogeochemical cycles. The major processes are shown in Figure 2, a chemical flow chart for nitrogen. Nitrogen occurs naturally in a number of different oxidative states (as in NH_3, N_2, NO, NO_2), giving great complexity to the system. The largest reservoir of nitrogen is the atmosphere, which is 78 per cent nitrogen gas. A portion of the elemental nitrogen is removed by various microbes and incorporated into living tissue by a variety of biochemical pathways. This process of nitrogen fixation permits the fixed nitrogen to be assimilated by plants and incorporated into protein. Then it is either returned directly to the environment or consumed by animals before its ultimate release. When this release occurs, biochemical degradation takes place and ammonia is released. This ammonia in turn is re-assimilated or changed by microorganisms into nitrate, which may then be leached to groundwater as part of the hydrological cycle. This process is called nitrification. Biochemical reactions resulting in a net change in the opposite direction occur in denitrification, the process in which nitrates are transformed to nitrogen or nitrous oxide. The atmospheric and biological cycles of

TABLE 2. Budget for the Nitrogen Cycle.*

	Land		Sea		Atmosphere	
	rate/yr	% error	rate/yr	% error	rate/yr	% error
Input						
Biological nitrogen fixation	–	–	10	50	–	–
Symbiotic (31)	14	25	–	–	–	–
Non-symbiotic (31)	30	50	–	–	–	–
Atmospheric nitrogen fixation (31)	4	100	4	100	–	–
Industrially fixed nitrogen fertilizer (31)	30	5	–	–	–	–
N-oxides from combustion*	14	25	6	25	20	25
Return of volatile nitrogen compounds in rain	?	–	?	–	–	–
River influx (31)	–	–	30	50	–	–
N_2 from biological denitrification (31)	–	–	–	–	83	100
Natural NO_2	–	–	–	–	?	–
Volatilization (HN_3)	–	–	–	–	?	–
Total Input	92+		50		103+	
Storage						
Plants (31)	12,000	30	800	50	–	–
Animals (31)	200	30	170	50	–	–
Dead organic matter (31)	760,000	50	900,000	100	–	–
Inorganic nitrogen (31)	140,000	50	100,000	50	–	–
Dissolved nitrogen (31)	–	–	20,000,000	10	–	–
Nitrogen gas (31)	–	–	–	–	3,800,000,000	3
NO + NH_4 (25)	–	–	–	–	Less than 1	50
NH_3 + NH_4 (17)	–	–	–	–	12	50
N_2O (33)	–	–	–	–	1,000	50
Total Storage	912,200		21,000,970		3,800,001,013	
Loss						
Denitrification (31)	43	–	40	100	–	–
Volatilization	?	–	?	–	–	–
River runoff (31, 32)(includes enrichment from fertilizers)	30	50	–	–	–	–
Sedimentation (31)	–	–	0.2	50	–	–
N_2 in all fixation processes	–	–	–	–	92	50
NH_3 in rain (17)	–	–	–	–	Less than 40	50
NO_2 in rain	–	–	–	–	?	–
N_2O in rain	–	–	–	–	?	–
Total Loss	73		40.2		132+	

*All numbers are in millions of metric tons. The error columns list plus-or-minus probable errors as a percentage of the estimate. From Institute of Ecology. 1971. Man in the Living Environment. Report of the Workshop on Global Ecological Problems.

nitrogen are so rapid and strong that, in contrast to phosphates, the geological aspects are trivial. The rates at which nitrogen moves to and from some of the major compartments in its cycle are shown in Table 2. Clearly, the values are rough estimates, and much work remains to be done.

LITERATURE CITED

Biggar, J. W., and Corey, R. B. 1969. Agricultural drainage and eutrophication. *In* Eutrophication: Causes, Consequences, and Correctives. International Symposium on Eutrophication. National Academy of Sciences, Washington, D.C. 661 pp.

Institute of Ecology. 1971. Man in the Living Environment. Report of the workshop on global ecological problems. 267 pp.

IV. Radiophosphorus Equilibrium with Mud, Plants, and Bacteria under Oxidized and Reduced Conditions[1]

F. R. Hayes and J. E. Phillips

Zoological Laboratory, Dalhousie University, Halifax, Nova Scotia

ABSTRACT

The phosphorus equilibration pattern and rate between mud and water was the same in natural Jenkin sampler cores, in artificial cores, and in bottles in which dredged surface mud was packed by centrifuge. Thus any specific natural physico-chemical or bacteriological layering of the surface muds of lakes is relatively unimportant in phosphorus exchange. Phytoplankton or bacterial cells equilibrate within a few minutes after addition. When antibiotics are used the P^{32} remains as inorganic PO_4 and is rapidly taken up by higher plants, or in the absence of plants, there is a rapid loss of P^{32} to the mud. With one exception, in over 100 artificial systems tested, the amount of P^{32} remaining in the water at equilibrium was greater in the presence of bacteria than where antibiotic had been added. This was true whether the system was treated with nitrogen or air, *i.e.*, was aerobic or anaerobic. After a week less than 10% of the P^{32} remained in the water of an antibiotic treated sample, while two thirds remained in the control.

The remarkable ability of bacteria to hold phosphorus in the water might be accomplished in two ways:

1) By an acceleration of the rate of P^{32} return from the sediment to the water by bacteria in the mud. In all experiments while the turnover time of water was generally of the same order of magnitude for all systems, the turnover time for mud was much shorter in the controls than in the antibiotic treated systems.

2) By the rapid uptake of radiophosphate by water bacteria and their ability to hold the radiophosphorus from the chemical or colloidal adsorption mechanism of the mud, which would be accomplished by incorporating the phosphate into non-participating organic compounds. An affinity, or holding back by water bacteria of P^{32} would be indistinguishable from an accelerated return to the water from the mud.

Dead plankton deposited on the mud decay and greatly increase the removal of P^{32} from the water. This reaction also is blocked by antibiotics. In bottle experiments there is a natural fallout of bacterial cells of about 4% per day. Neither the redox state of the system nor the level of lake productivity could be shown to influence either living or inorganic exchanges. The events following addition of radiophosphorus can be described as a modified first order consecutive reaction in which PO_4 yields organic P in the bodies of bacteria which in turn yields organic soluble phosphorus to the water.

The rate of exchange is measured as turnover time, which is the time required for the appearance or disappearance of as much phosphorus as is present in the test material, say phytoplankton or water or mud. Some turnover times are: water of a whole lake, one week; water in a bottle over mud, 0.5 week; return from lake sediment in nature including rooted aquatics, 1 month; from mud in a bottle, 0.5 week; from bottle mud without bacteria, 2 weeks. Equilibration of PO_4 between water and the inside of bacterial or phytoplankton cells is almost immediate, say 5 minutes, but conversion to the organic state is slower with average turnover time 0.3 days. Rooted aquatics, probably cannot take up organic P; with inorganic P their time is 0.5 week. Zooplankton are opposite in behavior, unable to use phosphorus until bacteria have made it organic; their time is then 1 day.

INTRODUCTION

When phosphorus is added to a lake as fertilizer nearly all of it disappears in a few days (Smith 1945, Orr 1947, Pratt 1949). The decline, taken together with the known effects of fertilization of agricultural land, might be attributed to new growth, although no immediate algal blooms appeared. The theory of growth stimulation demanded reconsideration when it was discovered that tracer phosphorus atoms, measured as radio-

[1] Acknowledgement is made of financial support by the Nova Scotia Research Foundation and the National Research Council of Canada.

Reprinted from *Limnology and Oceanography*, 3:459–475, 1958.

activity, declined in the water in the same way as large masses of fertilizer (Hayes *et al.* 1952, Rigler 1956). It now appears that there is a single pool of phosphorus belonging to lake water and solids, which is distributed between them in a dynamic equilibrium or steady state. Disappearance from water is a consequence of an experimental arrangement in which phosphorus is added to the water phase. Were the opposite technique followed, of taking up phosphorus from the lake water on, say, ion-exchange resins, a continuous replacement from the solids would be expected.

The foregoing interpretation does not deny the general observation that addition of fertilizers to lakes stimulates growth. Obviously after equilibration there will be more nutrient in the system than before. The point is that the decline of phosphorus in the water phase is not a measure of increased productivity, since its rate will tend to be independent of the quantity added.

The oxygen relations of red blood cells in their plasma provide a simple analogy to the lake nutrient system. Oxygen is distributed between the phases in dynamic equilibrium. Under ordinary conditions nearly all the oxygen is in the cells. Any disturbance of either phase will cause an appropriate rearrangement of the system. For example, the addition of oxygen to the plasma would be followed by a "loss" to the cells and vice versa. To a physiologist the plasma, or fluid phase, is merely an innocuous transporter between the active parts of the system.

The dynamic equilibrium of a lake might be represented as

Phosphorus in aqueous phase, a small fraction of the whole	$\rightleftharpoons$	Phosphorus in solid phase, a large fraction of the whole.

with a constant value for each phase but with a continuous exchange between them. The radioactive tracer technique makes it possible to measure the rate of exchange, which is given as the turnover time, defined by convention as the time for as many atoms to move through the phase as are present in the phase. To illustrate turnover time, one might imagine a country of 16 million people (say Canada) in tourist equilibrium with a country of 160 million people (say U. S. A.). Suppose there are 32 million people crossing the border each way each year. The turnover time for Canada would be $16/32$ yrs or 6 months; for the U. S. it would be $160/32$ or 5 yrs. It will be noticed that there is no uptake of tourists by the U. S. although there would erroneously appear to be such if the only observation made were a count of persons leaving Canada. Also, it is not implied that every Canadian moves across twice a year. Many persons will not cross at all, others several times per year and still others will go over once and stay. Turnover time is not obtainable by counting the populations of the two countries, however frequently, for the counts do not change. It is necessary to detect persons actually crossing the border.

In limnological work the P^{32} is an identifiable sample moving one way, like Rotarians leaving Canada. Extra mathematics, over and above tourist-like counts, are made necessary by lack of knowledge of the total populations of phosphorus atoms in the two phases, so that indirect means must be used to secure turnover times.

The purpose of this paper is to discuss the effect on the exchange reaction of some natural variants in lakes, namely the level of productivity, the state of oxidation or reduction, and the presence of green plants and bacteria. In a lake all these will be acting together to produce the observed equilibrium, but in the laboratory the components can be separated.

An account of preliminary experiments is given by Hayes (1955), and the results to follow are an amplification and in one respect a modification of that report. Whereas the dominant role in exchange was previously attributed to the state of oxidation or reduction with bacteria playing a supplementary part, the results to follow indicate that bacteria are decisive and can to a considerable degree suppress the classical inorganic mechanism.

MATERIAL AND METHODS

Of the lakes whose sediments were studied, one was judged eutrophic (Southport Pond), two were acid bog (Punchbowl and Silver), and five were unproductive or marginal (Copper, Black Brook, Grand, Lily, Bluff). For descriptions see Hayes and Anthony (1958).

Methods of handling and counting P^{32}, calculations of turnover times, handling of Jenkin sediment cores, etc. are given by Coffin *et al.* (1949), Hayes (1955), Hayes *et al.* (1952), and Harris (1957). Methods for measuring oxidation-reduction and for counting bacteria are given respectively in papers II and VI of this series.

In previous work (Hayes 1955) Jenkin core samples were collected to measure the radiophosphorus exchange between mud and water from lakes. This method is relatively exacting and inconvenient when many samples are required, and a simpler one was developed for the present investigation. Mud was collected in an Ekman dredge, the water was decanted off, and the surface mud was scooped into a sterile quart bottle. The remaining mud was put into a second sterile bottle. Lake water was also collected, and the samples were chilled and transported back to the laboratory as quickly as possible, generally within three or four hours. Here they were kept in the dark at 4°C until use, which was within a week or less. Before use the lake water was filtered through Whatman number 42 paper to remove large mud particles and plankton.

Artificial systems, as shown in Figure 1, were set up and maintained at 4°C. Before being used they were allowed to stand for five to seven days, to give the bacteria time to reach a stable number, under aerobic or anaerobic conditions, and to allow the whole system to reach biological and chemical equilibrium. When it was desired to inactivate bacteria the antibiotic terramycin, or later the more powerful tetracycline, was used. A quantity of 100 mg of terramycin was needed per bottle but only 15 mg of tetracycline. In practice 30 mg of the latter was used, to allow a safety factor. This concentration was found to inactivate the bacteria for at least a week. A further advantage of tetracycline was the absence of any observable deposit on the mud surface. The antibiotic was dissolved in a few ml of distilled water and then passed through a millipore filter to remove any resistant bacteria or moulds. It was added to the artificial system a day before the experiment was to begin. This interval was found to be sufficient to inactivate the bacterial population almost entirely.

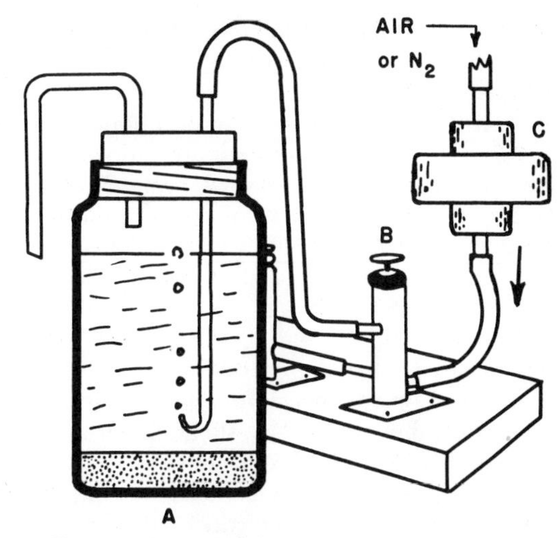

FIG. 1. An artificial mud-water system *A*, with gas jet connected to a needle valve *B*, and an aerosol bacterial filter *C*. Sterile apparatus was used and precautions against contamination taken so that the only opportunity for entrance of foreign bacteria was for about 15 sec. during collection of samples.

With a large syringe, about 20 ml of dredged mud was put into an 8-oz wide-mouthed bottle *A*, avoiding contamination of the sides. Twenty ml of surface mud was then added over the bottom mud and the bottles were capped and centrifuged for five minutes. The supernatant water was pipetted off, leaving a clean solid mud surface, over which, without disturbance, 150 ml of filtered lake water was added. Each bottle was closed with a rubber stopper through which passed a J-shaped jet with its tip turned upward and about 0.5 cm from the mud surface, and a U-shaped outlet tube. The jet was connected either to air or nitrogen which was allowed to flow at the rate of approximately one bubble per second, thus keeping the water in constant motion. The gas flow to each bottle was regulated by a needle valve *B*. The gas, air or nitrogen, was passed through an aerosol air filter *C* to remove any foreign bacteria.

In order to discover whether there was any appreciable uptake and exchange of phosphorus by bacteria on the glass surface of the bottle, water blanks were set up in each experiment without the addition of mud. There was no loss of P^{32} in the glass blanks to which tetracycline was added. Further evidence of the effectiveness of tetracycline in inhibiting bacterial activity is seen in the investigation of qualitative changes in the radiophosphate added to Grand Lake and Punchbowl systems (see below). The culturing of water micro-organisms by the millipore technique indicated that moulds are relatively unimportant numerically in lake water.

Attempts that were made to sterilize samples by heat under standard autoclaving conditions, were successful in killing all organisms, but produced such changes in the appearance and evident colloidal properties of the mud that they could not be taken as representative of natural conditions. Nevertheless, reactions with P^{32} agreed well with those of antibiotic sterilized samples.

Radiophosphate in dilute HCl, as obtained from Atomic Energy of Canada Ltd., was diluted with distilled water to give a count of 500,000 c.p.m. per ml and at once sterilized to kill bacteria and to permit storage for future use without the formation of organic radiophosphate by microorganisms. About one ml of the dilute radiophosphate was added to each artificial system, and duplicate samples were taken at intervals for counting.

From paper I, Table 2, the average total P in lakes is 18.5 ppb and the inorganic fraction amounts to 2.8 ppb. The added 500,000 counts in 150 ml water works out at 5.6×10^{-5} ppb. Thus the phosphorus increase in the samples by reason of added P^{32} amounted to 0.0003% of the total P already present or 0.002% of the inorganic P.

To separate the radiophosphate fractions a sample of water was passed through a millipore filter, which retained bacteria and particulate matter. The filtrate contained total phosphorus in solution, from which the inorganic P was precipitated. It was possible to find in a sample the c.p.m. per ml of radiophosphorus (a) in bacteria, (b) in particulate matter, (c) as inorganic P^{32} in solution, and (d) as total P^{32} in solution. The bacterial fraction could be estimated by comparing the counts in particulate matter with and without tetracycline treatment. There is a possibility that the small particulate fraction in the tetracycline-treated artificial systems might be incorporated in tetracycline-resistant bacteria or moulds, rather than in non-living particulate matter, although as already noted, moulds were not observed.

It was important to make sure that reduced conditions were secured in artificial systems by bubbling nitrogen through the water for a week. The cylinder nitrogen used was stated to contain not over 0.05% oxygen, which at equilibrium would give a dissolved oxygen level of 0.03 ppm. According to Mortimer (1941–2) ferrous iron was found in artificial systems below 0.1 ppm and in lakes below 0.5 ppm. Thus our oxygen level is theoretically low enough. Winkler tests on water through which nitrogen was bubbled were indistinguishable from zero oxygen.

Reduced conditions, however, are not attained merely by withdrawal of oxygen, but require the action of microorganisms, or of a suitable catalyst. The decisive test is the redox potential which, as paper II points out, is not easy to measure or interpret.

In an artificial Silver Lake oxidized mud-water system the initial potential was 0.33 volts for mud and the pH was 4.7. The water was then bubbled with nitrogen for four days, whereupon the potential of the mud fell to 0.14 volts. After bubbling with air for 12 hours, the mud potential rose to 0.40 volts. The same result was obtained with tetracycline treatment indicating probably that the antibiotic did not penetrate the mud to kill the bacteria there, but allowed them to reduce the system.

Further results on the Punchbowl are

TABLE 1. *Effect on potential of bubbling nitrogen or air in a Punchbowl artificial mud-water system*

The mud was initially reduced and the water oxidized. These results show that the mud-water system was well oxidized or reduced before the antibiotic was added and the experiments started.

Duration of bubbling in days	Potential in volts (Nitrogen)		Potential in volts (Air)	
	Water	Mud	Water	Mud
0	0.45	0.06	0.45	0.12
0.5	0.27	0.27		
3.0	0.17	0.14		
4.0			0.38	0.38

given in Table 1 showing that here also reduced conditions can be produced. With Grand Lake however, bubbling with nitrogen for over two weeks failed to bring about a corresponding reduction in four oxidized mud-water systems, to two of which antibiotic had been added. The mean initial potential of these systems was 0.42 volts for mud and 0.65 volts for water. The potential was measured every three or four days, the lowest reading after 15 days being 0.3 volts for mud and water. Other Grand Lake systems, when bubbled with air, remained oxidized at values between 0.4 and 0.5 volts for mud. It appears probable, in the light of these results and those in paper II, that the reduction achieved in these tests was as effective as would be found in nature in the sub-surface mud.

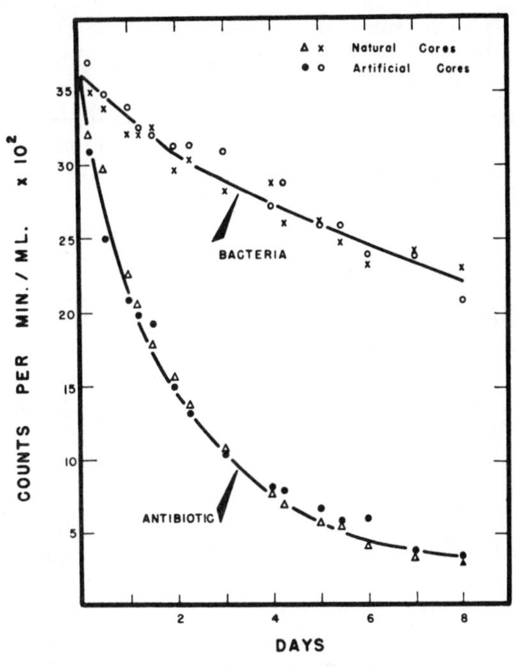

FIG. 2. The amount of radiophosphorus remaining in Grand Lake water over mud in natural and artificial Jenkin cores with and without antibiotic, *i.e.*, bacteria absent and present.

EXPERIMENTAL RESULTS

Natural and artificial cores compared

It was necessary for purposes of interpretation to know whether the mud-water radiophosphorus exchange in artificial systems represented the exchange under conditions more closely approaching those in the lake itself. The preparation of an artificial system involves the destruction of the natural physico-chemical and biological layering of the surface muds. If this layering were in any way peculiar and important to the phosphorus exchange in lakes, different results would be expected with natural Jenkin cores in which the mud-water interface is undisturbed. Two Jenkin cores were collected from Grand Lake and set up in the laboratory at 4°C. Water was siphoned off the cores until approximately 200 ml of water remained over the mud. Two artificial systems were set up by placing stirred bottom and surface mud in Jenkin cores and adding 200 ml of surface water over the mud in the usual manner. Tetracycline was added to the water of one natural and one artificial core. All cores were air bubbled.

The loss of P^{32} from the water is shown in Figure 2. It is apparent that natural and mocked-up Jenkin cores react similarly both in the presence and absence of bacteria. These results are almost identical to those for the eight-ounce bottle, artificial system results with the same lake. Any heterogeneous structure of the natural mud-water interface, if it is essential, was obviously re-established in the artificial systems before the P^{32} was added.

This experiment illustrates a general finding, namely that when bacteria are inactivated with antibiotic (or by heat sterilization) there is a rapid loss of P^{32} to the mud. With one exception, in more than 100 artificial mud-water systems tested the amount of P^{32} remaining in the water at equilibrium was greater in the controls in which bacteria were present than in the corresponding system to which antibiotic had been added. This was true whether the system was treated with nitrogen or air, *i.e.*, was aerobic or anaerobic. In the treated example under discussion, after a week less than 10% of the radiophosphate remained in the water. In the control two-thirds remained after the same period. This remarkable ability of bacteria to hold phosphorus in the water might be accomplished in two ways:

1) By an acceleration of the rate of P^{32}

return from the sediment to the water by bacteria in the mud. In all experiments, while the turnover time of water was generally of the same order of magnitude for all systems, the turnover time for mud was generally much shorter in the controls than in the antibiotic treated systems.

2) By the rapid uptake of radiophosphate by water bacteria and their ability to hold the radiophosphorus from the chemical or colloidal adsorption mechanism of the mud, which would be accomplished by incorporating the phosphate into non-participating organic compounds. An affinity, or holding back by water bacteria of P^{32} would be indistinguishable from an accelerated return to the water from the mud.

The marked atoms which were added follow along with the rest of the inorganic phosphate in solution. The test being in the dark, there is no green plant effect to consider, but only bacteria which are putting inorganic phosphate rapidly through their bodies, changing most of it enroute to the organic form. The organic fraction does not enter into non-living exchange reactions, of the kind described in earlier papers of this series. There is at the same time a regeneration of the inorganic phosphate by breakdown of the organic fraction, so that in the steady state of nature about 10 to 20% of the phosphorus in solution is inorganic.

When we prevent bacteria from entering the reaction by addition of antibiotic, the P^{32} remains in the inorganic state and is free to participate in, say, an adsorption equilibrium as discussed in paper III. This is the lower curve of Figure 2, showing the P^{32} atoms leaving the water. As each one leaves it is replaced by a conventional or P^{31} atom coming out of the mud, but these ordinary atoms have no identification marks.

A culture of phytoplankton

A side experiment with a culture of *Chlamydomonas dysosmos* illustrates the speed with which microorganisms can pass phosphorus through their bodies. It was a pure strain culture, bacteria free, and was well supplied, indeed over supplied, with nutrient in the medium so that no excess uptake due to possible starvation could occur. The phosphate concentration was 0.015% in the form of KH_2PO_4. The organisms made up 5% of the volume of the culture. Three-hundred counts per minute per ml of radiophosphorus was added to the medium, and its transfer from water to organisms followed, organisms being centrifuged from the medium before they were counted. In the final equilibrium the counts amounted to about 190 per minute per ml for the water and 1,760 counts per minute per ml for the cells. At the time of the first count, 2.5 minutes after the start of the experiment, equilibrium had already been established. On general grounds it is known that the turnover time for water is about one-third of the time for substantial completion of equilibration. Thus we can say that the turnover time for the phosphorus in the water was probably not more than 1 minute and may have been considerably less. At equilibrium the cells contained about half as much radiophosphorus as the water, and they had to return an atom to the water for every one they took out in exchange. It would thus appear that the turnover time for the phosphorus in *Chlamydomonas* cells must be not more than 0.5 minutes. Thus under natural conditions in a lake such microorganisms could very rapidly make over any inorganic phosphorus present into the organic form and return it to the water.

Competition between bacteria and plants

We turn now to the 8-oz bottle experiments, beginning with the competition for phosphorus between bacteria and higher plants. The mud layer at the bottom is omitted and the duplicates set up include a control of filtered lake water, the same plus one or other of the plants tested, and the same plus plant and antibiotic. Results on two lakes are shown in Figure 3.

The flowering plant *Eriocaulon*, which is very abundant over the bottom of Bluff Lake, was collected, and 1 g of sprigs was placed in each of two aerated bottles of Grand Lake water. The loss of P^{32} from the water is shown in Figure 3 at left. In the absence of bacteria *Eriocaulon* very rapidly takes up radiophosphate and reaches an equilibrium at which 10% of the P^{32} is left

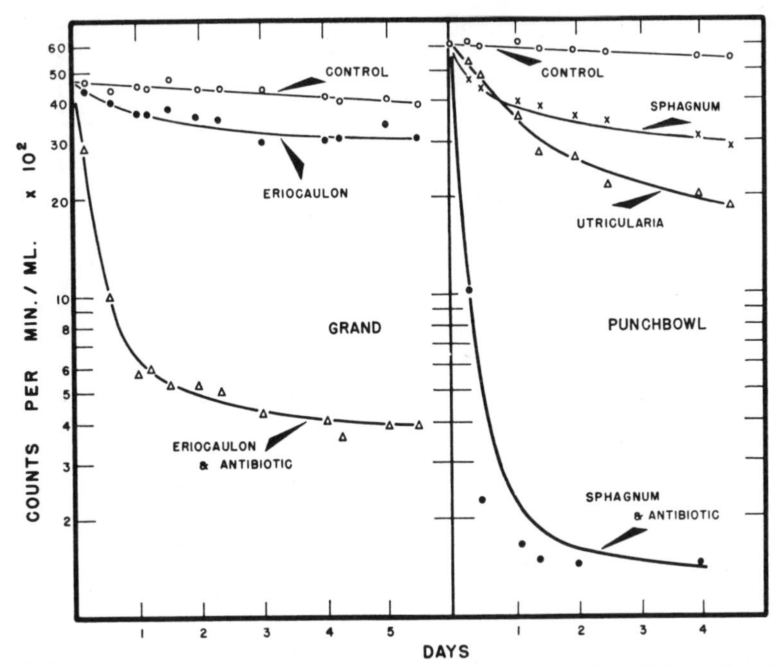

FIG. 3. Radiophosphorus remaining in aerated water (without added mud) from Grand Lake (primitive) and Punchbowl (acid bog) containing plants with and without antibiotic. The plants used were the pipewort (*Eriocaulon*), the bladderwort (*Utricularia*), and the peat moss (*Sphagnum*).

in the water. The phosphorus turnover times for *Eriocaulon* and water, in the absence of bacteria, were calculated to be 3.0 days and 0.34 days, respectively. The effect of bacteria is to hold phosphorus in the water and prevent the large loss to the bottom plants. When bacteria are present, the equilibrium between radiophosphorus in water and in mud was reached with two-thirds of the radiophosphorus in the water, including its bacterial population. This is because bacteria hold such large quantities of phosphorus in their cell bodies. A second cause is the production by the bacteria of soluble organic radiophosphorus, which the plant presumably cannot assimilate. This experiment was repeated with the plants *Sphagnum* and *Utricularia* in Punchbowl water, Figure 3 right. One and one-half grams of *Sphagnum* sprigs or 0.5 g of *Utricularia* sprigs were placed in 150 ml of water. Unfortunately tetracycline caused the *Utricularia* to turn brown, and there was little uptake by this plant in the presence of the antibiotic. Because of the adverse effect only the non-treated *Utricularia* result is plotted, which is similar to that for the other plants. In the absence of bacteria *Sphagnum* took up 97% of the radiophosphate in the water within half a day, the turnover times for *Sphagnum* and water being 3.5 days and 0.09 days, respectively. In the presence of bacteria 50% of the phosphorus remained in the water after 5 days. Although no turnover time for *Utricularia* and water in the absence of bacteria was obtained, the fact that the loss of radiophosphorus from the water in the presence of bacteria was greater for *Utricularia* than *Sphagnum* suggests that the former plant is as active as *Sphagnum* in exchanging phosphorus with water.

There was also a Grand Lake experiment including mud, to which *Eriocaulon* was added, and a Bluff Lake test including mud and *Utricularia*. This is placing all three components in competition. The two lake systems agreed well, that for Grand (Fig. 4A) showing that the addition of *Eriocaulon* to the control resulted in a loss of 70% of the radiophosphate from the water after six days. The ability of the plant to remove so

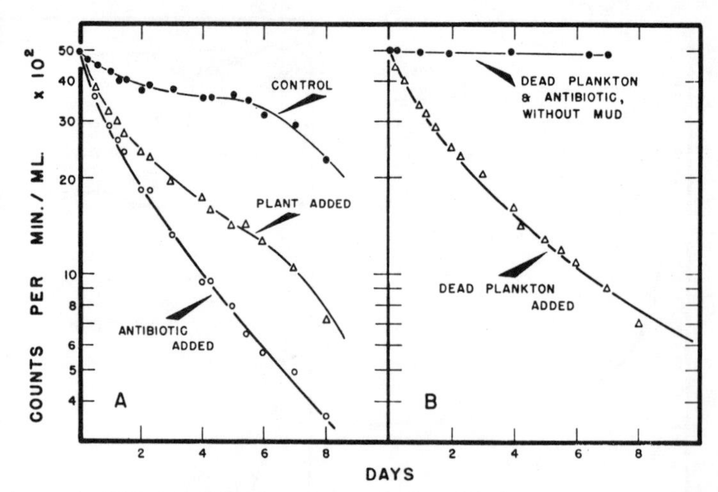

FIG. 4. *A*. Amount of P^{32} remaining in an aerated Grand Lake mud-water system to which 2 g of *Eriocaulon* sprigs or 30 mg of tetracycline was added. *B*. Effect of adding dead plankton in the presence and absence of bacteria. Comparing the lowest curve of *A* with the top one of *B* it is shown that mud is necessary for the inorganic exchange of P^{32} and that sterile plankton will not serve as a substitute for mud.

great a quantity of radiophosphate from the water when the mud control was unable to do so in the presence of bacteria indicates the affinity of *Eriocaulon* for phosphate. The control curve in Figure 4A is an extreme illustration of another phenomenon which was common to most experiments. While a control mud-water system seemed to reach equilibrium after three to eight days, a gradual loss of radiophosphorus continued over a period of weeks until practically none remained in the water. Evidence will be presented later which suggests this decline is due to a fall out of radiophosphorus in the bodies of bacteria. In the present example the turnover time would be calculated from the curve values before the sixth day.

In summary, when bacteria and higher plants compete for inorganic phosphorus the bacteria get there first and change a part of it to organic forms which are apparently unavailable to plant use. In the absence of bacteria the aquatic plants tested take up enormous quantities of P^{32} within 12 hr, indicating a rapid exchange and the establishment of an equilibrium with most of the phosphorus in the plant body.

Effect of plankton fallout

A large fallout of plankton following a bloom, might be expected to affect bacterial activity at the mud surface of a lake and thus alter the exchange of phosphorus. Plankton was netted from Grand Lake, autoclaved, and approximately 0.3 g was pipetted into duplicate mud-water systems from the same lake. The plankton sedimented immediately to the mud surface. Equal quantities of dead plankton were placed in two antibiotic treated glass blanks, omitting mud. Results are shown in Figure 4B. Obviously sterile dead plankton, of itself, is unable to adsorb significant quantities of phosphate by a colloidal mechanism or by other means. The addition of dead plankton to a mud-water system resulted in an enormous and rapid loss of P^{32} from the water. This greater uptake at the mud surface, as compared with the control in Figure 1A, was evidently caused by increased bacterial activity there, due to the large amount of readily decomposable organic matter. An uptake of phosphate by bacteria attached to falling organic matter in lakes is probably of importance in the natural cycle.

Effect of the state of oxidation

It is a matter of fact, repeatedly verified, that the removal of oxygen either over the lake bottom or in laboratory tubes will allow inorganic P to come up out of the mud. In nature this will take place in the dark below the green plant zone. A large fraction though presumably not all of the new P remains inorganic. The conditions of stagnation which are prerequisite to oxygen lack also operate to keep the new P in a close layer a few cm thick over the bottom where it may reach a high concentration. Any increase which might be observed if the new P were dispersed through the whole lake, would depend on both the volume of the lake and the quantity of P produced. The latter measurement has hitherto been too difficult to make (as far as we know). The new P may formerly have been adsorbed on a gel which was destroyed when reduced, (paper III) or (as Einsele stresses) it may exist as a salt of which the reduced form (ferrous phosphate) is more soluble than the oxidized (ferric phosphate).

How would the above considerations affect the behavior of P^{32} in the water? We know that under ordinary aerobic conditions the loss to the mud is kept small through the intervention of bacteria. The effect of removing oxygen is unpredictable because of competition between two processes: (a) bacterial conversion to organic P will be if anything diminished, and this would enhance removal of P^{32} from the water (see Fig. 2), and (b) the general flow of ordinary P out of the mud would so affect the exchange mechanism as to leave most P^{32} in the anaerobic water. Turning to actual results, in four out of ten experiments, the per cent of radiophosphorus in the water at equilibrium was higher under aerobic than under anaerobic conditions. The average difference between the equilibrium values in these four experiments was 5.5%, the highest being 11.0%. These differences are so small that the two conditions may be said to be indistinguishable. In the six experiments where there was more radiophosphorus in the water under anaerobic conditions, the average difference was 20% with a range from 32% to 5%. It is concluded from the whole series that the activities of microorganisms are dominant, so that the inorganic mechanism cannot be identified.

When antibiotics are added to prevent bacterial synthesis one might hope to observe

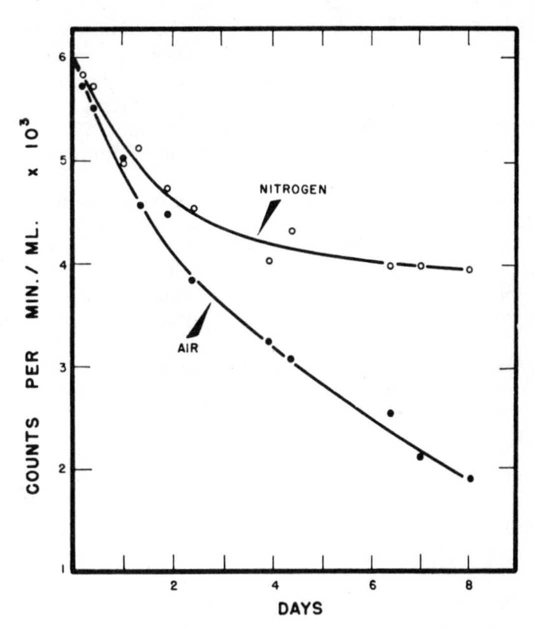

FIG. 5. The amount of radiophosphorus remaining in the water of antibiotic-treated Punchbowl mud-water systems under reducing (nitrogen) and oxidizing (air) conditions.

the inorganic mechanism, (b) of the previous paragraph. In ten out of twelve experiments the average per cent of radiophosphorus in the water at equilibrium was indeed higher under anaerobic conditions. In four out of these ten experiments however the difference was less than 5%. It was maximal in the two Punchbowl experiments in which there was twice as much radiophosphorus in anaerobic water at equilibrium (Fig. 5).

Effect of trophic level and lake type

Tests such as those described above have been carried out on eight lakes, with results expressed as turnover times for mud and for water and as P^{32} left at equilibrium. The data are in Table 2, in which the lakes are arranged in productivity groups. This judgment is in large measure qualitative, particularly as regards the separation between "marginal" and "unproductive."

A complete experiment would number 12 to 16 bottles each arranged as in Figure 1. Four artificial mud-water systems were bubbled with air and four with nitrogen. Antibiotic was added to two bottles in each

TABLE 2. *Turnover times for mud and water in 8-oz bottle experiments, and P^{32} concentration left at the equilibrium time*

The symbols are: A_c aerobic control; A_t aerobic test with tetracycline added to inactivate bacteria; N_c anaerobic or nitrogen control; N_t nitrogen with antibiotic added. Some heat sterilization results are also included, being placed for economy under the antibiotic values with which they are to be compared. The values are averages of duplicate tests or, in several instances, pairs of duplicates done at different times.

A mathematical limitation may be mentioned here, for illustration of which attention is directed to the upper curve of Figure 4. Suppose the flattening of this curve at 6 days and its subsequent decline were described by points with a little more scatter. The line marked "control" would then have to be drawn straight, a treatment which would make little difference to the turnover calculation for water, but would make the mud value read as infinity. The per cent left in water at equilibrium would come out to be zero. Such results, being meaningless, have been omitted from the table.

The treatment just described, involving the assumption that the decline, when plotted logarithmically, is linear, was the first way to calculate turnover time (Zilversmit *et al.* 1943). When applicable however, the complete method in which allowance is made for a two-way flow of phosphorus, yields more information.

Lake	Turnover time for mud Days				Turnover time for water Days				% P^{32} left in water at equilibrium			
	A_c	A_t	N_c	N_t	A_c	A_t	N_c	N_t	A_c	A_t	N_c	N_t
Productive												
Southport	3.2	13.8	2.1	8.6	3.8	3.0	1.7	2.1	53	17	44	20
Marginal												
Copper	3.4	11.3	2.5	22.0	2.0	1.7	3.7	2.0	36	13	65	9
Lily	4.4	26.5	5.0	10.7	1.9	3.3	2.0	2.5	31	10	28	19
ditto, heat sterilized				7.5		2.0		1.9		11		19
Black Brook	2.2	12.1	3.1	12.1	1.7	1.7	1.9	2.3	44	13	37	16
Unproductive												
Bluff	2.3	18.9	3.2	9.1	1.0	1.8	3.1	1.8	34	12	50	14
Grand	3.4	18.8	3.3	21.0	2.0	1.5	1.8	1.9	60	9	35	8
Acid Bog												
Punchbowl	6.3	7.0	1.6	3.7	3.8	3.1	2.7	4.1	41	30	62	56
ditto, heat sterilized		6.3		2.4		2.8		4.7		32		65
Silver			7.4		5.3	5.0	3.6	6.1			33	

group. In the Punchbowl and Lily Lake experiments heat-sterilized systems were also set up to compare with the effect of antibiotics. Each experimental arrangement was thus prepared in duplicate. Glass blanks (150 ml of water without mud) were set up under each condition in all experiments with the exception of Copper Lake and Bluff Lake. After the addition of P^{32} between eight and eighteen duplicate samples of water were taken from each bottle at intervals until equilibrium was reached or neared, and from these samples the turnover times for water and mud were calculated, as given in Table 2. They are averages of duplicate mud-water systems, the mean probable deviation of the two values corrected by the *t*-test being less than 20%. The per cent of the initial radiophosphorus concentration remaining in the water at equilibrium is also given.

The first conclusion to be drawn from Table 2 must be negative. Looking vertically at the columns, there is no clear relation between trophic level and the exchange reaction. Only the acid bog differs from the rest of the groups in the antibiotic or heat-treated "*t*" columns. Here the turnover time for mud is shorter and for water longer, and more P^{32} is left at equilibrium than in the others. Since the difference is in the sterile samples it is evidently related to the physical properties of the sediment rather than microorganisms. The brown water of bog lakes is carrying in and depositing large quantities of foreign organic material from surrounding swampy country which may account for the P^{32} effects.

The absence of differences in laboratory tests may not reflect behavior in nature, where phytoplankton enters the scene. These plants will not only compete with bacteria for phosphorus, but may suppress them by their own antibiotics (Steemann Nielsen 1955).

Table 2 shows that the turnover time for mud, *i.e.*, the rate at which mud can put phosphorus back into the water, becomes greatly lengthened when microorganisms are inactivated. We may postulate a normal cycle in which there is (a) an uptake and synthesis of organic P^{32} in the water, followed by (b) a fall out to the sediment surface, where (c) a breakdown to inorganic P^{32} occurs under bacterial action, so that (d) the P^{32} is restored to the water. Under the tetracycline treatment an inorganic mechanism is in operation with different time relations.

Looking at the center block of Table 2, the horizontal values have no trend. Thus the turnover time for water is not shown to be affected by bacteria or by the state of oxidation-reduction.

The right block however, does indicate bacterial action, in that the columns A_c and N_c have higher values respectively than A_t and N_t. This means that when bacteria are present they hold extra phosphorus in the water. To look for an oxidation-reduction effect we compare column A_c with N_c and A_t with N_t. No convincing behavior pattern can be discerned, *i.e.*, there is no indication of more P^{32} left in the water under one state than the other. The one lake, Punchbowl, which clearly obeys theoretical demands, has already been illustrated in Figure 5.

No appreciable loss of radiophosphorus was noted from the water of over thirty glass blanks, both anaerobic and aerobic, to which antibiotic had been added. This is evidence that the loss from the water of mud-water systems was not due to precipitation of phosphate as ferric phosphate.

The agreement between heat sterilization and antibiotic treatment of mud-water systems suggests that: (a) the effect of antibiotic on P^{32} exchange is due to the inactivation of bacteria; (b) moulds and other antibiotic-resistant microorganisms are relatively unimportant in the mud-water exchange; (c) the physico-chemical adsorption mechanism of the mud is not drastically affected by the very harsh treatment of heat sterilization.

A look at the whole of Table 2 suggests that the action of bacteria is to suppress the underlying inorganic exchange mechanism, which is thereby relegated to a subsidiary role. We may suppose that in nature the phytoplankton behaves in the same way.

Loss from the water of control glass blanks

There was a loss of radiophosphate from the water of over thirty control blanks, that is, blanks in which the bacteria had not been killed. This loss was generally much less and occurred at a slower rate than that from corresponding mud-water systems.

Some blanks showed an equilibrium with a solid phase, evidently periphytic or sedimented organisms. In most cases the decline was arithmetic during the time of sampling. Often there was no decline during the first few days followed by slow loss of radiophosphorus. Full data dealing with these results are not presented here because of their voluminous and diverse nature. The loss of radiophosphorus from Grand Lake and Punchbowl glass blanks is shown in Figure 3.

Heukelekian and Heller (1940) report that the bacterial population of stream water is increased by addition of clean sand, while ZoBell (1943) found that increased bacterial activity in stored sea water is directly proportional to the ratio of glass area to water volume. According to Taylor and Collins (1949) growth is stimulated by Bohemian glass but not by Pyrex or fused silica. These reports led us to enquire whether the loss of P^{32} in the control glass blanks was due to bacteria on the sides of the glass bottles. If this were so it would be a source of error in the mud-water systems. If however, the loss were due to fallout of bacteria from the water to the bottom of the bottle, or to accumulation by sessile bacteria on the bottom, there would be no error introduced.

Tests were conducted on Punchbowl mud-water systems in which the water volume was reduced from the usual 150 ml to 100 ml and 50 ml, and to which pieces of glass tubing were added to double the glass surface. Tabular data will be omitted since all results were negative, *i.e.*, there was no difference in behavior of P^{32} caused by changed surface-volume ratios.

It is known (Fig. 3, lines marked "control") that some loss occurs from water even when mud is not present. To discover whether such loss of radiophosphorus might be due to a fallout of bacteria, a glass filter disc was placed on the bottom of an aerobic Grand Lake preparation. After 15 days 50% of the radiophosphorus remained in the water. The sintered glass disc, when it was removed and its activity tested, accounted for 89% of the P^{32} which had disappeared from the water. Thus most of the loss of P^{32} is to the bottom of the bottles, and the effect is not a source of error where the mud surface replaces the glass bottom of the blank. It is concluded that periphytic bacteria, if they are present on the glass walls, are not active in taking up radiophosphorus.

Consecutive forms of phosphorus

It has already been suggested that tracer phosphate is soon taken up by bacteria and changed into the organic state, and that later some of it leaks out into the water again as soluble organic phosphate. We may now consider the time and equilibrium relations of the process which may be compared and contrasted with a consecutive reaction, $A \rightarrow B \rightarrow C$, as shown in the left block of Figure 6, in which the two reactions are proceeding simultaneously. If events were followed by analyzing for A, curve A would be obtained; if periodic measurements were made of the end product C, curve C would result; finally if only the intermediate product B were determined, the course of the reaction would rise to a maximum and fall off as shown by curve B.

The course of events with a P^{32} experiment differs from the reaction described in that there is a gradual loss of the total, $A + B + C$, from the water both by reason of fallout, etc. as already described, and by inorganic exchange with mud if mud is present. Also the lake water reaction does not proceed to virtual completion, but to an equilibrium in which appreciable quantities of all forms remain. As already mentioned general water analyses show 10 to 20% of the phosphate as inorganic, *i.e.*, corresponding to A.

Three aerated bottle series were set up, with and without mud, in which the P^{32} in several fractions was followed. Two were on Grand Lake material, collected in January and July and the third on Punchbowl collections made in July. Results agreed very well, and the Punchbowl, for illustra-

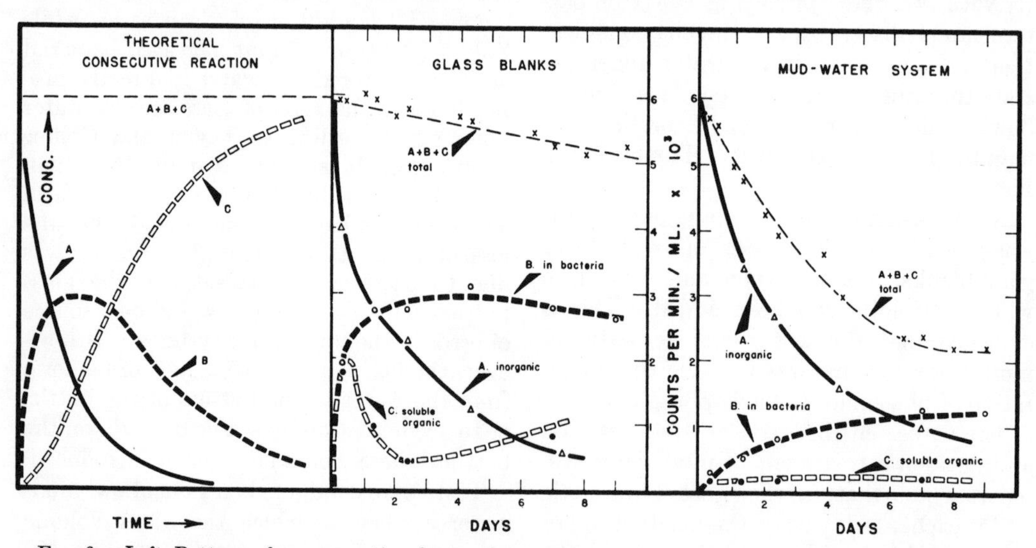

FIG. 6. *Left:* Pattern of a consecutive first-order reaction of type A→B→C. The second step is half as fast as the first. From Getman and Daniels 1937, p. 331. *Center:* Forms of P^{32} as they vary with time in an aerated bottle of Punchbowl water. *Right:* The same in a mud-water system, set up as shown in Figure 1.

tion, will be presented in detail, as shown in Figure 6.

Rapid changes were observed in the P^{32} which had been introduced as inorganic PO_4. Figure 6, center, shows the water blank for the first nine days. Water bacteria rapidly incorporated 50% of the P^{32} as part of their body protoplasm, and reached virtual equilibrium within 12 hours. (The same equilibrium at 12 hours was observed in a test on Black Brook water by which time two-thirds of the P^{32} was in the bacteria.)

There was a rapid formation of soluble organic radiophosphorus (C) reaching a maximum by the time of the first observation at twelve hours. This fraction then fell to an equilibrium value after two days. The maximum might be due to a very rapid production via bacteria and a slower loss from the water to mud or sessile microorganisms. The subsequent change in C was slight with a possible rise at right. While the increase at 7 days is not proven by the single analysis, it is made plausible by the theoretical necessity that the total A + B + C should sum up to the value for the top line. The turnover time for the bacteria was calculated at 5 hr.

When mud is added to the system (Fig. 6, right) the uptake of P^{32} by water bacteria is considerably diminished. The reason is apparent in the difference between the topmost or "total" curves showing the great loss from the water when mud is present. In other words there is introduced with the mud a great number of competing bacteria as well as an inorganic exchange mechanism. The soluble organic fraction (C) in the water is also smaller when mud is present. It appears that the two experiments are showing the same phenomena but that the mud one is modified by the steep down slope of the A + B + C curve. The turnover time for water bacteria worked out at four days, nearly an order of magnitude larger than when mud is absent. Evidently the mud is pulling for P^{32} against the water bacteria, which tend, as a consequence, to hold on to what they can get.

In Grand Lake tests the mud has diminished effectiveness as a P^{32} remover (see Fig. 4A, top line). For this reason the turnover time for water bacteria was about the same as that of the Punchbowl water blank, 4 hr or less. As a further consequence there was a greater uptake by the water bacteria (B). In other words the Grand Lake complete system was not unlike the Punchbowl water blank in behaviour.

INTEGRATION

There are now sufficient data, as assembled in Table 3, to warrant a preliminary assessment of the time relations of the several phosphorus reactions.

For the water of a whole lake, the turnover time averages about a week (center col., blocks A and B). For a bottle experiment it is nearly half a week (block C, 1st 4 entries). The bottle experiments lack phytoplankton and rooted aquatics. The two results are surprisingly alike, and they seem at first to contradict the tourist analogy of the introduction, according to which a lake of infinite depth (compared to a bottle) would have an infinite population of phosphorus atoms, hence an infinitely long turnover time. The fact seems rather to be that in these lakes the exchanging mechanism builds up with the water volume, perhaps because the littoral zone is large and phytoplankton is proportional to volume of water. In our analogy the tourists are proportional to the population of the country.

The return from solids (left col.) is an order of magnitude slower in lakes than in bottles, presumably because of extra competition from rooted aquatics, phytoplankton, and zooplankton in the natural state. We suggest provisional acceptance of the bottle results as describing the behavior of mud alone. The lack of difference between oxidized and reduced systems and the four-fold slowing of the response when bacteria are inactivated has already been discussed.

The remainder of Table 3 deals with reactions between aquatic life and water in absence of mud. With microorganisms a sharp time distinction exists between two processes. There is a quick reaction, evidence for which rests mainly on Rigler's results with freshly collected lake water in-

TABLE 3. *A collection of turnover times*

The first three lake experiments are with large quantities of phosphate as fertilizer; all the others are with tracer amounts of radiophosphorus. Although added as inorganic PO_4 the phosphorus is soon changed to organic compounds when bacteria are present. Times are given in days except for equilibria between outside water and cells, which are given in minutes (shown by italics in table).

Turnover time for 1st mentioned component	Equilibrium system	Turnover time for 2nd mentioned component	Lake, location, experiment	Reference
		A. Whole lakes. Addition of dry phosphate fertilizer		
176	Solids, *i.e.*, mud + plants, *vs.* water incl. bacteria	17	Unstrat. Crecy, N. B.	Smith 1945
	Solids, *i.e.*, mud + plants, *vs.* water incl. bacteria	3.2	Unstrat. marine Loch Craiglin, Scotland	Orr 1947
	Solids, *i.e.*, mud + plants, *vs.* water incl. bacteria	2.4	Unstrat. marine pond, Cohasset, Mass.	Pratt 1949
		B. Whole lakes. P^{32} tracer experiment		
37	Solids, *i.e.*, mud + plants, *vs.* water incl. bacteria	7.6	Punchbowl, N. S. epilimnion	Coffin *et al.* 1949
39	Solids, *i.e.*, mud + plants, *vs.* water incl. bacteria	5.4	Unstrat. Bluff, N. S.	Hayes *et al.* 1952
	Solids, *i.e.*, mud + plants, *vs.* water incl. bacteria	3.6	Toussaint, Ont. epilimnion	Rigler 1956
29	Solids, *i.e.*, mud + plants, *vs.* water incl. bacteria	10.2	Toussaint, recalculated by us from Rigler data.	
		C. Laboratory experiments with P^{32}		
3.6	Mud *vs.* water, with bacteria present, oxidized system	2.7	Average of values in Table 2	Orig.
3.5	Same, but reduced system	2.6	Average of values in Table 2	Orig.
15.5	Mud *vs.* water, with bacteria absent, oxidized system	2.6	Average of values in Table 2	Orig.
12.5	Same, but reduced system	2.9	Average of values in Table 2	Orig.
1.1	Inorg. phosphorus in sol. & in bact. *vs.* org. phosphorus in sol. & in bact.	0.79	Filtered surf. water from Chocolate, a polluted lake near Halifax	Harris 1957
0.21	Inorg. phosphorus in sol. & in bact. *vs.* org. phosphorus in sol. & in bact.	0.21	Same from Punchbowl N. S.	Orig.
0.2	Inorg. phosphorus in sol. & in bact. *vs.* org. phosphorus in sol. & in bact.	0.2	Same from Grand, N. S.	Orig.
	Bacteria & algal cells *vs.* water	*4.5*	Surf. water freshly collected from Toussaint, Ontario	Rigler 1956
	Bacteria & algal cells *vs.* water	*3.6*	Same from Oiseau, Ont.	Rigler 1956
	Bacteria & algal cells *vs.* water	*2.6*	Same from Maskinonge, Ont.	Rigler 1956
0.5	Phytoplankton culture *vs.* water	*1.0*	Pure *Chlamodomonas* in laboratory nutrient solution	Orig.
3.0	Flowering plant *Eriocaulon vs.* water	0.34	1 g sprigs in 150 ml lake water; bacteria absent	Orig.
3.5	Peat moss (*Sphagnum*) *vs.* water	0.09	1.5 g sprigs in 150 ml lake water; bacteria absent	Orig.
0.58	Brine shrimp (*Artemia*) *vs.* sea water		Naturally occurring bacteria present	Harris 1957
∞	Same, but bacteria absent		P^{32} remains in inorganic state	Harris 1957
1.8	Beach flea (*Gammarus*) *vs.* sea water		Naturally occurring bacteria present	Harris 1957
∞	Same, but bacteria absent		P^{32} remains in inorganic state	Harris 1957

cluding natural phytoplankton, and presumably some bacteria; there is also a slow reaction observed in our laboratory results with bacteria but not including phytoplankton. These are resolved below (Fig. 7) into a harmonious scheme, for the validity of which further proof is desirable. In the scheme it is supposed that PO_4 in water goes in and out of the bodies of phytoplankton and bacteria with a turnover time of the order of 5 min. Most of it is chemically unchanged. (One is tempted to think about the way glucose or urea crosses cell walls in a higher animal.) The process of conversion to the organic form is about two orders of magnitude slower, with a turnover time averaging 0.3 days.

Two plants tested had the same turnover times as natural mud. As previously noted they were apparently unable to take up organic phosphorus, although we have not yet proved this. Proof would consist in offering plants organic P^{32} in solution with no inorganic fraction present.

Finally, two marine invertebrates are reported which may be considered to indicate the expected behavior of zooplankton. They are unable to take up inorganic phosphorus, but when bacteria are present to serve as food, their turnover time is of the order of one day.

In Figure 7 the data on turnover are combined in a single plan. The numbers are all days except for one 5-min entry. Four kinds of line are used as qualitative indicators of time. It is seen that upon addition of phosphate to water the immediate reaction, within minutes, is a transfer through the bodies of unicellular floating forms of life (heavy lines to right). Next, as shown by the light solid lines to the left, there occurs within a matter of hours, *i.e.*, two orders of magnitude slower than the above, an exchange in which the floating cells and the higher aquatics compete on approximately equal terms for the PO_4. This they make into their own body structures, throwing some of it back to the water as soluble organic phosphorus. We have cancelled out the virtually instantaneous passage through floating cells and set the process down as an equilibrium between

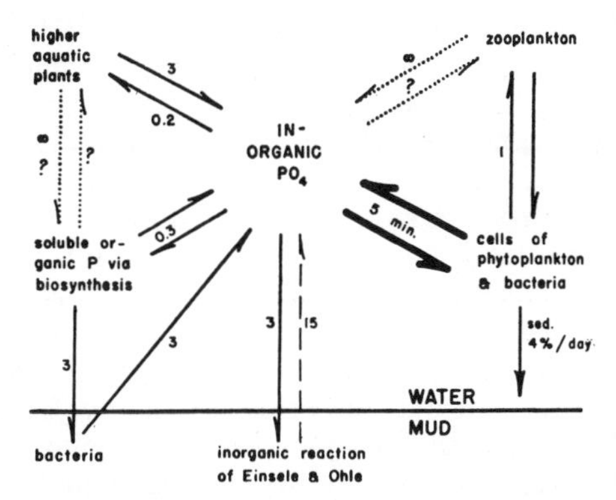

FIG. 7. Transformations of phosphorus in a lake with turnover times for different equilibria. Very heavy lines indicate the first reaction with floating cells, time in minutes. Other times in days. Lighter solid lines are reactions at intermediate speeds—two or three orders of magnitude slower than the initial one. Dashed line is the return from mud by inorganic release, a still slower turnover. Dotted lines at top indicate reactions too slow to measure, called infinitely slow by comparison with the rest.

inorganic and organic phosphorus in solution, for which the turnover time is 0.3 days.

At top left is indicated a doubt as to whether higher plants can utilize organic phosphorus, and at top right the feeding of zooplankton is given a turnover time, and their inability to utilize inorganic P is noted.

The lower part of Figure 7 brings in the sediment surface. At right bacteria are noted to fall out at a few per cent per day. This is in bottles and not to be read as net fallout in lakes, which is subject to wide variation. The fallout at right probably describes the same phenomenon as the line leading down at left, *i.e.*, the settling of organic matter to be reduced again by bottom microorganisms for regeneration to the water. The turnover time for leaving the water and for return is here 3 days, an order of magnitude slower than for exchanges with floating life.

The inorganic mechanism, which can be observed when bacteria are suppressed, is shown at lower center. There are variants for reduced and for oxidized conditions which have been described in papers II and III of this series. As already noted, the observed time relations are not affected by the redox state. In taking out inorganic P the mechanism has the same turnover as the bacterial action just described, namely 3 days. The return to water however is much slower, at 15 days. This means that there is about $15/3 = 5$ times as much participating inorganic P in the mud phase as in the water (as in our tourist example the U. S. has a larger population than Canada). Now as previously noted, the ratio of soluble organic to inorganic P in the water is also 5 to 1. Putting the ratios together we conclude that the P exchanging in the mud by the inorganic mechanism is about equal to the total P in the water. Thus if we could get it all released it would double the lake supply. (However the removal of the participating mud pool would trigger a further equilibration reaction by which more phosphorus in the mud would become available and so on *ad infinitum*, like the flow of ink out of a fountain pen.)

The final level of total exchangeable phosphorus in a lake is eventually determined by: (1) The net rate at which phosphorus in solution and in particulate matter enters the lake through drainage and rainfall. Ruttner (1953) states that the spring waters entering Lunz Lake contain 0 to 3γ/L of phosphate, while rain water occasionally contains 10γ/L of phosphorus originating in dust. This inflow is dependent on the geology, fertility, and size of the drainage area, and on the amount of rainfall. (2) The rate at which exchanging phosphorus is lost from further metabolism by incorporation into inorganic insoluble precipitates and undecayed organic matter in the lake bottom. Phosphorus in sediments amounts to about 0.05% dry wt of which two-thirds is organic. The largest concentrations and highest organic fractions are in surface sediments of recent geological formations (Kleerekoper 1957). (3) The morphometric properties of a lake. The volume determines the extent to which phosphorus is diluted. The area and depth influence the thermal stratification and relative size of the photosynthetic zone, which in turn affect the phosphorus cycle.

Within the limits set by these properties the various parts of the phosphorus cycle will operate. For example rooted aquatics are probably unimportant in large deep lakes where the littoral region is negligible compared to the total volume. In smaller and shallower lakes, in which the profundal zone is minimal, the growth and bacterial decay of rooted aquatics probably dominates the phosphorus exchange. It is even possible for the whole productive capacity to be occupied by rooted aquatics, leaving no nutrients to spare for support of phytoplankton (Prowse 1955).

REFERENCES

Coffin, C. C., F. R. Hayes, L. H. Jodrey, and S. G. Whiteway. 1949. Exchange of materials in a lake as studied by the addition of radioactive phosphorus. Can. J. Res., D, **27:** 207–222.

Getman, F. H., and F. Daniels. 1937. Outlines of theoretical chemistry. 6th Ed. Wiley, New York. 662 pp.

Harris, E. 1957. Radiophosphorus metabolism in zooplankton and microorganisms. Can. J. Zool., **35:** 769–782.

Hayes, F. R. 1955. The effect of bacteria on the exchange of radiophosphorus at the mud-water interface. Verh. Int. Ver. Limnol., **12:** 111–116.

Hayes, F. R., and E. H. Anthony. 1958. Lake water and sediment. I. Characteristics and water chemistry of some Canadian east coast lakes. Limnol. Oceanogr., **3:** 299–307.

Hayes, F. R., J. A. McCarter, M. L. Cameron, and D. A. Livingstone. 1952. On the kinetics of phosphorus exchange in lakes. J. Ecol., **40:** 202–216.

Hayes, F. R., B. L. Reid, and M. L. Cameron. 1958. Lake water and sediment. II. Oxidation-reduction relations at the mud-water interface. Limnol. Oceanogr., **3:** 308–317.

Heukelekian, H., and A. Heller. 1940. Relation between food concentration and surface for bacterial growth. J. Bact., **40:** 547–558.

Kleerekoper, H. 1957. Une étude limnologique de la chimie des sédiments de fond des lacs de l'Ontario mèridional, Canada. Thèses Fac. Sci. Univ. Paris, Sér. A, 3006. No. d'ordre 3878. 205 pp.

Macpherson, L. B., N. R. Sinclair, and F. R. Hayes. 1958. Lake water and sediment. III. The effect of pH on the partition of inorganic phosphate between water and oxidized mud or its ash. Limnol. Oceanogr., **3:** 318–326.

Mortimer, C. H. 1941–2. The exchange of dissolved substances between mud and water in lakes. J. Ecol., **29:** 280–329, and **30:** 147–201.

Orr, A. P. 1947. An experiment in marine fish

cultivation. II. Some physical and chemical conditions in a fertilized sea-loch (Loch Craiglin, Argyll). Proc. Roy. Soc. Edin. B, **68:** 3–20.

PRATT, D. M. 1949. Experiments in the fertilization of a salt water pond. J. Mar. Res., **8:** 36–59.

PROWSE, G. A. 1955. The role of phytoplankton in studies of productivity. Verh. Int. Ver. Limnol., **12:** 159–163.

RIGLER, F. H. 1956. A tracer study of the phosphorus cycle in lake water. Ecology, **37:** 550–562.

RUTTNER, F. 1953. Fundamentals of Limnology. Transl. by D. G. Frey and F. E. J. Fry. Univ. Toronto Press, Toronto. 242 pp.

SMITH, M. W. 1945. Preliminary observations upon the fertilization of Crecy Lake, New Brunswick. Trans. Amer. Fish. Soc., **75:** 165–174.

STEEMANN NIELSEN, E. 1955. The production of antibiotics by plankton algae and its effect upon bacterial activities in the sea. Deep Sea Res., Suppl. to vol. **3**(Papers in Mar. Biol. Oceanogr.): 281–286.

TAYLOR, C. B., AND V. G. COLLINS. 1949. Development of bacteria in waters stored in glass containers. J. Gen. Microbiol., **3:** 32–42.

ZILVERSMIT, D. B., C. ENTENMAN, M. C. FISHLER, AND I. L. CHAIKOFF. 1943. The turnover rate of phospholipids in the plasma of the dog as measured with radioactive phosphorus. J. Gen. Physiol., **26:** 333–340.

ZOBELL, C. E. 1943. The effect of solid surfaces upon bacterial activity. J. Bact., **46:** 39–56.

23 TRACER STUDY OF THE PHOSPHORUS CYCLE IN SEA WATER

W. D. Watt and F. R. Hayes
Institute of Oceanography, Dalhousie University, Halifax, N. S., Canada

ABSTRACT

When an aquatic biological system approaches a steady state the distribution of dissolved inorganic phosphorus (DIP), particulate phosphorus (PP), and dissolved organic phosphorus (DOP) may be dealt with as though it were the result of the equilibrium of a chemical exchange system. The standing stock and rate of exchange has been determined for coastal water collected during late summer from a site near Halifax, N. S.:

DIP (0.35 μg-atoms P/L)
⇅ 0.23 μg-atoms P/L/day, in each direction
PP (2.98 μg-atoms P/L)
⇅ 1.2 μg-atoms P/L/day, in each direction
DOP (0.60 μg-atoms P/L)

DOP is derived from dead and dying organisms in the water and is taken up and metabolized by bacteria which release DIP. There is no direct breakdown from DOP to DIP; filtered sea water showed no evidence of phosphatase activity. DOP labelled with P^{32} was resolved by paper chromatography into 6 components.

INTRODUCTION

The annual cycle of phosphorus in the sea and in lakes and its importance to plankton abundance has been noted by a number of workers (*e.g.*, Atkins 1925, 1928; Cooper 1933). Redfield, *et al.* (1937) did an extensive survey of the distribution of phosphorus in the Gulf of Maine, and prepared balance sheets describing the transport through three phases: dissolved inorganic phosphorus, dissolved organic phosphorus, and particulate phosphorus.

Pratt (1950) attempted to measure rates of phosphorus assimilation, of release as organic phosphorus, and of regeneration as dissolved inorganic phosphate in tanks of fertilized salt water. He was unable to determine the concentration changes of dissolved organic phosphorus because of the magnitude of variability.

Analytical techniques can measure only net changes in phosphorus and cannot provide information about an equilibrated cycle. A steady state may exist such that during the period between two sets of observations no change can be found in the phases of phosphorus. This does not mean that exchanges have not taken place, but rather that all changes are compensated, *i.e.*, the system is in dynamic equilibrium.

In analytical studies three forms of phosphorus are distinguished, and it is generally assumed that these form a transfer cycle:

Diss. Org. Phos.
↗ ↘
Partic. Phos. ← Diss. Inorg. Phos.

Hayes and Phillips (1958) have shown that a one-way cycle such as this is part of a system of dynamic equilibria. If P^{32} be added as tracer orthophosphate (DIP) to a sample of lake water it equilibrates with the particulate and dissolved organic phases (Harris 1957; Hayes and Phillips 1958). The same is true for sea water (Fig. 1). A stable biological system may be dealt with as though it were the result of a dynamic equilibrium of the sort:

$$\begin{array}{ccccc} \text{Dissolved} & & \text{Particulate} & & \text{Dissolved} \\ \text{Inorganic} & \rightleftharpoons & \text{Phosphorus} & \rightleftharpoons & \text{Organic} \\ \text{Phosphorus} & & & & \text{Phosphorus} \\ \text{(DIP)} & & \text{(PP)} & & \text{(DOP)} \end{array}$$

The reaction DIP $\rightarrow$ PP represents uptake of orthophosphate by living organisms. Zooplankton do not play a significant role in this reaction. Harris (1957) found that P^{32} uptake by *Gammarus* was inhibited by antibiotics and concluded that "*Gammarus* does not take up appreciable

[1] This work was supported financially by the National Research Council of Canada.

Reprinted from *Limnology and Oceanography*, 8:276–285, 1963.

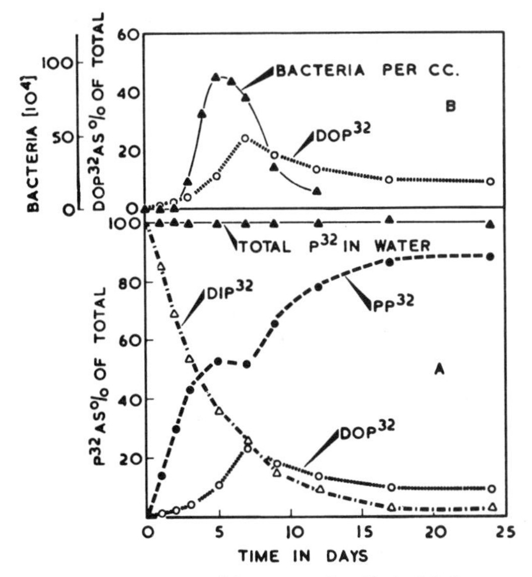

FIG. 1. A. Equilibration of P^{32} (added as a trace amount of dissolved inorganic orthophosphate) with the phosphate system of sea water collected near Halifax, N. S. in June 1960. The experiment was carried out in darkness and at the *in situ* temperature of the water. B. The change in bacterial numbers in the water during the equilibration shown in A. The dissolved organic P^{32} curve from A is shown again here for comparison with the curve of bacterial numbers.

quantities of phosphorus by adsorption through the body wall, intestine, or gills." Rigler (1961) drew a similar conclusion from his work on *Daphnia magna*. Bond (1933) has presented evidence that marine invertebrates in general are impermeable to water, salts, and organic solutes.

Harris (1957) suggests that bacteria in lake water will take up inorganic phosphate and incorporate it into organic compounds, and the same is doubtless true of phytoplankton.

The process PP → DOP represents the release of organic phosphorus from dead organisms presumably as a result of bacterial attack and leaching out by water.

Hayes and Phillips (1958) added tracer P^{32} to mud–water systems. They found that when antibiotics were used or the systems otherwise sterilized there was a marked increase in the P^{32} loss from water to mud. The plants *Sphagnum* and *Eriocaulon* also took up more P^{32} when antibiotics were added. The authors suggest that bacteria, when present, take up inorganic phosphorus and return it to the water in an organic form which is retarded from exchange reactions with the sediments and plants, and so tends to remain free in solution. They point out, however, that their results can be equally well explained by uptake and storage of P^{32} in particulate form by microorganisms in the water.

The return process DOP ⇀ PP represents the uptake of dissolved organic phosphorus compounds by bacteria, and possibly phytoplankton and certain protozoans. Krogh (1931) in his review of the subject concluded that no multicellular animals take up dissolved organic substances to any significant extent.

Finally, PP ⇀ DIP is the result of several simultaneous processes: (1) The release of DIP from organic compounds which have been broken down by bacteria. The source could be dissolved compounds taken up from the water or obtained from dead cells, organic detritus, or organic colloids. (2) The return of DIP to the water by phytoplankton and bacteria as a result of exchange in which inorganic phosphate is continually passing into and out of living cells. (3) Autodephosphorylation of labile organic phosphorus compounds contained within organisms which have died. Matsue (1949) killed *Skeletonema* cells with chloroform, which leaves the phosphatase active, and by heating, which destroys the enzyme. His results, given in Table 1, show that stored phosphate, *i.e.*, phosphate

TABLE 1. *Effect on organic cell phosphorus of killing* Skeletonema *with and without destruction of cell phosphatase. Figures show the percentage of stored organic cell phosphate released to water as orthophosphate after different treatments. Data of Matsue (1949)*

Cells killed by	After 2 hr	After 23 hr
Chloroform	75	98
Heating to 45°C with or without chloroform	62	83
Boiling	12	18

taken up by the cells and stored in an organic form, is readily attacked by the phosphatase enzymes in the cell.

Margalef (1951) and Rigler (1961) have shown that cladocerans release phosphatase into the water of the environment. Rigler found that the filtrate from a bacteria-free culture of *Daphnia magna* hydrolyzed glycerophosphate dissolved in Ottawa River water, but naturally occurring organic phosphorus compounds in the same water were unaffected. Rigler concluded that the "hydrolysis of naturally occurring organic phosphates does not take place, or takes place extremely slowly."

It may well be that all organic phosphorus compounds subject to enzyme hydrolysis are broken down by the organism's own enzymes as soon as it dies, and only compounds resistant to enzymatic attack are released to the water.

A phosphate system in sea or lake water approaches equilibrium when environmental conditions are constant, and no net phosphorus exchanges can be measured. When a steady state is approximated the system can be treated *as though* it consisted of two simultaneously occurring reversible first-order reactions:

$$\mathrm{DIP} \underset{\beta_2}{\overset{\alpha}{\rightleftharpoons}} \mathrm{PP} \underset{\gamma}{\overset{\beta_1}{\rightleftharpoons}} \mathrm{DOP}$$

where α represents the rate of loss of material from the DIP phase, β from PP, and γ from DOP. To identify whether loss is from PP to DIP or DOP respectively, subscripts are used (β_2 and β_1). Of course the exchanges occurring in a living system are not true first-order reactions; the calculated rates are applicable only to the particular conditions under which they are determined; they are ecological rather than chemical rate constants.

METHODS

Particulate P^{32} (PP^{32}) was taken as the difference between total P^{32} in the water and the P^{32} in the filtrate passed through a membrane filter of 0.5 μ pore diameter (Millipore® filters[2] were used). Tests showed that the filters adsorbed small amounts of P^{32}; hence they were always rinsed first with 5–10 ml of the solution to be tested.

Dissolved inorganic P^{32} (DIP^{32}) was precipitated from filtrates by a modification of the procedure described by Willard and Diehl (1943). Three 2-ml samples of the solution to be tested were placed in 15-ml graduated conical centrifuge tubes. To each tube there were added: 1 ml of a solution of KH_2PO_4 containing 1 mg of non-radioactive phosphorus per ml; 1 ml of 50% saturated NH_4NO_3; 2 ml of the molybdic acid reagent of Willard and Diehl (1943). The tubes were shaken vigorously, and the yellow precipitate allowed to settle for 3 min. Alcohol was layered on top to reduce flotation of the precipitate and the tubes were centrifuged for 3 min. The precipitates were washed twice in 7 ml of 0.25 N nitric acid, then dissolved in 2–3 drops of concentrated ammonium hydroxide, and diluted to 2 ml in the graduated centrifuge tubes. One ml was then pipetted into an aluminum planchet for estimation of the radioactivity.

A statistical analysis based on 14 determinations (each in triplicate) with this precipitation technique gave 96.8% recovery, the standard error of the mean being ±0.5%. Accordingly, all experimental values for DIP^{32} have been divided by 0.968.

In another measurement, untreated aliquots of the Millipore filtrate were evaporated and their radioactivity estimated. Dissolved organic P^{32} (DOP^{32}) was taken as the difference between the P^{32} content of the filtrate and of the precipitate, *i.e.*, what would remain in the water after precipitation.

Radioactivity was estimated on 1-ml aliquots air dried at room temperature in aluminum planchets 1 in. in diameter to which one drop of aerosol solution, of sufficient concentration to cause even spreading, was added. Counts were made with a Nuclear-Chicago Corp. model D-47 gas flow counter (with a "micromil" window in position). All samples were counted to a reliable error (90% confidence level) of

[2] Registered trademark, Millipore Filter Corp., Bedford, Mass.

±0.5%. Sampling and counting were done in triplicate.

During all experiments involving constant temperature the range was ±0.25°C. Hayes and Phillips (1958) found that loss of P^{32} from water in pyrex containers was largely the result of settling out of bacteria rather than growth of sessile forms. In all equilibration experiments here reported the loss of P^{32} from the water was rendered negligible by daily shaking. The effectiveness of this method is indicated by a plot of total P^{32} in the water as shown in the top line of Figure 1a.

The amount of non-radioactive phosphorus present in each phase was determined by the procedure of Strickland and Parsons (1960) for the development of the phosphomolybdate complex and its subsequent reduction to a highly colored blue compound; and for perchloric acid oxidation in determinations of total phosphorus. Optical density was measured using a Photovolt Corp. Lumetron Colorimeter with a 150-mm light path for DIP determinations and a 75-mm light path for total phosphorus. DIP was determined by direct analysis of sea water passed through a membrane filter of 0.5-μ pore diameter. DOP was taken as the difference between total phosphorus in membrane-filtered water and DIP. PP was taken as the difference between total phosphorus in the untreated water and total phosphorus in the water after membrane filtration.

Estimated errors, unless otherwise identified, are standard errors of the mean expressed as percentages of the mean.

RESULTS

The maximum in the DOP^{32} curve during exchange with DIP^{32}

The experimental curves of Figure 1 differ from theoretical curves in that the dissolved organic phosphate passes through a maximum. Such a maximum was also noted by Hayes and Phillips (1958, Fig. 6, center).

During the course of the equilibration shown in Figure 1a bacterial counts were made on the water using an agar plate culture technique and the medium proposed by ZoBell and Anderson (1936). The water was diluted by 10^4 before plating. The bacterial numbers showed no change during the first day, rose to a maximum by the 5th day, then declined. In Figure 1b these counts are plotted with the DOP^{32} curve from Figure 1a for comparison. The maximum in DOP^{32} occurs during the decline in bacterial numbers suggesting that it is the result of organic phosphorus compounds released from the cells of dead bacteria.

Sea water was collected and stored in 1-L flasks at 10°C for 14 days; then inorganic P^{32} was added and the equilibration followed at 10°, 15°, and 20°C (Fig. 2) in darkness. The 10° equilibration curves (Fig. 2a) show no maximum in DOP^{32}, whereas at higher temperatures maxima were observed (Figs. 2b and 2c). After 14 days at 10°C the bacterial population of

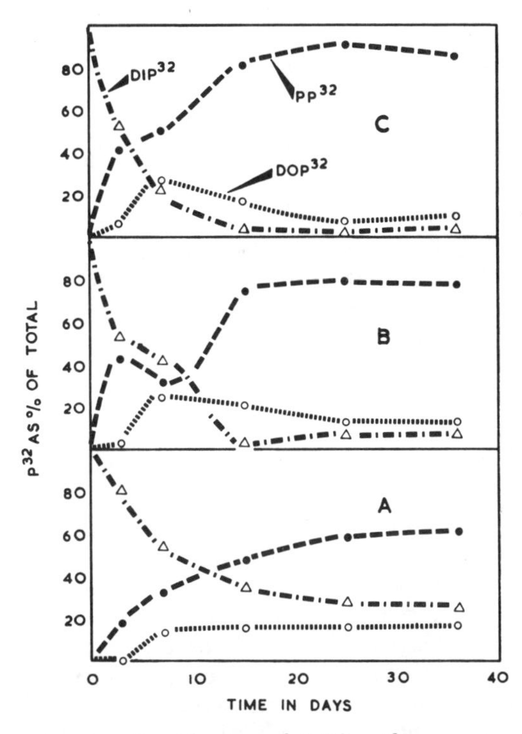

FIG. 2. Equilibration of DIP^{32} with sea water collected near Halifax and stored at 10°C for 14 days before the experiment. The equilibrations were conducted in darkness. Equilibration A was at 10°C, B at 15°C, and C at 20°C.

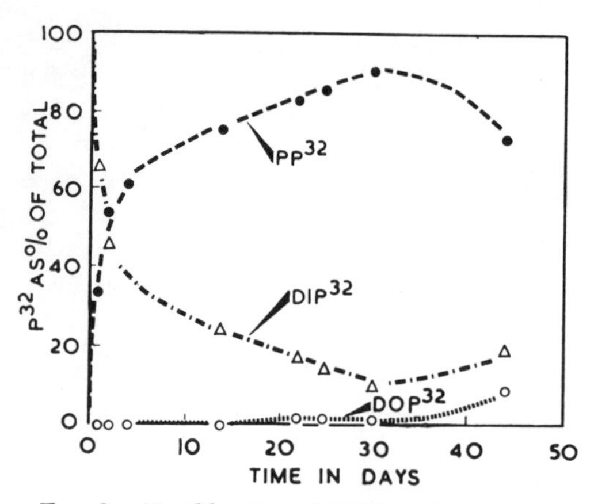

FIG. 3. Equilibration of DIP^{32} with a growing population of *Nostoc*. There was a noticeable death of algae after the thirtieth day.

the sea water must have passed through its maximum (as in Fig. 1b) and declined to what ZoBell and Anderson (1936) found to be a relatively constant number. The addition of DIP^{32} to the 10° system demonstrated equilibration curves (Fig. 2a) which are very similar to the theoretical curves of a three-phase system and without a DOP^{32} maximum. At higher temperatures (15° and 20°C) some new bacterial growth was induced as evidenced by the larger PP^{32} fractions. The growth and death of bacteria here led to a maximum in the DOP^{32} curve.

DOP^{32} is derived from PP^{32} and the equilibration curves for the particulate phase always show a dip associated with the organic maximum, *e.g.*, Figures 1, 2b, 2c, and 3.

Source of the DOP

The finding of Fogg (1952) that the Myxophyceae release organic compounds during growth led to a further experiment. A 2-gal pyrex bottle was filled with a vitamin-enriched (biotin, B_{12}, and thiamine) nutrient salt medium in sea water which had been stored at 4°C for two months. The medium was stirred continuously by air bubbling, and a fluorescent bulb was kept alight inside the bottle.

After two weeks the surface of the bottle was covered with a dense growth of the myxophyte *Nostoc*. At this point a trace amount of DIP^{32} was added and the culture was continued undisturbed except for intermittent sampling for a further 44 days.

The changes in DIP^{32}, PP^{32}, and DOP^{32} are shown in Figure 3. Because the *Nostoc* grew attached to glass, rather than in suspension, PP^{32} was taken as the difference between DIP^{32} initially added to the medium and the total remaining P^{32} in filtrates. The DIP^{32} was rapidly taken up by the alga, and the continued uptake after 14 days' equilibration reflects continued and noticeable growth of the *Nostoc*. No DOP^{32} could be detected until the 22nd day when a small amount appeared, presumably because, by this time, an appreciable number of cells were dead and dying. Between day 30 and day 44 growth ceased and there was noticeable cell death with a corresponding drop in PP^{32} and rise in both DIP^{32} and DOP^{32}.

A nutrient broth containing 3 g/L beef extract and 5 g/L peptone was made up with aged (4 months at room temperature) sea water and a trace amount of DIP^{32}. Aliquots of 12 ml were placed in 40 Erlenmeyer flasks each of 50 ml and equipped with side arm tubes. The flasks were stoppered with cotton and incubated at 20°C on a rotary shaker in the dark. The growth of the bacterial population in each flask was estimated by tipping the broth into the side arm and measuring the optical density with a Klett colorimeter. After the first 12 hr and again after 24 hr and thereafter at 24-hr intervals for 8 days the contents of 3 flasks, chosen at random, were removed for estimation of the 3 phases of P^{32}. The results are shown in Figure 4. The curve for PP^{32} parallels the growth curve for the population measured as optical density. Only after maximum growth was attained did significant amounts of DOP^{32} appear.

On the basis of experiments illustrated in Figures 1 to 4 it may be concluded that the flow of DIP^{32} into the pool of dissolved organic phosphorus compounds is primarily due to death and decay of microorganisms.

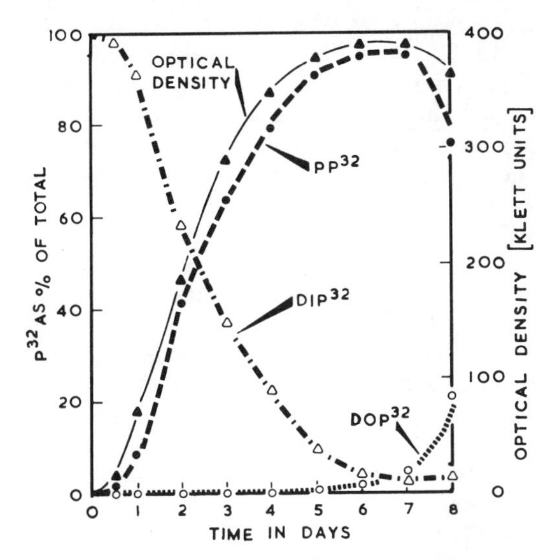

FIG. 4. Bacterial and phosphorus changes in aged sea water to which organic nutrients and DIP^{32} were added. The optical density is taken to be proportional to the number of bacterial cells.

Preparation of a DOP^{32} solution and its addition to fresh sea water

If the exchange system

$$DIP \rightleftharpoons PP \rightleftharpoons DOP$$

is applicable to natural aquatic systems, then when DOP^{32} is added to water the P^{32} should pass first into the PP phase, then to DIP.

To test this, a solution of naturally occurring organic phosphorus compounds partially labelled with P^{32} but free from PP^{32} and DIP^{32} was prepared. The method of preparation takes advantage of the fact that *Fucus vesiculosus* (a common seaweed) exchanges with DIP in sea water, but neither takes up nor releases DOP. This fact was established in the following experiments.

Twenty grams (fresh weight) of intact *F. vesiculosus* were shaken vigorously in 6 successive 500-ml portions of sea water sterilized by membrane filtration. The *Fucus* was then placed in 500 ml of filter-sterilized sea water to which a trace amount of DIP^{32} was added. The decline in total P^{32} and DIP^{32} in the water was followed at 4°C in darkness for 23.5 hr (Fig. 5). The control without *Fucus* showed no change in the overall radioactivity level or DIP^{32} level of the water.

It can be seen from Figure 5 that under the conditions of the experiment, uptake of DIP^{32} by microorganisms has been eliminated; neither PP^{32} nor DOP^{32} appeared in the water. The uptake of DIP^{32} by *Fucus* is an exchange reaction, and some of the DIP^{32} was returned from the plant to the water during the course of the experiment. DIP^{32} accounted for all P^{32} in the water. After 24 hr in sterile sea water containing DIP^{32} the *Fucus* was washed 3 times in sterile sea water and placed in another 500 ml of filter-sterilized sea water. The release of P^{32} was followed (in darkness) for 15 hr, and again only DIP^{32} could be found in the water. It is concluded that *Fucus vesiculosus* does not release DOP^{32} compounds.

DIP^{32} was allowed to equilibrate at the *in situ* temperature (17.6°C) for 3 days. The water was then sterilized by membrane filtration using a filter of 0.5-μ pore diameter, and 30% of the P^{32} in the filtrate was found to be in the organic form. Under the same conditions as were used in

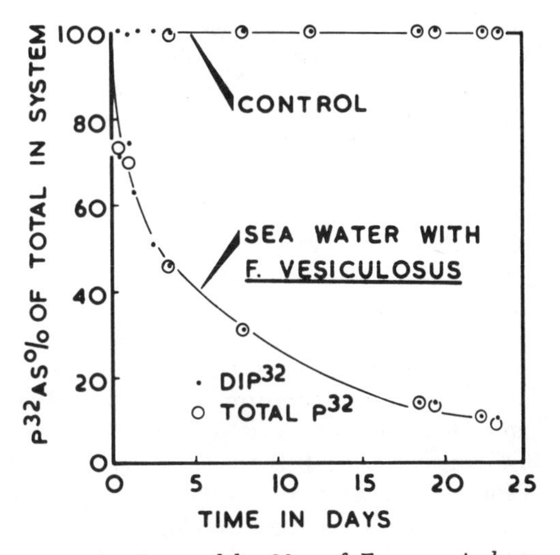

FIG. 5. Removal by 20 g of *Fucus vesiculosus* of DIP^{32} from 500 ml of sterile sea water at 4°C. The control (without *Fucus*) showed no change. DIP^{32} in the water always corresponded to total P^{32} in the water. The experiment was conducted in darkness.

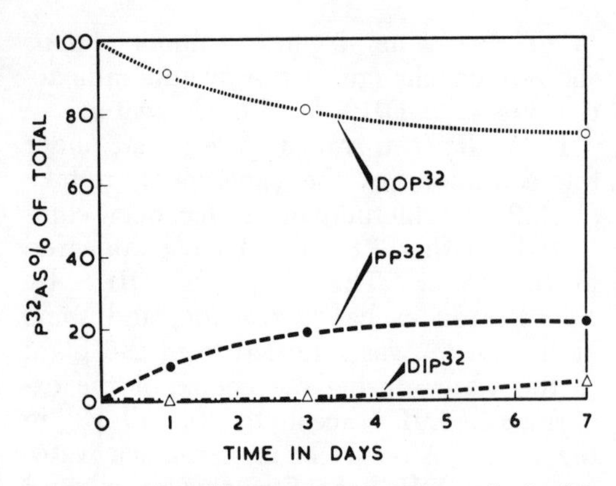

FIG. 6. Equilibration of DOP^{32} with freshly collected sea water (mean of 5 trials). The equilibrations were carried out in darkness, at the *in situ* temperature of the water.

obtaining the curves in Figure 5 the filtrate was 4 times treated with 20 g of washed *Fucus vesiculosus*. Of the P^{32} remaining in the filtrate 87% was then found to be in the organic form. There was no measurable loss of DOP^{32}. Subsequent treatments with *Fucus* did not increase the percentage of DOP^{32}. The solution was passed through a membrane filter of 100-mμ pore diameter and the radioactivity in the filtrate was then found to be 95% in the DOP^{32} form. After passage through a filter of 10-mμ pore diameter the DOP^{32} accounted for 99.2% of the remaining P^{32}. The small amount of precipitable P^{32} remaining may be phosphate adsorbed onto particles less than 10 mμ in diameter and so unavailable to the *Fucus*. It is also possible that this represents a portion of the DOP^{32} that is hydrolyzed in the preciptation procedure.

DOP^{32} prepared in this manner remains stable if kept sterile. After one month storage at room temperature, P^{32} in the filter-sterilized solution was still 99.2% in the DOP^{32} form, and there was no loss of radioactivity from the water. Autoclaving such a solution causes no loss of DOP^{32}, but may change the chemical nature of the compounds.

Fifty milliliters of the 99.2% DOP^{32} solution were added to 450 ml of freshly collected sea water and the flasks were kept in darkness at the *in situ* temperature (16.1°C) for 7 days. The distribution of P^{32} among the 3 phases was determined on days 1, 3, and 7. The average results of 5 replicates are shown in Figure 6. During the first day the value for DIP^{32} remained at zero (after 0.8% was subtracted) while PP^{32} rose to 9.8%. The process of conversion to PP^{32} continued on the 3rd-day observation, and by the 7th day there was an appreciable quantity of DIP^{32}. It is concluded that P^{32} in this system passed from DOP^{32} to PP^{32} but that there was no direct conversion of DOP^{32} into DIP^{32}.

On 1 October 1961, an uncontaminated sample of sea water was collected outside Peggy's Cove, N. S. DIP^{32} was added to 1 L of this and after being kept for 3 days, the water was sterilized by membrane filtration, and DIP^{32} was removed by treatment with *Fucus vesiculosus* and membrane filtration as described above. The radioactivity in the resulting solution was 99% in the DOP^{32} form. The P^{32}-labelled components of the solution were separated by unidimensional descending paper chromatography into 6 fractions. The solvent used was 25% trichloroacetic acid and acetone in a ratio of 1:3 of 25% TCA to acetone. A radiochromatogram scanner record for the sample is shown in Figure 7. Very similar results were obtained from water samples collected on 28 September and 3 October.

On 1 October 1961, the sample of water taken at Peggy's Cove was of salinity 30.0‰ and the temperature was 15°C. The chlorophyll *a* content was 1.1 μg/L as determined by the method of Richards with Thompson (1952). The results of a phosphorus analysis are given in Table 2.

To check whether any free phosphatase might be present in this sea water, two 500-ml samples were sterilized by membrane filtration. Before filtration 10 μg-at./L of phosphorus were added to one sample in the form of disodium-p-nitrophenyl phosphate. Both samples were saturated with chloroform, kept at 15°C for two weeks, and then analyzed for DIP. In neither sample was there a measurable

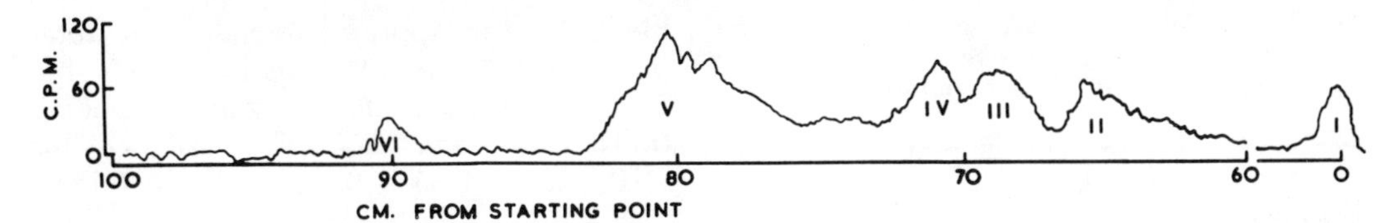

FIG. 7. Radiochromatogram scanner record of naturally occurring DOP^{32} compounds from sea water labelled with P^{32}.

increase in DIP in the water, a result which indicates absence of phosphatase.

The main experiment of this series was set up to determine the exchange rates of the phosphorus fractions. Within an hour of collection, quadruplicate 500-ml samples of the water collected on 1 October were treated with a trace amount of DIP^{32} as carrier-free H_3PO_4 and kept in darkness at 15°C for 3 days. Samples were taken at intervals to follow the equilibration, the results of which are shown in Figure 8.

From Figure 8 it appears that the DOP^{32} phase almost reached equilibrium during the first day, but after the second it shows a sharp increase. If the exchange system

$$\mathrm{DIP} \rightleftharpoons \mathrm{PP} \rightleftharpoons \mathrm{DOP}$$

still applies, and the system is at equilibrium, the changes in DIP^{32} and PP^{32} during the second day should be less than the changes during the first day. The equilibration curve for DOP^{32} would not be immediately affected by the growth of bacteria because during a phase of rapid growth bacteria do not release organic phosphorus compounds. The DOP^{32} curve from day 0 to day 2 may be taken as representative of the equilibrated system as it was *in situ*.

The peak in the PP^{32} curve indicates that a bacterial population was growing in the water during the second day, and this accounts for the continuing rapid change in DIP^{32} and PP^{32} during day 2. It is therefore best to use only the interval from day 0 to day 1 in calculating the exchange rates.

The rate of change equation for DIP^{32} is:

$$\frac{d(\mathrm{DIP}^{32})}{dt} = -\alpha(\mathrm{DIP}^{32}) + \beta_1(\mathrm{PP}^{32}).$$

For a finite time interval Δt this equation may be written approximately:

$$\frac{(\mathrm{DIP}^{32})_2 - (\mathrm{DIP}^{32})_1}{\Delta t} = -\alpha\frac{(\mathrm{DIP}^{32})_1 + (\mathrm{DIP}^{32})_2}{2} + \beta_1\frac{(\mathrm{PP}^{32})_1 + (\mathrm{PP}^{32})_2}{2}$$

TABLE 2. *The forms of phosphorus in Halifax coastal water, free from urban drainage. Collected at Peggy's Cove, N. S., 1 October 1961. Values are means of 4 analyses; errors are within ± 1%*

	DIP	PP	DOP
μg-at./L	0.35	2.98	0.60
%	8.9	75.8	15.3

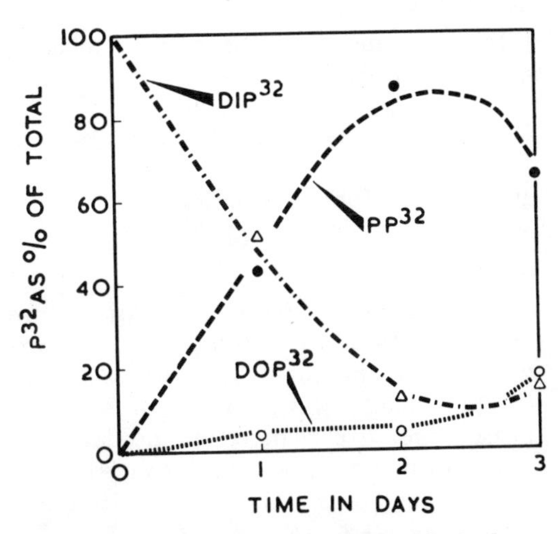

FIG. 8. Equilibration of DIP^{32} with fresh sea water (mean from 4 determinations). The water was taken at Peggy's Cove, N. S., and the equilibrations carried out 3–6 October 1961, at the *in situ* temperature of the water, and in darkness.

TABLE 3. *Calculated exchange rates (days⁻¹) for sea water collected 1 October 1961, at Peggy's Cove, N. S. for the equilibrium system*

$$\mathrm{DIP} \underset{\beta_2}{\overset{\alpha}{\rightleftharpoons}} \mathrm{PP} \underset{\gamma}{\overset{\beta_1}{\rightleftharpoons}} \mathrm{DOP}$$

	α	β_1	β_2	γ
Mean $\pm\sigma_m$	0.67±19%	0.078±19%	0.41±13%	2.0±13%

where $(DIP^{32})_1$ and $(PP^{32})_1$ each represent the percentage of P^{32} in the phase at the beginning of the time interval Δt and $(DIP^{32})_2$ and $(PP^{32})_2$ are the percentages after Δt. In this way an equation was written for each flask for the time interval from day 0 to day 1.

The amount of phosphorus in each phase (Table 2) was expressed as a percentage of the total phosphorus and these values for (DIP) and (PP) placed in the equilibrium equation:

$$\alpha(\mathrm{DIP}) = \beta_1(\mathrm{PP}),$$

a separate equation being written for each of the 4 analyses.

This yielded 4 pairs of equations: one member of each pair being an approximate rate of change equation and the other an equilibrium equation. Each pair of equations was solved simultaneously for values of α and β_1.

In the same way the rate of change equation for (DOP^{32}):

$$\frac{d(\mathrm{DOP}^{32})}{dt} = \beta_2(\mathrm{PP}^{32}) - \gamma(\mathrm{DOP}^{32})$$

and the equilibrium equation:

$$\beta_2(\mathrm{PP}) = \gamma(\mathrm{DOP})$$

were solved for values of β_2 and γ.

The results of these calculations are given in Table 3. Multiplying the exchange rates in Table 3 by 100 will give the percentage of the phosphorus in the phase which is exchanged in one day.

The turnover times are the reciprocals of the exchange rates. The turnover time for:

$$\mathrm{DIP} = 1/\alpha = 1/0.67 = 1.5 \text{ days}$$
$$\mathrm{PP} = 1/(\beta_1 + \beta_2) = 2.0 \text{ days}$$
$$\mathrm{DOP} = 1/\gamma = 0.5 \text{ day.}$$

Multiplying an exchange rate from Table 3 by the appropriate value for total phosphorus in the phase as given in Table 2 one obtains a value for μg-at. P/L being transferred each day. For instance for DIP, $0.67 \times 0.35 = 0.23$; similarly $0.78 \times 2.98 = 0.23$. With these values the exchanges in surface water from Peggy's Cove may be written:

DIP (0.35 μg-at. P/L)
⇃↾ 0.23 μg-at. P/L/day, in each direction
PP (2.98 μg-at. P/L)
⇂↿ 1.2 μg-at. P/L/day, in each direction
DOP (0.60 μg-at. P/L).

DISCUSSION

The experimental results may be summarized as follows. Organic phosphorus compounds are released into solution from dead or dying organisms. Rapidly growing populations of bacteria or green plants do not release organic phosphorus compounds. Dissolved organic phosphorus compounds are taken up by bacteria, broken down, and inorganic phosphorus is released.

That portion of the dissolved organic phosphorus phase into which P^{32} could be incorporated was resolved by paper chromatography into 6 components. This suggests that dead cells release 6 organic phosphorus compounds to the water. The relative proportions in which the DOP compounds exist are not necessarily the same as in living cells, nor need they be the same as the proportions in which the compounds are initially released to the water. It is to be expected that some of the DOP compounds released from dead cells will be more susceptible to bacterial decomposition than others. The calculated exchange rate for DOP in Peggy's Cove water (Table 3) is for the phase as a whole; *i.e.*, it is the average exchange rate of the 6 compounds.

There is nothing in the literature which enables one to predict exchange rates; until such predictions are possible separate exchange rates, and turnover times, will have to be measured for each biological system. Rigler (1956) found a turnover time of 3.6

min for DIP in Toussaint Lake water (exchange rate of 0.28 min^{-1}) in exchange with phytoplankton and bacteria in the water. If the DIP in the water had been greater, or the population of phytoplankton and bacteria smaller, the turnover time for DIP would have been longer. This may apply to Maskenonge Lake where Rigler found a turnover time of 26 min (almost one order of magnitude greater). Then too, in Peggy's Cove water the turnover time of DIP was here found to be 1.5 days. It is apparent that the exchange rates of one biological system can be quite different from those of other systems, or seasonally from the same body of water once the population has changed.

When there is more evidence some readily understood relationship may be found linking nutrient supply and population level and composition with exchange rates. At the present time no such relationship is apparent.

REFERENCES

ATKINS, W. R. G. 1925. Seasonal changes in the phosphate content of sea water in relation to the growth of the algal plankton during 1923 and 1924. J. Mar. Biol. Ass. U. K., **13**: 700–720.

———. 1928. Seasonal variations in the phosphate and silicate content of sea water during 1926 and 1927 in relation to the phytoplankton crop. J. Mar. Biol. Ass. U. K., **15**: 191–205.

BOND, R. M. 1933. A contribution to the study of the natural food cycle in aquatic environments. Bull. Bingham Oceanogr. Coll., **4**: 1–89.

COOPER, L. H. N. 1933. Chemical constituents of biological importance in the English Channel, Nov 1930 to Jan 1932. Part I. Phosphate, silicate, nitrate, nitrite, ammonia. J. Mar. Biol. Ass. U. K., **18**: 677–728.

FOGG, G. E. 1952. The production of extracellular nitrogenous substances by a blue-green alga. Proc. Roy. Soc. London, B, **139**: 372–397.

HARRIS, EUGENE. 1957. Radiophosphorus metabolism in zooplankton and microorganisms. Can. J. Zool., **35**: 769–782.

HAYES, F. R., AND J. E. PHILLIPS. 1958. Lake water and sediment. IV. Radiophosphorus equilibrium with mud, plants and bacteria under oxidized and reduced conditions. Limnol. Oceanogr., **3**: 459–475.

KROGH, AUGUST. 1931. Dissolved substances as food of aquatic organisms. Biol. Rev., **6**: 412–442.

MATSUE, Y. 1949. Phosphorus storing by a marine plankton diatom. J. Fish. Inst., **2**: 29–42. (Unpublished translation from Japanese, courtesy of Dr F. Uyeno.)

MARGALEF, R. 1951. Rôle des entomostracés dans la régéneration des phosphates. Verh. Internat. Verein. Limnol., **11**: 246–247.

PRATT, D. M. 1950. Experimental study of the phosphorus cycle in fertilized salt water. J. Mar. Res., **9**: 29–54.

REDFIELD, A. C., H. P. SMITH, AND B. H. KETCHUM. 1937. The cycle of organic phosphorus in the Gulf of Maine. Biol. Bull., **73**: 421–443.

RICHARDS, F. A., WITH T. G. THOMPSON. 1952. The estimation and characterization of plankton populations by pigment analysis. II. A spectrophotometric method for the estimation of plankton pigments. J. Mar. Res., **11**: 156–172.

RIGLER, F. H. 1956. A tracer study of the phosphorus cycle in lake water. Ecology, **37**: 550–562.

———. 1961. The uptake and release of inorganic phosphorus by *Daphnia magna* Straus. Limnol. Oceanogr., **6**: 165–174.

STRICKLAND, J. D. H., AND T. R. PARSONS. 1960. A manual of sea water analysis. Fish. Res. Bd. Canada, Bull., **125**: 1–185.

WILLARD, H. H., AND H. DIEHL. 1943. Advanced quantitative analysis. D. Van Nostrand Co., Inc., New York.

ZOBELL, C. E., AND D. Q. ANDERSON. 1936. Observations on the multiplication of bacteria in different volumes of stored sea water and the influence of oxygen tension and solid surfaces. Biol. Bull., **71**: 324–342.

24 NITRATE REDUCTION IN SEAWATER OF THE DEEP NITRITE MAXIMUM OFF PERU[1]

A. F. Carlucci and Hazel R. Schubert
Institute of Marine Resources, University of California, San Diego, La Jolla 92037

ABSTRACT

Nitrate-reducing bacteria were isolated from samples collected inside and outside the secondary nitrite maximum in oxygen-poor deep waters off the coast of Peru. Under anaerobic conditions at 20C in the laboratory, these bacteria reduced nitrate to nitrite in seawater enriched with 0.5 to 1.0 mg glucose/liter. Nitrate reduction also occurred in seawater supplemented only with 20 μg-atom NO_3^--N/liter and trace metals. Nitrite production generally was greatest during the first 24 hr while cell numbers were increasing. On continued incubation cells died, nitrate increased, and nitrite decreased. Dissolved organic carbon increased concurrently with the death of bacteria, indicating release of cell contents into the medium by lysis.

Microbiological evidence is presented to show that the high nitrite concentrations in oxygen-poor deep waters of the secondary nitrite maximum arise from nitrate reduction.

INTRODUCTION

Gilson (1937), working in the Arabian Sea, found nitrite concentrations greater than 7.5 μg-atom N/liter associated with low oxygen concentrations. Subsequently, high nitrite concentrations in oxygen-poor waters have been observed by Ivanenkov and Rozanov (1961) in the Arabian Sea, Wooster, Chow, and Barrett (1965) and Fiadeiro and Strickland (1968) in waters off the coast of Peru, and Brandhorst (1959) and Thomas (1966) in waters of the northeastern tropical Pacific Ocean.

Brandhorst (1959) believed that the nitrite of the primary nitrite maximum, generally found in the thermocline of most water columns, resulted from nitrification or oxidation of ammonium—a mineralization product of organic matter decomposition. Vaccaro and Ryther (1960) considered nitrite to be present as a result of excretion by phytoplankton.

It has been suggested that the nitrite in the secondary maximum of the oxygen-poor waters resulted from denitrification (Brandhorst 1959; Wooster et al. 1965; Thomas 1966). Goering and Dugdale (1966), in studies using ^{15}N, found no evidence for denitrification in these waters. They did not, however, investigate a depth profile and it is possible that active regions of denitrification were missed. Goering (1968) did find denitrification occurring in waters of the secondary nitrite maximum off the west coast of Mexico. Fiadeiro and Strickland (1968) concluded that nitrate reduction was responsible for the high concentration of nitrite in oxygen-poor waters.

None of the above workers made microbiological studies in the waters of the secondary nitrite maximum.

Several workers have isolated nitrifying bacteria from seawater collected at various depths (*see* Carlucci and Strickland 1968), but few of these studies have attempted to describe the role of the nitrifiers in nitrite maxima.

Marine denitrifying and nitrate reducing bacteria have been isolated from shallow waters off India (Sreenivasan and Venkataraman 1956*a*, *b*; Venkataraman and Sreenivasan 1956; Sreenivasan 1957). The denitrifying bacteria required a high concentration of peptone (1 g/liter) before gas production was demonstrated. Brisou and Vargues (1963) found bacteria capable of reducing nitrate to nitrite in coastal waters off the coast of Algeria. A high

[1] This work was supported in part by the Marine Life Research Program (Scripps Institution of Oceanography's component of the California Cooperative Oceanic Fisheries Investigation, a project sponsored by the State of California) and in part by U.S. Atomic Energy Commission Contract No. AT(11-1)GEN 10, PA 20.

Reprinted from *Limnology and Oceanography*, 14:187–193, 1969.

TABLE 1. *Composition of the media*

Constituent	Media (mg/liter of aged seawater)			
	HP	D*	DGL	DGH
Bacto-peptone	20	—	—	—
Glucose	—	—	0.5	1.0
KNO_3	10.1	2.02	2.02	2.02
$FePO_4$	20	—	trace	trace

* The medium contained 1 ml/liter of a chelated metal solution: $CoCl_2 \cdot 6H_2O$, 0.004 g; $CuSO_4 \cdot 5H_2O$, 0.004 g; $FeCl_3 \cdot 6H_2O$, 1.0 g; $ZnSO_4 \cdot 7H_2O$, 0.3 g; $MnSO_4 \cdot H_2O$, 0.6 g; $Na_2MoO_4 \cdot 2H_2O$, 0.15 g; ethylenediaminetetraacetate, 6.0 g; made to 1 liter with double glass-distilled water and *p*H adjusted to about 7.5 by adding sodium hydroxide solution.

concentration of peptone was also used in these experiments. ZoBell (1946) reported that about 5% of the marine bacteria that he isolated were capable of denitrifying activity (i.e., production of gas from nitrate). It seems, therefore, that both microbiological nitrate reduction and nitrification occur in the sea, but the role of the responsible microorganisms in the production of nitrite in waters such as the primary and secondary nitrite maxima needs to be evaluated.

During the cruise of B.A.P. *Unanue* (unpublished) the bacteriology, as well as a number of other factors, of waters possessing secondary nitrite maxima was studied. Heterotrophic bacteria were isolated from inside and outside the secondary nitrite maximum. Some of these isolates could reduce nitrate to nitrite under anaerobic conditions.

This communication describes the results of a number of experiments with these bacteria, mainly in low concentrations of oxygen and nutrients—conditions characteristic of the waters of the secondary nitrite maximum.

We wish to thank P. Robison and P. McNally for technical assistance. We are grateful to Mrs. I. H. Ji for determining dissolved carbon in some of the samples. We would also like to thank our colleagues at the Institute of Marine Resources and at the Instituto del Mar del Peru for assistance in the shipboard work and analytical determinations.

MATERIALS AND METHODS

The bacteria used in most of these studies are referred to as cultures 10 and 34 and were isolated from samples taken off the coast of Peru—10 from waters of 400 m having a relatively low nitrite concentration at 09° 04′ S lat, 83° 37′ W long and 34 from waters of 240 m in the secondary nitrite maximum at 16° 13′ S lat, 80° 50′ W long.

A 3-liter Niskin sampler was used to collect water at the desired depths. Standard dilution and plating procedures were used for the isolations. Subcultures were made on 2216E solid medium (0.5% peptone, 0.01% yeast extract, 0.01% $FePO_4$, and 1.5% agar in "aged" seawater) by spreading 0.1 ml of the original culture with a glass rod. After 48 hr at 20C, several colonies were picked and transferred to 2216E slants; cultures 10 and 34 were selected for further study.

The media are listed in Table 1. Nitrate, nitrite, ammonia, dissolved carbon, and dissolved oxygen were determined with the methods described by Strickland and Parsons (1968), modified to accommodate small samples. Dissolved carbon was measured after removing cells by centrifugation at about $16{,}000 \times g$ for 15 min. Bacteria were counted by spreading 0.1 ml of the culture or its dilution in triplicate on the surface of 2216E medium with a glass rod. Plates were incubated at 20C and counted after 48 hr.

Before all experiments, cultures were adapted to a low nutrient medium by at least two transfers of 48–72 hr in DGH medium. When they reached approximately 6×10^5 per ml, the cultures were used for inocula in the following experiments.

HP (high peptone) medium in tubes

Thirty ml of HP medium were put in each of a number of 18- × 150-mm screw-cap tubes; this amount nearly filled the tubes. They were autoclaved at 1.02 atm and 121C for 15 min, during which time the dissolved oxygen was driven off; while the medium was still hot the caps were tightened. Just before inoculation the caps were removed, then 0.1 ml of an active culture was added to each tube (filling it completely), the caps again secured tightly, and the tube contents uniformly dispersed with a mechanical mixer. Representative tubes showed no detectable dissolved oxygen (<0.1 ml O_2/liter). The tubes were transferred to anaerobic jars which were then filled with purified N_2 after 10–15 min of flushing with the gas. Incubation was at 20C. Uninoculated HP medium was treated in the same way and served as a control. Each treatment was replicated three times. Periodically, tubes from all treatments were analyzed for nitrate, nitrite, and numbers of viable bacteria. Values for replicate tubes were averaged.

D (no organic supplement), DGH, and DGL media (high and low glucose) in tubes

This series of experiments was essentially the same as above with the following exceptions. Instead of peptone, the media contained either low levels of glucose (sterilized separately and added after tubes were autoclaved) or no energy source other than the organic matter present in aged seawater. Before analysis, the contents of the tubes were pooled in a large flask and aliquots were used for nitrate, nitrite, and, in some cases, ammonia and dissolved carbon determinations.

DGH in flasks

A 2-liter erlenmeyer flask containing 1,100 ml of DGH medium was inoculated with 4 ml of an active culture of the test bacterium. The flask had a rubber stopper fitted with three lengths of glass tubing—one an N_2-inlet tube submerged in the medium to allow bubbles to agitate and flush it. The N_2 was passed through a tube of sterile cotton before entering the culture. A short tube plugged with cotton provided an "exhaust" for the system; a piece of silicone tubing with a pinch-clamp on its outer end was used to regulate gas flow. The third tube was also submerged in the medium and bent outside the flask, forming a siphon with a piece of silicone tubing and pinch-clamp at its outer end. This was the sampling tube and was closed during incubation. Incubation was at 20C and an uninoculated flask served as a control. Before inoculation the whole setup was autoclaved at 1.02 atm for 15 min. When a sample was required, the "exhaust" tube was closed and the sampling tube opened. Because of positive pressure in the vessel, the culture medium could flow out of the flask and be collected in a sterile glass-stoppered bottle. At various times during a 24-hr period about 100 ml was removed. The sample was analyzed for nitrate, nitrite, ammonia, dissolved oxygen, and bacterial numbers.

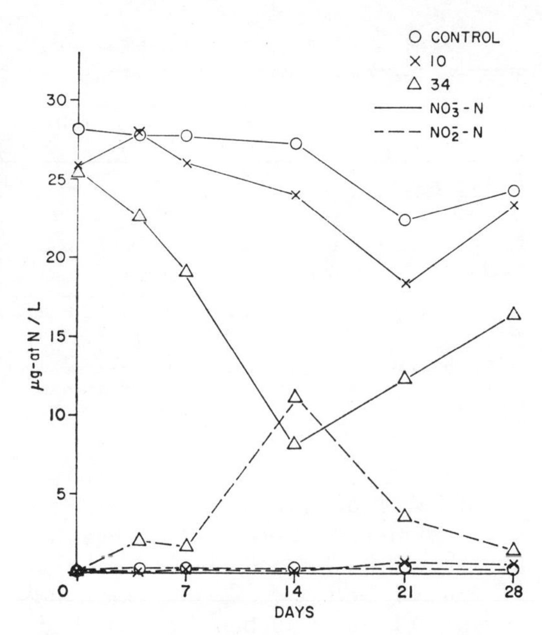

Fig. 1. Nitrate reduction and nitrite production by cells of cultures 10 and 34 in HP medium under anaerobic conditions.

RESULTS

Both cultures 10 and 34 reduced nitrate under anaerobic conditions in HP medium (Fig. 1). A high concentration of nitrite was produced by culture 34, but the sum of nitrate- and nitrite-nitrogen was not equal to the amount of nitrate-nitrogen initially present. This indicated that nitrite

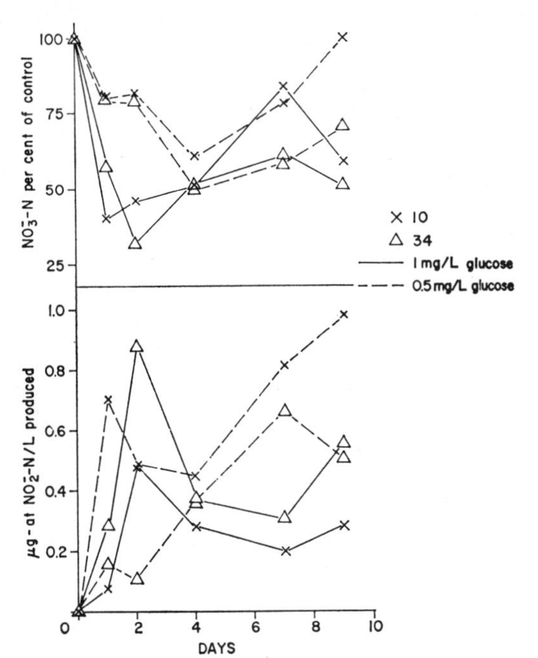

FIG. 2. Nitrate reduction and nitrite production by cells of cultures 10 and 34 in DGL and DGH media under anaerobic conditions.

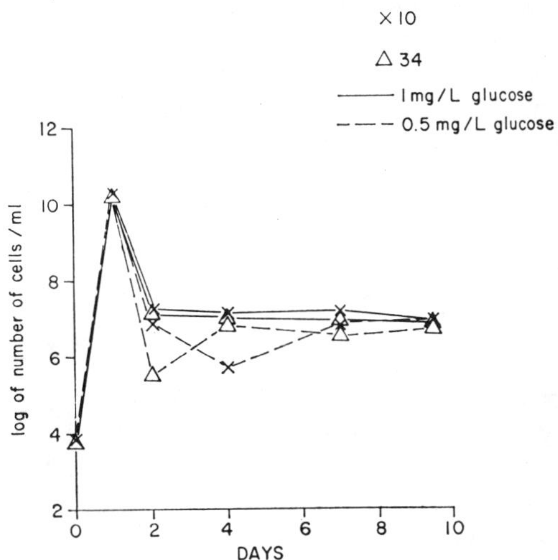

FIG. 3. Cell numbers for cultures 10 and 34 in DGL and DGH media under anaerobic conditions.

was further reduced. Ammonia was not determined in this experiment because peptone was present in such high concentration. Culture 10 did not produce significant amounts of nitrite, although some nitrate was reduced.

Media DGL and DGH (Table 1) were used in test tube experiments to see if nitrate would be reduced under low concentrations of glucose (Fig. 2). Nitrate was rapidly reduced during the first day by culture 10; it was also reduced by culture 34 to about 30% of its original concentration. With 0.5 mg/liter glucose, nitrate was reduced less and more slowly than with 1 mg/liter. Nitrate concentrations were increased by both cultures for the remainder of the experiment. This increase after the initial decrease was difficult to interpret until after a number of experiments had been done. It then became apparent that as nitrate was reduced and nitrite increased (Fig. 2), cells multiplied, generally over a two-day period (Fig. 3), then died rapidly; after this nitrate was no longer reduced, but probably increased in the medium as a result of cell lysis.

Nitrite was produced by cells of cultures 10 and 34 in D medium in test tubes (no added energy source, Fig. 4). The increase in nitrite occurred during the first day. Cell numbers increased proportionally to nitrite production (Fig. 4), but the cells began to die after 24 hr. Ammonia was not produced during the experiment. Dissolved organic carbon showed no increase for the first 24 hr, but as cells died and released organic materials into the medium it increased (Fig. 5).

All the previous experiments showed that most of the nitrate was reduced during the first day or two. In the next experiment (flask), samples were analyzed during the first 24 hr. Nitrate was rapidly reduced by cultures 10 and 34 during this time in DGH medium, and most of it could be recovered as nitrite (Fig. 6). Since there was some nitrite initially present in the medium and some nitrogen incorporated into cells, some nitrate may have been reduced beyond nitrite, although no increase in ammonia concentration was observed. Nitrate reduction or nitrite production was proportional to the number of cells present.

In a 12-hr experiment, under similar conditions, nitrate was again reduced to

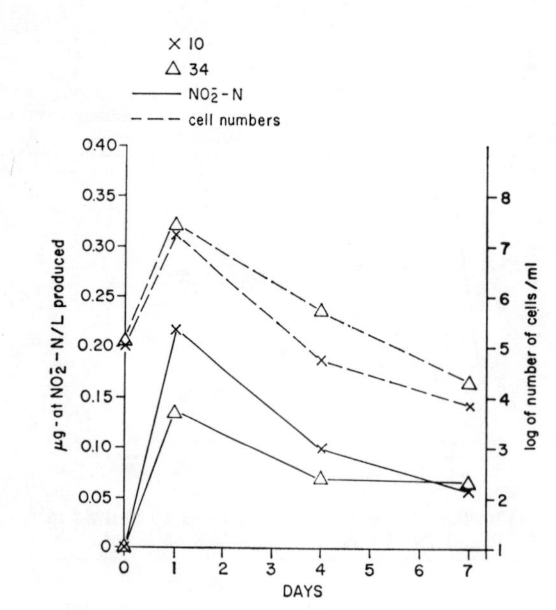

FIG. 4. Nitrite production and cell numbers of cultures 10 and 34 in D medium under anaerobic conditions.

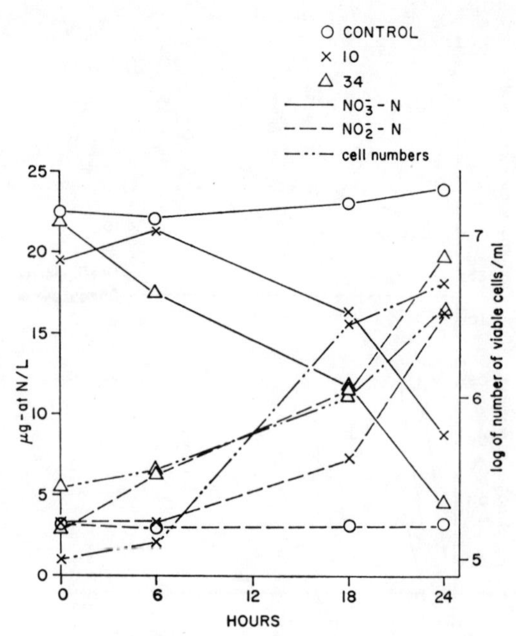

FIG. 6. Nitrate reduction, nitrite production, and cell numbers of cultures 10 and 34 in DGH medium during a 24-hr incubation.

nitrite. Here, as in the 24-hr experiment, the sum of nitrate- and nitrite-nitrogen was almost equal to the sum of the nitrate concentration initially present (Table 2). Given the experimental conditions, we do not know the reduction products beyond nitrite; no increase in ammonia concentration was observed.

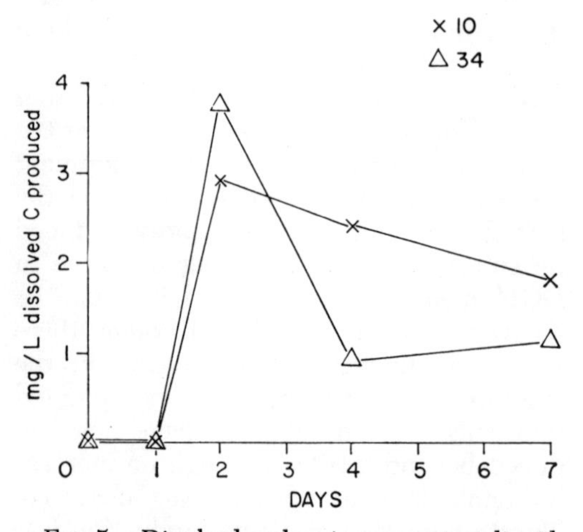

FIG. 5. Dissolved carbon in supernatant liquids of DGH media in which cultures 10 and 34 had been inoculated. Control value subtracted in each plotted concentration.

DISCUSSION

According to Verhoeven (1956), there are three types of nitrate reduction: 1) nitrate assimilation, in which nitrate is reduced only for the building of cell protein; 2) incidental dissimilatory nitrate reduction, in which nitrate is a nonessential hydrogen acceptor; and 3) true dissimilatory nitrate reduction, in which nitrate acts, at least under certain conditions, as the essential hydrogen acceptor enabling the organism to grow. When the last type leads to gaseous products (N_2 and N_2O), it is commonly referred to as denitrification. All three types of nitrate reduction, however, can result in the formation of nitrite in the medium, even though it may be a rapidly reduced intermediate.

Samples collected from above, below, and within the secondary nitrite maximum were analyzed for (in addition to temperature and salinity) oxygen, phosphate, silicate, nitrate, nitrite, and ammonia; dis-

TABLE 2. *Reduction of nitrate and nitrite production by cells of cultures 10 and 34 in DGH medium during a 12-hr incubation (0 = 0 time; 12 = 12 hr)*

Culture	No./ml 0	No./ml 12	NO_3^--N (μg-atom/liter) 0	NO_3^--N (μg-atom/liter) 12	NO_2^--N (μg-atom/liter) 0	NO_2^--N (μg-atom/liter) 12	N incorporated into cells (μg-atom/liter*)	N reduced further than nitrite (μg-atom/liter)
Control	—	—	24.1	24.4	4.8	4.8	—	—
10	2×10^5	3×10^6	24.8	13.2	5.5	11.2	4.0	1.9
34	2.7×10^5	2×10^6	24.8	13.6	5.4	9.9	3.9	2.8

* Calculated from the assumption that one bacterium weighs about 1 μμg, approx 80% of which is water, and that nitrogen is 10% of the dry weight.

solved and particulate organic carbon, nitrogen and phosphorus; particulate carbohydrate; biotin, thiamine, and vitamin B_{12}; adenosine triphosphate and bacterial count (Fiadeiro and Strickland 1968). None of these parameters could be related to high nitrite concentrations in the oxygen-poor waters. The total number of bacteria growing on 2216E medium did not differ greatly from the different waters. In six stations where there was a secondary nitrite maximum, there was a corresponding decrease in nitrate concentration.

Thomas (1966), working in waters off the west coast of Central America, reconstructed nitrate profiles from assumed nitrogen : phosphorus ratios. Where the observed nitrate plus nitrite-nitrogen concentrations did not equal the calculated values, he concluded that denitrification had taken place. Since he found little or no ammonia in these waters, it was presumed that gaseous products such as nitrogen and nitrous oxide had been produced. His reconstruction of nitrate profiles is subject to criticism.

Fiadeiro and Strickland (1968) were more cautious and said that biological nitrate reduction was responsible for the formation of the secondary nitrite maximum in waters off the coast of Peru. They studied the relationship between phosphate and nitrate and apparent oxygen utilization (AOU). From such relationships, they concluded that nitrate reduction, most likely denitrification, was occurring in these waters.

In the process of heterotrophic denitrification, organic matter is oxidized biochemically by nitrate. The results of our experiments with either low or no glucose additions indicated that nitrate reduction was occurring. No ammonia was observed and the sum of nitrate- and nitrite-nitrogen did not equal the amount initially present. Calculations from cell counts and nitrate "lost" (Table 2) indicated that all the unaccountable nitrogen had not been assimilated. It was, therefore, possible that gaseous end products could have been produced, and hence that denitrification had occurred. Gaseous end products were not observed in Durham tube experiments (*see* Sreenivasan and Venkataraman 1956*a*), but it was not unreasonable to assume that, as a result of low concentrations of organic matter, they were produced in amounts not exceeding their solubilities in seawater. Sreenivasan and Venkataraman (1956*a*) could not demonstrate denitrification (gas production in Durham tubes) by nearshore bacterial isolates unless peptone was present in amounts greater than 1,000 mg/liter. The bacteria, however, grew with lesser amounts. The amount of dissolved organic matter in most oceans rarely exceeds 0.5 mg C/liter.

It seems that nitrate reduction was responsible, in part, for the high concentrations of nitrite in oxygen-poor waters off Peru. This conclusion is supported by the results of our microbiological experiments, since culture 34 was isolated from these waters.

REFERENCES

BRANDHORST, W. 1959. Nitrification and denitrification in the eastern tropical north Pacific. J. Conseil, Conseil Perm. Intern. Exploration Mer, **25**: 3–20.

BRISOU, J., AND H. VARGUES. 1963. Proteolysis and nitrate reduction in seawater, p. 410–414.

In C. H. Oppenheimer [ed.], Marine microbiology. Thomas, Springfield, Ill.

CARLUCCI, A. F., AND J. D. H. STRICKLAND. 1968. The isolation, purification, and some kinetic studies of marine nitrifying bacteria. J. Exptl. Marine Biol. Ecol., **2:** 156–166.

FIADEIRO, M., AND J. D. H. STRICKLAND. 1968. Nitrate reduction and the occurrence of a deep nitrite maximum in the ocean off the west coast of South America. J. Marine Res., **26:** 187–201.

GILSON, H. C. 1937. The nitrogen cycle. John Murray Exped., Sci. Repts., **2:** 21–81.

GOERING, J. J. 1968. Denitrification in the oxygen minimum layer of the eastern tropical Pacific Ocean. Deep-Sea Res., **15:** 157–164.

———, AND R. C. DUGDALE. 1966. Denitrification rates in an island bay in the equatorial Pacific Ocean. Science, **154:** 505–506.

IVANENKOV, V. N., AND A. G. ROZANOV. 1961. Hydrogen sulphide contamination of the intermediate waters of the Arabian Sea and Bay of Bengal. Okeanologiya, **1:** 443–449.

SREENIVASAN, A. 1957. Nitrate-reducing bacteria from marine environment. Current Sci. India, **26:** 392–393.

———, AND R. VENKATARAMAN. 1956*a*. Marine denitrifying bacteria from south India. J. Gen. Microbiol., **15:** 241–247.

———, AND ———. 1956*b*. *Pseudomonas tuticorinensis* n. sp., a marine denitrifying bacterium. Current Sci. India, **25:** 294–295.

STRICKLAND, J. D. H., AND T. R. PARSONS. 1968. A practical handbook of seawater analysis. Bull., Fisheries Res. Board Can. No. 167. 311 p.

THOMAS, W. H. 1966. On denitrification in the northeastern tropical Pacific Ocean. Deep-Sea Res., **13:** 1109–1114.

VACCARO, R. F., AND J. H. RYTHER. 1960. Marine phytoplankton and the distribution of nitrite in the sea. J. Conseil, Conseil Perm. Intern. Exploration Mer, **25:** 260–271.

VENKATARAMAN, R., AND A. SREENIVASAN. 1956. The occurrence and activities of certain bacterial groups in off-shore marine environment of Gulf of Mannar, off Tuticorin. Proc. Natl. Inst. Sci. India, Pt. B, **22:** 357–367.

VERHOEVEN, W. 1956. Some remarks on nitrate and nitrite metabolism in microorganisms, p. 61–85. *In* W. D. McElroy and B. Glass [eds.], Inorganic nitrogen metabolism. The Johns Hopkins Press, Baltimore.

WOOSTER, W. S., T. J. CHOW, AND I. BARRETT. 1965. Nitrite distribution in Peru Current waters. J. Marine Res., **23:** 210–221.

ZOBELL, C. E. 1946. Marine microbiology. Chronica Botanica Co., Waltham, Mass. 240 p.

NUTRIENT LIMITATION IN THE SEA: DYNAMICS, IDENTIFICATION, AND SIGNIFICANCE[1]

25

R. C. Dugdale[2]

Institute of Marine Science, University of Alaska, College

ABSTRACT

A mathematical model has been constructed to provide a theoretical framework for the investigation and discussion of nutrient limitation in the sea. The euphotic zone is divided into nutrient-limited and light-limited regions. For simplicity, the regeneration term is ignored, and a steady state is assumed. The uptake of major nutrients is assumed to follow the Michaelis-Menton expression, and some data indicate that the assumption is probably correct. Also assuming a dominant phytoplankton population, with characteristic Michaelis-Menton kinetic parameters, it is demonstrated that the size of the population, fractional growth rate, the rate of production, and the concentration of limiting nutrient are determined by the sum of the fractional loss rates for the phytoplankton. The role of micronutrient components of enzymes is developed theoretically using the effect of molybdenum on a nitrate-limited system as an example.

For the sea, the significance of nutrient limitation theory is primarily in the study of phytoplankton competition and succession. Applications to lakes are discussed, and it appears that the effects of micronutrient deficiencies are more likely to be observed there than in the sea.

INTRODUCTION

The role of nutrients in limiting primary production in the sea is widely acknowledged. A nutrient term was incorporated into the productivity models of Riley (1963*a*) and Steele (1962), while Harvey (1963) wrote of threshold levels of a limiting nutrient. However, there has been no serious attempt to develop a model for the dynamics of nutrient-limited productivity in the sea. As a result, there is confusion regarding the design of adequate experiments and the interpretation of the results. The purpose of this paper is to present theoretical considerations based on one possible model for nutrient limitation. Virtually no data are included, partly because very little relevant data exist, but also for another more important reason: my conviction that biological oceanography suffers from a lack of theoretical work conducted on a "best guess" basis in advance of at-sea or laboratory experiments.

Thus, this paper is an attempt to elaborate a framework for understanding the processes of growth and production under nutrient limitation sufficiently well to allow the intelligent design of experiments. Therefore, no attempt is made to predict standing crops or productivity in particular regions of the sea as Riley, Steele, and others have done with much success.

In an attempt to reach a larger audience, no mathematics more complicated than elementary algebra have been employed. This procedure lacks rigor in the derivations, but does not affect steady-state solutions for the equations. Another major simplification has been made by lumping all losses from the phytoplankton population into a single loss rate. By doing so, the nature of stability and restoring reactions becomes easier to understand.

THEORY

Primary nutrients

In subtropical and tropical regions, a thermocline at approximately the same depth as the bottom of the euphotic zone restricts the movement of nutrients from

[1] Contribution No. 33 from the Institute of Marine Science, University of Alaska, College. I am greatly indebted to Drs. John Ryther, John Goering, Don Button, and Holger Jannasch for their helpful suggestions and especially for their endless patience in discussing these ideas in agreement and disagreement with me.

[2] Present address: Department of Oceanography, University of Washington, Seattle 98105.

Reprinted from *Limnology and Oceanography*, 12:685–695, 1967.

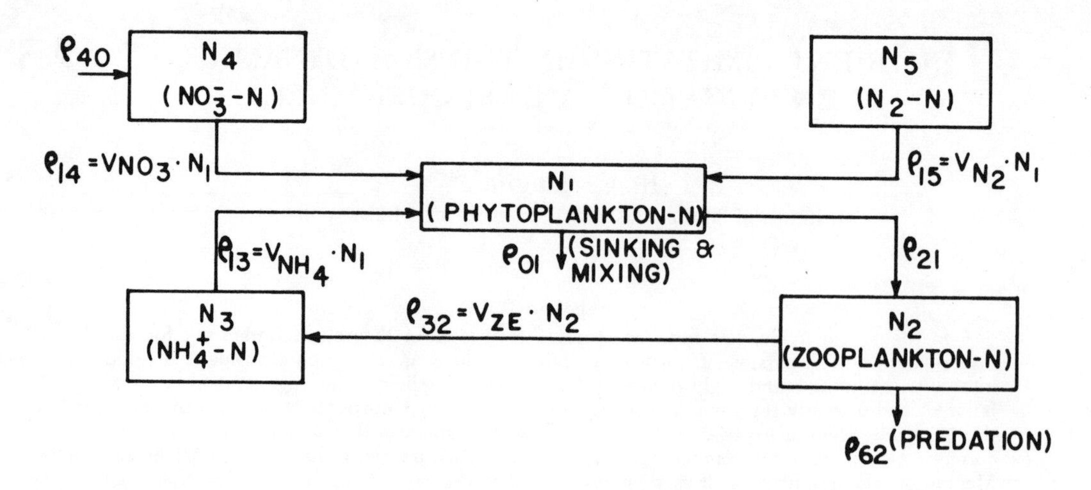

FIG. 1. Flow of nitrogen in the euphotic zone.

the rich layers below the thermocline upward to replace those used by the phytoplankton. A gradient in nutrient concentration is thus developed with sign opposite to that of light. Making the initial assumption that, at low concentrations, some positive correlation exists between phytoplankton productivity and the concentration of a limiting nutrient, two regions can be defined: 1) a lower euphotic region where light is less available than nutrients and is, therefore, the limiting factor, and 2) an upper euphotic region where adequate light is available but where nutrient concentration has been sufficiently reduced to become the factor limiting growth. These will be referred to as the light-limited and the nutrient-limited regions. Later, some important differences will be pointed out.

The productivity models of Riley (1963*a*, 1965) and Steele (1962) include terms for the effect of nutrients, and both have been remarkably successful in predicting populations and productivity for particular regions of the sea. These models have been constructed using carbon as a base, and much of the information regarding nutrient limitation has been obscured. There are significant advantages in constructing a model for nutrient limitation based on the flow of the limiting nutrient and this is the approach used here.

Dugdale and Goering (1967) proposed the measurement of primary production in the sea by measuring the flow of nitrogen using the stable isotope ^{15}N as a tracer. Their diagram for the flow of nitrogen in the euphotic zone (Fig. 1) designates transport rates by ρ's with subscripts indicating the direction of flow from one compartment to another, after the terminology of Sheppard (1962). For example, ρ_{14} is the rate of nitrogen transport [weight or atoms of N (unit time)$^{-1}$ (unit volume)$^{-1}$] into compartment 1 from compartment 4, in this case the rate of nitrate uptake by the phytoplankton population. Velocities, designated as V's, are transport rates reduced to unit nitrogen concentrations. For example, $V_{NO_3^-} = \rho_{14}/N_1$ and has units of NO_3^--N taken up (unit time)$^{-1}$ (unit N in the phytoplankton)$^{-1}$. The units reduce to time^{-1}. $V_{NO_3^-}$ can therefore be considered to be a growth rate in terms of nitrogen. The nitrogen available to the phytoplankton is designated as new (for example, nitrate from below the thermocline or formed by nitrogen fixation, or regenerated, such as decomposition of organic nitrogen into ammonia).

The model (Fig. 1), while representing quite accurately the processes occurring in nature, is more complicated than necessary for an initial approach to nutrient limitation theory. Steele and Menzel (1962) in

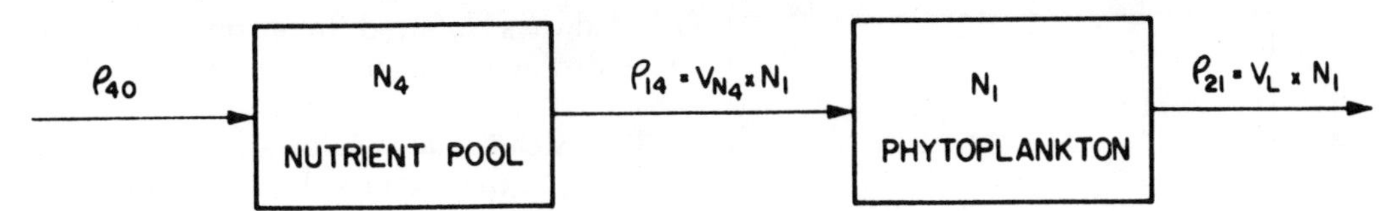

FIG. 2. Simplified flow-diagram of a major nutrient in the euphotic zone with regeneration pathways omitted.

their model for the Sargasso Sea near Bermuda neglected nutrient regeneration, using only nitrate concentrations for the nutrient term. This simplifying procedure will be followed and, in addition, the discussion will be restricted to those processes occurring in the nutrient-limited region of the euphotic zone. Thus, in its simplest form the primary production system is diagrammed as in Fig. 2, the notation remaining unchanged. N now stands for a limiting nutrient rather than specifically for nitrogen, and ρ_{40} is the new productivity, the absolute rate of supply of limiting nutrient, assumed here to be supplied by mixing from below the nutrient-limited region; N_1 is the phytoplankton concentration; V_{N_4} is the growth rate of the phytoplankton in terms of the limiting nutrient, units t^{-1}; and V_L is the loss rate of the phytoplankton per unit phytoplankton, principally from grazing, sinking, and mixing, units of t^{-1}.

Then

$$\rho_{14} = V_{N_4} \times N_1 , \qquad (1)$$

the absolute transport rate of limiting nutrient into the phytoplankton, and

$$\rho_{21} = V_L \times N_1 , \qquad (2)$$

the absolute transport rate of limiting nutrient out of the phytoplankton, that is, the sum of all losses. At steady state, all ρ values must be equal. Thus,

$$\rho_{14} = \rho_{21} ,$$

and from equations (1) and (2)

$$V_{N_4} \times N_1 = V_L \times N_1 ,$$

and

$$V_{N_4} = V_L . \qquad (3)$$

The growth rate of the phytoplankton is therefore determined by the sum of the loss rates due to sinking, grazing, and mixing. The nature of the term, V_L, is crucial to an understanding of the implications of the model. The sinking rate of the phytoplankton appears to be a function of species, age, and physiological state. The inclusion of a separate term for zooplankton has been avoided deliberately for simplicity, and it should be emphasized that a solution for the size of the zooplankton population is not necessary for the purpose of this model. Different phytoplankton-zooplankton communities will probably stabilize in different relative proportions to give varying values for the grazing component of V_L. The effect of mixing in removing the phytoplankton is less in this model than in those of Steele (1958) and Riley (1965). This results from consideration here of a nutrient-limited layer underlain by a light-limited layer containing a finite population, whereas Steele (1958) and Riley (1965) treat the euphotic zone as a whole with a phytoplankton population exchanging with an underlying zone containing no phytoplankton. In mathematical terms:

$$\rho_{21}(\text{mixing}) = m[N_1 - N_{1(0)}] ,$$

where $N_{1(0)}$ is the concentration of phytoplankton in the light-limited layer.

A mechanism for achieving a steady state can be demonstrated if a positive relationship exists between the concentration of limiting nutrient and the velocity of uptake of the nutrient by the phytoplankton. Work with chemostats has depended on the extension of Michaelis-Menton enzyme kinetics to whole organisms, usually bacteria. In that case the following expression holds (Monod 1942):

$$V_{N_4} = N_4 \frac{V_{\max}}{K_s + N_4}, \qquad (4)$$

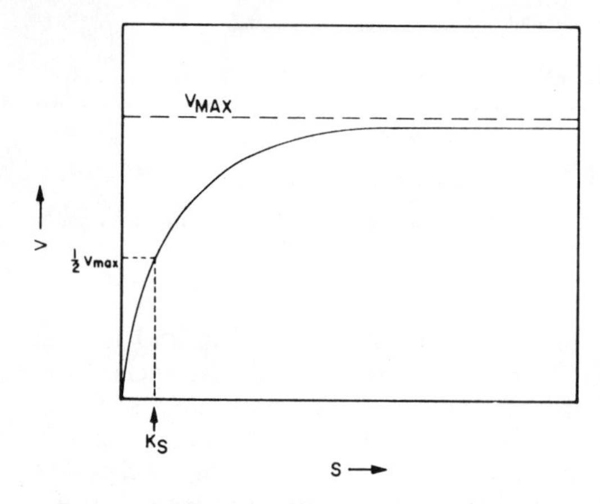

FIG. 3. Nutrient uptake as a function of nutrient concentration, according to the Michaelis-Menton expression.

where V_{max} is the maximum uptake velocity for the organisms under the experimental conditions, and K_s is the substrate concentration at which $V_{N_4} = V_{max}/2$ (Fig. 3).

Evidence for the validity of Michaelis-Menton kinetics for algae is beginning to accumulate. Caperon (1965), using the chemostat, showed Michaelis-Menton kinetics to apply to nitrate-limited growth of *Isochrysis galbana*. MacIsaac (unpublished), using batch methods and ^{15}N for a tracer, has evidence that the same is true for ammonia uptake by *Chlorella* sp. The data of Harvey (1963) for the uptake of phosphate by *Phaeodactylum tricornutum*, when replotted (Fig. 4), also show Michaelis-Menton kinetics. Although the data are scanty at present, the preliminary assumption that the nutrient uptake behavior of algae is described by equation (4) appears to be upheld.

To apply this function in the model, it must be assumed either that all species of algae have an identical K_s value or that one species or a group of species with similar K_s values are in sufficient majority to produce the effect of a unialgal population. Wright and Hobbie (1966) have studied the kinetics of glucose uptake by natural populations of microorganisms in lakes. They designate the K value so obtained as K_t, a coefficient for transport, to distinguish it clearly from K_s, which has in the past always referred to a single species. The distinction is important and will be observed here.

The steady-state value for N_4 corresponding to the required V_{N_4}, which has the same value as V_L, can be found from a curve similar to Fig. 3 plotted with the appropriate values of K_t and V_{max}. From equations (3) and (4), an expression can be obtained for the concentration of limiting nutrient:

$$N_4 = K_t \Big/ \frac{V_{max}}{V_L} - 1. \qquad (5)$$

The important point is that the concentration of limiting nutrient is set by the magnitude of the loss rate and by the characteristic uptake kinetic parameters of the population, V_{max} and K_t. This simple system is passively regulated according to control theory (Milsum 1966). In qualitative terms, the system can be seen to regulate through the interaction between phytoplankton and the nutrient concentration. For example, an increase in the loss rate would result in a decrease in the size of the population, N_1, thereby reducing ρ_{14}. N_4 increases as a result and forces an increase in uptake rate until V_{N_4} again equals V_L.

The size of the standing crop of phytoplankton can be obtained from equation (2):

$$N_1 = \rho_{21}/V_L.$$

At steady state $\rho_{21} = \rho_{40}$, and then

$$N_1 = \rho_{40}/V_L. \qquad (6)$$

The size of the phytoplankton population therefore varies directly according to the rate of nutrient supply and indirectly as the loss rate. In the two-layered system, the rate of nutrient supply is, according to Riley (1965),

$$\rho_{40} = m(N_0 - N_4), \qquad (7)$$

where m is the mixing coefficient between the nutrient-limited region and that lying just below, and N_0 is the concentration of limiting nutrient just below the nutrient-limited region.

N_4 is a function of the kinetic character-

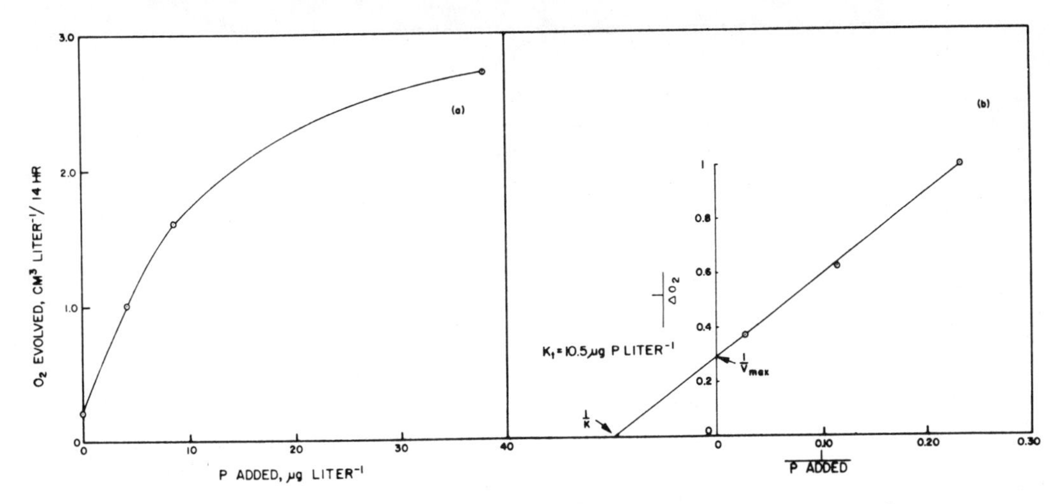

FIG. 4. The data of Harvey (1963) replotted to show the response of *Phaeodactylum tricornutum* according to Michaelis-Menton kinetics. (a) Oxygen evolved as a function of added phosphorous. (b) Reciprocal plot of the same data according to the Lineweaver-Burke modification to obtain values for V_{max} and K_t.

istics of the dominant phytoplankton species as given in equation (5) and

$$\rho_{40} = m\left[N_0 - \left(K_t \Big/ \frac{V_{max}}{V_L} - 1\right)\right]. \quad (8)$$

Combining equations (6) and (7),

$$N_1 = \frac{m}{V_L}\left[N_0 - \left(K_t \Big/ \frac{V_{max}}{V_L} - 1\right)\right]. \quad (9)$$

The size of the phytoplankton population and the productivity of the system, therefore, depend upon the size of the term $K_t/[(V_{max}/V_L) - 1]$ occurring in equations (8) and (9). That is, $N_1 \rightarrow 0$ and $\rho_{40} \rightarrow 0$ as

$$K_t \Big/ \left(\frac{V_{max}}{V_L} - 1\right) \rightarrow N_0, \quad (10)$$

while for maximum productivity

$$\rho_{40} \rightarrow mN_0 \quad \text{and} \quad N_1 \rightarrow mN_0/V_L \quad \text{as}$$

$$K_t \Big/ \left(\frac{V_{max}}{V_L} - 1\right) \rightarrow 0. \quad (11)$$

Thus, low loss rates or the presence of a dominant species with high V_{max} or low K_t values are conditions under which a large fraction of the maximum production will be realized. It is also apparent that the maximum allowable loss rate is set at some value below V_{max} by the same mechanism. This can be seen from equations (3) and (4):

$$V_L = V_{max}N_4/(K_t + N_4),$$

and since $N_4 \rightarrow N_0$ as a maximum value,

$$V_{L_{max}} = V_{max}N_0/(K_t + N_0). \quad (12)$$

Equation (12) sets the maximum loss rate beyond which the population cannot respond with an increased growth rate.

The reactions of a nitrate-limited system to changes in loss rates can be seen in Fig. 5 for the following conditions: $V_{max} = 0.05$/hr, $K_t = 2.5$ μg-at./liter, $m = 0.001$/hr, and $N_4 = 10.0$ μg-at. NO_3^--N/liter. A common feature of the curves is the rapid change that occurs as the limits are approached, for example, the high rate of change in ambient nitrate concentration, N_4, and in productivity, ρ_{40}, in the region where V_L increases from 0.03 to 0.04. The standing crop increases rapidly as $V_L \rightarrow m$. In the case $V_L = m$, from equation (9),

$$N_1 = \left[N_0 - \left(K_t \Big/ \frac{V_{max}}{V_L} - 1\right)\right].$$

At low V_L values, $N_1 \simeq N_0$. Fig. 5 also illustrates the principle that high phytoplankton growth rates result in low pro-

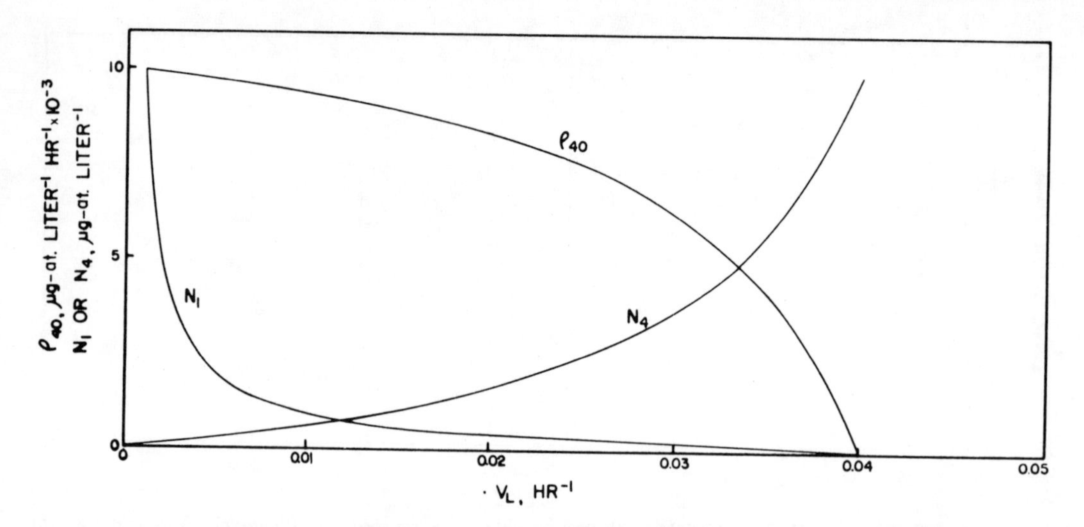

FIG. 5. Size of phytoplankton population, N_1, rate of production, ρ_{40}, and concentration of limiting nutrient, N_4, as functions of fractional loss rate, V_L, in the model for a nutrient-limited system without regeneration, for assumed values of K_t, V_{max}, N_0, and m.

ductivity; that is, the nutrient available is wasted as a result of high ambient nutrient concentration. In the case illustrated, 90% or more of the potential productivity can be realized only when $V_L < 0.3\,V_{max}$.

Micronutrients

Metal constituents of enzymes such as iron and molybdenum have been named as limiting nutrients. For example, Goldman (1960) believed molybdenum to be limiting primary productivity in Castle Lake, California, and Menzel and Ryther (1961) added iron to Sargasso Sea water and obtained enhanced photosynthetic activity for 24 hr. In view of these findings, it is appropriate to examine theoretically the role of this class of micronutrients.

Molybdenum is a constituent of nitrate reductase and might reasonably be expected to affect the uptake of nitrate if it were to be in short supply. The question then revolves about the manner in which the concentration of the enzyme affects the conversion of nitrate to ammonia within the algal cell. Some insight is provided if the reactions forming the basis for the Michaelis-Menton theory of enzyme kinetics are examined. The derivations can be found in White et al. (1959) and elsewhere. In their notation, the basic reaction is written:

Enzyme (E) + substrate (S) ⇌ enzyme-substrate complex (ES) → Enzyme (E) + Products (P)

and

$$E + S \underset{K_2}{\overset{K_1}{\rightleftharpoons}} ES \overset{K_3}{\rightarrow} P + E\,.$$

K_1, K_2, and K_3 are first order rate constants. The final expression is as equation (4):

$$V = V_{max} S/(K_m + S),$$

where V is the velocity of conversion of substrate to product; $K_m = (K_2 + K_3)/K_1$ is the Michaelis constant; and

$$V_{max} = K_3 E\,. \qquad (13)$$

A short supply of molybdenum might be expected to decrease the concentration of the enzyme nitrate reductase, thereby reducing linearly the magnitude of V_{max}.

Assuming that V_{max} for the cellular uptake of nitrate is set in this manner, the effect of a reduced V_{max} can be seen from equation (5) to result in an increased value of ambient nitrate concentration, N_4. Increased N_4 results in turn, from equations

(6) and (7), in a decrease in the supply rate ρ_{40} and in the size of the phytoplankton population, N_1. The productivity of the system has thereby been reduced. The maximum allowable loss rate, $V_{L_{max}}$, set by equation (12) is also reduced. Although the system is molybdenum limited, regulation or accommodation to changes in V_L is still dependent upon nitrate at the new value of the kinetic parameter, V_{max}. This is the case when the molybdenum requirement of the population is small compared to the amount available, a condition met in the sea where Barsdate (1963) has shown that the concentration of molybdenum varies only slightly.

The effects of reduced molybdenum concentration in the model are shown in Fig. 6a and 6b for the following conditions:

$V_{max} = 0.05/\text{hr}$, $K_t = 2.5$ μg-at./liter, $m = 0.001/\text{hr}$, and $N_4 = 10$ μg-at. NO_3^--N/liter.

Fig. 6a was obtained from equation (13) and the additional assumption that the concentration of nitrate reductase will be directly proportional to the molybdenum concentration in the surrounding water according to the expressions:

$$V_{max} = 0.5 \times \text{Mo concentration}, \quad \text{Mo} < 0.1\ \mu\text{g-at./liter};$$

$$V_{max} = 0.05/\text{hr}, \quad \text{Mo} \geqslant 0.1\ \mu\text{g-at. Mo/liter}.$$

The curves for ambient nitrate, N_4, productivity, ρ_{40}, and phytoplankton, N_1, show high rates of change when the molybdenum concentration falls to about one-half the saturating concentration, assumed here to be 0.1 μg-at./liter. At one-quarter of the saturating molybdenum concentration, $V_{max} = V_{L_{max}}$, and the population collapses. The existence of a critical range may be of particular interest in lakes where relatively large changes in concentration of molybdenum are likely to be observed (Barsdate 1963).

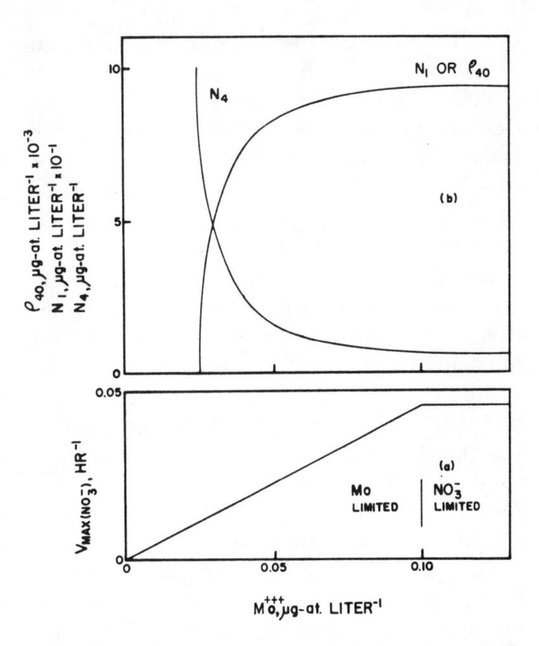

FIG. 6. Effects in the model of a diminished supply of molybdenum on an initially nitrate-limited system for assumed values of K_t, V_{max}, N_0, and m. (a) Assumed relationship between molybdenum concentration and V_{max}. (b) Size of phytoplankton population, N_1, rate of production, ρ_{40}, and concentration of initially-limiting nitrate, N_4, as functions of molybdenum concentration.

IDENTIFICATION

From the foregoing discussion it should be clear that the investigation of nutrient limitation should be made by considering the production system as a whole and with some model in mind. If the model proposed here were to be used to design experiments, some technique or combination of techniques for the measurement of uptake rates must be chosen. The necessity for obtaining results quickly calls for the use of tracer methods. Radioisotopes or stable isotopes are available for measuring the uptake of common nutrients or micronutrients (for example, ^{15}N, ^{32}P, ^{99}Mo, and ^{30}Si). It should be obvious that the kinetics of a particular nutrient should be measured directly, if possible, with a labeled form of the nutrient, for example, $^{15}NO_3^-$ and $^{32}PO_4^{3-}$.

Indirect methods, in which a nutrient that may be limiting is added and the response measured by the reaction of some other system in the cell, have also been used. The ^{14}C method of Steemann Nielsen (1952), providing a highly sensitive assay of the photosynthetic uptake of carbon by phytoplankton, was used by Goldman (1960, 1961) for his investigations of the effects of trace metals on primary production in Brooks Lake, Alaska, and Castle Lake, California; by Menzel and Ryther (1961) in studies of nutrient limitation in the Sargasso Sea; and by Becacos (1962) on Lago Maggiore, Italy.

Results obtained with the ^{14}C technique are subject to uncertainties of interpretation since additions of suspected limiting nutrients may influence the photosynthetic system specifically. This is especially likely to occur with micronutrients; Fogg (1965) has pointed out that the rate of photosynthesis can be changed simply by altering the divalent : monovalent ion ratio. In a nutrient-limited system, such a disturbance would be followed by an eventual return to the prexisting photosynthetic rate. An observation of enhanced photosynthesis under these conditions has no validity.

The possibility of confusing transient and steady-state response to experimental additions of limiting nutrient also exists when using the direct method, that is, observing the uptake of limiting nutrient. It has been shown in equation (5) that the concentration of the limiting nutrient depends upon the growth rate V_{N_4} and upon the parameters V_{max} or K_t. Thus, any experimental manipulation, accidental or deliberate, that affects either V_{max} or K_t will result in a temporary change in V_{N_4} and ρ_{21}, followed by eventual restoration of the previously existing rates at a new ambient concentration of nutrient. No data for algae appear to be available for the effect of environmental conditions on K_t. Vaccaro and Jannasch (1966) have shown that, for a species of marine bacteria resembling *Achromobacter aquamarinus*, the value of K_t is temperature dependent. A large number of factors may be important in setting the value of V_{max} and changing the concentration of any one of them may be expected to throw the system into transient oscillation through its effect on V_{max}. No information appears to be available regarding the time-scale for these perturbations. The magnitude and period of oscillation probably depend on the size of the population and the rate of flow of limiting nutrient.

Uptake velocities for nitrate and ammonia in seawater have been obtained by Dugdale and Goering (1967) using ^{15}N. The method has the advantage that growth rates in terms of nitrogen, $V_{NO_3^-}$ and $V_{NH_4^+}$, can be obtained. Although the accuracy of a single measurement is influenced by the amount of detritus present, relative velocities can be obtained from a variety of experiments made upon a single sample of seawater. For example, the search for a limiting nutrient at an oceanic station might proceed by a series of experiments in which serial additions of phosphate and nitrate were made to seawater samples with $^{15}NO_3^-$ added as a tracer. After incubation for a few hours in saturating light, the samples would be filtered and the filters processed to obtain $V_{NO_3^-}$. If a kinetic response is observed for nitrate but not for phosphate, the system apparently would be nitrogen limited. In another series of experiments, the influence of molybdenum could be demonstrated if its addition resulted in altered nitrate kinetics, possibly as an increase in V_{max} for nitrate uptake.

SIGNIFICANCE OF NUTRIENT LIMITATION

Redfield, Ketchum, and Richards (1963) pointed out that inorganic nitrate and phosphorus occur in deep oceanic water in approximately the same ratio that these elements occur in the phytoplankton. In general, both become exhausted through uptake by phytoplankton at about the same time. Since both nutrients are supplied from below the thermocline by upwelling or mixing processes, it is nearly as correct to say that primary production is limited by the rate of supply of both nutrients as it is to name the nutrient that actually limits the growth rate of the phytoplankton cells at any given moment. The importance of nutrient limitation theory lies, rather, in the dynamic behavior of the system.

One of the most interesting aspects of nutrient limitation theory is concerned with phytoplankton succession and competition.

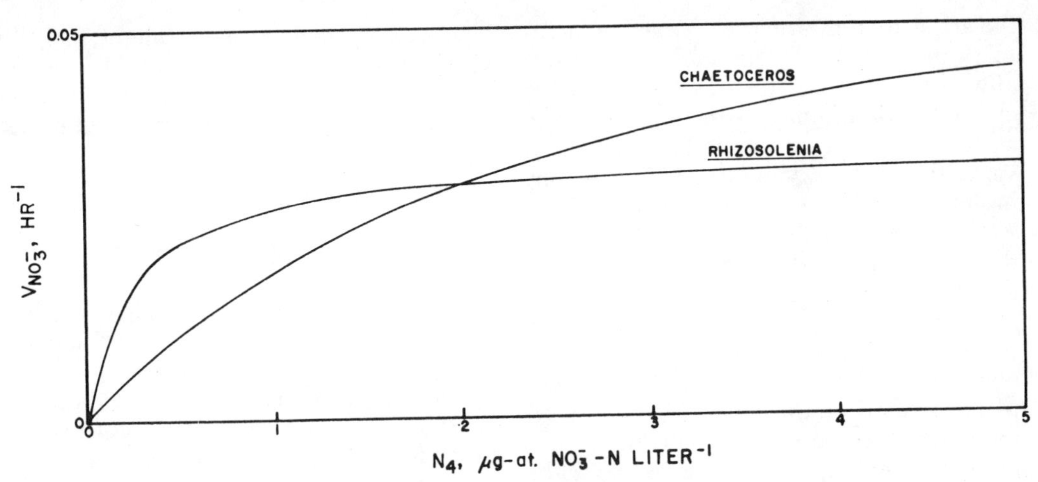

FIG. 7. Hypothetical curves for nitrate uptake vs. nitrate concentration for two species of algae. Values for V_{max} were taken from Riley (1963*b*); hypothetical values of K_t for nitrate uptake were assigned. For *Chaetoceros socialis*, $V_{max} = 0.068$/hr, $K_t = 2.50$ µg-at. NO_3^--N/liter; for *Rhizosolenia alata*, $V_{max} = 0.034$/hr, $K_t = 0.25$ µg-at. NO_3^--N/liter.

The model proposed in this paper requires the assumption that a single species of phytoplankton or a group of species with similar kinetic characteristics dominates the population.

The effect of a dominant population, N_1, will be to fix the level of dissolved nutrient concentration, N_4, since the sum of $\rho_{14}' + \rho_{14}'', \ldots \rho_{14}^n$ is assumed to be small compared to ρ_{14}. The minor populations N_1', $N_1'', \ldots N_1^n$ will have cellular growth rates set by the existing level of N_4 and their uptake characteristics K_t', $K_t'', \ldots K_t^n$, and V_{max}', $V_{max}'', \ldots V_{max}^n$. The fate of these minor populations will be determined by the relative size of the loss rates V_L', $V_L'', \ldots V_L^n$, in comparison to the uptake rates V_{N_4}', $V_{N_4}'', \ldots V_{N_4}^n$. Since these populations have no feedback response system while they remain small, they are likely to be growing or diminishing in size.

Values for V_{max} are available for some species under culture conditions. Data of Lanskaya are given in Riley (1963*b*) for generation times of diatoms obtained from the Black Sea and from these a value of V_{max} for growth can be computed. These values range from 0.226/hr for *Skeletonema costatum* (Grev) Cl. to 0.034/hr for *Rhizosolenia alata* Bright and 0.068/hr for *Chaetoceros socialis* f. *autumnalis* Pr.-Lavr. The important question is whether the species having lower V_{max} values have lower K_t values, enabling them to compete with species having higher V_{max} values. For example, in Fig. 7 the instantaneous growth rate is plotted against nitrate concentration for *R. alata* and *C. socialis* with the values given above for V_{max}. The effect of a low value of K_t in compensating for low maximum growth rates is shown. *R. alata* with a maximum growth rate half that of *C. socialis* would show higher instantaneous growth rates at any nitrate concentration below 2.0 µg-at./liter than would *C. socialis*. The situation represented in Fig. 7 may be extreme, but it does serve to illustrate the potential importance of the kinetic characteristics for nutrient uptake.

Competition for available nutrients at low concentrations, such as those prevailing in the Sargasso Sea, appears likely to result in an evolution of species with low K_t values, and it is possible that this cannot be achieved and still maintain a high value of V_{max}. In that case, species characteristic of low productivity tropical regions can be expected to show low values for both V_{max} and K_t. Other consequences follow, especially the necessity to reduce loss rates correspondingly in a variety of ways, for example, by reduction of sinking rates and the acquisition of armor to reduce vulnerability to predation. If diatoms are eventually shown to be organisms that have evolved with a high maximum growth rate at the cost of a correspondingly high K_t value, then the puzzle of diatom dominance in nutrient-rich areas may be explained.

Although a steady state was assumed in developing the model, that condition probably never occurs in nature. The existence of a control mechanism, however, tends to restore equilibrium when the system is perturbed. The feeding activities of zooplankton may produce perturbations with interesting consequences. Grazing is not carried on at a steady rate over the 24-hr period but rather is concentrated during the dark hours when at least parts of the population migrate toward the surface and feed. A daily cycle in ammonia concentration may be expected to occur because the phytoplankton population will be reduced allowing ammonia to accumulate temporarily. An increase in supply rate for ammonia will probably also occur as a result of its excretion by the zooplankton. The increased concentration of ammonia during the night hours will, however, result in an increase in $V_{NH_4^+}$, offset in part by a decreased uptake rate in the dark. Cycles of this nature have been described at Bermuda by Goering, Dugdale, and Menzel (1964) and by Beers and Kelly (1965). In the first case, the results were interpreted to mean an inherent daily rhythm of nitrogen uptake by the plant cells; however, it appears to be more likely that the phytoplankton were responding to the increased ammonia concentration.

With sufficient experience, it may be possible to apply nutrient limitation theory to situations where natural populations are perturbed deliberately through the addition of nutrients, for example, by the addition of sewage effluent to lakes or to coastal regions of the sea. The usefulness of an adequate theory may prove especially valuable in lakes where the ratio between nutrients in the water may vary widely from the ratio required by the phytoplankton. A detailed knowledge of the uptake kinetics for common algae might make it possible to control fishponds to obtain desired algae if a continuous addition of nutrients were to be made in place of the usual batch additions.

SUMMARY

The model proposed provides only a beginning towards understanding the dynamic behavior of nutrient-limited production. Regeneration terms for zooplankton grazing and bacteria should be added, but when this is done, the model becomes considerably more complex. The existence of additional forms of nitrogen, especially ammonia, must be taken into account. The possibility of interaction between nutrient concentrations and K_t values must be taken into account, particularly in the case of nitrate uptake, which is likely to be influenced by the ambient concentration of ammonia. The study of nitrate and phosphate uptake by Ketchum (1939) shows evidence for phosphate-nitrate interaction also. For simplicity, silicate has been ignored here but must also be considered in any generalized model.

The model provides a framework for the investigation of nutrient limitation in the sea and should be helpful in experimental design and interpretation. There is an urgent need to discover the important kinetic parameters for the uptake of nutrients (if other than Michaelis-Menton kinetics as postulated in this paper) and to measure them for phytoplankton algae characteristic of different productivity regimes. If this can be done, valuable new information about the adaptations of these organisms to their environment will become available, and a new tool will be provided to help in the prediction of their occurrence and behavior under nutrient-limited conditions.

REFERENCES

Barsdate, R. J. 1963. The biogeochemistry of molybdenum in lakes and in the sea. Ph.D. Thesis, Univ. of Pittsburgh. 92 p.

Becacos, T. 1962. Azione di alcune sostanze inorganiche sull'attività fotosintetica in due laghi dell'alta Italia (Lago Maggiore e Lago di Mergozzo). Mem. Ist. Ital. Idrobiol., **15:** 45–68.

BEERS, J. R., AND A. C. KELLY. 1965. Short-term variation of ammonia in the Sargasso Sea off Bermuda. Deep-Sea Res., **12:** 21–25.

CAPERON, J. W. 1965. The dynamics of nitrate limited growth of *Isochrysis galbana* populations. Ph.D. Thesis, Scripps Institution of Oceanography, La Jolla, Calif. 71 p.

DUGDALE, R. C., AND J. J. GOERING. 1967. Uptake of new and regenerated nitrogen in primary productivity. Limnol. Oceanog., **12:** 196–206.

FOGG, G. E. 1965. Algal cultures and phytoplankton ecology. Univ. Wisconsin Press, Madison. 126 p.

GOERING, J. J., R. C. DUGDALE, AND D. W. MENZEL. 1964. Cyclic diurnal variations in the uptake of ammonia and nitrate by photosynthetic organisms in the Sargasso Sea. Limnol. Oceanog., **9:** 448–451.

GOLDMAN, C. R. 1960. Molybdenum as a factor limiting primary productivity in Castle Lake, California. Science, **132:** 1016–1017.

———. 1961. Primary productivity and limiting factors in Brooks Lake, Alaska. Verhandl. Intern. Ver. Limnol., **14:** 120–124.

HARVEY, H. W. 1963. The chemistry and fertility of sea waters. Cambridge Univ. Press, London. 240 p.

KETCHUM, B. H. 1939. The absorption of phosphate and nitrate by illuminated cultures of *Nitzschia closterium*. Am. J. Botany, **26:** 399–407.

MENZEL, D. W., AND J. H. RYTHER. 1961. Zooplankton in the Sargasso Sea off Bermuda and its relation to organic production. J. Conseil, Conseil Perm. Intern. Exploration Mer, **26:** 250–258.

MILSUM, J. H. 1966. Biological control systems analysis. McGraw-Hill, New York. 466 p.

MONOD, J. 1942. Recherches sur la croissance des cultures bacteriennes. Hermann, Paris. 210 p.

REDFIELD, A. C., B. H. KETCHUM, AND F. A. RICHARDS. 1963. The influence of organisms on the composition of seawater, p. 26–77. *In* M. N. Hill [ed.], The sea, v. 2. Interscience, New York.

RILEY, G. A. 1963*a*. Theory of food-chain relations in the ocean, p. 438–463. *In* M. N. Hill [ed.], The sea, v. 2. Interscience, New York.

——— [ED.]. 1963*b*. Marine biology I, Proc. 1st Intern. Interdisciplinary Conf., p. 77. AIBS, Washington.

———. 1965. A mathematical model of regional variations in plankton. Limnol. Oceanog., **10**(Suppl.): R202–R215.

SHEPPARD, C. W. 1962. Basic principles of the tracer method. Wiley, New York. 282 p.

STEELE, J. H. 1958. Plant production in the North Sea. Scot. Home Dept., Marine Res., No. 7, p. 1–36.

———. 1962. Environmental control of photosynthesis in the sea. Limnol. Oceanog., **7:** 137–150.

———, AND D. W. MENZEL. 1962. Conditions for maximum primary production in the mixed layer. Deep-Sea Res., **9:** 39–49.

STEEMANN NIELSEN, E. 1952. The use of radioactive carbon (C^{14}) for measuring organic production in the sea. J. Conseil, Conseil Perm. Intern. Exploration Mer, **18:** 117–140.

VACCARO, R. F., AND H. W. JANNASCH. 1966. Studies on heterotrophic activity in seawater based on glucose assimilation. Limnol. Oceanog., **11:** 596–607.

WHITE, A., P. HANDLER, E. L. SMITH, AND D. STETTEN, JR. 1959. Principles of biochemistry. McGraw-Hill, New York. 1149 p.

WRIGHT, R. T., AND J. E. HOBBIE. 1966. Use of glucose and acetate by bacteria and algae in aquatic eco-systems. Ecology, **47:** 447–464.

Part 6

AQUATIC POLLUTION PROBLEMS

A pervasive and alarming side effect of man's developing technology and increasing population size is the ease with which he can modify terrestrial and aquatic ecosystems. The problem is compounded greatly by the fact that in most cases ecologists have not yet developed the detailed understanding of these ecosystems required to set limits on such modifications or even to predict their specific effects. Growing public concern and serious evaluations of the problem (see, for example, Tukey, 1965; Daddario, 1966) provide hope that these modifications can be minimized.

Both a working definition of pollution and a concise summary of the practical realities we face are provided in a statement by the National Research Council Committee on Pollution (1966):

> Pollution is an undesirable change in the physical, chemical or biological characteristics of our air, land and water that may or will harmfully affect human life or that of desirable species, our industrial processes, living conditions, and cultural assets; or that may or will waste or deteriorate our raw material resources. Pollutants are residues of the things we make, use and throw away. Pollution increases not only because as people multiply the space available to each person becomes smaller, but also because the demands per person are continually increasing, so that each throws away more year by year. As the earth becomes more crowded, there is no longer an "away." One person's trash basket is another's living space.

Aquatic ecosystems possess several characteristics which have made them seemingly ideal disposal sites for a wide variety of by-products from human activities. Most are relatively large in water volume and have a depth and opacity which help conceal waste accumulations. Solids and gases are dissolved readily, and in most aquatic systems mixing and transport processes are relatively rapid and efficient. Revelle (1960) and others suggest that part of the problem may stem from human attitudes about the aquatic portions of the earth. As terrestrial animals, we have tended to view oceans and lakes both as alien environments and as giant "holes in the ground" which, until recently, seemed logical places to dump our most undesirable wastes. All these factors contribute to the insidious nature of aquatic pollution, for usually it has been difficult to demonstrate harmful effects until they have progressed quite far.

Warren (1971) suggests that aquatic pollutants can best be considered in terms of the kinds of direct or indirect adverse effects they produce in the receiving waters. In this context, it is important to recognize that similar adverse changes in the aquatic environment occur as the result of natural processes, although generally they are much more limited in scope. Because they are a part of normal environmental change, such "natural pollution" effects (Hynes, 1960) generally are considered separately from those induced by man.

Organic materials which require oxidation by heterotrophic microorganisms to effect their decomposition form one such major functional category. Resulting depletion of dissolved oxygen to low levels can have lethal or sublethal effects on organisms inhabiting aquatic systems which receive large quantities of organic materials in the form of untreated sewage or industrial wastes.

Accumulation of nitrates, phosphates, and other inorganic plant nutrients from treated and untreated sewage effluents, agricultural run-off, and other sources allows increased primary production by algae and other autotrophic microorganisms. In lakes and other aquatic systems in which the exchange of water is limited, resulting blooms of these organisms can cause major physical and biological changes within a relatively short period of time. This process of nutrient accumulation, or eutrophication, also occurs naturally, but at a much slower rate, as such bodies of water age. The marked acceleration of the eutrophication process and its damaging effects by indiscriminate discharge of chemical by-products from human activities is a subject of major concern (National Academy of Sciences, 1969). Papers in this collection (Edmondson; Ryther and Dunstan) demonstrate, however, that adverse effects of nutrient accumulation can be alleviated or prevented if the cycling and utilization processes involved are clearly understood and if meaningful regulatory measures are adopted.

The effects of toxic and radioactive wastes from a wide variety of industrial, agricultural, and domestic sources are among the most serious problems facing the aquatic ecologist. These problems are compounded by the fact that little is known about the transfer, biological concentration, and specific effects of most such pollutants in aquatic ecosystems. Recent studies indicate that selective biological concentration effects involving radioactive isotopes and chlorinated hydrocarbons (such as DDT), all of which have come into extensive use since the 1940's, represent serious aquatic pollution problems. Similarly, widespread use of mercury in agricultural applications and of lead as a gasoline additive has produced apparently dangerous accumulations of these persistent heavy metals in both marine and freshwater systems. Toxic pollution resulting from careless industrial and shipboard handling of petroleum chemicals has long been a significant and obvious problem, particularly in developed bays and rivers.

Suspended solids, such as those resulting from the erosion of land areas disturbed by man or from dredging operations, can have adverse effects on both freshwater and coastal marine ecosystems. High turbidity caused by the suspension may lead to the partial or complete elimination of the natural aquatic vegetation and its associated animal species because light conditions are no longer adequate for photosynthesis. Subsequent deposition of these solids may alter the character of the bottom sediments, thereby causing other changes in the benthic community.

Temperature is considered to be one of the most important environmental factors controlling life processes. This is particularly true for most aquatic organisms, in which body temperature is determined by that of the surrounding water. Thermal pollution, involving unnatural changes in the aquatic temperature regime which lead to adverse biological or physical effects, is caused by such human activities as restriction or impoundment of streams and rivers, irrigation practices, and industrial use of water as a heat transfer medium (Krenkel and Parker, 1969). Cooling water effluent from the steam condenser systems of electric power generating plants is the primary cause of thermal pollution in most areas. Rapid expansion of the industry to meet current and projected needs for electrical power indicates that unless effective engineering solutions are forthcoming, thermal pollution will continue to be a major concern.

As the papers in this collection indicate, an ecologist's approach to a typical study of aquatic pollution differs little from that he might use in considering an unaffected natural system. Indeed, one of the major shortcomings of many pollution monitoring studies is that they do not incorporate the kinds of quantitative approaches to sampling and data analysis which characterize a modern ecological field study.

LITERATURE CITED

Daddario, E. Q. 1966. Environmental pollution. A challenge to science and technology. Rept. of Subcommittee on Science, Research, and Development, 89th Congress, U.S. Government Publication, 70–1770.

Hynes, H. B. N. 1960. The Biology of Polluted Waters. Liverpool University Press, Liverpool. 202 pp.

Krenkel, P. A., and Parker, F. L. 1969. Biological Aspects of Thermal Pollution. Proceedings of the National Symposium on Thermal Pollution, Vanderbilt University Press. 407 pp.

National Academy of Sciences. 1969. Eutrophication: Causes, Consequences, and Correctives. International Symposium on Eutrophication, Washington, D.C. 661 pp.

National Research Council. Committee on Pollution. 1966. Waste Management and Control. National Academy of Sciences (NAS Publication 1400), Washington, D.C. 257 pp.

Revelle, R. 1960. Welcoming address. *In* E. A. Pearson (ed.), Proceedings of the First International Conference on Waste Disposal in the Marine Environment. Pergamon Press, New York. 569 pp.

Tukey, J. W. (ed.). 1965. Restoring the quality of our environment. Report of the Environmental Pollution Panel, President's Science Advisory Committee, The White House (available from the Superintendent of Documents, Washington, D.C.). 317 pp.

Warren, C. E. 1971. Biology and Water Pollution Control. W. B. Saunders Company, Philadelphia. 434 pp.

26 Changes in Lake Washington following an increase in the nutrient income[1]

W. T. EDMONDSON (WASHINGTON, USA)

With 5 figures and 2 tables in the text

When, in 1955, Lake Washington developed a dense bloom of *Oscillatoria rubescens*, it was instantly obvious that an interesting limnological problem existed (EDMONDSON, ANDERSON and PETERSON 1956). It was known that the growing human population in the Lake Washington watershed was contributing nutrients to the basically relatively unproductive lake. The fact that *Oscillatoria rubescens* had similarly appeared in other polluted lakes which later developed nuisance conditions because of excess alga production suggested that Lake Washington should be studied (HASLER 1947, THOMAS 1957).

About 1930, the last major source of raw sewage was removed from the lake, although untreated sewage still enters the lake in relatively small quantities through storm sewer overflows which operate during heavy rain. Also, material drains into the lake and its tributaries from septic tanks. In 1957 ten sewage treatment plants, serving about 64 000 people, were putting treated effluent directly into Lake Washington (Fig. 1). Full information is given by BROWN and CALDWELL (1958). Since the initial report, a program of observation has been made between October 1956 and July 1959 by Dr. JOSEPH SHAPIRO and several assistants. The lake continued to exhibit progressively higher concentrations of nutrients, to produce larger crops of phytoplankton and zooplankton, and to change in other ways that indicate an increased productivity (Fig. 2).

The mean concentration of chlorophyll in the epilimnion during the summer was 4.7 times as high in 1958 as in 1950. Oxygen concentrations in the metalimnetic minimum and in the deepest part of the lake were much lower in 1957 than in any previous year; values less than 1.0 mgm/l occurred for the first time in September 1957, and, as would be expected, release of phosphate from the sediments was greatly increased after that time.

Because of public interest in the lake, an engineering survey was made to establish the magnitude of the sewage effluent and other sources of nutrients (BROWN and CALDWELL 1958). The increase in nutrients between 1950 and 1955 looks relatively small, but the limnological condition changed in an important way.

While it is obvious that the productivity of a lake will be affected by changes in the total amount of nutrients entering it, it is difficult to establish a simple basis on which to make quantitative comparisons. For this reason it now seems useful to review the concept of nutrient budget, especially income, to see how a useful

[1] The data on the conditions in Lake Washington reported here have been obtained with financial help from the following agencies, to which grateful acknowledgement is made: National Institutes of Health (Grant RG 4623), The National Science Foundation (Grant G 6167), and the State of Washington Fund for Research in Biology and Medicine (Initiative 171).

Reprinted from *Vehrhandlungen der internationalen Vereinigung für theoretische und angewandt Limnologie*, 14:167–175, 1961.

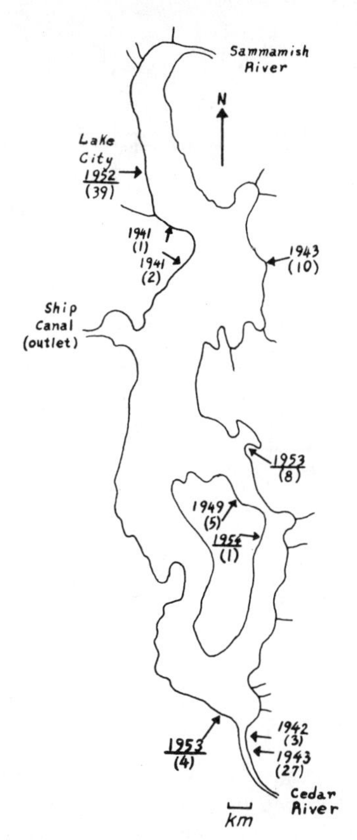

Fig. 1. Map of Lake Washington showing the location of sewage treatment plants and the date of establishment of each. The number in parentheses is the percentage of phosphate phosphorus in sewage effluent contributed by the plant in 1957.

comparison among lakes may be established. It may be that no simple numerical comparison will be possible, but some way must be developed for properly comparing the nutrient status of different lakes. Because of lack of space the present note is limited to a small part of the problem, a few points about the comparison of the income of phosphorus among lakes.

Obviously the annual nutrient income must be related to some property of the lake. A common way of figuring nutrient income is to divide the total amount of a nutrient annually entering the lake by the area of the lake, and to state the income as mass per unit area of lake (gm/m^2 or lb/acre) (Sawyer 1947, Thomas 1957). This will be called "areal income".

Computation of an areal income implies that the nutrients will have their effect through processes that are proportional to the area of the lake, and it makes no allowance for the depth of the lake, hence the degree of dilution and the resulting concentration of the nutrients in the water. Undoubtedly some of the important productive processes are related to area, but the concentration of the nutrients can have an effect of its own as shown by a large number of culture studies (e. g., Rodhe 1948, Gerloff and Skoog 1954, 1957).

It would seem useful, therefore, to base the income on the volume of the lake to give the potential concentration that would be achieved if dilution were the only process involved, ignoring for the moment utilization and regeneration. This

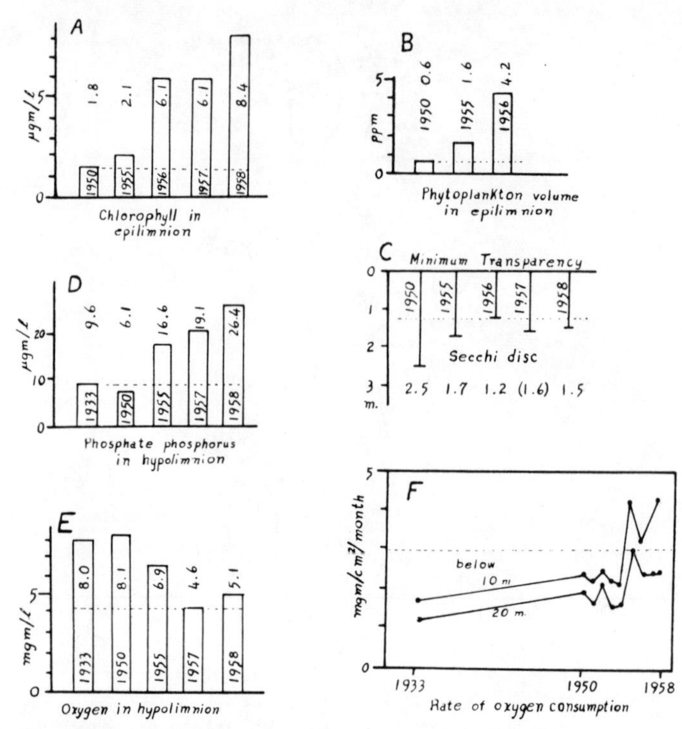

Fig. 2. Limnological conditions in Lake Washington. A. Mean concentration of chlorophyll in the top 10 meters between June 20 and August 20. B. Mean volume of phytoplankton in top 10 meters during the same period. Data for 1957 and 1958 not yet available. C. Minimum transparency observed during each year. D. Mean concentration of phosphate-phosphorus in the layer below 20 meters late in the summer. The date varies slightly from year to year. E. Same for oxygen. F. Rate of development of oxygen deficit between June 20 and August 20. Because of the oxygen minimum at about 12 meters, the deficit has been calculated both for the layer below 10 meters and that below 20 meters.

gives a base figure with which consideration of various processes can be developed. Inflowing water is not immediately diluted to the volume of the lake, and some algae will grow with nutrients in more nearly entering concentration.

If the inflowing water has the same or smaller concentration of nutrients as that in the lake, when it displaces water through the outlet, the potential concentration developed will not be increased by increasing the rate of flow of the water, all other things being equal, although the areal budget would be increased and give a false impression of the biological consequences. When lake is more dilute than the average concentration of the inflows, increasing the rate of flow increases the potential concentration. Utilization causes the actual concentration achieved to be less. A simple diagram shows some of the numerical relationships involved to substitute for further discussion (Fig. 3).

The potential concentration is related to the areal income and the mean depth of the lake.

$$\frac{\text{areal income}}{\text{mean depth}} = \text{potential concentration} \quad \frac{\text{gm/m}^2}{\text{m}} = \text{gm/m}^3$$

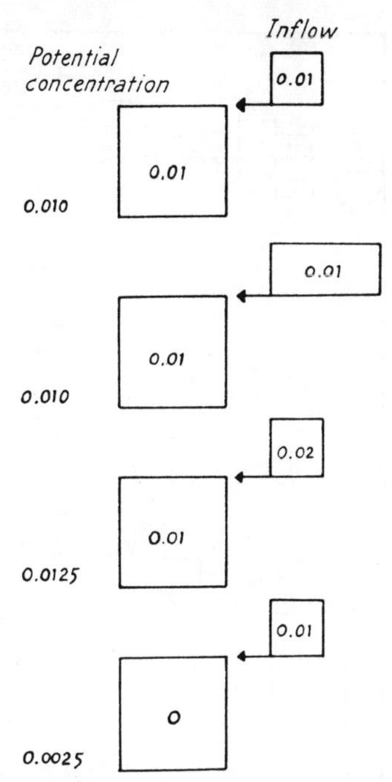

Fig. 3. Diagram to suggest the results of different conditions of nutrient income. For each figure in the center is a square whose area represents the volume of the lake. The number shows the concentration of nutrient before the addition of the annual increment of water. At the right is a rectangle whose area represents proportionally the volume of the annual increment of water. The mean concentration of nutrient in the increment is shown. At the left is shown the resulting potential concentration assuming that the increment displaces lake water and is then mixed evenly with the remaining water.

We can now consider the phosphorus income of Lake Washington (Table 1, Fig. 4). The Cedar River, which brought in 54% of the water income of the lake in 1957, is the most dilute source according to measurements made upstream from the City of Renton where a large sewage treatment plant empties into the river. The other inlets have higher concentrations; all of these streams run to a greater or less extent through settled, partly cultivated, areas served by septic tanks which contribute nutrients since the soil is in general poor for septic tanks. The highest monthly average obtained for any of the streams entering Lake Washington was 0.40 mgm/l of phosphate-P. The concentration in the effluent of sewage treatment plants is about 100 times larger than that encountered in even the most contaminated inlets. The concentration that would be achieved in the lake by diluting all these sources of nutrients with water having the average concentration observed in the center of the lake is 0.037 mgm/l.

Computations have been made for the lakes at Madison, Wisconsin and six European lakes, using published information (Table 2, Fig. 5). In a general way, the lakes which produce more severe alga nuisances have high incomes of phosphorus, and those which are not noteworthy for algae have lower incomes. Thus,

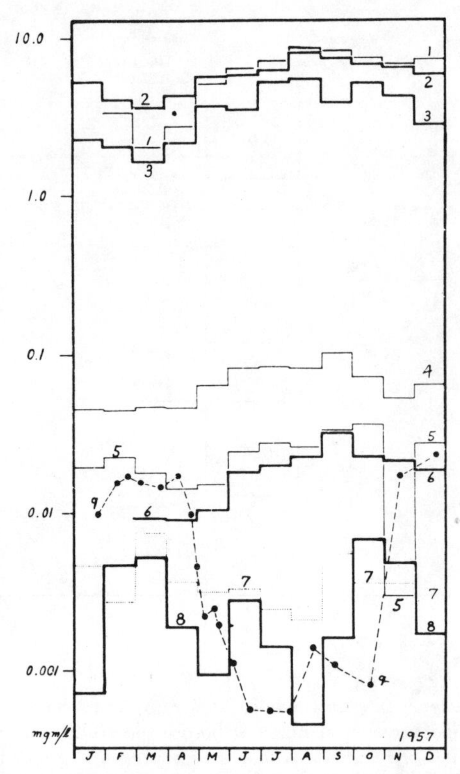

Fig. 4. Monthly averages of phosphate phosphorus concentration, 1957 on a logarithmic scale.

1.—3. Sewage treatment plant effluents: 1, Kirkland; 2, Renton; 3, Lake City.

4.—7. Inlets to Lake Washington (chosen to show the extremes in concentration): 4, Thornton Creek; 5, Samamish River; 6, Coal Creek; 7, Cedar River at Landsburg.

8. Ship canal (outlet of Lake Washington).

9. Individual measurements made in surface water near center of lake.

Data for 1—8 from Hollis M. Phillips, Seattle Department of Engineering; for 9, by Joseph Shapiro.

of the Wisconsin series, Waubesa, Kegonsa and Monona are famous for their alga nuisances and have been heavily treated with copper sulfate. Of the Swiss series, Pfäffikersee and Greifensee are described by Thomas (1957) as eutrophic. Zürichsee stands above Türlersee in areal income, but slightly less in potential concentration which is consistent with the relative status of their algal crops and other conditions. Türlersee develops anaerobic conditions and hydrogen sulfide in the hypolimnion, and Thomas (1957) comments that Zürichsee would be worse than it is if it were not for its large volume. Actually it is difficult to rate the potential concentration for Zürichsee since the lake is meromictic, or at least has reduced mixing, and the industrial contribution has not been measured, so the nutrients in the mixolimnion actually must be more concentrated than shown in the graph. Lake Washington stands high in the areal budgets, and although it is relatively lower in potential concentration is still not far below Furesø which has begun to show changes toward nuisance conditions in recent years.

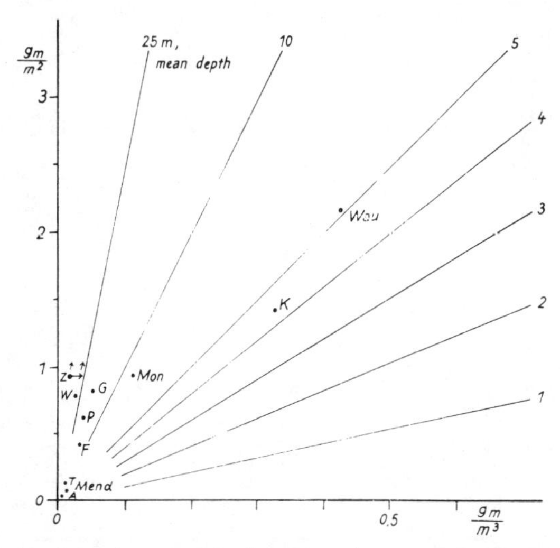

Fig. 5. Comparison of areal income (ordinate) with potential concentration (abscissa) of phosphate-phosphorus for the lakes given in Table 2. The lines radiating from the origin show the mean depth of the lakes. Since the annual increment of water to Lakes Monona, Kegonsa and Waubesa is larger than the volumes of the lakes, the quantities plotted are those contained in a volume of incremental water equal to the volume of the lake (annual income × replenishment time).

Lake Washington is close to Monona in the areal income, far below in potential concentration. Thus it seems that the potential concentration is more closely related to algal standing crop than is the areal income. The annual inflows into Lakes Waubesa and Kegonsa are larger than the volumes of the lakes, and therefore the areal expression of income greatly exaggerates the fertilization of the lakes although high concentrations are achieved. Unfortunately, direct measurements of the primary productivity are not available, and it is possible to make comparisons only of features which are indirectly related to primary productivity.

Table 1. Known contributions of phosphorus to Lake Washington (kgm per year of P, 1957).

	A Phosphate P	B Total Dissolved P	C Particulate P	D Total P (= B + C)
Sewage Treatment Plants	37 696	41 842	5 156	46 998
Streams[1]	18 474	33 731	6 816	40 547
Industrial Waste	7 774	10 585	[3]	(10 585)
Sewage overflow	5 823[2]	6 470	[3]	(6 470)
Total	69 767	92 628	11 972	104 600

[1] Measurements in Cedar River made upstream from Renton Sewage Plant.
[2] Only Total dissolved P measured: assume 90% as phosphate.
[3] Not measured.
Note: The figures are subject to some modification as more information becomes available. The inlets contributed 886 million m³ of water in 1957 and the sewage plants 9.7 million.

Table 2 Phosphorus and water income of lakes.

	Washington	Mendota	Monona	Waubesa	Kegonsa	Furesø	Zürichsee[1]	Türlersee	Aegerisee	Päffikersee	Greifensee
A. Area of lake, millions m^2 (= km^2)	87.615	39.38	14.10	8.23	12.73	9.4	68.42	0.48	7.19	3 25	8.38
B. Volume of lake, millions m^3	2 884.2	478.4	119.0	40.2	59.1	126.4	3 720	6.7	351.7	56.5	155.0
C. Mean depth (B/A), m	32.9	12.1	8.4	4.9	4.6	13.4	54.4	14.0	48.9	17.4	18.5
D. Area: Volume (A/B)	0.030	0.082	0.118	0.204	0.215	0.074	0.018	0.716	0.020	0.056	0.054
E. Annual increment of water, millions m^3	895.9	39.9	98.0	136.4	172.0	3.997	2 938	3.15	39.6	12.4	88.7
F. Water increment: Lake volume (E/B)	0.311	0.084	0.823	3.386	2.911	0.031	0.790	0.470	0.113	0.219	0.572
G. Replenishment time, years (B/E)	3.2	12.0	1.2	0.30	0.35	31.6	1.3	2.1	8.9	4.6	1.7
H. Phosphate-phosphorus											
1. Annual increment, kgm	69 767	2 575	13 427	58 680	53 922	4 031	65 000	61	258	1 999	6 926
2. Concentration in water increment, mgm/l (HI/E)	0.078	0.065	0.120	0.4226	0.321	1.008	0.022	0.019	0.007	0.161	0.078
3. Concentration in lake, mgm/l (HI/B)	0.024	0.005	0 113	(1.456)[2]	(0.912)[2]	0.032	0.017	0.009	0.0007	0.035	0.045
4. Areal income, gm/m^2 (HI/A)	0.796	0.065	0.952	(7.130)[2]	(4.236)[2]	0.427	0.950	0.127	0.036	0.615	0.827
I. Total dissolved phosphorus[3]											
1. Annual increment, kgm	92 628	6 881	30 169	81 736	84 554	—	119 000	143	1 174	4 499	13 366
2. Concentration in water increment, mgm/l (II/E)	0.1057	0.172	0.270	0.5886	0.504	—	0.041	0.045	0.030	0.363	0.151
3. Concentration in lake, mgm/l (II/B)	0.032	0.014	0.254	(2.028)[2]	(1.431)[2]	—	0 032	0.021	0.003	0.080	0.086
4. Areal income, gm/m^2 (II/A)	1.034	0.174	2.140	(9.931)[2]	(6.642)[2]	—	1.739	0.297	0.163	1.384	1.600

[1] Data are given for the lower basin of Zürichsee, and the phosphorus income of the whole lake applied to this basin.

[2] Since the annual increment of water is larger than the volume of the lake, this concentration is not achieved, and the values in Row 2 should be used for comparison.

Source of data: Lake Washington (Table 1), HOLLIS M. PHILLIPS, Seattle Engineering Department; Mendota, SCHNEBERGER, et al; Monona, Waubesa, Kegonsa, SAWYER, 1947; Furesø, BERG, et al, 1958; Swiss lakes, THOMAS 1955, 1956, 1956—57. The value for Zürichsee does not include industrial waste.

[3] For the Swiss lakes, this value includes particulate P.

The potential concentrations, as presented in Table 2, are, of course, unrealistic as descriptions of conditions since the incoming water is not uniformly and completely mixed with the entire volume of the lake and nutrients are withdrawn by biological activity and released from the bottom. However, existence of very concentrated sources of nutrients can have biological consequences, and these must be considered.

To summarize, it is desirable to find, if possible, a simple numerical basis on which to compare the rate of nutrient supplies of lakes from outside sources. A system which incorporates a consideration of the potential concentration has advantages over a system which relates income to area. Effects of nutrient storage by algae, exchange with sediments, rates of turnover, dilution and distribution need, of course, to be studied separately.

Discussion

MORTIMER: How far is production controlled by the depth ratio of the photic layer and the mixing layer? The ratio the Fursö will be very different to that in the much deeper Lake Washington.

EDMONDSON: The compensation depth is almost always within the epilimnion. Therefore light is one of the limiting properties.

MORTIMER: Has the "nutrient income" been corrected for loss of total and dissolved nitrogen and phosphorus to the outflow and sediments?

EDMONDSON: In a proper budget one has separate numbers for income and spending (income, outflow and decomposition). This paper is limited to a consideration of one aspect of the income of phosphorus, although measurements are available for the outlet and sediments.

LUND: 1. When the speaker mentions nutrients what nutrients is he referring to? 2. I think in these very rich lakes we may reach a situation where nitrogen and phosphorus are in excess and other nutrients may be limiting.

EDMONDSON: 1. In general, any substance required or used by algae as a component for growth. In particular, the emphasis in this work has been on phosphorus and nitrogen. 2. See Dr. SHAPIROs comment.

SHAPIRO: With regard to Dr. LUNDs suggestion that perhaps too much attention is being given to nitrogen and phosphorus, Lake Washington, in spite of the large amount of sewage entering it, is still at the stage where addition of nitrate and phosphate results in increased algal growth. Addition of trace elements or soil extract had no effect.

References

BERG, K., et al. 1958. Furesøundersøgelser 1950—1954. — *Folia Limnol. Scand.,* **10,** 1—89.

BROWN & CALDWELL. 1958. (Civil and Chemical Engineers.) *Metropolitan Seattle Sewerage and Drainage Survey.* A Report for the city of Seattle, King County, and the State of Washington.

EDMONDSON, W. T., ANDERSON, G. C., and PETERSON, D. R. 1956. Artificial eutrophication of Lake Washington. — *Limnol. and Oceanogr.* **1,** 47—53.

GERLOFF, G. C., and SKOOG, F. 1954. Cell contents of nitrogen and phosphorus as a measure of their availability for growth of *Microcystis aeruginosa.* — *Ecology,* **35,** 348—353.

— 1957. Nitrogen as a limiting factor for the growth of *Microcystis aeruginosa.* — *Ecology,* **38,** 556—561.

HASLER, A. D. 1947. Eutrophication of lakes by domestic drainage. — *Ecology*, **28**, 383—395.

RODHE, W. 1948. Environmental requirements of fresh-water plankton algae. — *Symb. Bot. Upsaliensis.*, **10**, 1—149.

SAWYER, C. N. 1947. Fertilization of lakes by agricultural and urban drainage. — *J. New England Waterworks Assoc.*, **61**, 109—127.

THOMAS, E. A. 1948. Limnologische Untersuchungen am Türlersee. — *Schweiz. Z. Hydrobiol.*, **11**, 90—177.

— 1955. Stoffhaushalt und Sedimentation im oligotrophen Aegerisee und im eutrophen Pfaffiker- und Greifensee. — *Mem. Ist. Ital. Idrobiol., Suppl.* **8**, 357—465.

— 1957. Der Zürichsee, sein Wasser und sein Boden. — *Jb. vom Zürichsee*, **17**, 173—208.

Wisconsin Committee on Water Pollution. 1949. *Report on Lake Mendota studies concerning conditions contributing to occurrence of aquatic nuisances* **1945—1947**. — Published by Wisconsin Committee on Water Pollution (hectographed).

PHOSPHORUS, NITROGEN, AND ALGAE IN LAKE WASHINGTON AFTER DIVERSION OF SEWAGE

W. T. EDMONDSON

Department of Zoology, University of Washington, Seattle 98105

Abstract. *After diversion of sewage effluent from Lake Washington, winter concentrations of phosphate and nitrate decreased at different rates. From 1963 to 1969, phosphate decreased to 28 percent of the 1963 concentration, but nitrate remained at more than 80 percent of the 1963 value. Free carbon dioxide and alkalinity remained relatively high. The amount of phytoplanktonic chlorophyll in the summer was very closely related to the mean winter concentration of phosphate, but not to that of nitrate or carbon dioxide.*

Lake Washington has responded promptly to major changes in its nutrient income. In 1955 the lake was unmistakably deteriorating, for in that year *Oscillatoria rubescens* became prominent in the plankton (*1*). The lake was receiving effluent from ten secondary biological treatment plants amounting to about 24,200 m³ [6.4 million gallons (1 gallon = 3.8 liters)] per day from a tributary population of about 68,000. In 1957, sewage effluent was contributing about 56 percent of the phosphorus and 12 percent of the nitrogen income of the lake (*2*). A program of diversion was voted by public action. Because of the magnitude of the project, diversion of all plants was not simultaneous. The first diversion of about one quarter of the effluent occurred in February 1963, by which time 11 treatment plants were enriching the lake with a total capacity of about 75,600 m³ per day. In 1965 the volume was reduced to 55 percent of the original, and the final diversion took place in February 1968, but about 99 percent of it had been diverted by March 1967 (*3*).

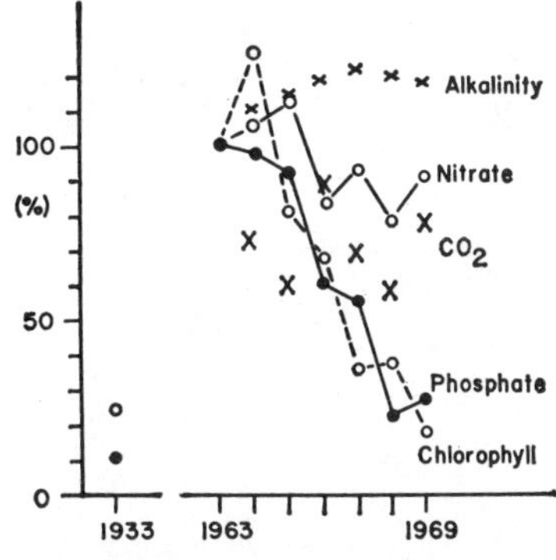

Fig. 1. Mean winter (January to April) values in surface water of phosphate-phosphorus and nitrate-nitrogen, and mean summer (July and August) values of chlorophyll in surface phytoplankton. The 1963 values, plotted as 100 percent, were (in micrograms per liter): P, 57; N, 428; and chlorophyll, 38. Unconnected points show winter means (January and February) of bicarbonate alkalinity and free carbon dioxide in surface water (25.3 and 3.2 mg/liter in 1963).

The purpose of this report is to describe some of the changes in chemical conditions and algae following diversion of sewage effluent.

After the first diversion the condition began to improve steadily as indicated by the phosphorus content, abundance of algae, and transparency of the lake. In Lake Washington, concentrations of phosphate and nitrate are at a maximum in winter and decrease to low values during the spring growth of phytoplankton (*2*). During the winter of 1969, the concentration of phosphate was only 28 percent of that in 1963, the year of maximum phosphate. The content of nitrate has decreased much less than that of phosphate, remaining 80 percent or more of the 1963 value. The winter mean free carbon dioxide has been rather variable, fluctuating around 75 percent of the 1963 value. The winter mean alkalinity (bicarbonate) increased by about 20 percent from 1963. At present, an explanation for the details of the changes in carbon dioxide and alkalinity is not available.

During the years of heaviest enrichment, the number of phytoplankton has fluctuated around relatively high values during the summer (*4*), and the phosphorus content of particulate matter (seston) in summer corresponds well to the concentration of dissolved phosphate the previous winter (*5*). The abundance of phytoplankton has decreased since diversion started. The nuisance condition of a lake is well measured by Secchi disk transparency; the mean summer transparency has increased from 1.0 m in 1963 to 2.8 m in 1969. In Lake Washington, changes in transparency are dominated by changes in phytoplankton rather than in silt. Phytoplankton counts have not been completed, but data are available on the chlorophyll content of the phytoplankton in the epilimnion. The summer mean (July and August) is strikingly related to the concentration of phosphate in the surface water during the previous winter but not to nitrate or carbon dioxide or alkalinity (Fig. 1). This relation strongly suggests that phosphorus is the most important limiting element in Lake Washington. This, of course, does not mean that other elements are not important to phytoplankton, but only that they are present in excess relative to phosphorus, and the abundance of algae, therefore, varies in proportion to the phosphorus content. While it is conceivable that some other element, not measured in our study, is varying in exact proportion to phosphorus, it seems very unlikely.

Reprinted from *Science*, Vol. 169, No. 3946, August 14, 1970. Copyright 1970 by the American Association for the Advancement of Science.

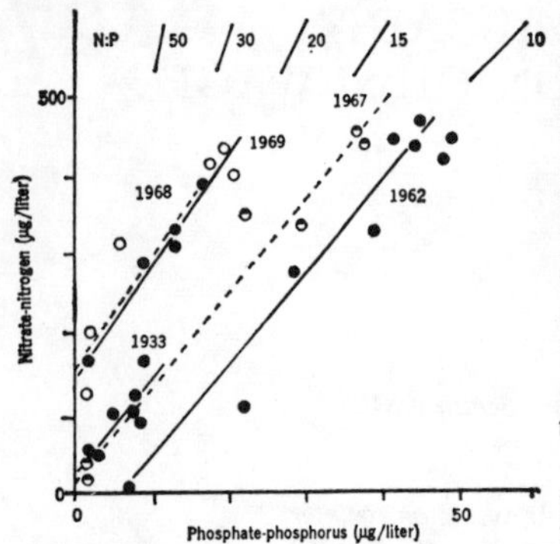

Fig. 2. Correlation between surface values of phosphate and nitrate during the spring increase of phytoplankton when the concentrations of nutrients are decreasing. In 1962 a distinct excess of phosphate occurred when nitrate had been exhausted, but in the other years shown nitrate was in excess, especially 1968 and 1969. The slopes of the lines, fitted by eye, vary from 11.9 (1962) to 14.8 (1968 and 1969). Ratios of nitrogen to phosphorus are shown by the numbered lines radiating from the origin. [Compare with fig. 5 of (*2*).]

The difference in the rates of decrease of nitrogen and phosphorus can be attributed at least in part to the fact that the water supply to the lake through the two major inlets is relatively much richer in nitrogen than in phosphorus, so that the lake water since 1963 has been progressively diluted with water that is poorer in phosphorus relative to nitrogen than is the water in the lake (*2*). Whether nitrogen fixation by blue-green algae has contributed to keeping the lake nitrogen elevated has not yet been determined.

The question has frequently been raised as to the effect of removal of phosphate from sewage effluent or of reducing the concentration of phosphate in the effluent by controlling the character of the input. Concern has been expressed that heavy enrichment with nitrate may cause alga problems in waters with adequate natural supplies of phosphorus (*5*). While this might be true in some regions for geochemical reasons, it appears that Lake Washington in its unpolluted condition has more than enough nitrogen relative to phosphorus, and phosphorus is the dominating limiting element. Lake Washington probably represents a large class of lakes in which phosphorus is the dominating element (*6*).

An indication of the relative importance of these two elements can be seen by measuring their decreasing concentrations during the growth of phytoplankton in the spring (Fig. 2). In 1933, when the lake was less polluted, a small concentration of nitrate was left over when phosphate was nearly exhausted. In strong contrast, after many years of enrichment with sewage effluent rich in phosphorus, an excess of phosphate occurred when nitrate was exhausted in 1962; this excess phosphate was almost as much as the winter maximum of 1933. During diverison of sewage, the condition returned to resemble that of 1933 but continued to change, and in 1968 and 1969 a large excess of nitrate occurred. This kind of analysis should be generally useful in other lakes that have a winter maximum and spring decrease in nutrient concentration. It is obvious that the fact that phosphorus is in excess in a lake does not mean that control of phosphorus will be ineffective as long as it can be brought to low concentrations. Possibly in some regions excess phosphorus demonstrated this way can be regarded as evidence that the lake has been affected by sewage effluent and its quantity used as a measure of the magnitude of the effect.

Thus, Lake Washington has responded promptly and sensitively to changes in its nutrient income. While the changes during increase are not as well documented by direct limnological data as those during decrease, they are recorded by paleolimnological evidence (*7*). On the basis of the present data and existing knowledge, it seems valid to predict that noticeable improvements can be made in similar lakes even by partial limitation of phosphorus. Total elimination of phosphorus is impractical, of course. The worst sources are the most concentrated ones. The popular system of calculating total loadings on an areal basis (kilograms per hectare) without accounting for concentration can be misleading when dealing with lakes that receive large volumes of dilute drainage. The annual total of phosphorus can be very large, but if it is dispersed in a large volume of water it cannot generate as dense concentrations of algae as can the same total in much more concentrated sewage effluents and some kinds of agricultural drainage. Diversion of sewage from Lake Washington reduced the phosphorus income by only about half, but the sewage effluent was nearly 200 times as concentrated in phosphate as the influent streams (*2*), and the effect has been great.

References and Notes

1. W. T. Edmondson, G. C. Anderson, D. R. Peterson, *Limnol. Oceanogr.* **1**, 47 (1956).
2. W. T. Edmondson, in *Eutrophication: Causes, Consequences, Correctives*, Publ. 1700 (National Academy of Sciences, Washington, D.C., 1969), pp. 124–149.
3. In (*2*) the final diversion date was given as 1967 in the legend of fig. 4, rather than 1968, as correctly given on p. 136.
4. W. T. Edmondson, *Verh. Int. Verein. Theoret. Angew. Limnol.* **16**, 153 (1966).
5. ———, in *Water Resources Management and Public Policy*, T. H. Campbell and R. O. Sylvester, Eds. (Univ. of Washington Press, Seattle, 1968), pp. 161–163.
6. E. A. Thomas, in (*2*), pp. 29–49.
7. W. T. Edmondson, *Mitt. Int. Verein. Theoret. Angew. Limnol.*, No. 17 (1969), p. 19; J. C. Stockner and W. W. Benson, *Limnol. Oceanogr.* **12**, 513 (1967).
8. Supported by grants from the National Science Foundation. Most of the field and chemical work in recent years has been carried out under the supervision of David E. Allison. Many of the data reported here have been obtained with the cooperation of Shirley M. Clark, Diane Egan, Donald J. Hall, Lois Kiehl, and Michael Parker.

NITROGEN, PHOSPHORUS AND EUTROPHICATION IN THE COASTAL MARINE ENVIRONMENT

28

JOHN H. RYTHER
WILLIAM M. DUNSTAN

Woods Hole Oceanographic Institution, Woods Hole, Massachusetts 02543

Abstract. *The distribution of inorganic nitrogen and phosphorus and bioassay experiments both show that nitrogen is the critical limiting factor to algal growth and eutrophication in coastal marine waters. About twice the amount of phosphate as can be used by the algae is normally present. This surplus results from the low nitrogen to phosphorus ratio in terrigenous contributions, including human waste, and from the fact that phosphorus regenerates more quickly than ammonia from decomposing organic matter. Removal of phosphate from detergents is therefore not likely to slow the eutrophication of coastal marine waters, and its replacement with nitrogen-containing nitrilotriacetic acid may worsen the situation.*

The photosynthetic production of organic matter by unicellular algae (phytoplankton) in the surface layers of the sea is accompanied by, is indeed made possible by, the assimilation of inorganic nutrients from the surrounding water. Most of these substances are present at concentrations greatly in excess of the plants' needs, but some, like nitrogen and phosphorus, occur at no more than micromolar levels and may be utilized almost to the point of exhaustion by the algae. It is, in fact, the availability of these nutrients that most frequently controls and limits the rate of organic production in the sea.

Harvey (*1*) was among the first to point out that phytoplankton growth caused the simultaneous depletion of both nitrate and phosphate from the ambient seawater. Much has since been written about the interesting coincidence that these elements are present in seawater in very nearly the same proportions as they occur in the plankton (*2–4*). For example, Redfield (*3*) reported atomic ratios of available nitrogen to phosphorus of 15 : 1 in seawater, depletion of nitrogen and phosphorus in the ratio of 15 : 1 during phytoplankton growth, and ratios of 16 : 1 for laboratory analyses of phytoplankton. This relationship may have resulted from adaptation of the organisms to the environment in which they live, but Redfield suggested a mechanism, the microbial fixation of elementary nitrogen, which could regulate the level of fixed nitrogen in the sea relative to phosphorus to the same ratio as these elements occur in the plankton. In other words, any deficiency of nitrogen could be made up by nitrogen fixation.

Such a process could, in times past, have adjusted the oceanic ratio of nitrogen to phosphorus to its present value, and it may be important in regulating the level or balance of nutrients in the ocean as a whole and over geological time. It is certainly not effective locally or in the short run. As analytical methods have improved and as the subject has been studied more intensively, it has become increasingly clear that the concept of a fixed nitrogen to

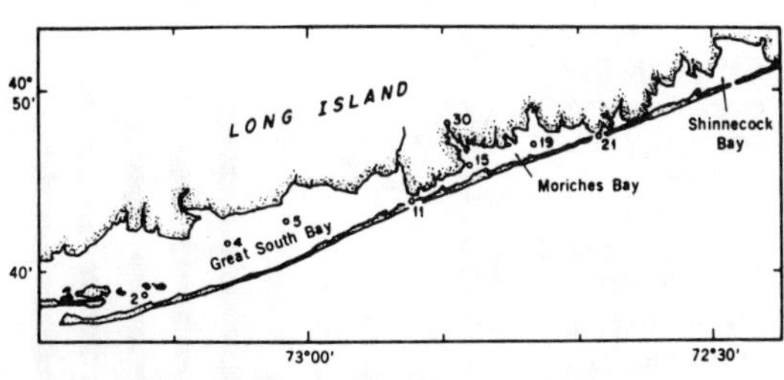

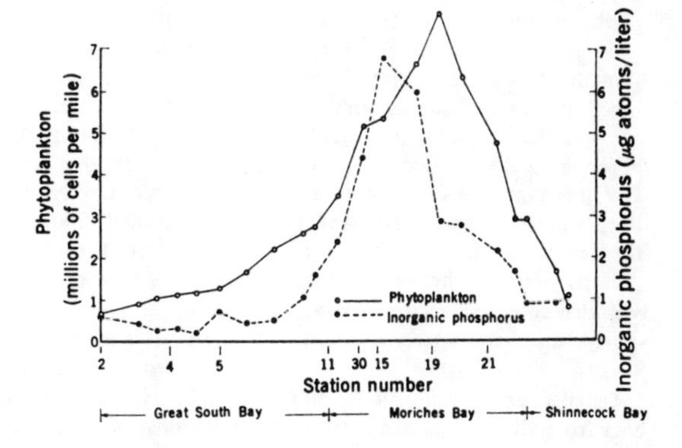

Fig. 1. The distribution of phytoplankton and inorganic phosphorus in Great South Bay, Moriches Bay, and Shinnecock Bay, Long Island, in the summer of 1952. Station numbers on the map (above) correspond to station numbers on the abscissa of the figure (right).

Reprinted from *Science*, Vol. 171, No. 3975, March 12, 1971. Copyright 1971 by the American Association for the Advancement of Science.

phosphorus ratio of approximately 15 : 1, either in the plankton or in the water in which it has grown, has little if any validity.

As early as 1949 Ketchum and Redfield (*5*) showed that deficiencies of either element in culture mediums may drastically alter their ratios in the algae. They reported (*5*) nitrogen to phosphorus ratios by atoms in cultures of *Chlorella pyrenoidosa* of 5.6 : 1 for normal cells, 30.9 : 1 for phosphorus-deficient cells, and 2.9 : 1 for nitrogen-deficient cells. A number of subsequent studies of both algal cultures (*6, 7*) and oceanic particulate matter (*8, 9*) have reported highly variable ratios of nitrogen to phosphorus. These ratios are somewhat difficult to interpret in oceanic particulate matter, since living algae may comprise a very small fraction of the total particulate organic matter collected by the usual sampling methods, and the origin and nature of the remaining material are largely unknown. On the other hand, the chemical composition of algae grown in the usual culture mediums may differ significantly from that of naturally occurring organisms. Despite these uncertainties, the following generalizations may be made: (i) ratios of nitrogen to phosphorus from less than 3 : 1 to over 30 : 1 (by atoms) may occur in unicellular marine algae; (ii) the ratio varies according to the kind of algae grown and the availability of both nutrients; and (iii) although there is no indication of any "normal" or "optimum" nitrogen to phosphorus ratio in algae, values between 5 : 1 and 15 : 1 are most commonly encountered and an average ratio of 10 : 1 is therefore a reasonable working value.

In seawater, a 15 : 1 atomic ratio may be typical of the ocean as a whole. But since 98 percent of its volume lies below the depth of photosynthesis and plant growth, such mean values have little relevance to the present discussion. If one considers only the remaining 2 percent of the ocean's volume, the so-called euphotic layer, high ratios appoaching 15 : 1 occur only at the few times and places where relatively deep water is mixed or upwelled into the euphotic layer (*9*). Over the greater part of the sea surface, the two elements appear to bear no constancy in their interrelationship (*5, 9–12*).

Detailed examination of the nutrient data from the sea surface reveals that, as the two elements are utilized, nitrogen compounds become depleted more rapidly and more completely than does phosphate. This is particularly true when only nitrate-nitrogen is considered. Both Vaccaro (*9*) and Thomas (*13*) have pointed out, however, that ammonia may often be quantitatively a more important nitrogen source than is nitrate in surface ocean waters, particularly when nitrogen levels in general become reduced through plant growth. But even when all known forms of available nitrogen are considered together, they are often found to be reduced to levels that are undetectable in the euphotic layer. In this event, almost invariably a significant amount of phosphate remains in solution. There is, in short, an excess of phosphate, small but persistent and apparently ubiquitous, in the surface water of the ocean, relative to the amount of nitrogen available to phytoplankton nutrition. This is true in both the Atlantic (*9, 10, 12*) and Pacific oceans (*11, 13, 14*). Thus, the ratio of nitrogen to phosphorus in surface seawater may range from 15 : 1, where subeuphotic water has recently been mixed or upwelled to the surface, to essentially zero when all detectable nitrogen has been assimilated. Since most of the surface waters of the ocean are nutrient deficient most of the time, nitrogen to phosphorus ratios appreciably less than 15 : 1 are the common rule.

A puzzling question remains to be answered. If the ultimate source of nutrients is deep ocean water containing nitrate and phosphate at an atomic ratio of 15 : 1 and if the average phytoplankton cell contains these elements at a ratio of about 10 : 1, why is it that nitrogen compounds are exhausted first from the water and that a surplus of phosphate is left behind? How can nitrogen rather than phosphorus be the limiting factor? Before turning to this question, we will present two examples of nitrogen as a limiting factor to phytoplankton growth.

Long Island bays. Great South Bay and Moriches Bay are contiguous and connected embayments on the south shore of Long Island, New York, formed by the barrier beach that extends along much of the East Coast of the United States. They are shallow, averaging 1 to 2 m in depth, have a hard sandy bottom, and traditionally have supported a productive fishery of oysters and hard clams. Introduction and growth of the Long Island duckling industry, centered along the tributary streams of Moriches Bay, resulted in the organic pollution of the two bays and the subsequent development of dense algal blooms in the bay waters, to the detriment of the shellfisheries.

Table 1. Regeneration of nitrogen and phosphorus accompanying the decomposition of mixed plankton [after Vaccaro (*18*)]. Excess phosphorus (last column) was calculated on the assumption that all nitrogen was assimilated as produced and that phosphorus was assimilated at a nitrogen to phosphorus ratio of 10 : 1.

Days	NH_3 + NO_2 + NO_3 (μg atoms N/ liter)	PO_4 (μg atoms P/ liter)	N : P (by atoms)	Excess P (μg atoms P/ liter)
0	0.00	0.80	0.0	0.80
7	0.86	0.79	1.1	0.70
17	2.81	0.84	3.3	0.56
30	3.08	1.05	2.8	0.74
48	3.86	0.98	3.9	0.59
87	4.14	1.04	4.1	0.63

As a result of studies by the Woods Hole Oceanographic Institution during the period 1950–55, the ecology of the region and the etiology of its plankton blooms were described in some detail (*15*). The situation has since changed, but certain of the unpublished results of the study are especially pertinent to this discussion and will be reviewed here.

During the period of dense phytoplankton blooms, the peak in the abundance of phytoplankton occurred in Moriches Bay in the region nearest the tributaries, where most of the duck farms were located. The algal populations decreased on either side of this peak in a manner that suggested dilu-

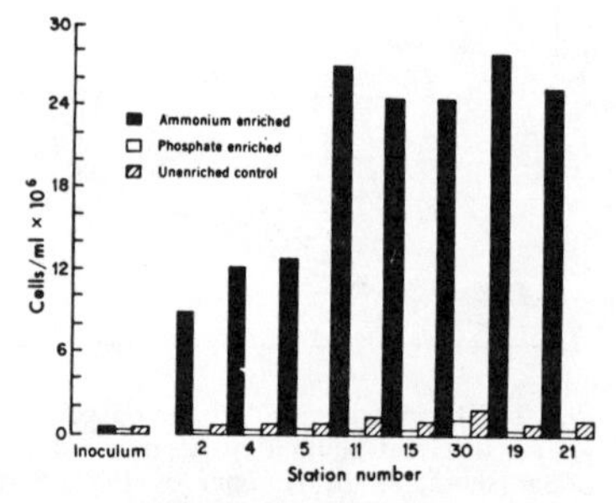

Fig. 2. Growth of *Nannochloris atomus* in unenriched, ammonium-enriched, and phosphate-enriched water collected from Great South Bay and Moriches Bay at the station locations indicated by number in Fig. 1.

tion from tidal exchange via Great South Bay to the west and Shinnecock Bay to the east (Fig. 1). Further study suggested that growth of the phytoplankton was actually confined to the tributaries themselves and that the algae in the bays represented a nongrowing population that was able to persist for long periods of time, during which they became distributed in much the same way that a conservative oceanographic property (for instance, freshwater) would behave.

Roughly coincident with the distribution of the phytoplankton was that of phosphate, which reached a maximum concentration of 7.0 μmole per liter in eastern Moriches Bay and fell to levels of about 0.25 μmole per liter at the eastern and western ends of the region (Fig. 1). Phosphate, in fact, was used throughout the study as the most convenient and diagnostic index of pollution from the duck farms.

Analyses were also made for nitrogen compounds, including nitrate, nitrite, ammonia, and uric acid (uric acid is the nitrogenous excretory product of ducks). Except in the tributaries that were in direct receipt of the effluent from the duck farms, no trace of nitrogen in any of the above forms was found throughout the region studied. It was tentatively concluded that growth of the phytoplankton was nitrogen-limited and that the algae quickly assimilated nitrogen in whatever form it left the duck farms, exhausting the element from the water well up in the tributaries before it could reach the bay.

To confirm this theory, water samples were collected from a series of stations (Nos. 2, 4, 5, 11, 15, 30, 19, and 21) in Great South Bay and Moriches Bay and in the Forge River, one of the tributaries of Moriches Bay on which several duck farms were located. These station locations are indicated by number in Fig. 1. The water samples were Millipore-filtered, and each was then separated into three 50-ml portions. The first of these served as a control while the other two received separately NH_4Cl and $Na_2HPO_4 \cdot 12H_2O$ at concentrations of 100 and 10 μmole per liter, respectively. All flasks received an inoculum of *Nannochloris atomus*, the small green alga that was the dominant species in the blooms. The cultures were then incubated for 1 week at 20°C and approximately 11,000 lu/m² of illumination, after which the cells were counted (Fig. 2).

The algae in the unenriched controls increased in number by roughly two- to fourfold, the best growth occurring in the water collected from within (station 30) or near (stations 11 and 15) the Forge River. No growth occurred in any of the samples enriched with phosphate, with the exception of that taken from the Forge River (station 30), in which the cell count increased about threefold. In fact, the addition of phosphate seemed to inhibit the growth of the algae relative to that observed in the unenriched controls. In contrast, all the water samples to which ammonia-nitrogen were added supported a heavy growth of *Nannochloris*, resulting in cell counts an order of magnitude greater than were attained in the control cultures. About twice as many cells were produced in the samples from Moriches Bay as were produced in samples from Great South Bay, which suggests that some other nutrient became limiting in the latter series. One might surmise, from the distribution of phosphate in the two bays (Fig. 1), that phosphorus was the secondary limiting factor in Great South Bay, but this possibility was not investigated. There can be little doubt, however, that nitrogen, not phosphorus, was the primary limiting factor to algal growth throughout the region.

New York bight and the eastern seaboard. In September 1969, an oceanographic cruise (R.V. *Atlantis II*, cruise 52) was undertaken along the continental shelf of the eastern United States between Cape Cod and Cape Hatteras. The primary objective of the cruise was to study the effects of pollution of various kinds from the population centers of the East Coast upon the

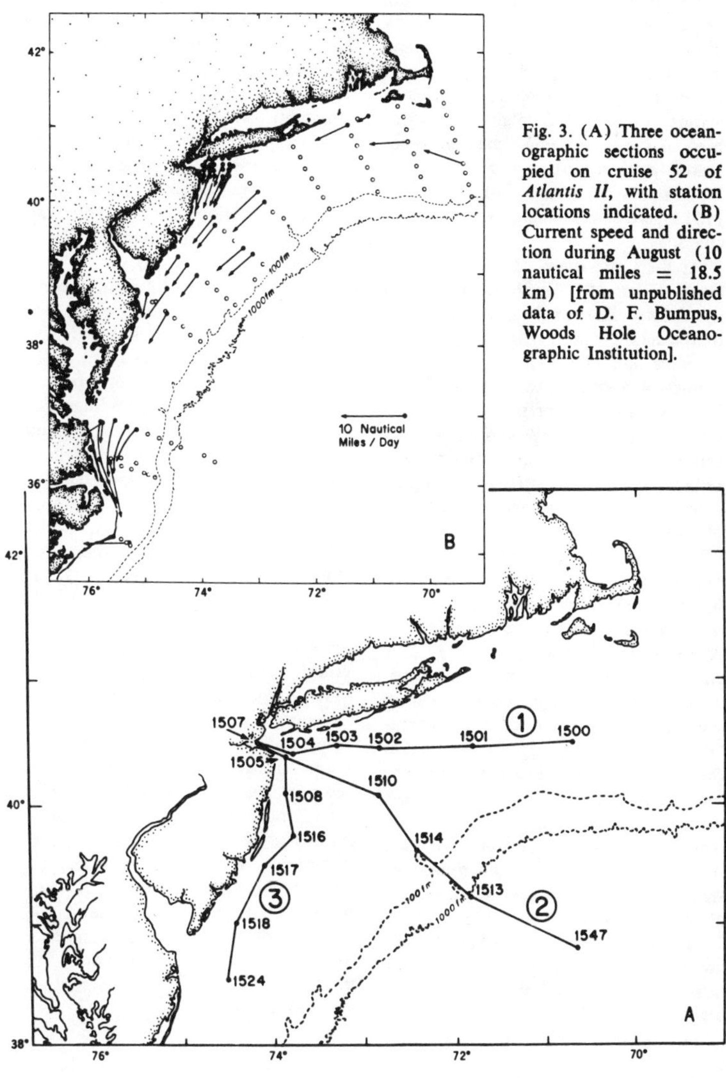

Fig. 3. (A) Three oceanographic sections occupied on cruise 52 of *Atlantis II*, with station locations indicated. (B) Current speed and direction during August (10 nautical miles = 18.5 km) [from unpublished data of D. F. Bumpus, Woods Hole Oceanographic Institution].

productivity and cycles of organic matter in the contiguous coastal waters. To obtain input data, samples were collected inside the New York bight, at the locations where sewage sludge and dredging spoils from New York City are routinely dumped, as well as from up the Raritan and Hudson rivers.

Of the 52 stations occupied during the cruise, 16 will be discussed here. These stations constitute three sections originating in New York Harbor, one extending eastward along the continental shelf south of Long Island and the New England coast, one extending southeasterly along the axis of Hudson Canyon and terminating at the edge of Gulf Stream, and one running southerly along the coast of New Jersey to the mouth of Delaware Bay (Fig. 3A). The nearshore nontidal currents of the region are predominantly to the south (Fig. 3B), so that pollution emanating from New York Harbor would be expected to spread in that direction roughly along the axis of section 3. Section 1 and the inshore stations of section 2 might be considered as typical of unpolluted or moderately polluted coastal waters, whereas the two distal stations of section 2 (1513 and 1547) should be oceanic in character.

At the inshore end of the three sections, station 1507 was located in heavily polluted Raritan Bay. The water at that location was a bright apple-green in color and contained nearly a pure culture of a small green alga identified in an earlier study of the area by McCarthy (*16*) as *Didymocystis* sp. Karshikov. Stations 1505 and 1504 are the respective dumping sites for dredging spoils and sewage sludge for the city of New York.

The distribution of total particulate organic carbon at the surface are shown for the three sections in Fig. 4. The measurements were made by the method of Menzel and Vaccaro (*17*). It is clear from the two sets of measurements that living algal cells made up only a small fraction of the particulate organic content of the water. On the basis of either criteria, one can see that the high content of particulate organic matter characteristic of the New York bight extends seaward for less than 80 km to the east and southeast, whereas evidence of pollution occurs at least 240 km to the south (section 3), along the New Jersey coast to Delaware Bay, presumably the direction of flow of the water flushed out of the bight.

The distribution of inorganic nitrogen, as combined nitrate, nitrite, and ammonia, in the surface water of the three sections is shown in Fig. 5. If one considers that the two terminal stations of section 2 (1513 and 1547) are oceanic in character, it is clear that the level of inorganic nitrogen immediately outside New York Harbor, even at the two dumping sites, are as low as, if not lower than, those found in the open sea and that such low values are, in fact, characteristic of the entire continental shelf. Phosphate (Fig. 6), however, presents a quite different picture. Its surface concentrations throughout the shelf area and particularly in the water south of the New York bight are appreciably higher than those observed at the two oceanic stations. As in the Great South Bay–Moriches Bay situation, the available nitrogen from the center of high pollution within the New York bight seems to be utilized by microorganisms as quickly as it becomes available, but there is a surplus of phosphate, which is carried seaward and is distributed throughout the continental shelf.

Experiments similar to those described above were carried out with surface water collected from the 16 stations of sections 1 to 3. This water was Millipore-filtered immediately after its collection and was stored frozen in polycarbonate bottles until used. In the laboratory, the water from each station was divided into three portions, as in the experiments described earlier; one served as a control, one received 10 μmole of sodium phosphate, and one received 100 μmole of ammonium chloride. In this instance, the mediums were inoculated with the common coastal diatom *Skeletonema costatum*, the cells of which had been washed and nutrient-starved in sterile, unenriched Sargasso Sea water for 2 days prior to their use. Growth of the cultures at 20°C and 11,000 lu/m^2 of illumination proceeded for 5 days; the results are shown in Fig. 7.

As in the earlier experiment, there was some growth in most of the unenriched controls, and this varied from station to station. Growth in the samples enriched with phosphate was no better and, in several cases, not so good as growth in the control cultures. In contrast, and again as in the bay experiments, heavy growth of *Skeletonema* occurred in most of the samples enriched with ammonia; in several cases growth was ten times or more the growth in the controls and phosphate-enriched samples.

Variability of the growth in the NH_4-enriched samples from station to station again reflects the differential development of secondary nutrient deficiencies, perhaps of phosphate but possibly of one or more other nutrients. This would be expected to occur first in the more offshore stations (see Fig. 6), and such an explanation is consistent with the experimental results. Currently unexplained, however, is the poor growth in the ammonia-enriched samples from stations 1507 and 1505 (two of the heavily polluted stations from within the New York bight), particularly at station 1505, where the best growth occurred in the unenriched and phosphate-enriched series. With the exception of the anomalous results from those two stations, the generalization can be made, consistent with the nutrient distribution picture, that also in these waters nitrogen, not phosphorus, is the primary limiting factor to algal growth.

Sources and mechanisms. To return to the question of why and how nitrogen can limit the growth of phytoplankton when the amount of phosphorus

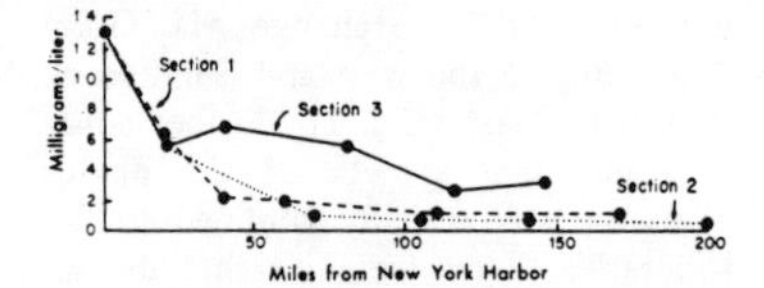

Fig. 4. The distribution of particulate organic carbon along the three sections shown in Fig. 3 (50 miles = 80 km).

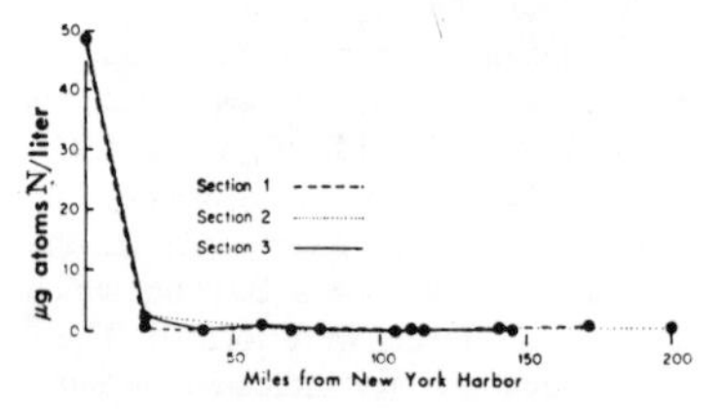

Fig. 5. The distribution of inorganic nitrogen ($NO_2 + NO_3 + NH_3$) along the three sections shown in Fig. 3 (50 miles = 80 km).

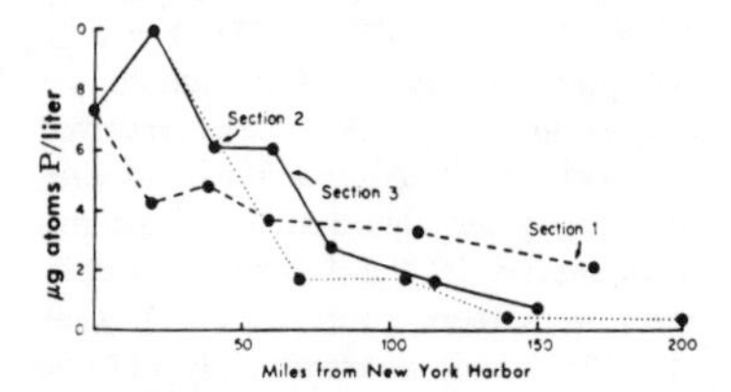

Fig. 6. The distribution of phosphate along the three sections shown in Fig. 3 (50 miles = 80 km).

relative to nitrogen in the plants is greater than it is in seawater, there are probably two explanations. One explanation applies to the ocean in general; the other, to coastal waters and estuaries specifically.

Seasonally or aperiodically, as a result of surface cooling, wind mixing, or other processes leading to vertical instability, the surface layers of the ocean are recharged with nutrients from subeuphotic depths. This mechanism, important though it is as the ultimate source of enrichment of the open sea, probably occurs infrequently. Most of the time, in the thermally stratified, nutrient-impoverished surface waters of the open ocean, organic production is maintained largely through recycling. The supply of nutrients by vertical transport from beneath the thermocline is relatively insignificant. Under these circumstances plant production is limited by the rate of regeneration of the nutrient that is mineralized most slowly. Table 1 lists the relative rates of mineralization of inorganic nitrogen compounds and of phosphate from a mixed plankton tow (*18*). The excess phosphate left in the water is also shown; the amount was calculated on the assumption that all the nitrogen is assimilated as quickly as it is formed and that phosphate is used at a ratio of one atom of phosphorus for each ten atoms of nitrogen assimilated (see above). Even if nitrogen and phosphorus were assimilated at a ratio of 5 : 1, an appreciable amount of phosphate would still be left unassimilated. This mechanism is probably responsible for the small but persistent supply of dissolved phosphate observed in surface waters throughout most of the open ocean environment.

The situation is quite different in coastal waters and estuaries. Here the surplus of phosphate may be quite large, as we have seen, and its source is unquestionably the land.

In Great South Bay and Moriches Bay it was pointed out that phosphate could be used as a tracer of the pollution originating from the duck farms located on the tributaries to Moriches Bay. Nitrogen and phosphorus are contained in duck feces in the ratio of 3.3 : 1 by atoms. Total nitrogen and phosphorus analyses of dissolved and suspended matter in the tributaries and in Moriches Bay itself gave nitrogen to phosphorus ratios of 2.3 : 1 to 4.4 : 1, consistent with the presumed origin of this material. About half of the total phosphorus was present as dissolved, inorganic phosphate, with the remainder being tied up in the algae and other particulate matter. All the nitrogen occurred in the latter form. As mentioned earlier, no inorganic nitrogen in any form and no uric acid could be detected anywhere in the water (*15*).

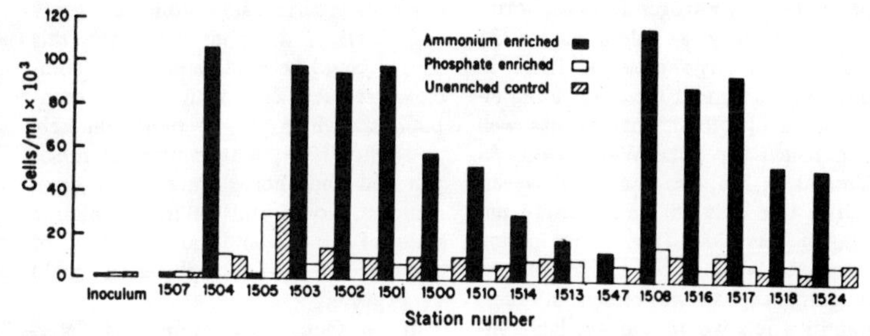

Fig. 7. Growth of *Skeletonema costatum* in unenriched, ammonium-enriched, and phosphate-enriched water from the New York bight collected from the stations shown in Fig. 3.

According to the above data, the ratio of nitrogen to phosphorus in the plankton would be about 6.6 : 1. The population of algae in the area consisted of an almost pure culture of two species of green algae, which were identified at the time as *Nannochloris atomus* and *Stichococcus* sp. The green algae (Chlorophyceae) are characterized by a low nitrogen to phosphorus ratio (*5, 6*). This fact was believed to be partly responsible for their presence in the bay waters, though other factors such as low salinity and high temperature were also shown to be important selective mechanisms (*15*).

In the New York bight and the contiguous coastal waters, a high level of phosphate was again measured (Fig. 6). From its distributional pattern there can be little doubt that this material originated in New York Harbor and its tributaries. In Raritan Bay, as mentioned, the phytoplankton consisted almost exclusively of a small green alga, owing presumably to a combination of ecological conditions similar to those that obtained in Great South Bay and Moriches Bay. In the New York bight and the waters farther offshore, conditions more typical of the marine environment prevailed, and the plankton flora consisted of a mixture of diatoms, flagellates, and other forms (*19*). What is the origin of the surplus phosphate in this case?

From data provided by Pearson *et al.* (*20*) one can calculate that the nitrogen to phosphorus ratio by atoms in domestic wastes that have been subjected to primary sewage treatment is 5.8 : 1. In wastes that have undergone secondary treatment, the ratio, according to Weinberger *et al.* (*21*), is 5.4 : 1. On the assumption that the ~4 billion kl per day of domestic wastes entering the New York bight from the New York–New Jersey megalopolis have been subjected to something intermediate between primary and secondary treatment, some 90 metric tons of nitrogen and 36 metric tons of phosphorus are discharged into these waters each day. If the phytoplankton that inhabit the area assimilate nitrogen and phosphorus in the ratio of 10 : 1 by atoms (4.5 : 1 by weight), nearly half the phosphate entering the system is in excess of the amount that can be used by the plants.

Eutrophication. As we have seen, phosphate is a convenient index or tracer of organic pollution. Its analysis by conventional colorometric techniques is quick, accurate, and highly sensitive and is far easier than analysis of other chemical nutrients. Furthermore, it persists when other products of organic decomposition, such as nitrogenous compounds, have disappeared from solution. Thus, domestic wastes can be tracked longer and farther from their source of input by looking at the distribution and concentration of phosphate than by using almost any other criteria. From this fact, it is a short and easy step to the conclusion that phosphate is the causative agent of algal growth, eutrophication, and the other adverse effects associated with organic pollution. In the sea, such is far from true.

There is the possibility, alluded to briefly above, that blue-green algae, and possibly other microorganisms capable of fixing atmospheric nitrogen, may by this process bring enough nitrogen into the biological cycle to balance the surplus of phosphate. Filamentous blue-green algae are common in freshwater lakes, and their ability to fix nitrogen is well demonstrated (*22*). For this reason, or simply because of a high natural ratio of nitrogen to phosphorus, there is probably, as Edmondson (*23*) suggests, "A large class of lakes in which phosphorus is the dominating element," a hypothesis that he has well documented for Lake Washington. As Edmondson has also shown, however, such is true only in the relatively unpolluted condition. During the period when Lake Washington received sewage effluent, phosphate was present in excess quantities relative to the available nitrogen.

In the open tropical ocean, there are also filamentous blue-green algae, of the genus *Trichodesmium*, that are capable of fixing nitrogen, though the process is so slow and inefficient as to be almost undetectable (*24*). In the more eutrophic coastal waters and estuaries, such algae are almost unknown, and nitrogen fixation has not been demonstrated. Here, as we have shown, it is unquestionably nitrogen that limits and controls algal growth and eutrophication.

Much of the phosphate in domestic waste has its origin in detergents. The fraction of the land-derived phosphate in our coastal waters that can be attributed to this source is difficult to assess but has been estimated to be 25 to 50 percent of the total (*25*). The total land-derived phosphate also includes human excreta, agricultural runoff, industrial wastes, and other material, all of which vary greatly from place to place. As shown earlier, the nitrogen to phosphorus ratio in domestic waste is slightly higher than 5 : 1 by atoms. Even if as much as half of the phosphate in sewage came from detergents and if all of the phosphate from this source could be eliminated by its complete replacement with other compounds, which is a most unlikely possibility (*26*), the amounts of nitrogen and phosphorus entering the environment would still be in the atomic ratio of 10 : 1, and no reduction of algal growth or eutrophication could be expected.

If, in fact, the phosphate in detergents is replaced with nitrilotriacetic acid (NTA), as is the current trend in the industry (*26*), the net effect could be an acceleration and enhancement of the eutrophication process. In sewage treatment (and presumably in nature, if more slowly), NTA undergoes biodegradation and probably yields glycine and glycolic acid as intermediate decomposition products (*27*). These compounds may be used directly as a nitrogen source by at least some species of unicellular algae (*15*), or they may be deaminated to ammonia, which is universally available to phytoplankton.

Coastal waters already receive the sewage of roughly half the population of the United States. To replace a portion of the phosphate in this sewage with a nitrogenous compound and to then discharge it into an environment in which eutrophication is nitrogen-limited may be simply adding fuel to the fire.

References and Notes

1. H. W. Harvey, *J. Mar. Biol. Ass. U.K.* **14**, 71 (1926).
2. A. C. Redfield, *James Johnston Memorial Volume* (Univ. of Liverpool Press, Liverpool, 1934), p. 176.
3. ———, *Amer. Sci.* **46**, 205 (1958).
4. H. U. Sverdrup, M. W. Johnson, R. H. Fleming, *The Oceans* (Prentice-Hall, New York, 1942), p. 236; L. H. N. Cooper, *J. Mar. Biol. Ass. U.K.* **22**, 177 (1937).
5. B. H. Ketchum and A. C. Redfield, *J. Cell. Comp. Physiol.* **33**, 281 (1949).
6. J. D. H. Strickland, *Bull. Fish. Res. Bd. Can. No. 122* (1960).
7. T. R. Parsons, K. Stephens, J. D. H. Strickland, *J. Fish. Res. Bd. Can.* **18**, 1001 (1961); C. D. McAllister, T. R. Parsons, K. Stephens, J. D. H. Strickland, *Limnol. Oceanogr.* **6**, 237 (1961).
8. D. W. Menzel and J. H. Ryther, *Limnol. Oceanogr.* **9**, 179 (1964).
9. R. F. Vaccaro, *J. Mar. Res.* **21**, 284 (1963).
10. E. Harris and G. A. Riley, *Bull. Bingham Oceanogr. Collect. Yale Univ.* **15**, 315 (1956).
11. U. Stafánsson and F. A. Richards, *Limnol. Oceanogr.* **8**, 394 (1963).
12. B. H. Ketchum, R. F. Vaccaro, N. Corwin, *J. Mar. Res.* **17**, 282 (1958).
13. W. H. Thomas, *Limnol. Oceanogr.* **11**, 393 (1966).
14. ——— and A. N. Dodson, *Biol. Bull.* **134**, 199 (1968); W. H. Thomas, *J. Fish. Res. Bd. Can.* **26**, 1133 (1969).
15. J. H. Ryther, *Biol. Bull.* **106**, 198 (1954).
16. A. J. McCarthy, thesis, Fordham University (1965).
17. D. W. Menzel and R. F. Vaccaro, *Limnol. Oceanogr.* **9**, 138 (1964).
18. R. F. Vaccaro, in *Chemical Oceanography*, J. P. Riley and G. Skirrow, Eds. (Academic Press, New York, 1965), vol. 1, p. 356.
19. E. M. Hulbert, unpublished data.
20. E. A. Pearson, P. N. Storrs, R. E. Selleck, *Ser. Rep. 67-3* (Sanitary Engineering Research Laboratory, Univ. of California, Berkeley, 1969).
21. L. W. Weinberger, D. G. Stephan, F. M. Middleton, *Ann. N.Y. Acad. Sci.* **136**, 131 (1966).
22. R. C. Dugdale, V. A. Dugdale, J. C. Neess, J. J. Goering, *Science* **130**, 859 (1959); D. L. Howard, J. I. Frea, R. M. Pfister, P. R. Dugan, *ibid.* **169**, 61 (1970).
23. W. T. Edmondson, *ibid.* **169**, 690 (1970).
24. R. C. Dugdale, D. W. Menzel, J. H. Ryther, *Deep-Sea Res.* **7**, 298 (1961).
25. F. A. Ferguson, *Environ. Sci. Technol.* **2**, 188 (1968).
26. *Chem. Eng. News* **1970**, 18 (17 August 1970).
27. R. D. Swisher, M. M. Crutchfield, D. W. Caldwell, *Environ. Sci. Technol.* **1**, 820 (1967).

28. Supported by the Atomic Energy Commission, contract AT(30-1)-3862, ref. NYO-3862-40, and by NSF grant GB 15103. Contribution 2537 from the Woods Hole Oceanographic Institution.

31 August 1970; revised 3 December 1970

THE EFFECTS OF THERMAL EFFLUENTS ON MARINE ORGANISMS*† 29

J. E. WARINNER and M. L. BREHMER

Virginia Institute of Marine Science Gloucester Point, Virginia, U.S.A.

1. INTRODUCTION

URBANIZATION and industrialization along with the rapid increase in population since World War II have placed a great demand on our waterways. The water available must serve several purposes including domestic water supplies, water for industrial consumption, sewage disposal, irrigation, commercial fishing, transportation, and recreation. A number of industries utilizing water must locate on rivers and streams and along the estuaries. Those which use water for cooling purposes do not consume appreciable amounts of water nor contaminate the water supply but they do return the water to the stream at a higher temperature. The largest use of cooling water is by the rapidly expanding steam electric generating industry. The thermal efficiency of steam turbines which turn the generators is increased by condensing the steam on the exhaust side of the turbines, thus creating a partial vacuum. The temperature of the cooling water is raised 6° to 9.5° C generally but may be raised under some circumstances to as much as 24.0° C.

One of the earliest investigations into the effects of thermal discharges upon aquatic life was reported by VAN VLIET (1957). MOORE (1958) used the term "thermal pollution" in 1958 "Pollution" of rivers was discussed by HERRY (1959); MARKOWSKI (1959) published the first of a series of papers on the thermal discharges of a number of power stations in Great Britain. Since that time research has been continued at an accelerated pace by ecologists, government, and industry in order to evaluate the effects of thermal elevation on aquatic life. Limitations on maximum temperatures and maximum allowable areas of thermal elevation have been established by several states in the interest of pollution abatement.

Most of the work has been done on fresh-water streams because of the possibility of low flow conditions and the subsequent use of water for domestic purposes. The volume of water in the typical coastal plain estuary and the presence of tidal currents favor the utilization of these waters for the discharge of thermal effluents. However, there is a fundamental difference between inland streams and tidal estuaries. The introduction of a biocide, whether chemical or physical, into an inland stream may destroy aquatic organisms in a given section of the stream, but the stream may be classified as "healthy" upstream from the point of discharge and below the zone of recovery. A similar introduction into a restricted estuary could easily affect populations in the entire estuary and contiguous waters. This is because many of the animals have retained a link somewhere in their life histories between marine and fresh water.

* Contribution No. 171, Virginia Institute of Marine Science, Gloucester Point, Virginia 23062, U.S.A.

† This paper was presented at the 19th Annual Industrial Waste Conference, Purdue University, Lafayette, Indiana, U.S.A., 5th May, 1964, and is reproduced by permission of the authors and Professor D. E. Bloodgood.

Reprinted from *Air and Water Pollution International Journal*, 10:277–289, 1966.

The Atlantic and Pacific salmon are classical examples of fish that require a healthy aquatic environment from the open sea all the way to fresh headwaters in order to reproduce and survive as a species. The striped bass, shad, herring and smelt, to name a few of the better known species, have similar life histories. Less generally known are those species which spend part of their lives in the estuaries but migrate to higher salinities to spawn. The young of these species return to the estuaries in the post-larval stage and traverse the entire length of the estuary to reach nursery grounds just below the fresh-salt water transition zone. Examples are the Atlantic croaker, spot, gray sea trout, and the commercially important menhaden. The immature stages of the blue crab also migrate the length of the estuary before reaching adult form. Other invertebrates such as the oysters and clams produce planktonic larvae which may travel a fifty mile oscillatory course before settling down as sedentary adults.

An estuary must be considered a *biological zone* and the establishment of sub-minimal conditions across any point in the estuary may affect valuable aquatic forms in the entire estuary.

A study was undertaken by the Virginia Institute of Marine Science in 1963 to determine the effects of thermal effluents on a community of marine organisms. Both laboratory and field observations were made. The studies reported here are divided into two parts, primary productivity and the effects on benthic invertebrates.

2. LOCATION OF STUDY

The field investigations were made at the Virginia Electric and Power Company's steam electric generating station at Yorktown, Virginia. The cooling water is taken from the York River via a canal. The entrance to the canal is protected by jetties which extend into the river approximately 360 m. The intake facility supplies water both to the power plant and to the American Oil Company refinery on adjacent property downstream from the power plant. The outfall canal from the power plant empties into the river at shore line some 450 m east of the intake. The flow of water through the single pass condenser is 220,000 gal/min at peak loading with an average temperature rise of 8° C. The depth of the water at the intake is 3 m at mean low water. The outfall canal continues offshore in a trench through the inshore shallows. The subtidal zone has a very gradual slope and is not more than 1 m deep at mean low water 300 m from shore. It is over these shallows that the effluent spreads after leaving the outfall. The direction and extent of the dispersion pattern are determined by tidal currents and wind.

The distribution of thermal effluents is depicted in FIG. 1 at surface and bottom during maximum flood current and high slack water. FIGURE 2 shows the distribution during maximum ebb current and at low slack water.

The York River is wide and deep in this reach and there is no possibility of a thermal barrier existing. The figures above also show that the thermal layer rises to the surface in a fairly well defined layer. During flood current a discrete stream of warm water is directed into the river by the intake jetty. There is evidence of turbulence in the effluent zone as shown by the patchiness in isotherms during slack water.

The study was made from June 1963 through May 1964. The maximum and minimum weekly temperatures during this period, taken from a recording thermograph at the Institute of Marine Science across the York River from the power

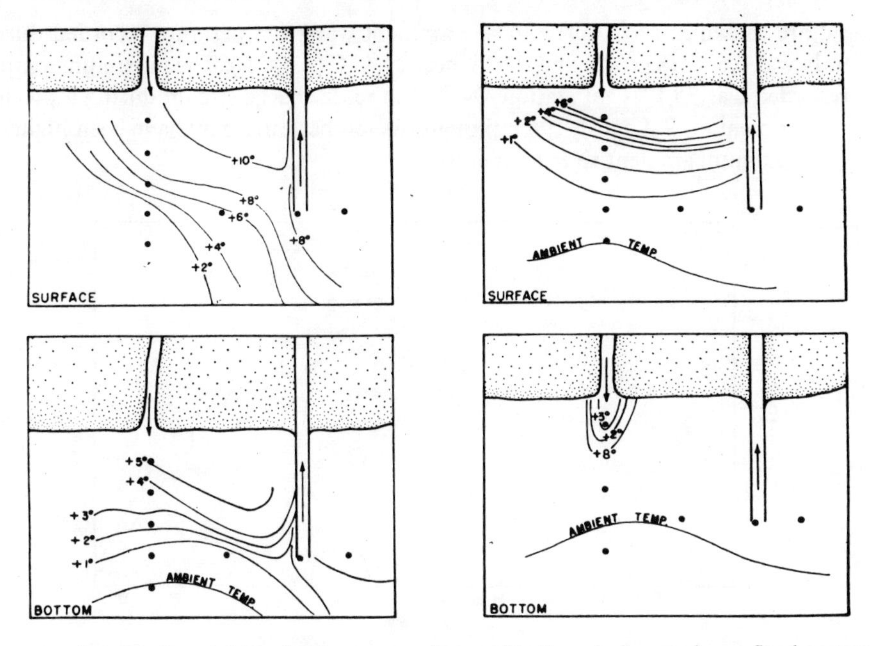

FIG. 1. Distribution of thermal effluents at surface and bottom during maximum flood current (left) and high slack water (right).

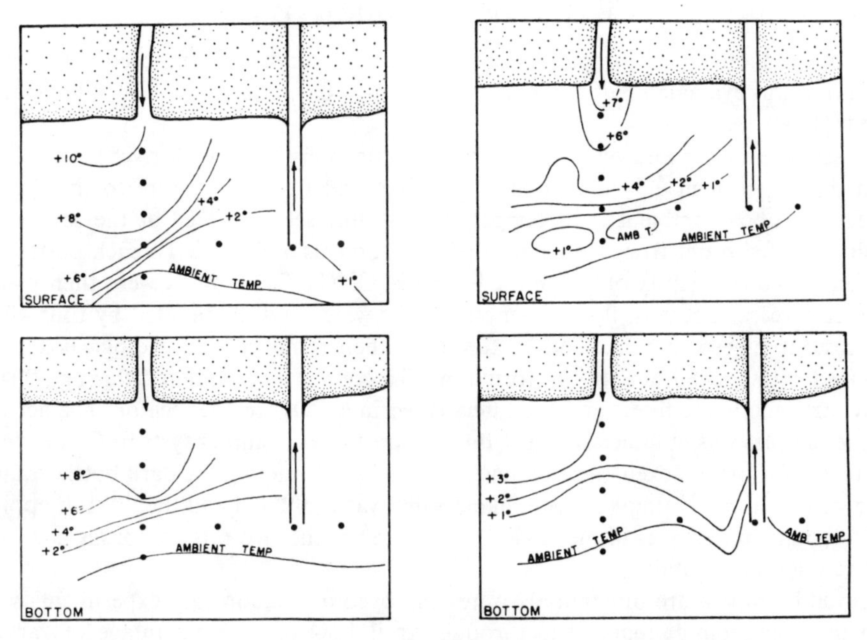

FIG. 2. Distribution of thermal effluent at surface and bottom during maximum ebb current (left) and low slack water (right).

station, are shown in Fig. 3. The lowest ambient water temperature recorded during the study was 0.6° C (33.0° F) and the highest 29.4° C (85.0° F). The maximum temperture recorded was 35.1 ° C at station D-100 in the discharge stream during a productivity experiment on 9 August, 1963, though the temperature may have been higher at times of maximum ambient river temperatures.

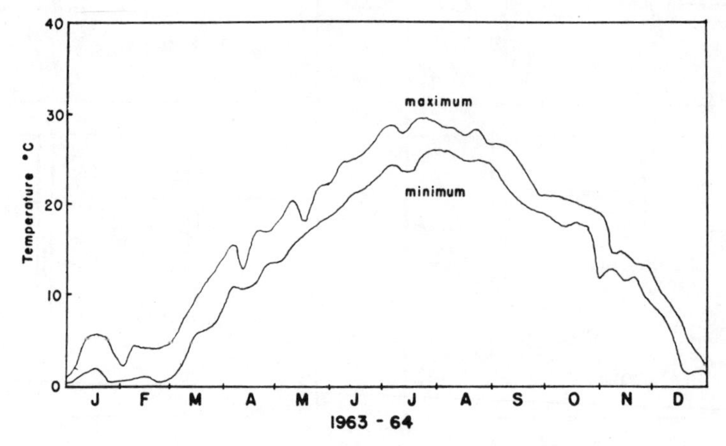

Fig. 3. Weekly maximum and minimum water temperatures recorded at the Virginia Institute of Marine Science, Gloucester Point, Virginia.

3. PRIMARY PRODUCTIVITY

The productivity of phytoplankton under the conditions imposed by thermal elevation was evaluated by the use of the carbon-14 technique as described by Strictland (1960).

A set of ten clear and two opaque BOD bottles were filled with river water which had passed through a heat exchanger which raised the temperature to the desired increment above ambient river temperature. A like set was filled at the same time with unheated water from the same supply to serve as a control. To each bottle was added the same quantity of C^{14} in the form $Na_2C^{14}O_3$. The bottles were then placed in a culture cabinet with flowing ambient river water and illuminated by four 40 W fluorescent tubes for 5 hr. The bottles were then removed and biological activity stopped with a small quantity of formalin. The contents were filtered through 0.80 μ pore size membrane filters and the filters dried in a CO_2 free desiccator. The activity of the samples was obtained by a gas flow proportional counting system. Counts were corrected for carbon exchange and background by subtracting the dark bottle counts. The variation in C^{14} uptake within like bottles was evaluated and the "t" test applied to determine differences in mean carbon-uptake for the two sets of bottles at the 95 per cent level of confidence.

Several temperature differentials were employed and laboratory experiments were run monthly from January 1963 through April 1964 to take advantage of varying ambient temperatures and seasonal plankton populations. Assuming the same number of degrees of freedom, the magnitude of the difference in carbon assimilation should be reflected by the size of the "t" value. The dependent variable "t" is plotted against the temperature difference, ΔT, for several ranges of ambient water temperature. This, incidentally, represents different seasons and plankton populations. When ambient water temperatures fell between 0° and 5° C a significant increase in carbon assimilation occurred at a temperature rise, ΔT, of 8° C. Temperature rise in this section refers to the increase in temperature as the river water passed through the laboratory heat exchanger.

At ambient water temperatures of 5°–10° C significant increases in production were found whenever the thermal elevation was over 2.5° C to as much as 14° C which was the largest temperature elevation used. When the temperature of the river water was between 10° and 15° C there was an interesting reversal in results.

During November, a temperature rise of 3° C stimulated production but production was depressed at a rise of 5.6° C and 14° C. The actual temperatures of the heated water were 17.8° C and 20.6° C respectively when production was depressed. When ambient river temperatures were between 15° and 20° C all temperature elevations of 5.6° C or more resulted in significant depression of production. The actual temperatures of the heated water were above 23.5° C. During July and August, the period of maximum water temperatures, temperature elevation of 3.5° C and above depressed production. The actual temperature of the heated water was 30.5° C.

Field studies using the dark and light bottle technique of GAARDER and GRAN (1927) were attempted but the results were not considered trustworthy over an extended period of culture such as 24 hr. Apparently during the periods of rapid production the water became supersaturated and bubbles of O_2 were evolved. Since the method depends upon the measure of dissolved oxygen change, the results were not entirely accurate. Several experiments were run in June 1963 with conflicting results, but generally showed a substantial net increase over 24 hr in the heated water. During August 1963 a 24 hr and a 4 hr culture were made both of which showed a net decrease in dissolved oxygen in the outfall water.

4. BENTHOS

One of the best methods of evaluating environmental conditions is to observe the kinds and numbers of organisms actually living in the environment. Consequently a year-round survey from June 1963 through May 1964 was made at selected stations representing areas influenced by the heated discharge water and areas free of the influence. The stations selected were on two transects. The first transect ran directly off the discharge canal and contained five stations at 100 m intervals, designated D-100, D-200, etc., the numerals indicating the distance from shore. The second transect consisted of a line of stations parallel to the shoreline and 400 m from shore. The stations were located as follows:

(A) 100 m upriver from the intake, (B) at the intake cribbing, and (C) halfway between the intake and the discharge transect.

At monthly intervals, two bottom grabs were taken at each station with a 1/20 m^2 Petersen grab. The composite samples from each station were sieved through screens of 1 mm mesh. The invertebrate specimens were preserved in formalin for counting and identification.

Sampling was limited to two grabs at each station because of the time required to take samples and to sort and identify the organisms, but an evaluation of the adequacy of the samples was made by taking several grabs at a single station on one occasion and analyzing the results. A trellis diagram such as that used by KULCZYNSKI (1927), and many others to compare fauna at different localities is used here to show the correlation coefficients for all possible combinations of the five samples taken at a single location. (TABLE 1).

TABLE 1. CORRELATION COEFFICIENTS

	1	2	3	4	5
1	100				
2	70	100			
3	71	84	100		
4	74	70	69	100	
5	69	82	82	65	100

The correlations were obtained by listing all the species obtained in the five samples. In each column representing a separate sample the abundance of a particular organism was recorded as percentage of the total number of all individuals in that sample. To compare any two samples as to species affinity, a sum was made of the lesser percentages for each species common to both samples. In this correlation method the relative abundance of a species in two samples and the presence or absence is reflected in the final percentage value derived.

The lowest correlation between single grab samples by this method was 65 per cent and the highest correlation nearly 84 per cent. The species composition of the two samples with the lowest correlation were combined to make a composite sample and this list compared with two samples having the next lowest correlation. That is, samples 3 and 4 were combined and compared with combined samples 1 and 5. The resulting correlation coefficient was 81.6 per cent. The same comparison was made with the two highest correlations with resulting coefficient of 82.9 per cent. It was felt that correlations of this magnitude and consistency were sufficient to limit sampling to two grabs per station.

From the standpoint of representing the total number of species at a station, two grabs are certainly not sufficient. Two grabs represent between 70 and 90 per cent of the total number of species taken in the five grabs at a single station. Undoubtedly additional sampling would have turned up additional species. However, these species are represented by only a few individuals each. A sparsely distributed organism such as the hard clam may contribute greatly to the biomass of a sample but little numerically.

A number of indices have been developed by ecologists to express the relationship of numbers, species, and distributions of organisms or, in other words, to quantify a community of organisms. By these methods, areas, samples, or time intervals can be compared with respect to faunal (or floral) composition. The indices are so designed that each parameter carries a certain weight and obviously one index will not serve under all conditions. Accordingly two indices were selected to express the community relationships at each of the stations selected for comparison. One of these indices was developed from information theory by MARGALEF (1956) to describe plankton communities. The redundancy index of Margalef is a measure of the position of the sample between two extremes, that of maximum diversity of species and that of minimum diversity. The former is exemplified by a case in which all the species are represented by an equal number of individuals, while minimum diversity would be a community in which one species is dominant over all the others which are represented by only one individual each. Minimum diversity is expressed as maximum redundancy (R = 1) and maximum diversity is expressed as a low redundancy (R = 0).

The redundancy index values for all stations on the transect off the discharge canal, for samples taken in February and August 1964 are shown in FIG. 4. Two monthly

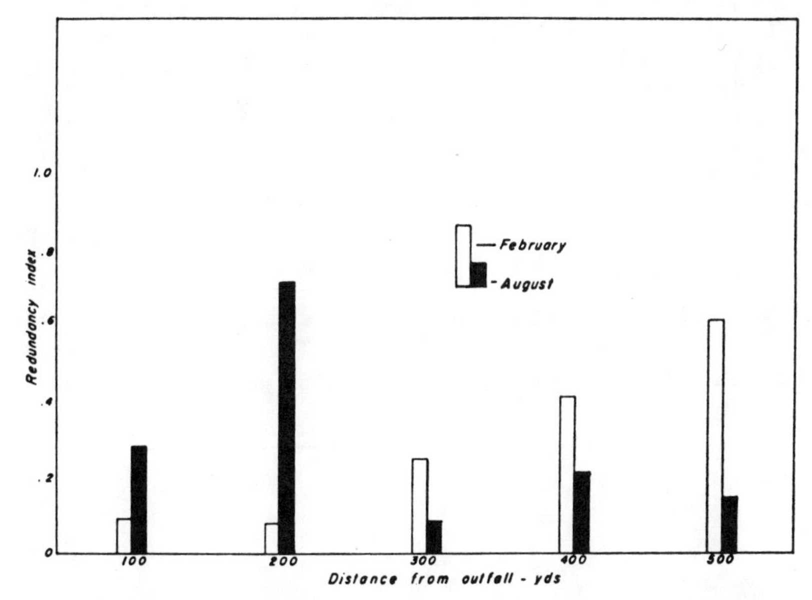

FIG. 4. Redundancy index for benthic fauna at stations opposite the discharge canal.

samples are shown here for comparison. Notice that during the winter months when the water is cold that lowest redundancy (maximum diversity) occurs at the inshore stations, that is, where the influence of warm water would be the greatest. In the summer months maximum redundancy is a characteristic of the inshore stations. Temperatures in the 32–38° C range are prevalent at stations D-100 and D-200 in the summer months.

Redundancy indices for the long-shore stations 400 m from shore are represented in FIG. 5. During winter minimum redundancy of species is found at station A-400—away from the thermal effluent. Redundancy is near the middle of the scale at the other stations. During summer there is a low redundancy at all stations 400 m from shore and only a slight increase in redundancy opposite the outfall.

Another index applied to these same samples weights the number of species somewhat more heavily. This index of diversity was discussed by Jones (1961).

$\frac{S-1}{\ln N}$ where S = number of species taken, $\ln N$ = natural logarithm of the number of individuals in the sample.

The histogram in FIG. 6 shows the indices of dispersion at the stations off the discharge for the same two months, February and August. Reflected in the bars is the fact that species become more numerous away from the discharge point.

During the warm months when river temperatures are normally high, the numbers of individuals and the number of species both declined. For example, the number of

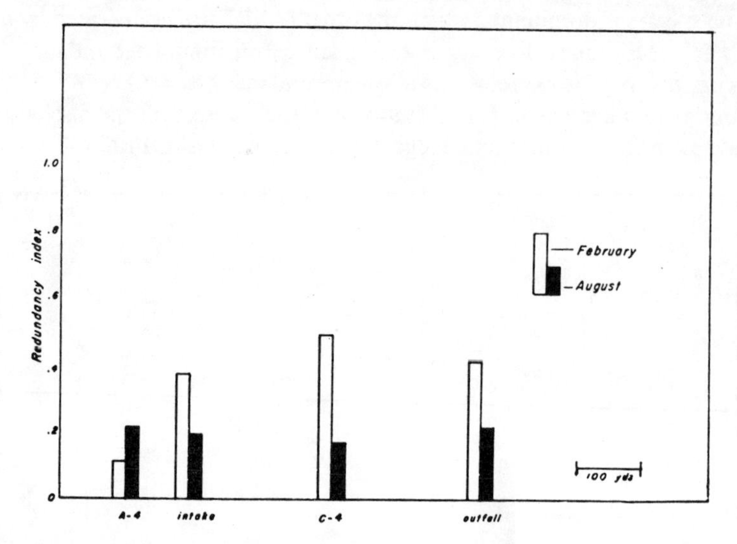

FIG. 5. Redundancy index for benthic fauna at longshore stations, 400 m offshore.

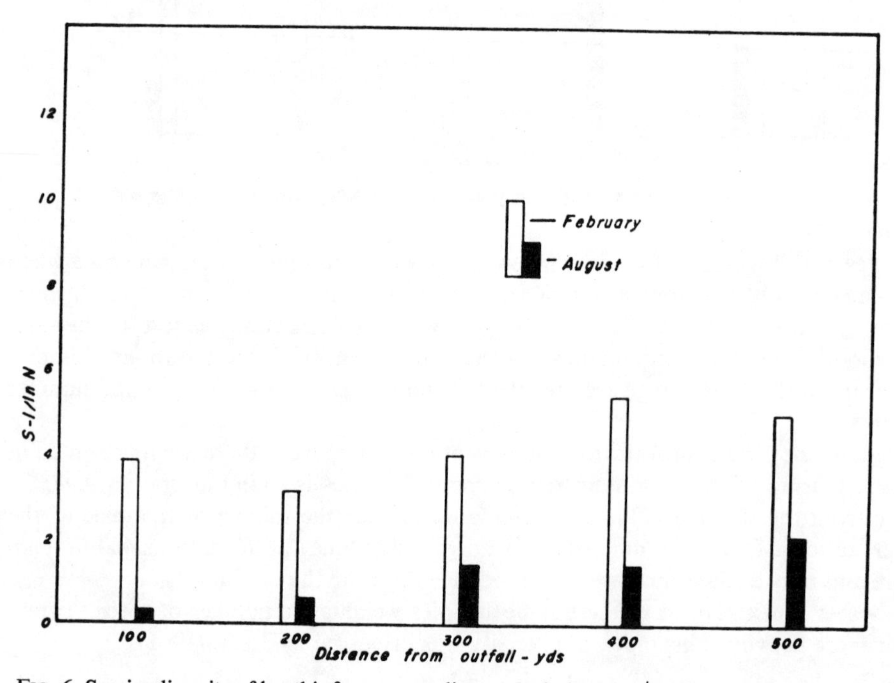

FIG. 6. Species diversity of benthic fauna according to the index $S-1/\ln N$ at stations opposite the discharge canal.

individuals taken at Station A-400 in August 1963 was about one fifth the number taken in February 1964. The number of *species* taken in August was likewise about one fifth the number taken in February. Since the number of species is weighted more heavily than the number of individuals in this index, the August indices are all depressed. While the reduction is evident at all stations it is more pronounced at the stations affected by the thermal discharge.

The species taken at stations D-200, D-400 and A-400 are shown in TABLE 2, for

TABLE 2. SPECIES TAKEN DURING AUGUST AND FEBRUARY IN 1/10 m² AT STATIONS D-100, D-400, AND A-400

		February			August		
		D-100	D-400	A-400	D-100	D-400	A-400
Coelenterata	*Diadumene leucolena*			2			
	Edwardsia leidyi		24	9			
Nemertea	*Amphiporus ochraceus*	1					
	Micrura leidyi		1				
	Tubulanus pellucidus		1				
Phoronida	*Phoronis architecta*	1	9				
Annelida	*Capitella capitata*			2			
	Cirratulidae unid.			1			
	Eteone lactea		4	1			
	Eumida sanguinea		1	2			
	Glycera americana		14	4		9	1
	Glycinde solitaria	1	11	1			
	Heteromastus filiformis	6	7		2		
	Nephtys picta						1
	Nereis succinea	3	3	12	32		
	Oligochaeta unid.		2	2			
	Paraprionospio pinnata	2	52	1			
	Pectinaria gouldi		6	5			
	Phyllodoce arenae	1	4				
	Pista palmata			1			
	Platynereis dumerilii			8			
	Podarke obscura			1			
	Polydora ligni	1	63	1			3
	Prionospio heterobranchia		1				
	Sabella microphthalma			13			2
	Scolelepis bousfieldi	1					
	Scoloplos robustus	1	3	1		1	
	Spiochaetopterus oculatus		13	1			
	Sthenelais boa			1			
	Tharyx setigera		1				
Mollusca	*Amygdalum papyria*			1			
	Anadara transversa			2			
	Bittium varium			9			
	Coryphella rufibranchialis		1				
	Crepidula convexa		2	88			20
	Epitonium rupicola		2	1			
	Haminoea solitaria		55				
	Lucina multilineata		2				
	Lyonsia hyalina		1	1			

(*continued overleaf*)

TABLE 2—*Continued*

		February			August		
		D-100	D-400	A-400	D-100	D-400	A-400
	Macoma phenax			4		2	
	Mangelia plicosa			1			
	Mitrella lunata			1			
	Mulinia lateralis		18	3		3	3
	Mya arenaria	2	1	4			
	Nassarius vibex	3	6	5			5
	Odostomia bisuturalis	1	2	8			
	O. impressa			5			
	Retusa canaliculata	3	345	23			
	Turbonilla interrupta		14				
Arthropoda	*Ampelisca macrocephala*		5	12		3	
	A. vadorum		22	25			
	Caprella geometrica			3			
	Carinogammarus mucronatus			23			
	Cyathura polita			1			
	Cymadusa filosa			23			
	Edotea triloba		11	13			
	Elasmopus pocillimanus		1	27			
	Erichsonella attenuata		2	21			
	Ericthonius brasiliensis						1
	Eurypanopeus depressus			1			5
	Idotea baltica			1			18
	Leptocheirus plumulosus	1					
	Libinia dubia			1			
	Listriella clymenellae		1				
	Monoculodes edwardsi		6	1			
	Neomysis americana			1			
	Oxyurostylis smithi	4	18	14			
	Sarsiella zostericola		1				
	Sphaeroma quadridentatum	1					
	Unid. amphipod			1			
Echinodermata	*Leptosynapta inhaerens*		1	1		1	
Chordata	*Molgula manhattensis*			2			4
	No. species	17	41	52	2	6	11
	No. individuals	33	737	334	34	19	63

comparison during the months of low and high water temperature. The only two species taken during August at the inshore stations were two polychaete worms, *Nereis succinea* and *Heteromastus filiformis*. The list of invertebrates taken in samples from the unaffected station A-400 is more diversified, having polychaete worms, pelecypods, gastropods, decapods, isopods, amphipods, and chordates. One reason for this is that eel grass grows at station A-400, but not in the discharge area.

During February station D-400 off the discharge canal had a few less species than A-400 but had a larger population even when the aggregation of the tectibranch, *Retusa canaliculata*, is omitted. Of course, the eel grass had died back and the community had changed at A-400 during the winter.

The abundance of species at each station during an annual cycle is shown in FIG. 7. The summer low is seen at every station but is more pronounced at stations D-100, D-200, and D-300.

A particle size analysis was done on the sediments from all stations to see if differences in bottom type might account for differences in populations. The results are shown in TABLE 3. Silt and clay fractions were determined by the pipette method.

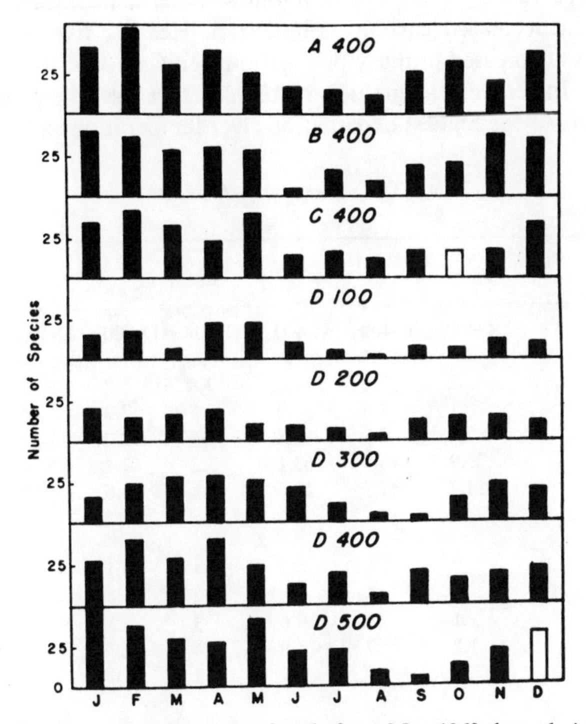

FIG. 7. Species abundance during an annual cycle from May 1963 through April 1964. Open bars are estimated values.

SANDERS (1958) suggests that the distribution of deposit feeders is correlated with the clay fraction of the sediments and the distribution of filter feeders may be correlated with either the silt-clay fraction or with the current conditions which determine the sediment type. In either case the sediment types at all the stations studied here do not differ appreciably from one another, thus it is unlikely that the differences in numbers or species found at the stations near the discharge (stations D-100, D-200, D-300) are related to differences in particle size distribution. However, in October and November a mat of detritus was found at stations D-100 and D-200 which was the result of the deposition of floating bits of eel grass and other vegetation which normally dies back in the fall of the year. Bottom samples at this time were black and smelled of hydrogen sulfide. This undoubtedly limited the numbers and species during this period of decomposition. The general depression of species year round is associated only with those stations within 200–300 m of the discharge.

5. SUMMARY AND CONCLUSIONS

The assimilation of carbon by the natural phytoplankton populations of the York River is affected by an artificial increase in water temperature. During the winter months temperature changes, which are normally experienced during passage through a power plant condenser, enhance primary production. When river water reached a temperature of 10 to 15° C, a 3 degree rise in temperature through the condenser resulted in an increased carbon assimilation, but a higher temperature rise (more than 5.5° C) resulted in decreased carbon assimilation. Heating river water which was above 15° C always depressed primary production significantly if the temperature rise was above 5.5° C. In general, the greater the temperature rise, the greater the depression of production. At the highest experienced river temperatures a 3.5° C increase in

TABLE 3. PARTICLE SIZE ANALYSIS

Size	Stations							
	A-400	B-400	C-400	D-100	D-200	D-300	D-400	D-500
Sand				(%)				
1000μ	.1	.2	.1	7.8	1.2	.3	.2	.2
500μ–1000μ	.1	2.6	.3	29.5	2.9	2.4	.5	.2
250μ–500μ	2.4	8.5	7.8	49.4	12.0	14.5	10.4	2.5
125μ–250μ	79.9	46.8	63.1	5.2	47.4	53.6	58.9	73.6
62.5μ–125μ	14.7	34.1	22.6	5.5	32.6	24.7	24.7	18.0
Silt								
32μ–62.5μ	.2	1.1	.7	.4	.4	.5	.4	.4
16μ–32μ	.1	.5	.4	.2	.2	.3	.4	.4
8μ–16μ	.1	.8	.6	.3	.3	.4	.5	.5
4μ–8μ	.2	.9	.8	.4	.3	.6	.7	.1
2μ–4μ	.3	.9	.7	.3	.4	.5	.6	1.4
Clay								
1μ–2μ	.3	.8	.6	.2	.4	.5	.6	.6
.5μ–1μ	.4	.6	.8	.3	.6	.7	.8	.8
.5μ	1.2	2.2	1.4	.5	1.3	1.0	1.3	1.3
Sand	97.2	92.2	93.9	97.4	96.1	95.5	94.7	94.5
Silt	.9	4.2	3.2	1.6	1.6	2.3	2.6	2.8
Clay	1.9	3.6	2.8	1.0	2.3	2.2	2.7	2.7
Silt and clay	2.8	7.9	6.0	2.6	3.9	4.5	5.3	5.5

temperature was sufficient to depress production. This evidence suggests that the range of tolerance to temperature narrows towards the higher natural river temperatures, since only a small temperature rise results in a great reduction in phytoplankton production at river temperatures above 25° C. This is to say that whenever the river temperature is above 15° C, a temperature rise of 8° C, which is typical of the power plant studied, depresses production and the depression becomes greater at higher ambient river temperatures.

The community composition and abundance of marine benthic invertebrates in the river were affected by the thermal discharge over a distance of 300–400 m from the discharge. The affected area is minimized by the fact that the heated water is less dense than the river water and rises above the bottom as it flows offshore. Both the remote sampling stations and the stations within 300 m of the discharge canal showed a marked seasonal change in numbers and species, reaching a minimum in August 1963 and a maximum in February 1964. Diversity indices showed that the lowest diversity of species was found at the stations within 300 m of the discharge during the summer months. This is an indication of stress on the population. During the winter however, diversity of community composition was much higher at these stations. The redundancy index suggests that the greatest diversity occurs in the heated water even though the number of species was only half of the number taken at the remote stations. There is clear evidence of stress on the benthic population over a limited area during the months of high normal river temperatures.

Acknowledgments—This investigation was supported in part by Public Health Service Grant No. WP00092 from the Division of Water Supply and Pollution Control. The authors gratefully acknowledge the assistance of MARVIN L. WASS, Associate Marine Scientist; JAMES KERWIN; and JOHN McCAIN in identifying and counting the benthic organisms.

REFERENCES

GAARDER T. and GRAN H. H. (1927) Investigations of the production of plankton in the Oslo fjord, *Rapp. P.-v. Reun Cons. perm. int. Explor. Mer.* **42,** 127–136.

HERRY S. (1959) Pollution of rivers by heated discharges. *Bull. Cent. belge Étud. Docum. Eaux,* **46,** 226–235.

JONES M. L. (1961) A quantitative evaluation of the Benthic fauna off Point Richmond, California, University of California Publications in Zoology, **67,** 219–320.

KULCZYNSKI St. M. (1927) Die Pflanzenassoziationen der Pienen, *Bull. int. Acad. sci. Lett. Cracovie,* **2,** 27–204.

MARGALEF R. (1956) Informacion of diversidad especifica en las communidades de organismos, *Investigación pesqu.* **3,** 99–106.

MARKOWSKI S. (1959) The cooling water of power stations; a new factor in the environment of marine and freshwater invertebrates. *J. Anim. Ecol.* **28,** 243–255.

MOORE E. W. (1958) Thermal "pollution" of streams. *Ind. Engng. Chem. ind. Edn.* **50,** 1–4, 87A.

SANDERS H. L. (1958) Benthic studies in Buzzards Bay.—I Animal-sediment Relationship. *Limnol. Oceanogry,* **3,** 245–258.

STRICKLAND J. D. H. (1960) *Measuring the Production of Marine Phytoplankton,* Bulletin 122, Fisheries Research Board of Canada.

VAN VLIET V. (1957) Effect of heated condenser discharge water upon aquatic life. *Am. Soc. Mech. Engng* **57,** 1–10.

Index